Lecture Notes in Electrical Engineering

Volume 641

Series Editors

Leopoldo Angrisani, Department of Electrical and Information Technologies Engineering, University of Napoli Federico II, Naples, Italy
Marco Arteaga, Departament de Control y Robótica, Universidad Nacional Autónoma de México, Coyoacán, Mexico
Bijaya Ketan Panigrahi, Electrical Engineering, Indian Institute of Technology Delhi, New Delhi, Delhi, India
Samarjit Chakraborty, Fakultät für Elektrotechnik und Informationstechnik, TU München, Munich, Germany
Jiming Chen, Zhejiang University, Hangzhou, Zhejiang, China
Shanben Chen, Materials Science and Engineering, Shanghai Jiao Tong University, Shanghai, China
Tan Kay Chen, Department of Electrical and Computer Engineering, National University of Singapore, Singapore, Singapore
Rüdiger Dillmann, Humanoids and Intelligent Systems Laboratory, Karlsruhe Institute for Technology, Karlsruhe, Germany
Haibin Duan, Beijing University of Aeronautics and Astronautics, Beijing, China
Gianluigi Ferrari, Università di Parma, Parma, Italy
Manuel Ferre, Centre for Automation and Robotics CAR (UPM-CSIC), Universidad Politécnica de Madrid, Madrid, Spain
Sandra Hirche, Department of Electrical Engineering and Information Science, Technische Universität München, Munich, Germany
Faryar Jabbari, Department of Mechanical and Aerospace Engineering, University of California, Irvine, CA, USA
Limin Jia, State Key Laboratory of Rail Traffic Control and Safety, Beijing Jiaotong University, Beijing, China
Janusz Kacprzyk, Systems Research Institute, Polish Academy of Sciences, Warsaw, Poland
Alaa Khamis, German University in Egypt El Tagamoa El Khames, New Cairo City, Egypt
Torsten Kroeger, Stanford University, Stanford, CA, USA
Qilian Liang, Department of Electrical Engineering, University of Texas at Arlington, Arlington, TX, USA
Ferran Martin, Departament d'Enginyeria Electrònica, Universitat Autònoma de Barcelona, Bellaterra, Barcelona, Spain
Tan Cher Ming, College of Engineering, Nanyang Technological University, Singapore, Singapore
Wolfgang Minker, Institute of Information Technology, University of Ulm, Ulm, Germany
Pradeep Misra, Department of Electrical Engineering, Wright State University, Dayton, OH, USA
Sebastian Möller, Quality and Usability Laboratory, TU Berlin, Berlin, Germany
Subhas Mukhopadhyay, School of Engineering & Advanced Technology, Massey University, Palmerston North, Manawatu-Wanganui, New Zealand
Cun-Zheng Ning, Electrical Engineering, Arizona State University, Tempe, AZ, USA
Toyoaki Nishida, Graduate School of Informatics, Kyoto University, Kyoto, Japan
Federica Pascucci, Dipartimento di Ingegneria, Università degli Studi "Roma Tre", Rome, Italy
Yong Qin, State Key Laboratory of Rail Traffic Control and Safety, Beijing Jiaotong University, Beijing, China
Gan Woon Seng, School of Electrical & Electronic Engineering, Nanyang Technological University, Singapore, Singapore
Joachim Speidel, Institute of Telecommunications, Universität Stuttgart, Stuttgart, Germany
Germano Veiga, Campus da FEUP, INESC Porto, Porto, Portugal
Haitao Wu, Academy of Opto-electronics, Chinese Academy of Sciences, Beijing, China
Junjie James Zhang, Charlotte, NC, USA

The book series *Lecture Notes in Electrical Engineering* (LNEE) publishes the latest developments in Electrical Engineering - quickly, informally and in high quality. While original research reported in proceedings and monographs has traditionally formed the core of LNEE, we also encourage authors to submit books devoted to supporting student education and professional training in the various fields and applications areas of electrical engineering. The series cover classical and emerging topics concerning:

- Communication Engineering, Information Theory and Networks
- Electronics Engineering and Microelectronics
- Signal, Image and Speech Processing
- Wireless and Mobile Communication
- Circuits and Systems
- Energy Systems, Power Electronics and Electrical Machines
- Electro-optical Engineering
- Instrumentation Engineering
- Avionics Engineering
- Control Systems
- Internet-of-Things and Cybersecurity
- Biomedical Devices, MEMS and NEMS

For general information about this book series, comments or suggestions, please contact leontina.dicecco@springer.com.

To submit a proposal or request further information, please contact the Publishing Editor in your country:

China

Jasmine Dou, Associate Editor (jasmine.dou@springer.com)

India, Japan, Rest of Asia

Swati Meherishi, Executive Editor (Swati.Meherishi@springer.com)

Southeast Asia, Australia, New Zealand

Ramesh Nath Premnath, Editor (ramesh.premnath@springernature.com)

USA, Canada:

Michael Luby, Senior Editor (michael.luby@springer.com)

All other Countries:

Leontina Di Cecco, Senior Editor (leontina.dicecco@springer.com)

** **Indexing: The books of this series are submitted to ISI Proceedings, EI-Compendex, SCOPUS, MetaPress, Web of Science and Springerlink** **

More information about this series at http://www.springer.com/series/7818

Andrey A. Radionov · Alexander S. Karandaev
Editors

Advances in Automation

Proceedings of the International Russian Automation Conference, RusAutoCon 2019, September 8–14, 2019, Sochi, Russia

Set 1

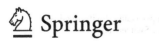
Springer

Editors
Andrey A. Radionov
Rectorate
South Ural State University
(National Research University)
Chelyabinsk, Russia

Alexander S. Karandaev
Nosov Magnitogorsk State
Technical University
Magnitogorsk, Russia

ISSN 1876-1100 ISSN 1876-1119 (electronic)
Lecture Notes in Electrical Engineering
ISBN 978-3-030-39224-6 ISBN 978-3-030-39225-3 (eBook)
https://doi.org/10.1007/978-3-030-39225-3

© Springer Nature Switzerland AG 2020
This work is subject to copyright. All rights are reserved by the Publisher, whether the whole or part of the material is concerned, specifically the rights of translation, reprinting, reuse of illustrations, recitation, broadcasting, reproduction on microfilms or in any other physical way, and transmission or information storage and retrieval, electronic adaptation, computer software, or by similar or dissimilar methodology now known or hereafter developed.
The use of general descriptive names, registered names, trademarks, service marks, etc. in this publication does not imply, even in the absence of a specific statement, that such names are exempt from the relevant protective laws and regulations and therefore free for general use.
The publisher, the authors and the editors are safe to assume that the advice and information in this book are believed to be true and accurate at the date of publication. Neither the publisher nor the authors or the editors give a warranty, expressed or implied, with respect to the material contained herein or for any errors or omissions that may have been made. The publisher remains neutral with regard to jurisdictional claims in published maps and institutional affiliations.

This Springer imprint is published by the registered company Springer Nature Switzerland AG
The registered company address is: Gewerbestrasse 11, 6330 Cham, Switzerland

Preface

International Russian Automation Conference (RusAutoCon) took place on September 8–14, 2019, in Sochi, Russian Federation. The conference was organized by South Ural State University (National Research University). The international program committee has selected 355 reports.

The conference was divided into 13 sections, including:

1. Process automation;
2. Modeling and simulation;
3. Control theory;
4. Machine learning, big data, Internet of things;
5. Flexible manufacturing systems;
6. Industrial robotics and mechatronic systems;
7. Computer vision;
8. Industrial automation systems cybersecurity;
9. Diagnostics and reliability of automatic control systems;
10. Communication engineering, information theory and networks;
11. Control systems;
12. Energy systems, power electronics, and electrical machines;
13. Instrumentation engineering;
14. Signal, image, and speech processing.

The International Program Committee has selected totally 125 papers for publishing in Lecture Notes in Electrical Engineering (Springer International Publishing AG).

The Organizing Committee would like to express our sincere appreciation to everybody who has contributed to the conference. Heartfelt thanks are due to authors, reviewers, participants, and to all the team of organizers for their support and enthusiasm which granted success to the conference.

Andrey A. Radionov
Conference Chair

Contents

Identification of Engineering Thermal Physics Objects Based
on Inverse Heat Conduction Problems Solving by Using
Parametric Optimization Methods 1
A. N. Diligenskaya, S. A. Kolpashchikov, and A. G. Mandra

Wireless Telemetry System for Gas Production 9
M. Yu. Prakhova, E. A. Khoroshavina, and A. N. Krasnov

Control of Linear Servo Pneumatic Drive Based on Fuzzy
Controller and Knowledge Base 17
E. L. Khaziev

Method for Controlling Tracking Actuator 26
E. V. Zubkov and M. L. Khaziev

Method for Composition of Two, Three and More
Multistructural Petri Nets into Single Equivalent Control Model
of Technological Process 35
M. P. Maslakov, S. V. Kulakova, and A. Z. Dobaev

Control System of Tension Device in Winding Fabrics
from Composite Materials 42
A. Mikitinskiy, B. Lobov, and P. Kolpachshan

Analysis of Drive of Mechatron Control System of Rotary Trencher ... 50
M. E. Shoshiashvili, I. S. Shoshiashvili, and T. P. Kartashova

Digital Platform for Non-destructive Testing 58
A. Kosach, I. Moskvicheva, and E. Kovshov

Intelligent Support System of Batch Selection for Electric Arc
Furnace: Consolidation of Empirical and Expert Information 66
O. S. Logunova and N. S. Sibileva

Automatic Control of Emulsifier Consumption in Oil Treatment Unit 74
K. V. Arzhanov and A. V. Arzhanova

Comparative Analysis of Automatic Methods for Measuring Surface of Threads of Oil and Gas Pipes 83
D. S. Lavrinov, A. I. Khorkin, and E. A. Privalova

Influence of Induction Heating Modes on Thermal Stresses Within Billets 97
Yu. E. Pleshivtseva and E. A. Yakubovich

Propellant Consumption Optimal Adaptive Terminal Control of Launch Vehicle 107
A. F. Shorikov and V. I. Kalev

Analysis of Adaptive Machines with Two Process-Connected Working Movements 120
M. A. Lemeshko, M. D. Molev, and A. G. Iliev

Approach to Building an Autonomous Cross-platform Automation Controller Based on the Synthesis of Separate Modules 128
G. M. Martinov, I. A. Kovalev, and A. S. Grigoriev

The Controlled Proportional Electromagnet-Driven Devices for Axial Rotor Loading in Test Rigs for Studying Fluid-Film Bearings 137
A. V. Sytin, V. V. Preys, and D. V. Shutin

Research of Process of Automatic Opening of Air-Penetrating Flexible Containers for Free-Flowing Products 147
A. M. Makarov, O. V. Mushkin, M. A. Lapikov, and Yu. P. Serdobintsev

Automatic Control of the Oil Production Equipment Performance Based on Diagnostic Data 158
K. F. Tagirova and A. R. Ramazanov

Technology of Building Movement Control Systems of Products in Warehouse Areas of Industrial Enterprises Using Radio Frequency Identification Methods 170
A. V. Astafiev, A. A. Orlov, and T. O. Shardin

Approach to Development of Specialized Terminals for Equipment Control on the Basis of Shared Memory Mechanism 181
P. A. Nikishechkin, N. Yu. Chervonnova, and A. N. Nikich

Control System for Gas Transfer by Gas Gathering Header in the Context of Water Accumulation 189
M. Yu. Prakhova, E. A. Khoroshavina, and A. N. Krasnov

Active Vibration Protection System with Controlled Viscosity of Working Environment 197
B. A. Gordeev, S. N. Okhulkov, A. B. Darenkov, and D. Yu. Titov

Increasing Precision of Absolute Position Measurement in NC Machine Tools by Means of Diminishing Inertia Loads in Measurement Unit 208
Ya. L. Liberman and K. Yu. Letnev

Calculating Lifting Capacity of Industrial Robot for Flexible Lathe System at the Design Stage 216
Ya. L. Liberman and K. Yu. Letnev

Method to Control Quality of Assembly of Induction Motors 225
K. E. Kozlov, V. N. Belogusev, and A. V. Egorov

Calculation and Analytical Module "Risk" for Selective Diagnostics and Repair of Main Gas Pipelines with Account of Technogenic Risks 233
Y. A. Bondin, N. A. Spirin, and S. V. Bausov

Energy-Saving Oriented Approach Based on Model Predictive Control System 243
T. A. Barbasova, A. A. Filimonova, and A. V. Zakharov

Using Method Frequency Scanning Based on Direct Digital Synthesizers for Geotechnical Monitoring of Buildings 253
D. I. Surzhik, O. R. Kuzichkin, and A. V. Grecheneva

Structural and Functional Model of the On-board Expert Control System for a Prospective Unmanned Aerial Vehicle 262
P. I. Tutubalin and V. V. Mokshin

Simulation Modelling of the Adaptive System of Structurally and Parametrically Indefinite Object with Control Lag 273
L. V. Chepak and Z. D. Pikul'

Experimental Verification of Flux Effect on Process of Aluminium Waveguide Paths Induction Soldering 282
A. V. Milov, V. S. Tynchenko, and A. V. Murygin

Intellectualization of the Induction Soldering Process Control System Based on a Fuzzy Controller 292
V. E. Petrenko, V. S. Tynchenko, and A. V. Murygin

Internal Combustion Engines Fault Diagnostics 305
L. A. Galiullin and R. A. Valiev

Optimization of Control for the Pneumatic Container Transport System 315
V. V. Khmara, A. M. Kabyshev, and Yu. G. Lobotskiy

Construction and Movement Control of Integrated Executive Device of Laser-Robot 323
I. N. Egorov, A. N. Kirilina, and V. P. Umnov

Design for Adaptive Wheel with Sliding Rim and Control System 331
A. Ivanyuk, D. Marchuk, and Yu. Serdobintsev

Improvement of Acoustic-Resonance Method and Development of Information and Measuring Complex of Location of Deposited Pipelines 339
S. O. Gaponenko, A. E. Kondratiev, and N. K. Andreev

Fuzzy Control of Underwater Walking Robot During Obstacle Collision Without Pre-defined Parameters 347
V. V. Chernyshev, V. V. Arykantsev, and I. P. Vershinina

Complex Application of the Methods of Analytical Mechanics and Nonlinear Stability Theory in Stabilization Problems of Motions of Mechatronic Systems 357
A. Ya. Krasinskiy and E. M. Krasinskaya

Determining Magnetorheological Coupling Clutch Damping Characteristics by the Rotary Shafts Shear Deformation 371
S. N. Okhulkov, A. I. Ermolaev, A. B. Daryenkov, and D. Yu. Titov

Automation in Foundry Industry: Modern Information and Cyber-Physical Systems 382
M. V. Arkhipov, V. V. Matrosova, and I. N. Volnov

Research of Non-sinusoidal Voltage in Power Supply System of Metallurgical Enterprises 393
R. V. Klyuev, I. I. Bosikov, and A. D. Alborov

Choice of Wind Turbine for Operation in Conditions of Middle Ural 401
A. Valtseva and K. Karamazova

Development Trend of Electrification and Small-Scale Power Generation Sector in Russia 409
K. V. Selivanov

Principles of Energy Conversion in Thermal Transformer Based on Renewable Energy Sources 417
Y. A. Perekopnaya, K. V. Osintsev, and E. V. Toropov

Poultry Wastes as Source of Renewable Energy................... 424
M. V. Zapevalov, N. S. Sergeev, and Yu. B. Chetyrkin

Application Features of Mathematical Model of Power System for Analysis of Technical and Economic Indicators of Reactive Power Compensation Device 434
D. V. Ishutinov, V. I. Laletin, and E. N. Malyshev

On Technogenic Impact of Electromagnetic Components of Rectified Current and Voltage on Environment................. 444
K. Kuznetsov and A. Zakirova

Engineering Solutions for the Use of Air Enriched with Oxygen in Power Engineering Complexes of Power Plants Operating on Secondary Gases...................................... 452
E. B. Agapitov, M. S. Sokolova, and A. E. Agapitov

Renewable Sources of Energy for Efficient Development of Electricity Supplies for Agriculture in Chechen Republic 460
A. R. Elbazurov and G. R. Titova

Renewable Energy Potential of Russian Federation 469
E. Solomin, A. Ibragim, and P. Yunusov

Study of Spiral Air Accelerators for Wind Power Plants Using a Vertical Rotation Axis................................ 477
A. A. Bubenchikov, T. V. Bubenchikova, and E. Yu. Shepeleva

Generating Gas from Wood Waste as Alternative to Natural Gas in Package Boilers 492
E. M. Kashin, R. R. Safin, and V. N. Didenko

Investigating the Effectiveness of Solar Tracking for PV Facility in Chelyabinsk 501
A. A. Smirnov, A. G. Vozmilov, and O. O. Sultonov

Optimization of Diesel-Driven Generators in Continuous Power Systems of Essential Consumers.......................... 509
V. V. Karagodin, K. A. Polyansky, and D. V. Ribakov

Study of Operating Characteristics of Pellets Made from Torrefied Wood Raw Material.................................... 519
R. R. Khasanshin, R. R. Safin, and A. R. Shaikhutdinova

Development and Research of Automated Compact Cogeneration Plant..................................... 526
A. M. Makarov, P. V. Dikarev, and V. V. Lazarev

CAD in Electrical Engineering: New Approaches to an Outdoor
Switchyard Design 536
E. A. Panova, A. V. Varganova, and N. T. Patshin

Increasing Efficiency of Water and Energy Supply Systems
in Southern Region Environment of Russian Federation 545
D. V. Kasharin

Electron-Ion Technology as Protection of Solar Modules
from Contamination 554
I. M. Kirpichnikova and V. V. Shestakova

Plating Technology for Improvement of Moveable Contact Joint
Performance in Power-Distribution Equipment 563
V. Goman and S. Fedoreev

Refinement of Transmission Line Equivalent Circuit Parameters
at Power System State Estimation by PMU Data 571
I. Kolosok and E. Korkina

Exergy Analysis of Single-Stage Heat Pump Efficiency Under
Various Steam Condensation Conditions 581
S. V. Skubienko, I. V. Yanchenko, and A. Yu. Babushkin

Comprehensive Comparison of the Most Effective Wind Turbines 588
E. Solomin, X. Lingjie, H. Jia, and D. Danping

The Development of Cost Assessment Models for Overhead
Power Transmission Lines 596
V. V. Cherepanov, N. S. Bakshaeva, and I. A. Suvorova

Energy Efficiency of Electric Vacuum Systems: Induction
Motor – Water Ring Pump with an Ejector 606
A. R. Denisova, A. I. Rudakov, and N. V. Rozhentcova

Cooling System Oil-Immersed Transformers with the Use
of a Circulating Sulfur Hexafluoride 613
M. G. Bashirov, A. S. Khismatullin, and E. V. Sirotina

Assessment Method of Technical Condition of Small Refrigerating
Machine Using Programmable Controller 622
M. A. Lemeshko, S. R. Urunov, and A. V. Kozhemyachenko

Research and Mathematical Modeling of the Thermal and Power
Performance of Resistance Furnaces at Metallurgical Enterprises 630
R. V. Klyuev, I. I. Bosikov, and A. D. Alborov

Comparative and Optimizing Calculations of Energy Efficiency Indicators for Operation of CHP Plants Using the Normative Characteristics and Mathematical Models 637
N. V. Tatarinova and D. M. Suvorov

Description of Complex Hierarchical Systems by Matrix-Predicate Method 647
V. S. Polyakov and S. V. Polyakov

Modeling and Automation of the Hydro-Transport System of Water-Coal Fuel at Negative Ambient Temperatures 657
K. V. Osintsev, O. G. Brylina, and Yu. S. Prikhodko

Modelling and Controlling the Temperature Status of the Turbine T-125/150 CCGT 450 Flow Part at the CCGT Operation in the GTU Based CHP Mode with Steam Turbine in the Motoring Drive Mode 667
E. K. Arakelyan, K. A. Andryushin, and F. F. Paschenko

Analysis of the Secure Data Transmission System Parameters 675
M. O. Tanygin, M. A. Efremov, and Ya. A. Hyder

Geometric Modeling and CAD System to Solve Tectonics-Related Tasks Using Core Pole 684
T. S. Guriev, A. V. Kalinichenko, and M. M. Tsabolova

Computer-Aided Design and Construction Development of the Main Elements of Aviation Engines 693
D. A. Akhmedzyanov and A. E. Kishalov

Complex Engineering System Acceptability Domain: Worst-Case Analysis via Fuzzy Forecast Technique 703
P. A. Zinovev and I. I. Ismagilov

AR Guides Implementation for Industrial Production and Manufacturing 715
A. Ivaschenko, V. Avsievich, and P. Sitnikov

The Modelling and Optimization of Machine Management System with Computer Numerical Control 724
I. M. Yakimov, A. P. Kirpichnikov, V. V. Mokshin, and Z. T. Yahina

Measuring Systems Application Analysis in Engineering 732
M. I. Kovalev

Test Bench with Controlled Impact Action for Analyzing Rotor-Bearing Assemblies 744
A. V. Gorin, A. V. Sytin, and A. Y. Rodichev

**Method of Lathe Tool Condition Monitoring Based
on the Phasechronometric Approach** 753
D. D. Boldasov, A. S. Komshin, and A. B. Syritskii

Depersonalization of Personal Data in Information Systems 763
D. V. Primenko, A. G. Spevakov, and S. V. Spevakova

**Complex Evaluation of Information Security of an Object
with the Application of a Mathematical Model for Calculation
of Risk Indicators** 771
A. L. Marukhlenko, A. V. Plugatarev, and D. O. Bobyntsev

**Application of Simulink and SimEvents Tools in Modeling
Marketing Activities in Tourism** 779
A. N. Kazak, D. V. Gorobets, and D. V. Samokhvalov

**Cascade Windows in Intellectual Agents of Multichannel
Images Classification** 787
I. A. Malyutina, S. A. Filist, and A. R. Dabagov

**Developing a Technical Diagnostic Systems for Internal
Combustion Engines** 797
L. A. Galiullin and R. A. Valiev

Automation of the ICE Testing Process 806
L. A. Galiullin, R. A. Valiev, and D. I. Valieva

**Noise-Robust Method to Determine Speech Prosodic Characteristics
to Assess Human Psycho-Emotional State in Free Motor Activity** 816
A. K. Alimuradov, A. Yu. Tychkov, and P. P. Churakov

Automatic Temperature Control System for a Bee Hive 827
S. V. Oskin, N. I. Bogatyrev, and A. A. Kudryavtseva

**Software Application for Determining Comfortable Conditions
of the Human-Computer Interaction** 838
A. A. Popov and A. O. Kuzmina

**The Control-and-Measuring System Built-in Automatic Control
System by the Technical Casting Process with Piezocrystallization** 852
M. Denisov

**Statistical Simulation and Probability Calculation of Mechanical
Parts Connection Parameters for CAD/CAM Systems** 861
S. Skvortsov, V. Khryukin, and T. Skvortsova

**Evaluation of the Harmonic Locus of the Milling Technological
System Based on the Analysis of the Vibro-Acoustic Signal** 871
R. M. Khusainov, A. R. Sabirov, and D. D. Safin

Contents

An Adaptive Speech Segmentation Algorithm to Determine
Temporal Patterns of Human Psycho-Emotional States 879
A. K. Alimuradov, A. Yu. Tychkov, and A. V. Ageykin

Development of Algorithms for the Correct Visualization
of Two-Dimensional and Three-Dimensional
Orthogonal Polyhedrons.................................. 891
V. A. Chekanin and A. V. Chekanin

Long-Term Digital Documents Storage Technology 901
A. V. Solovyev

Automation of the Process a Comprehensive Assessment
of Educational Organization............................. 912
L. A. Ponomareva, O. N. Romashkova, and E. N. Pavlicheva

Fuzzy Modeling of the Assessment of Using an Educational
Audience in Order to Improve the Quality of Training
of the Educational Process 923
R. U. Stativko and A. I. Rybakova

Process Modeling for Energy Planning of Technological Systems 933
A. Sychugov, Yu. Frantsuzova, and V. Salnikov

Problems and Solutions of Automation of Magnetron Sputtering
Process in Vacuum...................................... 944
S. V. Sidorova, A. D. Kouptsov, and M. A. Pronin

Information and Analytical Support for the Protection
of Important Critical Information Infrastructure Objects 953
V. Berdyugin and L. Dronova

The Artificial Neural Network Application for Service-Oriented
Evaluation of the Used Cars............................. 965
A. N. Guda and A. N. Tsurikov

Analysis of the Problems of Industrial Enterprises Information
Security Audit 976
I. I. Barankova, U. V. Mikhailova, and O. B. Kalugina

The Method of Automated Configuration Objects of the WinCC
Project for the Oil and Gas Industry 986
Sh. Khuzyatov and R. Valiev

Optimization the Process of Catalytic Cracking Using Artificial
Neural Networks 994
E. Muravyova

Simulation of a Multi-connected Process in iThink Program 1005
E. Muravyova and Y. Stolpovskaya

Optimization of the Process of Acoustic and Magnetic Geothermal Water Treatment Through Simulation in CoDeSys 1019
A. V. Korzhakov, V. E. Korzhakov, and S. A. Korzhakova

Estimating the Cost of Implementing Virtual Desktops as a Stage of Project Management in the Field of Cloud Technologies 1034
K. Makoviy and Yu. Khitskova

Automation of the Opal Colloidal Films Obtaining Processes 1044
E. V. Panfilova and V. A. Dyubanov

An Intelligent Automated Control System of Micro Arc Oxidation Process 1053
P. Golubkov, E. Pecherskaya, and T. Zinchenko

Improving the Parametric Reliability of Automatic Control Systems in Electric Drives 1062
A. V. Saushev, P. V. Adamovich, and O. V. Shergina

Transportation Management Systems for Airport Ground Handling 1071
A. Dorofeev and O. Nastasyak

Automated Text Classification System Based on Statistical Unified Model 1079
S. Skorynin and A. Surkova

Fundamentals for the Automation of Information Processing in the Identification of Chemical-Technological Systems 1088
I. V. Germashev, E. V. Derbisher, and T. P. Mashihina

R&D in Collection and Representation of Non-structured Open-Source Data for Use in Decision-Making Systems 1098
A. I. Martyshkin, I. I. Salnikov, and E. A. Artyushina

Modeling of Interaction of the Mechatronic Unit Segments in the Adaptable Part of an Aircraft Wing 1113
N. Sharonov, A. Makarov, A. Ivchenko, and A. Gorelova

Modelling Steel Casting on a Continuous Unit 1124
A. Galkin, P. Saraev, and D. Tyrin

Deployment of Intelligent Tools into the Distributed Translation System of Models 1138
M. Polenov, V. Guzik, and A. Kurmaleev

Enterprise Information Security Assessment Using Balanced Scorecard 1147
R. Fatkieva and A. Krupina

**Bottom Induction Stirrer for Induction Crucible Furnace
with Graphite Crucible** 1158
K. Bolotin and D. Brazhnik

**Approximation Method for Probability Density Function
of Non-Gaussian Random Processes in Telecommunication
and Data-Measuring Systems** 1167
A. B. Semenov and V. M. Artushenko

**System to Capture Movements of Buyers and to Determine Quality
of Store Employees** 1175
A. D. Ulyev, V. L. Rozaliev, Yu. A. Orlova, and A. V. Alekseev

**Automating the Detection of Sarcastic Statements in Natural
Language Text** 1185
A. V. Dolbin, V. L. Rozaliev, Yu. A. Orlova, and A. D. Ulyev

**Required Coke Quality Influence on a Coking Coal Mix Price
Research. Linearization of a Coking Coal Procurement
Optimization Model** 1195
A. Lipatnikov and D. Shnayder

Identification of Engineering Thermal Physics Objects Based on Inverse Heat Conduction Problems Solving by Using Parametric Optimization Methods

A. N. Diligenskaya, S. A. Kolpashchikov, and A. G. Mandra[✉]

Samara State Technical University, 244 Molodogvardeyskaya Street,
Samara 443100, Russian Federation
amandra@mail.ru

Abstract. The paper presents the techniques for the analysis of inverse heat conduction problems which allow one to restore the characteristics and parameters of the objects of technological thermophysics and those founded on the methodology of the optimal control theory of objects with distributed constants. The analysis of monodimensional inverse heat conduction problem provides contraction of a set of desired solutions up to the accuracy class. As a result of this, a conditionally correct problem in an extreme formulation is defined, where the identifiable characteristics act as an optimal control action. To switch to parametric optimization, it is necessary to parameterize the desired control and use a uniform metric for estimating the variation of the derived temperature state from a given one. The solution of the obtained problem towards the optimum values of the sought parameters is based on the analysis of minimax programming that considers the ultimate solutions with their alternating properties. Further, the authors give an illustration of a problem solution of identifying the power of internal heat sources during the heat of induction with technological restriction.

Keywords: Inverse heat conduction analysis · Problem of parametric optimization · Minimax-programming · Internal heat sources power

1 Introduction

The quality of the technological equipment operation depends mainly on the degree of the initial information authenticity and the problem of improving the production processes efficiency is impossible without system analysis, identification and optimization.

Identification of mathematical models of transient thermal physics processes characterized by a complex character of heat transfer means obtaining information about the thermal and physical characteristics of the material, internal or boundary actions such as the initial state, the function of distribution of internal heat sources or the values of the convective heat transfer coefficients, each of which can be a complex function depending on a combination of a large number of factors, and the error in their values can be a significant source of error.

The problems of determining unknown parameters or characteristics of the process can be formulated as inverse problems $f(u) = y$ [1, 2], where u are the unknowns of the heat exchange process that determine the behavior of the system, y are the observed values characterizing the thermal state of the system (temperature field), and the operator f is establishing the cause-and-effect relationship between them.

One of the effective approaches to inverse problems solution is their formulation in variational statement [3, 4] and the subsequent solution as optimal control problems [5]. In this case when the inverse heat conduction problems (IHCP) are being solved the heat condition can be regarded as the state space. A set of the temperature field, the thermophysical characteristics and the boundary conditions setting the heat transfer conditions is considered as a state vector that determines the thermal system state at any time. The set of observed values, which in typical cases are the temperature-time dependences, obtained at a finite number of thermometry points, represents the measurement vector. Inverse problems assume the restoration of the components of the state vector by using the measurement vector.

In general, the transient heat transfer processes are systems with distributed parameters, variable in time and space, and are described by parabolic partial differential equations.

The difficulty of IHCP solution lies in their specific features that means providing the causal characteristics by the measured effect.

This violation of the cause-effect relationship leads to instability of IHCP solution, expressed in the fact that big perturbations of the identifiable state vector can correspond to small perturbations of the input parameters. Thus, the IHCP relates to ill-posed problems [6, 7].

The special approaches that allow to find stable solution of ill-posed problems are required. To solve IHCP a large number of methods with different techniques are developed. Ill-posed problems can be solved either by using methods for solution regularization or by reducing them to conditionally well-posed problems. The contraction of a set of feasible solutions to the compact one which is proper class allows to frame stable approximate solutions for ill-posed problems [6]. Here, set selection is made according to the physical meaning of each specific problem.

In this paper to obtain the conditionally well-posed statement of IHCP the set of control activities is restricted on the basis of their smoothness requirements. The problem of minimizing the error of uniform approximation model temperature to the given one is formulated, which leads to the parametric optimization problem. the solution of which is in the methodology application of the systems optimal control having distributed parameters [8–10].

2 Formulation of the Problem

A wide class of transient heat transfer processes in typical formulations is described by one-dimensional linear heterogeneous equation of heat transfer for a temperature field $T(x, t)$.

$$\frac{\partial T(x,t)}{\partial t} = a\nabla^2 T(x,t) + \frac{1}{c\gamma}F(x,t), \qquad (1)$$

where $F(x,t)$ is the power function of internal sources of heat; ∇^2 is the Laplacian; spatial coordinate $x \in \Omega \subset R^1$, $\partial\Omega$ is boundary of domain; time $t \in [0, t^0]$; a, c, γ are the averages of temperature conductivity, specific heat and specific density of heated material respectively.

Appropriate initial conditions

$$T(x,0) = f_0(x), \ x \in \Omega \qquad (2)$$

and boundary conditions

$$\frac{\partial T(x,t)}{\partial x} = f_1(t), \ (x,t) \in \partial\Omega \times [0, t^0], \qquad (3)$$

are also required.

For inverse heat conductivity problems solution the sought spatial distributed $u(x)$ or lumped $u(t)$ function can be boundary control $f_1(t)$ or internal one $[w(x), p(t)]$ in case of $F(x,t) = w(x)p(t)$, initial state $f_0(x)$ or thermal physical parameters a, c, γ [11, 12].

The function $u = [u(x), u(t)]$ is subject to identification, all other parameters and characteristics of process are known functions. The information on the measurement vector is given in the form of a temperature dependence $T^*(t)$, at a certain fixed point $x^* \in \Omega$ on the identification interval $[0, t^0]$.

The sought control action is subject to the restriction belonging to a given set V of accepted values u, known, for example, in the form of known boundaries

$$u \in V, \ t > 0, \qquad (4)$$

To evaluate the absolute deviation $|T(x^*,t) - T^*(t)|$ from the accurate solution $T(x^*,t)$ of BVP (1)–(3) conforming to the required function u from the appointed temperature $T^*(t)$ we apply the uniform metric [13, 14] and formulate the extreme statement of IHCP.

For the unit (1)–(3), it is required to determine the control activity u subordinate to constraint (4) providing minimax relations at a specified time frame $t \in [0, t^0]$

$$I(u) = \max_{t \in [0, t^0]} |T(x^*, t) - T^*(t)| \to \min_{u \in V}, \qquad (5)$$

3 Method of Solution

For the resolution of the issue (1)–(3) the reducing set of feasible solutions is applied according to which the sought functions are approximated in polynomial form

$$u(t) \in V_1 = \left\{\sum_{n=0}^{\infty} \Delta_n t^n,\ t \in [0, t^0]\right\}, \tag{6}$$

or

$$u(x) \in V_2 = \left\{\sum_{n=0}^{\infty} \Delta_n x^n,\ x \in \Omega\right\}, \tag{7}$$

When the degree N of the approximating polynomial (6) or (7) is given it corresponds to the parametric representation of sought optimal control $u = u(t, \Delta)$ which is determined by the parameter vector $\Delta = \{\Delta_n\}$, $n = \overline{0, N}$.

In this case the assumed heat level $T(x, t) = T(x, t, \Delta)$, computed as a response for controlling $u = u(t, \Delta)$, additionally acts as a parameters vector function Δ, involving parametric form on a closed bounded set $G_{n+1} \ni \Delta$. By the obtained parametric depiction of the control activity $u(t, \Delta)$ and the thermal state $T(x, t, \Delta)$, it is possible to carry out the reduction of the initial problem in the context of the parameters vector Δ [8, 9]:

$$I_0(\Delta) = \max_{t \in [0, t^0]} |T(x^*, t, \Delta) - T^*(t)| \to \min_{\Delta}, \tag{8}$$

The received minimax-optimization problem result (8) is in the known quality features of optimal temperature distributions $T(x^*, t, \Delta^0)$ possessing at $\Delta = \Delta^0$ the most uniform approximating to a given state $T^*(t)$ [13, 15]. These properties establish the character of the curve of the approximation error $T(x^*, t, \Delta^0) - T^*(t)$ of the given temperature.

At $\Delta = \Delta^0$ on the required segment $t \in [0, t^0]$ at certain points the number of which exceeds by one the number of the sought parameters, alternating maximum deviations $I_0(\Delta^0)$ in absolute magnitude are achieved. On the basis of this property, it is possible to establish a closed system for equations to limit temperature differences with respect to all the unknowns: the elements belonging to the parameter vector Δ^0 and the value $I_0(\Delta^0)$.

4 Results of the Numerical Experiment

According to the above method, numerical experiments were carried out, the results of which are presented below. The following problem comes down to defining the function $u(t) = p(t)$ of the internal heat sources rate [16, 17] during the process of the cylindrical bodies periodic inductive heating if the following functions are identified: the dimensional component data, thermophysical parameters and edge conditions. The following model written with respect to relative units is presented [18]

$$\frac{\partial T(x,t)}{\partial t} = a\left(\frac{\partial^2 T(x,t)}{\partial x^2} + \frac{1}{x}\frac{\partial T(x,t)}{\partial x}\right) + w(x)p(t), \qquad (9)$$

$$0 < x < 1;\ 0 < t \leq t^0$$

$$T(x,0) = 0_0;\ \frac{\partial T(0,t)}{\partial x} = 0;$$

$$\frac{\partial T(1,t)}{\partial x} + Bi\big(T(1,t) - T_{am}(t)\big) = 0.$$

Here B_i is the Biot number, T_{am} determinates a specific value of outside temperature.

The function $w(x)$ of spatial internal heat conduction can be obtained with the help of Maxwell's equation solution for electromagnetic field and possesses the common form of a work in the shape of a cylinder

$$w(x) = \frac{ber'^2(\zeta x) + bei'^2(\zeta x)}{ber\,\zeta\,ber'\,\zeta + bei\,\zeta\,bei'\,\zeta}\zeta,$$

where $ber(\bullet)$, $bei(\bullet)$, $ber'(\bullet)$, $bei'(\bullet)$ are Kelvin's functions together with first-order derivatives; ς – is a characteristic feature, that depends on a frequency of coil current (here $\varsigma = const$).

Nominal values of the input temperature were found from the heat conductivity results as well as the time-optimal control constrained on the state.

The general view of the typical criteria optimum control contains nonuniform heating intervals with ultimate output $p(t) = p_{max}$ and temperature equalization $p(t) = 0$ [18]. The requirements of real induction heating most often have to take into account phase restrictions, for example, limitation on the maximum temperature $T_{max}(t)$ in the volume of the heated workpiece. The optimal control algorithms are supplemented by stabilization sections of the maximum temperature on the accepted level T_{adm}. Thus, the optimal control algorithm consists of a heating interval with maximum power and a stabilization one with the maximum temperature on the accepted level (interval of movement along the constraint) [18]. As a result, the typical algorithm of the time-optimal heat power control with consideration of the technological constraint can be represented by the following expression:

$$p^0(t) = \begin{cases} p_{max}, & t \in (0, t_1) \\ p^{\lim it}, & t \in (t_1, t^0). \end{cases}$$

Modeling on the interval $t \in (0, t_1)$ takes into account the influence of the perturbation effect distributed over the harmonical law with amplitude $\sigma = 0.05 p_{max}$. Thus, the desired optimal control has the following form

$$p^0(t) = \begin{cases} p_{\max} + \sigma \cos \beta_1 t, & t \in (0, t_1) \\ \alpha_{21} + \alpha_{22} e^{-\beta_2 t}, & t \in (t_1, t^0). \end{cases} \quad (10)$$

The procedure for the power $p^0(t)$ identification (10) consists of the solution of two IHCP. The first of the problems includes the interval $(0, t_1)$ (where the switching point t_1 is considered known from the optimal control algorithm) and is solved in the class of

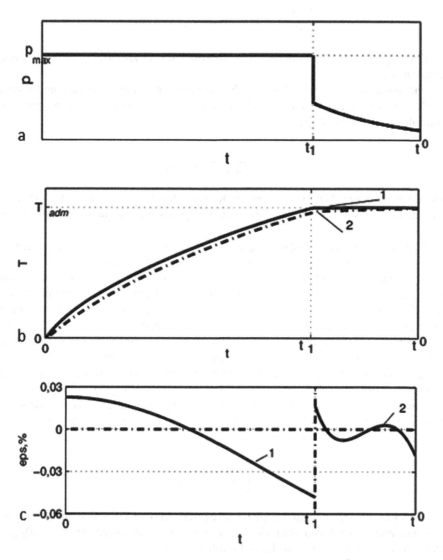

Fig. 1. The main characteristics of the problem: (a) optimal control $p^0(t)$; (b) temperature dependences $T_{\max}(t) - (1)$ and $T^*(t)$ (2); (c) error $\varepsilon = p^0(t) - p^*(t)$ of approximation of the identified action: 1 – on the interval $[0, t_1]$, 2 – on the interval $[t_1, t^0]$; $p^*(t)$ – computed optimal solution.

constant functions and provides recovery of $p^*(t) = p_{\max} = const$. The second problem is solved on an interval (t_1, t^0) in the class of polynomial functions of the form (6) for $N = 2$. The initial condition for the second problem is the heat distribution derived from the results of the first task for $t = t_1$.

The results of the solution of both IHCPs are presented in Fig. 1.

The analysis of the solution results shows the high accuracy of the recovery of the sought characteristic $p(t)$ on the basis of the IHCP solution by the methods of parametric optimization. If necessary, the approximation error $\varepsilon = p^0(t) - p^*(t)$ can be reduced by increasing the order N of the approximating polynomial (6) or (7).

5 Conclusion

The proposed methodology can be applied for a wide range of problems of identifying unknown parameters and characteristics in engineering thermal physics objects. The method is based on reducing the set of desired solutions to well-posedness class, evaluation of the approximation error of sought characteristic in the uniform metric, reducing to a parametric-optimization problem as well as applying optimal control methodology directed to systems having distributed parameters for its solution.

Acknowledgments. This work was supported by RFBR grants № 18-08-00048 and 18-08-00565.

References

1. Alifanov, O.M.: Inverse Heat Transfer Problems. Springer, Berlin (1994)
2. Beck, J.V., Blackwell, B., Clair, C.R.S.: Inverse Heat Conduction: Ill-Posed Problems. Wiley, New York (1985)
3. Alifanov, O.M., Artjuhin, E.A., Rumjancev, S.V.: Extreme Methods for Solving Ill-Posed Problems. Science, Moscow (1988)
4. Alifanov, O.M., Nenarokomov, A.V.: Boundary inverse heat conduction problem in extreme formulation. In: Proceedings of the 1st International conference on inverse problems in engineering: Theory and Practice, New York, pp. 31–37 (1993)
5. Butkovskii, A.G.: Optimal Control Theory for Systems with Distributed Parameters. Science, Moscow (1965)
6. Tihonov, A.N., Arsenin, V.Y.: Methods of the Solution of Incorrect Problems. Science, Moscow (1979)
7. Ivanov, V.K., Vasin, V.V., Tanana, V.P.: Theory of Linear Ill-Posed Problems and Its Applications. Science, Moscow (1978)
8. Pleshivceva, Y.E., Rapoport, E.Y.: Successive parameterization method of control actions in boundary value problems of optimal control of distributed parameter systems. J. Comput. Syst. Sci. Int. **3**, 22–33 (2009)
9. Diligenskaya, A.N., Rapoport, E.Y.: Analytical methods of parametric optimization in inverse heat-conduction problems with internal heat release. J. Eng. Phys. Thermophys. **87** (5), 1126–1134 (2014)

10. Diligenskaya, A.N., Rapoport, E.Y.: Method of minimax optimization in the coefficient inverse heat-conduction problem. J. Eng. Phys. Thermophys. **89**(4), 1008–1013 (2016)
11. Korotkii, A.I., Mikhailova, D.O.: Reconstruction of boundary controls in parabolic systems. Proc. Steklov Inst. Math. **280**(1), 98–118 (2013)
12. Bockstal, K.V., Slodicka, M.: Recovery of a space-dependent vector source in thermoelastic systems. Inverse Proble. Sci. Eng. **23**(6), 956–968 (2015)
13. Rapoport, E.Y.: Alternance Method for Solving Applied Optimization Problems. Science, Moscow (1986)
14. Rapoport, E.Y.: Minimax optimization of the stationary states in distributed parameter systems. J. Comput. Syst. Sci. Int. **2**, 3–18 (2013)
15. Collatz, L., Krabs, W.: Approximations Theorie. Tschebysheffsche Approximaton mit Anwendungen. B.G. Feubner, Stuttgart (1973)
16. Yang, L., Dehghan, M., Yu, J.-N., et al.: Inverse problem of time-dependent heat sources numerical reconstruction. Math. Comput. Simul. **81**(8), 1656–1672 (2011)
17. Shi, C., Wang, C., Wei, T.: Numerical solution for an inverse heat source problem by an iterative method. Appl. Math. Comput. **244**, 577–597 (2014)
18. Rapoport, E.Y., Pleshivtseva, Y.E.: Optimal Control of Induction Heating Processes. CRC Press/Taylor and Francis group, London, New York (2007)

Wireless Telemetry System for Gas Production

M. Yu. Prakhova(✉), E. A. Khoroshavina, and A. N. Krasnov

Ufa State Petroleum Technological University, 1 Kosmonavtov Street,
450062 Ufa, Russian Federation
prakhovamarina@yandex.ru

Abstract. When developing automated information and measurement systems (AIMS) for scattered automated or controlled facilities, choosing the type of communication channels for system components is an essential issue. The efficiency of communication channels determines the efficiency of the AIMS as a whole, thus affecting the quality of controls. This is especially relevant for gas-production facilities, which are mostly located in the Far North. The unprofitability or immaturity of wired connectivity enforces the exclusive use of wireless channels. Russia has a bandwidth that can be used for unlicensed radio communication. However, this bandwidth is very small (868.7 to 869.2 MHz), which is why various measurement channels might use very close frequencies; this in its turn means that the transmitters and receivers of crystal oscillators must produce very stable frequencies. The temperature and the "aging" of a crystal affect its frequency. The existing frequency stabilization technology only compensates the temperature-related frequency drifts. This paper proposes use of GPS data for frequency adjustment. This method can compensate the frequency drift of a crystal oscillator regardless of why it has occurred, which helps make use of a greater number of channels with the unlicensed bandwidth while keeping them reliably separated. Besides, a GPS receiver does not consume a lot of energy. The usage of the offered system allows to optimize the inhibitor's flow, to increase reliability of any gas field exploitation, even without full electricity supply system.

Keywords: Crystal oscillator · Parameter recorder · Gas-production telemetry system · Radio communication channel · GPS · Carrier frequency adjustment

1 Introduction

Major gas fields use complex and extensive gas-production and inter-facility gas transport systems [1]. The clustered location of wells, a distributed system of well-to-CGTP flowlines, a complex networking structure of the inter-facility reservoirs – these features of such production and transport systems justify the use of automated information-measurement systems (AIMS) as a part of process control systems (PCS). Given the scattered location of gas-production facilities, AIMS are better built with wireless technology [2]. Sometimes, wireless AIMS is the only option due to lack of wired connectivity; this applies to gas-production facilities located in the Far North (thus, most of Russia's gas-production facilities).

Provided a line-of-sight coverage and use of directed antennae, stable facility-to-BS connection might be established at a distance of up to 12 km. A RTP-4 [3] transducer lives two years or even more on a single battery provided a five-minute polling cycle and an annual average temperature of −5 °C. The system has certain drawbacks, such as the volatile frequency of its transceivers, which makes data transmission less efficient and may result in packet loss. Let's explain why.

In Russia, non-specialized wireless devices can officially use only two bandwidths: 864.0 to 865.0 MHz with an activity period of 0.1% or less, not allowed for use close to airports; and 868.7 to 869.2 MHz, no limitations (the latter bandwidth is referred to as 868 MHz) [4]. Thus, an AIMS developer only has a bandwidth of 500 kHz to deal with. To maximize the number of channels within this bandwidth, they have to minimize channel-specific bandwidths, esp. given the use of inter-channel guard bands. At the same time, reliable radio operation requires both the receiver and the transmitter to be set to a specific, fixed frequency.

However, some external factors such as the temperature may affect the transmitter's carrier frequency. This widens the range of frequencies produced, necessitating a reduction in channels in this bandwidth. Otherwise, the carrier frequency might deviate to the values of another transmitter, which will result in interference. The widened range of transmitter frequencies necessitates a wider bandwidth at the receiver output. It results in a worse signal-interference ratio at the receiver output as wider bandwidth means greater interference, thus lower quality of received signals.

To produce the same signal-interference ratio at the receiver output, one could use a more powerful transmitter, which would increase power consumption. This solution is not acceptable for gas-production AIMS as gas-production facilities lack mains electricity. Let's exemplify this by referring to the process parameter control system (PPCS) designed to control the pressure and temperature at the heads of gas and gas condensate wells, pipelines, and equipment [5]. This system is in use at all the gas condensate-production facilities of OOO Gazprom dobycha Urengoy. Measurement data are wirelessly transmitted to the base station (BS), either directly or via the parameter recorders (PR) of the network, see Fig. 1.

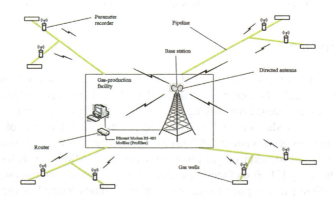

Fig. 1. Structural diagram of the process parameter control system.

Note that the frequency of high-frequency electromagnetic oscillations used to generate a radio signal is determined by the generator (the driver), whereas its instability can be eliminated in the subsequent cascades of the device. This is why it is the driver that should meet the frequency requirements first and foremost.

2 The Main Part

2.1 Relevance and Overview of the Existing Solutions

The operating frequency of a transmitter depends on the crystal oscillator. Quartz crystals are used as quartz is the only known material that combines the piezoelectric effect, low losses (thus a high Q-factor of up to 1,000,000), is naturally abundant, cheap, easy to produce en mass, as well as relatively highly stable.

This is why it is the instability of quartz frequencies due to various destabilizing electric, climatic, and mechanical factors that results in a significant drift of transmitter frequencies from the estimated value [6, 7]. Paper [8] cites data that enable a calculation of deviations. Thus, a typical 39 MHz quartz crystal has a frequency stability of ±25 ppm; thus, when using an 869 MHz channel, the real deviations of the carrier frequency may reach ±21.725 kHz. The experience of using a process parameter control system at the Urengoy oil and gas condensate field (OGCF), only two radio channels can be set up in a 500 kHz bandwidth to keep data transmission stable. This is due to the fact that the initial quartz frequency dispersion of 20 to 30 kHz is 'supplemented' by another 20 kHz due to temperature-induced drifts (the temperature dependency of quartz is non-linear, and precise estimation of such deviations requires special measures). Time-induced drifts accumulate and reach up to 20 kHz over a year. Thus, the transducers in the system only enabled temperature-versatile stable connection with a deviation of 60 kHz using a bandwidth of 120 kHz. Given a double-deviation guard band (and the fact that 5 deviations are necessary to ensure quality connection), the specified bandwidth can only contain two frequency channels: one in the upper half of the bandwidth, and one in the lower one. As such, only two channels may operate in parallel in a given area, which limits data transmission and is not acceptable for most AIMS. For example, the Urengoy OGCF gas-production AIMS has to receive temperature and pressure data from each well at least every five minutes [9, 10].

According to Gazprom dobycha Urengoy Communication Department (CD) specialists, the width of a channel should not exceed 24 kHz. As this parameter depends mainly on the stability of the crystal oscillator operation frequency, this problem remains relevant. Many papers have been written and published on the problem of improving this operating frequency, e.g. [7, 11–20].

The problem can be solved by improving the quality of the oscillator itself, or by temperature adjustment, using analog (thermistors and temperature sensors) and digital (microcontroller) technology.

Temperature-adjusted oscillators stabilize their frequency by means of a more complex circuitry and bigger size, hence greater power consumption and costs. Their basic solution mainly affects the pulsation of the resulting temperature frequency response and the efficiency of the oscillator configuration.

2.2 Statement of Problem

This paper discusses the GPS-based crystal oscillator operating frequency stabilization method.

2.3 Theory

The authors believe that the most optimal temperature adjustment method consists in using a GPS receiver as a component of the transceiver microchip; the device receives GPS signal from a satellite and provides timestamping with a picosecond precision. Based on GPS-generated time and comparing the size of a packet received (the no. of pulses) to the size of the packet to be sent over the same period of time, one can calculate the deviation of the crystal oscillator carrier frequency and adjust it in the algorithm. This method is not calculation-intensive and does not require any considerable upgrades in transceiver hardware.

Low-energy modules designed for precision clocks have recently entered the market. An example of those is Telit SL869-T [15], which is highly sensitive, consumes low power, and features a short cold start. One of its main features is TRAIM, a technology that filters out incorrect satellite signals; the devices produces a reference frequency synchronized to the Universal Coordinate Time with a deviation of less than $20 * 10^{-9}$ s. Capturing a single satellite is enough to generate the reference frequency. Figure 2 shows the functional diagram of the suggested device.

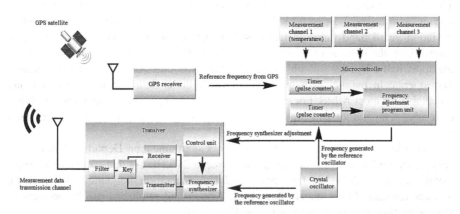

Fig. 2. Functional diagram of the GPS-based crystal oscillator operating frequency adjustment device.

The carrier frequency is adjusted when the recorder-to-BS connection is lost, or when the outer temperature deviates by 10° from the value used in calibration or previous adjustment. The microcontroller commands to enable the GPS receiver, which in its turn starts to generate reference frequency, after which the program unit runs the frequency adjustment algorithm.

GPS and crystal oscillator signals are recorded by respective microcontroller timers within 500 ms; the results are recorded in the adjustment unit.

Upon completion the microcontroller communicates the calculated adjustment for the frequency synthesizer to the radio module control unit. The calculated adjustment is recorded in the flash memory of the microcontroller, which in 60% of cases eliminates the need for GPS signals for a subsequent adjustment procedure.

The control unit registers can adjust the built-in synthesizer frequency win increments of 500 Hz. The built-in synthesizer generates both a heterodyne signal for the receiver and an FSK-modulated signal for the transmitter.

2.4 Experimental Results and Practical Significance

Figure 3 shows the structural diagram of the experimental bench for testing the GPS-adjusted crystal oscillator frequency. Figure 4 shows the results. One of the PR of the AIMS was placed in a temperature chamber. The signal spectrum was obtained by means of a spectrum analyzer.

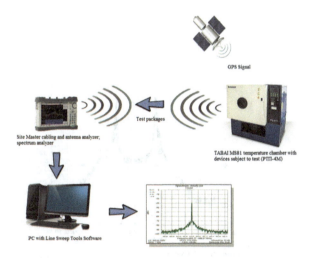

Fig. 3. Structural diagram of the experimental bench for testing the thermal stabilization of the crystal oscillator frequency.

Spectrum (a) corresponds to the initial crystal spectrum with a deviation of 12 kHz at a normal temperature. Spectrum (b) was generated in a temperature chamber at –27 °C, the temperature at which the connection to the PR was lost. As seen from the spectrum, the connection was lost due to a significant frequency drift. Spectrum (c) was generated in the temperature chamber at –30 °C, GPS-adjusted. No frequency drift, thus the PR connection was restored.

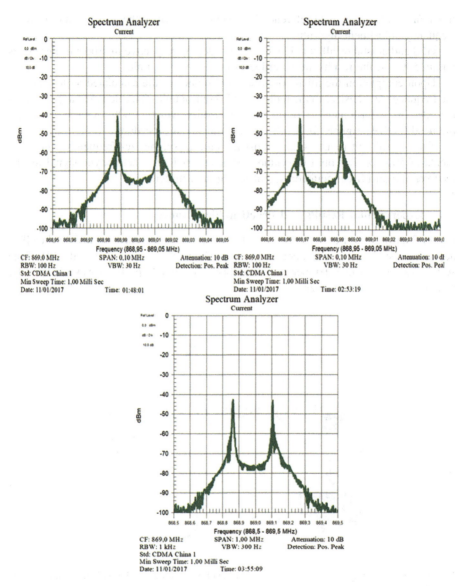

Fig. 4. Test results of the correction system of the clock frequency of the crystal oscillator by adjusting the GPS signal.

GPS adjustment results in greater power consumption. Paper [18] gives a formula for calculating the power consumption of a recorder. Taking into account the additional frequency adjustment channel, the formula (1) is written as:

$$\begin{aligned} e_i = {} & T_1 \cdot e_s + n \cdot t_m \cdot (e_m + e_w - e_s) + n \cdot t_c \cdot (e_w - e_s) + n \cdot t_w \cdot (e_w - e_s) + n \cdot t_p \cdot (e_w + e_r - e_s) \\ & + ((2n-1) \cdot t_r) \cdot (e_r + e_w - e_s) + ((2n-1) \cdot t_i) \cdot (s_i + n(e_w - e_s)) + ((2n-1) \cdot t_t) \cdot (z_i + e_w - e_s) \\ & + k_{gps} \cdot t_{gps} \cdot (e_{gps} + e_w) \end{aligned}$$

(1)

where t_m is the time the analog-digital converter spends to measure all the necessary parameters; t_c is the time the microprocessor spends to process the ADC-generated values; t_w is the time the microprocessor spends to switch from hibernation to the operating mode; t_p is the time interval from the moment the receiver is switched on to the moment the transmitter starts transmitting data; t_r is the time the radio module spends to switch to the receiver mode; t_t is the time the radio module spends to switch to the transmitter mode; t_{gps} is the time to adjust the radio module frequency; e_s is the microprocessor current consumption in hibernation; e_m is the ADC current consumption when taking measurements; e_w is the microprocessor current consumption in the operating mode; e_r is the radio module current consumption in the receiver mode; e_{gps} is the active current consumption of the GPS receiver; T_l is the real duration of a data acquisition cycle; n is the average expected number of connection attempts; k_{gps} is a 0 to 1 coefficient that indicates whether it is possible for adjust the frequency based on flash-stored data, i.e. without using the GPS receiver; s_i is the total power consumed by the transmitter in the ith unit to transmit responses to all the units that transmit data directly to the ith unit, with due account of necessary power; z_i is the power consumer by the transmitter in the ith unit to transmit measurement data packets, with due account of necessary power.

The operation of a recorder with an additional frequency adjustment channel was simulated in Castalia, a low-power network simulator. For power supply, we chose the regular lithium-thionyl chloride battery. Simulation showed that the recorder would operate stand-alone for 14,016 h with a five-minute polling cycle.

The analysis revealed that according to the experiments on test specimen, the proposed crystal oscillator carrier frequency adjustment method enables the operation of up to seven parallel channels; the additional GPS power consumption increased the total power consumption of the recorder, and thus reduced its stand-alone operating time, by 20% compared to the estimated values, which is acceptable given how greatly it improves the efficiency of wireless data transmission.

3 Conclusions

Precise maintenance of a carrier frequency reduces its deviation, thus enabling a greater number of channels in the unlicensed bandwidth, which is very important for high-volume AIMS; the method also makes data transmission and receiving more accurate and reliable.

References

1. Semukhin, M.V.: Multi-level system of models for calculating the parameters of a networking inter-facility reservoir and gas collection networks. In: Oil Industry. Ufa State Petroleum Technological University (2007). http://www.ogbus.ru/article/view. Accessed 25 Oct 2017
2. Wireless technology in information-measurement systems (2015). https://rtlservice.com/ru/company/blog/mesto_besprovodnyh_tehnologij_v_strukture_informacionnoizmeritelnyh_sistem. Accessed 27 Oct 2017
3. Parameter Recorder PTII-04 (2012) Registration no. 29581-12, 18 April 2012
4. Verkhulevsky, K.: Specifics and trends in the development of LoRaWAN. Wirel. Technol. **1–17**, 26–32 (2017)
5. Krasnov, A.N., Fyodorov, S.N.: Process parameter control telemetry system for the gas condensate wells and flowlines of the Urengoy OGCF. In: Proceedings of the All-Russian Scientific and Technical Conference Problems of Process and Production Control and Automation, Ufa (2010)
6. Improving the accuracy of a crystal oscillator. http://www.semtech.com/images/datasheet/xo_precision_std.pdf. Accessed 01 Nov 2017
7. Rubiola, E., Groslambert, J., Brunet, M., et al.: Flicker noise measurement of HF quartz resonators. IEEE Trans. Ultrason. Ferroelectr. Freq. Control **47**, 361–368 (2000). https://doi.org/10.1109/58.827421
8. Notes on usage. E-counter series. http://leapsecond.com/hpan/an200.pdf. Accessed 25 Oct 2017
9. Novikov, V.I.: Method of optimizing the parameters of gas and gas condensate wells. RU Patent 607326, 25 Oct 2015
10. Kolovertnov, G.Y.: Method of using a gas-production facility at a nearly-depleted field with a reservoir-radial collection system. RU Patent 2597390, 15 June 2015
11. Smith, S.F.: Frequency synchronization system for improved amplitude modulation and television broadcast reception. US Patent 6563893 B2, 15 September 2017
12. Fitasov, Y.E., Ivlev, D.N., Morozov, N.S., et al.: Wireless network-based time synchronization and local positioning system. Sens. Syst. **8–9**, 20–26 (2017)
13. Kosykh, A.V.: Adaptive dynamic ajustment of drifted crystal oscillator frequencies. Omsk Sci. Bull. Issue **1**, 168–169 (2008)
14. Levchenko, D.G., Nosov, A.V., Paramonov, A.A., et al.: AIMS information timestamping device. Sci. Instr. Making **17**(3), 88–95 (2007)
15. Telit wireless solutions. SL869 Product description. https://jt5.ru/files/pdf/Telit_SL869_Product_Description_r1.pdf Accessed 25 Oct 2017
16. Pavlova, Z.K., Baltin, R.R., Krasnov, A.N., et al.: The basics of designing wireless networks for monitoring the parameters of remote facilities. Int. Res. J. **12**(54), 161–164 (2016). Pt. 3
17. Krasnov, A.N., Prakhova, MYu., Khoroshavina, Y.A.: Wireless networks in oil-field automation systems. Oil Gas Bus.: e-J. **4**, 205–221 (2016)
18. Krasnov, A.N., Yefimova, V.N.: Optimizing the collection of data from scattered wireless sensors located at oil and gas condensate fields. Electric drive, electric technology, and industrial electrical equipment. In: Proceedings of the II International Research and Practice Conference, pp. 44–47. USPTU Publishing House (2015)
19. Improving the accuracy of a crystal oscillator. Revision 1. http://www.semtech.com. Accessed 20 Sept 2017
20. Bagayev, V.P., Kosykh, A.V., Lepetayev, A.N., et al.: Double-mode crystal oscillator with a digital adjustment for temperature. Electr. Conn. **3**, 48–51 (1988)

Control of Linear Servo Pneumatic Drive Based on Fuzzy Controller and Knowledge Base

E. L. Khaziev[✉]

Naberezhnye Chelny Institute, 68/19, Prospect Mira,
Naberezhnye Chelny, Russian Federation
emilius@yandex.ru

Abstract. The study is devoted to the control system of a linear pneumatic actuator based on fuzzy logic with the ultimate task of improving the quality of control indicators, namely the positioning accuracy at intermediate points of possible displacements, while maintaining high performance. Fuzzy inference techniques are widely used in the development of fuzzy controllers. The primary purpose of the fuzzy controller is to control an external object where the behavior of the managed object is described by fuzzy rules. Fuzzy logic controllers are the most important application of the fuzzy set theory. Their functioning differs from that of ordinary controllers by the fact the knowledge is used to describe the system instead of differential equations. The automated pneumatic actuator control system based on a fuzzy controller should be based on a knowledge base containing fuzzy frames (rules). The formation of this base is carried out on the basis of the knowledge of experts or the method of direct measurement with the help of control equipment.

Keywords: Pneumatic actuator · Control system · Servo mode · Adaptive mode · Fuzzy logic · Fuzzy controller · Knowledge base

1 Introduction

The construction of a system tracking control based on fuzzy logic allows to simplify the structure elements of the system, and also gives a discrete nature control method some continuous properties, which should enhance governance and reduce the volume of the knowledge base.

Fuzzy logic allows you to enter the tracking management of a certain well-known (a priori) information about the object in the form of fuzzy frame control, proximity to the natural shape of the language allows you to easily get the required knowledge from the experts. A priori information provides one of the main initial conditions of a system constructed according to the method of tracking control, the maximum condition of the initial adaptation.

2 Relevance, Scientific Significance of the Question

In modern engineering production of electromechanical actuators to actively replaced with pneumatic [1]. This is due to economic and operating advantages of pneumatic actuators [2].

The main drawback, which formerly hindered the development of such devices is the lack of an integrated management system that allows more flexibility to adjust and control pneumatic drives of machine tools or robots in the range of their possible movements [3–7].

The proposed control system pneumatic actuator is a hardware and software system that includes independent machine learning [8] software module based on fuzzy logic. This element of the system allows you to control patented device control – crane pneumatic dispensers [9]. With this solution the accuracy of the process control is improved [10–12].

Fuzzy control type is the method for approximation of data obtained as a result of experience or knowledge of experts. This issue is described in detail in the literature, for example [13, 14].

3 Statement of the Problem

Control of a linear pneumatic actuator in intelligent systems in the implementation of a tracking or adaptive modes of operation can be implemented using the methods of fuzzy inference using fuzzy controllers. The main task of the controller is the management of the external object at which the behavior of a managed object is described by a set of fuzzy rules. Controllers fuzzy logic is the most important application of fuzzy set theory, the functioning of which is significantly different from other types of controllers that the knowledge of experts instead of the data obtained by solutions of complex differential equations [15].

Control system linear pneumatic actuator on the basis of a fuzzy controller must rely on a knowledge base of fuzzy frames (rules). The formation of this database is based on the knowledge of experts or the method of direct measurements using monitoring equipment [16].

4 The Theoretical Part

Control system of pneumatic drive based on fuzzy controller can be represented in a diagram, shown in Fig. 1.

The system includes the following elements: the block set modes of operation of pneumatic actuators, the knowledge base of the programs of the task of operating modes of pneumatic actuators, unit of learning, the knowledge base of fuzzy control frames, the forming unit control programs, XML unit, CNC unit, compressor, crane valves, pneumatic actuator, the feedback sensors for position.

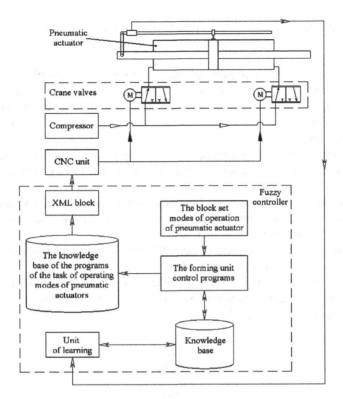

Fig. 1. Control system of the pneumatic drive based on fuzzy controller.

The initial stage of operation of the control system of the pneumatic drive starts with a knowledge base of fuzzy control frames on the basis of which is the creation of a control program [17].

At the stage of setting parameters of working modes of the actuators, the operator sets the expert through the modes characteristic of the methodology of the actuator on the basis of the elementary control programs.

Further, based on the characteristic of the methodology, the operator creates a sequence of changes of parameters with time. This produces the initial control program which is recorded in the "knowledge base of the programs of the task of operating modes of pneumatic actuators".

In this multilevel database is a sequential conversion of the input parameters of the operation modes in the control actions through fuzzy controller, which is located in the "the forming unit control programs".

For output variables the fuzzy controller uses linguistic variables and fuzzy frames, which are located in the "the knowledge base of fuzzy control frames". The result is a sequence of control parameters that is stored in the "the knowledge base of the programs of the task of operating modes of pneumatic actuators".

The resulting sequence is written to an XML file [16]. Based on the resulting data a "CNC unit" carries out direct control of stepper motors pneumatic crane valves and the block of preparation of air. The main parameters is the rotation angles of the stepping motors (distributors) and gas flow depending on the external load (for the block of preparation of air).

During operation of the pneumatic actuators using magnetostrictive sensors feedback (measuring path) obtain information about the current position of the object, each point of the measuring path (depending on the accuracy lies in the range 1–100 μ) shows the rate of change of the parameters of the two distributors, responsible for the operation of a single motor.

To prevent emergency situations, such as a short circuit or when the pressure increases and the failure of the safety valves, the sensor data is also directly received in the "CNC unit" to quickly disable the drive.

The "unit of learning" receives information from sensors in feedback mode. This is necessary in order to implement the adaptive control mode, and fill the knowledge base upon which are generated control frames.

The most important component of any intelligent system is a knowledge base. The knowledge base is a structure that contains a set of facts, rules or frames (object description in the form of attributes and their values) of output allowing logical inference and meaningful information processing [18]. This database consists of three levels and is presented in Fig. 2.

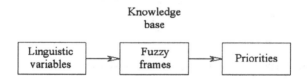

Fig. 2. Knowledge base of a fuzzy controller and methods for its completion

On the first level there are linguistic variables. These variables are compiled by the operator at the stage of formation of the program control to a specific type of pneumatic actuator, proceeding from the tasks of his job. the loading and unloading operations or functions tracking. They should include parameters which can be used to manage and set modes of operation. Linguistic variables are populated with experts in the field and methods of direct measurements. At this level is called the basic range, and the number of fuzzy labels, and the type of the membership function. The number of fuzzy labels depends on the accuracy of control. The more you have, the more accurate control, but this increases the time required for filling of the knowledge base. At this level there are also syntactic and semantic rules of linguistic variables [19].

On the second level applied a fuzzy control frames are used to convert the parameters in the control. The frames are made up of linguistic variables defined on the level above. They can also ask the experts, by direct measurements or by using intelligent approaches, such as self-learning neural network. In the latter case, full automation of the process of filling knowledge at this level. At this level there are also semantic rules that define the effect of linguistic variables on each other.

On the third level are the vectors of priorities, is intended to rank the characteristics. The vectors of priorities is composed of the specified parameters. On their basis it is possible to specify a special mode of operation of the drive, and they help to overcome the mid-section of the magnetic circuit of the sensor, it is necessary that the drive is not stopped at this point. This level is set by the method of expert estimates.

Important role in the knowledge base is played by the feedback. It is used not only for recording the results and adjusting management actions, but also to populate the knowledge base using the method of precedents.

Obtaining output data based on precedent is a decision-making method that uses knowledge about previous facts, situations or cases (precedents). When considering a new problem (the current case) sought a similar precedent as analogue. Instead of looking for the solution from the beginning every time, you can try to use the decision in a similar situation, perhaps adapting it to the changed circumstances of the current case. After the current event is processed, it is entered into the database of precedents, together with its decision for its possible use in the future [20, 21].

The precedent includes:

- A description of the task operation of an actuator
- The solution of the problem
- Outcome of applying the solution

The task description should contain all the information necessary to achieve the goal of optimal output (selection of the most appropriate solutions).

Basic requirements for finding the solution of the problem - the optimality of time finding the most accurate solution.

Description output result contains the rotation angle of governors of the stepper motors of the valves and the pressure and flow at a specific point in time for each point of the magnetic sensor feedback with the aim of securing the achievement of the specified point or continue movement. Description of results may also include links to other precedents, additional text information.

The method of precedents includes the following stages [19]:

- Retrieving the most relevant precedents for the current case from the library of precedents
- Adapting the chosen solution for the current case, if necessary
- Application solutions
- Application evaluation (validation) [22]
- Save, add the current case into the database of precedents

Consider filling the knowledge base of such frame to the operating mode of the pneumatic actuator. In this case, the control frame will look like the following:

$$IF\ L_i\ THEN\ \alpha_{1i}\ AND\ \alpha_{2j}, \tag{1}$$

where: L – it is a linear length measuring element - waveguide (magnetic) sensor meter path, it shows a specific value corresponding to the desired point; α_{1i} – the angle of rotation of the first stepper motor directional control valve; α_{2j} – the angle of rotation of the second stepper motor directional control valve.

On the basis of such frames can be the knowledge base for the operating mode of the pneumatic actuator. when driving the pneumatic actuator from one extreme to commit to some intermediate point, i.e. to a point with a fixed value L_0, then there is no further change of the parameters α_{li} and α_{2j}. Then base modes with fixing at some point can be represented as follows:

$$IF\ L_0\ THEN\ \alpha_{li}\ AND\ \alpha_{2j}, \qquad (2)$$

Fill values are not fixed points of L_n occurs with the use of precedents.

Assume that the movement of the actuator occurs from left to right ($L_{sign(+)}$) starting with an extreme point, then the control frame will look like this:

$$IF\ L_{sign(x+n)}\ THEN\ \alpha_{1x}\ AND\ \alpha_{2x}, \qquad (3)$$

where: $L_{sign(x+n)}$ – a linguistic variable is the length of the magnetic sensor showing the character areas is defined as the difference between the two values passed the last two points of the magnetic: $L_{sign(x+n)} = \alpha_i + \alpha_{(i+1)}$; α_{ix} – unknown increasing rotation angle of the first stepper motor directional control valve; α_{2x} – unknown increasing rotation angle of the second stepper motor directional control valve.

To fill this rule (namely, the movement from left to right) are looking for the nearest precedent, it is a frame:

$$IF\ L_{sign(x+1)}\ THEN\ \alpha_{1k+2}\ AND\ \alpha_{2m+1}, \qquad (4)$$

where: α_{1k+2} – a fixed increase of the rotation angle of the first stepper motor directional control valve with a big step; α_{2m+1} – a fixed increase of the rotation angle of the second stepper motor directional control valve with a smaller step.

On this basis create two control frame precedent:

$$IF\ L_{sign(x+1)}\ THEN\ \alpha_{1k+2}\ AND\ \alpha_{2m+1}, \qquad (5)$$

$$IF\ L_{sign(x-1)}\ THEN\ \alpha_{1m+1}\ AND\ \alpha_{2m+2}, \qquad (6)$$

The first frame shows an increase of the parameter L (the movement from left to right), the second frame shows the decrease of the parameter L (the movement from right to left).

For both frames true increase parameter change $L_{sign(x+n)}$, for (5) in the direction of increasing values of the nearest fuzzy label for (6) in the direction of decreasing values of the nearest fuzzy labels.

The development of a knowledge base based on frames-precedents (2), (5), (6) allows you to configure the actuator for different operations tasks for the implementation of the tracking or adaptive modes.

The steps of creating control programs of the pneumatic drive based on fuzzy logic. The process of creating control programs of pneumatic actuator based on fuzzy logic is the following sequence of steps presented in Fig. 3.

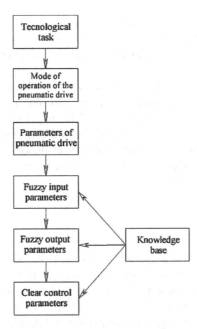

Fig. 3. Stages of creation of control programs of the pneumatic drive on the basis of fuzzy logic.

The upper level is the instruction of the process engineer on the task of carrying out certain technological operations to support a certain production process.

The next step is to select the desired mode of operation of the pneumatic actuator, the main indicators for the selection-load, positioning accuracy, speed and possibly tracking function.

The third step is to convert the parameters of the modes into specific numerical values of the parameters of the elements of the pneumatic actuator control system, incrementally, depending on the data from the sensors.

The fourth stage is the fuzzification, that is, the transformation of clear experimental values of input variables into fuzzy ones using the linguistic description of parameters (L, α_1, α_2). On the basis of linguistic variables we form fuzzy control frames:

$$IF\ L\ THEN\ \alpha_1\ AND\ \alpha_2, \qquad (7)$$

And then the knowledge base is formed on the basis of the received frames.

The final stage is the defuzzification, that is, the bringing of fuzzy frames of actual control commands that are recorded and stored in the database programs set modes of operation of pneumatic actuators.

5 The Results of Experimental Studies

Based on the theoretical study was developed a software module that allows for control of pneumatic actuator using FreeCAM [16].

The tests confirmed the performance of the model; an experimental setup was developed on the basis of pneumatic robot MP-9S with the application of the proposed hardware-software control system, and experimental study of its dynamic characteristics [23, 24].

A comparative study with the existing analog, which showed that, using the proposed control system and depending on the specified program of work, increases the performance when reaching the end positions of the output links of the robot, a possible adjustment of the speed and positioning of the working body of the robot at intermediate positions due to program control of stepper motors; it is determined that the performance of the proposed control system is higher by 16.7% compared to the system using the closest analogue [5].

6 Conclusion

The proposed control system pneumatic linear drive based on fuzzy logic allows to manage the pneumatic actuator via the differential method enable the crane control devices through software control of their actuators, electric stepper motors using fuzzy logic methods; to improve the quality of process control of pneumatic robot through the controlled regulation of the stages of acceleration and deceleration when reaching the intermediate position drives its actuators with an accuracy better than 0.1%.

References

1. Lovin, D.: Create an android robot with your own hands. Publishing House DMK-Press, Moscow (2007)
2. Pashkov, E.V., Kramar, V.A., Kabanov, A.A.: Servo drives, industrial process equipment: Training manual. Lan Publishing House, St. Petersburg (2015)
3. Khaziev, E.L.: Mathematical modeling of control system of pneumatic manipulator of an industrial robot. Sci. Tech. Gaz. Volga Region **3**, 173–177 (2011)
4. Khaziev, E.L.: Control system pneumatic industrial robot. Sci. Tech. Gaz. Volga Region **4**, 216–222 (2012)
5. Khaziev, E.L.: The calculation of the main parameters of the crane valve system of an industrial robot. Sci. Tech. Gaz. Volga Region **4**, 223–226 (2012)
6. Khaziev, E.L.: Simulation industrial pneumatic robot. In: Information Technologies. Automation. Updating and Solving Problems of Training of Highly Qualified Personnel, pp. 230–238 (2014)
7. Isaev, G.N.: Design of information systems: studies. Publishing House Omega-L, Moscow (2013)
8. Nikolenko, S., Kadurin, E., Arkhangel'skaya, E.: Deep Learning. Piter, St. Petersburg (2018)
9. Khaziev, E.L., Dmitriev, S.V.: Pneumatic dispenser. RU Patent 158927, 20 October 2016 (2016)
10. Litvinenko, A.M., Vasiliev, M.A.: Industrial robot with parallel kinematic chains. Mach. Eng., 46–48 (2007)
11. Kaliaev, I.A., Lokhin, V.M., Makarov, I.M., et al.: Intelligent robots: textbook for universities. Mashinostroenie, Moscow (2007)

12. Akimenko, T.A., Arshakyan, A.A., Budkov, S.A., et al.: Industrial robot with a management information system. Proc. Tula State Univ. **4**, 133–138 (2013)
13. Yarushkina, N.G.: The foundations of fuzzy and hybrid systems: study guide. Finance and Statistics, Moscow (2004)
14. Yarushkina, N.G.: Applied intelligent systems based on soft computing. ULSTU, Ulyanovsk (2004)
15. Khaziev, E.L., Khaziev, M.L.: Control system pneumatic robot based on fuzzy logic. Mod. High Technol. **3**(1), 74–78 (2016)
16. Khaziev, E.L., Khaziev, M.L.: Fuzzy control for pneumatic feed milling-boring machine with the use of the XML specification. Mod. High Technol. **9**(1), 84–88 (2016)
17. Khaziev, E.L.: Fuzzy control of pneumatic subsystems of the machine tool during loading and unloading operations. Dissertation, Naberezhnye Chelny (2017)
18. Gavrilova, T.A., Khoroshevskiy, V.F.: Knowledge base of intelligent systems. Piter, St. Petersburg (2000)
19. Zubkov, E.V., Dmitriev, S.V., Khayrullin, A.K.: Algorithmization of technological processes automated testing of diesel engines. Kazan University, Kazan (2011)
20. Solodovnikov, I.V., Rogozin, O.V., Shuruev, O.V.: Implementing logical inference mechanism for the prototype expert system. New Information Technologies, Moscow (2004)
21. Batyrshin, I.Z., Nedosekin, A.O., Stecko, A.A.: Fuzzy hybrid system. Theory and practice. Fizmatlit, Moscow (2007)
22. Leonenkov, A.V.: Fuzzy modeling in MATLAB and fuzzyTECH. BKhV Petersburg, St. Petersburg (2005)
23. Sidnyaev, N.I.: Theory of experiment planning and statistical data analysis. Urait, Moscow (2012)
24. CAMOZZI: Pneumatics for everyone. Tutorial Company (2015)

Method for Controlling Tracking Actuator

E. V. Zubkov[✉] and M. L. Khaziev

Kazan Federal University Naberezhnye Chelny Institute, 68/19 Mira Avenue,
Naberezhnye Chelny 423812, Russian Federation
eugen_z@mail.ru

Abstract. The work belongs to the field of automation and can be used in automatic control systems, servo drives in various industries. The technical result is the expansion of the control range of automatic systems without loss of stability when changing the properties of the controlled object and/or when there are significant disturbances both in the load and in the control channel. It is achieved through the correction of the proportional and integral components of the regulating signal, and corrective action occur in accordance with the values of the exponential function for the proportional component of the PI-controller and integral – using the inversely exponential dependence, the argument of which is the current error regulation. The proposed method of regulation will meet the requirements for the stability margin for automatic control systems. The control technique can be used to solve the problems of static accuracy of controlled objects and issues of quality assurance and dynamic accuracy of the system as a whole.

Keywords: Servo drive · Automatic control system · The method of extension of the control range · PI controller

1 Introduction

The servo drives operate in accordance with the built-in program which allows for detection of deviations of controlled variable from the control action and, ultimately, the impact on the object of regulation (the working body), with the aim of information specified deviations to zero [1].

As the requirements to the behavior of the servo actuator dynamics, depend on their purpose, principle of action, the nature of the external influences, specific conditions, etc., they can be very different, but they can be summarized in four categories: requirements for stability margin; static accuracy; requirements for system behavior in the transition process (terms of quality); requirements for dynamic precision of the system [2].

Create a servo drive that meets all of these requirements, according to the authors, possibly on the basis of code-switching (digital) nature of the control signals of the controllers implemented in the programmable controllers or digital computers.

This paper considers a new way to extend the control range of the automatic control systems without loss of stability.

2 Methodology

The vast majority of manufactured and used by regulators in automatic control systems, implemented in the program the PID control law. Controllers with the PID control are basically the only controllers used in practice in automatic control systems of technological processes. Wide application of the PID act is due to the fact that the algorithm of their functioning successfully imitates the work of an experienced human operator.

The regulators executed on a microprocessor basis, provides such a service, as the setting parameters by the operator with the ability to programmatically set these settings.

The disadvantage of these regulators is the immutability of settings in large and small perturbations, which leads to limited sustainable operation of automatic systems of regulation, and indicators of the quality of the transition process worse. The most significant deterioration regulation occurs when the change in parameters of object of regulation.

When you change the properties of the object of regulation and the appearance of significant disturbances as load, and the control channel, the coefficients of the system configuration needs to change, but formation of the control action controller remains the same as in steady and in transient modes, which leads to an increase of oscillating, overshoot and time regulation.

Known methods of adjusting the controller parameters with the identifier that includes identification of the control object by applying an object perturbations of a certain type, for example speed, latching of the response object to these perturbations, the calculation of the optimal settings of the regulator according to the obtained dynamic model of the object and compare the found parameters of the already established previously, and, if the compared parameters are different, then set the new settings and re-performed the identification of the object, and if not, the process correction parameters stop and translate the system in operational mode.

A disadvantage of the known method is the use of models with insufficient number of parameters that do not account for all properties of the object, which leads to the installation not corresponding to the setting object and, as a consequence, poor quality control.

There are also ways of self-adjustment, based on the classical method of Ziegler-Nichols [3] and its modifications. The essence of these methods is that a closed system is in an oscillatory mode, determined values of the critical gain of K_{cr} and critical period of oscillations of T_{cr}. Then, according to [3], determined the optimal settings for a typical linear regulator, expressed through K_{cr} and T_{cr}:

- for P-controller:

$$K = 0,5 \cdot K_{cr}, \tag{1}$$

- for PI controller:

$$K = 0,45 \cdot K_{cr}; \; T_i = T_{cr}/1,2, \tag{2}$$

- for PID controller:

$$K = 0,6 \cdot K_{cr}; \ T_i = T_{cr}/2; \ T_d = T_{cr}/8, \quad (3)$$

Methods of determining K_{cr} and T_{cr} can be different. The critical gain coefficient K_{cr} is determined by the output of the system at the border of stability [3] with variation of the gain knob. In [4, 5], the system enters a two-position control, wherein the system having oscillations with the parameters used to configure the controller. The input object is served trial harmonic signal with a variable frequency fluctuation [6]. The oscillation frequency is chosen to provide the critical frequency at which phase shift between input and output is equal to 3.14 rad.

The disadvantage of these methods of self-adjustment is the duration of the identification process associated with the statistical analysis of several periods of oscillation. To improve the accuracy of determining the K_{cr} and T_{cr} may require multiple iterations, which also delays the process of self-adjustment. The method is not suitable for objects that have dynamic characteristics which depend on the sign of the error regulation. These objects include, for example, a wide class of thermal objects (stoves, heaters, sterilizers, etc.) for which the processes of heating and cooling may have very different characteristics and require different controller settings for heating and cooling. In the present methods, the parameters of self-oscillations are averaged for positive and negative half waves and the result is determined by some average values of settings that do not match with the heating process or the cooling process. In all considered methods after determining the two parameters K_{cr} and T_{cr} used formulas Ziegler-Nichols to calculate controller settings. These formulas are empirical and are designed for objects with the ratio τ/T is from 0 to 0.3 so they do not guarantee quality control for high delay. In addition, in the case of PID controller for three settings, T_i, T_d are determined only by two parameters K_{cr} and T_{cr}, which indicates the inadequacy of the received settings of the actual control object.

In low-inertia objects (relative to the speed of movement of the regulatory body) under large perturbations close to the speed, control value changes fast enough. The situation arises when the integral component of control action accumulates faster than regulatory body has time to work out. As a result, the system is uncertain or even buckling.

The known method of correction of automatic control systems [7], in which, when reducing the current errors of regulation to values equal to or less than the specified, the change of the signal in the feedback circuit, which is formed as the product of the rate of change of the output parameter of the object at a reduced fixed ratio that translates the transition in monotone.

This technique increases the stability of the system within the established deviation of the stabilized parameter from the specified object and speeds up the elimination of deviation in case of exceeding set value, but weakens the proportional and the differential component of the control signal, increasing the accuracy of stabilization of the controlled parameter in the steady state.

A method of expanding the boundaries of the stable operation of automatic control system implemented with the configuration of the controller for the minimum time regulation [8].

Transient regulation of regulators, as configured by the method of minimal control time, become aperiodic in nature. If the perturbation at the load it is characterized by a greater stability of the automatic regulation system. The duration of a single half-wave (when the perturbation in load) substantially longer time the action of the first half-wave other types of settings. The transition of the aperiodic characteristic is the increase in the accuracy of stabilization of the controlled parameter in the steady state.

Thus, at the moment, there are technical solutions that provide a result that would be aimed at expanding the boundaries of the stable operation of automatic system of regulation and maintaining valid indicators of the quality of management of the automatic system when changing the characteristics of the object, and the appearance of increased perturbing effects on the stabilization system controlled variable as transient and in steady-state mode, not by a General "weakening" of the controller settings, and the adjustment of the proportional and integral components of the control signal as a function of the current error regulation in a selected field values from a preset to zero.

3 Results

The analysis of existing methods of control automatic control systems it was decided to look for a new method of regulation, which required a series of tests.

The test stand includes six parallel connected pumps driven by asynchronous electric motor. Five of them are not controlled pumps and one pump is frequency regulation of the electric motor. All pumps have an adjustable check valve. Pipe, which connected the pumps equipped with a pressure sensor, a flow meter and a valve to release pressure in the system. The stand simulates the flow of water to the consumer.

The need for joint inclusion of the pumps occurs in cases where one pump cannot provide the proper flow or pressure, or the required reserve to ensure continuity of water supply.

Control of the pump was carried out using programmable logic controller SIMATIC S7-300.

During the tests were obtained the statistical number of adjustable values of the pressure needed to create a simulation model identifying the main characteristics of the transition process.

To create a simulation model Fig. 1 package was used Scicos is included in Scilab mathematical program. Scicos – a tool for editing block diagrams and simulations (similar to simulink in MATLAB).

In the course of the simulation model has been applied to the way of setting up a PI-regulator which is not in the General weakening of the settings of the controller, and in the correction of the proportional and integral components of the control signal depending on the current error Δ in the regulation of a wide range of values of the current error regulation when changing the coefficients of the controller, in the process of his work.

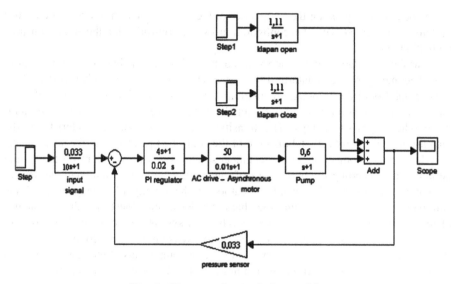

Fig. 1. Diagram of a simulation model.

The proposed method is illustrated in the block diagram of automatic system of regulation "Fig. 2" that implements a PI control law with the correction current error regulation of Δ in the proposed method.

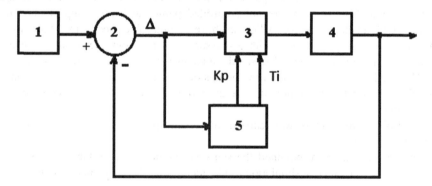

Fig. 2. A block diagram of an automatic control system.

The system contains an element of comparison 2, the inputs of which are connected to the master device 1 stabilized parameter and to the output of the control object 4. The output of the comparison member 2 is connected to the inputs of the controller unit 3 and unit 5 of the correction parameters K_p and T_i relative to the current regulator error control of nonlinear functions that characterize the object of regulation.

The action block of the correction current error regulation may be explained as follows. For example, finding, through the required step value changes adjustable values within a specific range and the optimum control for these differences, for which known methods of approximation can be represented by the following exponential dependence:

$$T_i = A \cdot \Delta^{-B}, \tag{4}$$

$$K_p = C \cdot \Delta^{D}, \tag{5}$$

Using the found approximation of the power-like dependencies B and D, the calculated parameters of PI controller according to the formulas (4), (5) and are able to determine close to optimal controller parameters which allow the desired correlation coefficient in the whole range of regulation.

On "Fig. 3" presents the pressure changes (trend 1), obtained under the action of external perturbations (trend 2) without adjusting the regulator output pressure.

Six parallel pumps, in the course of the experiment, experienced dynamic hydraulic loads in the form of different values of pressure drops. The dynamics of the loading pump was held not periodic laws, in this case, the loss of stability of an adjustable pump of the system was observed.

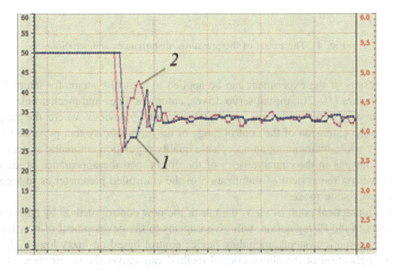

Fig. 3. Trends of pressure and disturbance.

As a result of correction of the parameters of the PI controller using the presented technique, the pump station gave out sustainable indicators Fig. 4 the required output pressure (trend 3) and medium pressure pumping stations (trend 2), regardless of the magnitude of external disturbances (trend 1), while retaining stability throughout the control range close to a predetermined value (trend 4).

Upon completion of the survey they obtained the patent for the invention [9] a method of expanding the control range of the automatic control systems without loss of stability.

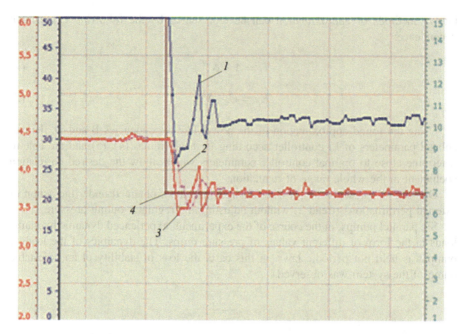

Fig. 4. The trends of the pressures, disturbances and setpoint.

The results of the experiment can be applied to control systems for internal combustion engines [10–26], digital servo drives, automatic lines and mechatronic devices that are managed by controllers or digital computing devices, where the proposed method is the extension of the control range of the automatic system prevents the loss of stability while maintaining an acceptable quality regulation automatic system in the event of changes in the characteristics of the object, and the appearance of increased disturbances on the system of stabilization of the controlled parameter as in transients and in steady-state mode.

From a practical point of view, the use of the new control method by adjusting the parameters of the PI-regulation, will give impetus to the development of new types of adaptive automatic systems, including device control based on fuzzy logic [27, 28].

The application of the above control method can satisfy the requirements for stability margin for control systems. The proposed method can be prospectively applied to solve problems on demand to the static accuracy of the regulated objects, as well as to the requirements in terms of quality and dynamic accuracy of the system as a whole.

References

1. Chemodanov, B.K., Blaze, E.S., Zimin, A.V., et al.: Servo Drives. Publishing House of BMSTU, Moscow (1999)
2. Pashkov, E.V., Kramar, V.A., Kabanov, A.A.: Servo Drives, Industrial Process Equipment: Training Manual. Publishing House Hind, St. Petersburg (2015)

3. Ziegler, J.G., Nichols, N.B.: Optimum settings for automatic controllers. Trans. ASME **64**, 759–768 (1942)
4. Pevsner, V.V., Lakhova, N.V., Nikolskaya, L.V., et al.: Microprocessor Controller Remikont R-130. Printing House of Research Institute Teplopribor, Moscow (1990)
5. Semenets, V.P.: Method of automatically setting system of regulation. RU Patent 2002289, 30 October 1993 (1993)
6. Mazurov, V.M.: Self-tuning control system. RU Patent 2068196, 20 October 1996 (1996)
7. Lubentsova, E.V.: Proportional-integral-differential controller. RU Patent 2234116, 27 August 2004 (2004)
8. Varlamov, L.G.: What criteria optimization of transient processes in automatic control systems it is more efficient to use in practice. Ind. Autom. Control Syst. Controll. **5**, 56–57 (2005)
9. Zubkov, E.V., Khaziev, M.L.: Method to expand the range of regulation is an automatic control system without loss of stability. RU Patent 2619746, 24 March 2016 (2016)
10. Zubkov, E.V., Dmitriev, S.V., Hajrullin, A.H.: Algorithmization of Technological Processes of the Automated Tests of Diesel Engines. Kazan University, Kazan (2011)
11. Zubkov, E.V., Makushin, A.A., Ilyukhin, A.N.: Rules, functions, and systems for the formation of models of the test conditions for truck and tractor combustion engines. Traktors Selkhozmash **5**, 17–20 (2009)
12. Zubkov, E.V., Novikov, A.A.: Regulation of the crankshaft speed of a diesel engine with a common rail fuel system. Russ. Eng. Res. **32**(7–8), 523–525 (2012)
13. Zubkov, E.V., Makushin, A.A., Novikov, A.A., et al.: Imitating modeling of testing of the diesel with common rail fuel supply system. Assem. Mech. Eng. Instrum.-Mak. **8**, 29–31 (2011)
14. Galiullin, L.A.: Automated test system of internal combustion engines. Proceedings of the International scientific and technical conference innovative mechanical engineering technologies, equipment and materials **86**, 12–18 (2014)
15. Galiullin, L.A., Valiev, R.A.: Automated system of engine tests on the basis of Bosch controllers. Int. J. Appl. Eng. Res. **10**(24), 44737–44742 (2015)
16. Zubkov, E.V., Mochalov, D.I., Hajrullin, A.H.: Imitating modeling of technological processes of the automated tests of diesels at dynamic loads at machine-building enterprise. Sci. Tech. Volga Reg. Bull. **1**, 274–277 (2013)
17. Galiullin, L.A., Zubkov, E.V., Mochalov, D.I.: Mathematical modeling of the modes of tests of diesel engines. St. Petersburg State Polytech. Univ. J. **5**, 77–81 (2011)
18. Yarushkina, N.G.: Fundamentals of fuzzy and hybrid systems. Finance and Statistics, Moscow (2004)
19. Makushin, A.A., Zubkov, E.V., Ilyukhin, A.N.: Fuzzy logic in the modeling of tests for internal combustion engines. Assembl. Mech. Eng. Instrum.-Mak. **12**, 39–44 (2009)
20. Ilyukhin, A.N., Zubkov, E.V.: Modernized algorithm of neural network initial weighting factors during the diagnosis of diesel engine faults. Int. J. Appl. Eng. Res. **10**(24), 44848–44854 (2015)
21. Biktimirov, R.L., Valiev, R.A., Galiullin, L.A., et al.: Automated test system of diesel engines based on fuzzy neural network. Res. J. Appl. Sci. **9**(12), 1059–1063 (2014)
22. Zubkov, E.V., Galiullin, L.A.: Hybrid neural network for the adjustment of fuzzy system when simulating tests of internal combustion engines. Russ. Eng. Res. **31**(5), 439–443 (2011)
23. Galiullin, L.A., Zubkov, E.V.: Neuro and indistinct setup of the automated system of tests of engines. Assembl. Mech. Eng. Instrum.-Mak. **7**, 26–31 (2011)

24. Zubkov, E.V., Galiullin, L.A.: A hybrid neural network for modeling of the modes of tests of internal combustion engines. Sci. Tech. Statements St Petersburg Polytech. Univ. **1**, 245–250 (2011)
25. Galiullin, L.A., Zubkov, E.V.: Intellectual Setup of the Automated Systems of Tests of Diesels. LAMBERT Academic Publishing, Saarbrücken (2011)
26. Zubkov, E.V.: The module of imitating modeling of modes of behavior of diesels of the automated system of tests. Basic Res. **11**(1), 49–53 (2015)
27. Khaziev, E.L., Khaziev, M.L.: Control system of the pneumatic robot based on fuzzy logic. Mod. High Technol. **3**(1), 74–78 (2016)
28. Khaziev, E.L., Khaziev, M.L.: Fuzzy control for pneumatic feed milling-boring machine with the use of XML. Mod. High Technol. **9**(1), 84–88 (2016)

Method for Composition of Two, Three and More Multistructural Petri Nets into Single Equivalent Control Model of Technological Process

M. P. Maslakov[✉], S. V. Kulakova, and A. Z. Dobaev

North Caucasian Institute of Mining and Metallurgy, 44 Nikolayeva Street,
Vladikavkaz 362021, Russian Federation
kalbash1@mail.ru

Abstract. This paper proposes a new approach to the construction of control models of complex technological processes consisting of sub-processes (operations) presented in the form of models described in terms of differential structural modified Petri networks. Method implementation procedures are defined and implementation algorithm is described. Two types of composition are presented when implementing this construction approach: serial and parallel. This approach is recommended when obtaining a single control model by a complex technological process, if the operations implementing it are presented in terms of one mathematical apparatus, but in various modifications thereof. The Cartesian product of graphs modified under Petri nets taking into account their different structure and requirements to composition types is the basis for the offered approach. The offered approach - a composition method the multi-structural of the modified Petri nets, provides increase in extent of synchronization of the technological operations making difficult technological process. The received managing models with use of the presented approach provide required extent of synchronization at the smallest labor costs, the conversions caused by need the multistructural of the models making difficult technological process of operations to models of identical structure. The offered algorithm is illustrated on a specific example.

Keywords: Technological process control model · Modified Petri nets · Method for multistructural Petri nets composition · Complicated transition

1 Introduction

The issues of automating the design of the processes to develop the systems controlling various complicated technological processes (CTP) as well as introducing the methods and algorithm for their efficient control have been studied by such scientists as: Sleptsov, Yurasov, Andriyevskiy, Megerhut, Konyukh, Zaitsev, Dedegkaev, Rutkovskiy, Bosikov, Braginsky, Dragomir, However, their papers [1–13] do not consider the issues of the development of equivalent control models in terms of increasing the synchronization degree of the operations performed in CTP.

The development of control models for complicated technological processes shall take into account many factors which will afterwards determine the efficiency of the control system as well. One can identify the equivalence of a control model to the manufacturing instructions implemented at the managed production facility as a main factor. The papers [14–17] proposed a number of methods and ways for building control models equivalent to complicated technological processes on the basis of the proposed Petri net modifications. However, none of them is suitable for the implementation of the composition of multistructural Petri net modifications.

This article proposes the method for the composition of multistructural Petri net modifications describing the operations included into a technological process. The rationale for this method development is explained by the fact that the development of a single model for the complicated technological process control does not always provide for the opportunity to represent the process operations in terms of a single mathematical apparatus.

The proposed composition method is more efficient in relation to the enhancement of the CTP operation synchronization degree.

The control models obtained with the help of this method provide for an undoubted necessary degree of synchronization at the lowest labor efforts which can be caused by the necessity to transform multistructural models making the CTP operations into unistructural models.

The results of the early research [15–21] became the basis for the development of the method for the composition of multistructural Petri net modifications, in particular, the authors used the Petri net modifications and composition methods from the papers [18–21].

2 Description of the Proposed-Composition Method

Let us illustrate the proposed method for the composition of multistructural Petri net modifications by the following example.

Given two modified Petri nets: $N1 = <P_v, T, I, O, \mu_0, P_i, P_o, I_i, O_o>$ (Fig. 1) and $N2 = <P_v, T, I, O, T_n^m, I_n^m, O_n^m, \mu_0, P_i, P_o, I_i, O_o>$ (Fig. 2). In Figs. 1 and 2 dot-and-dash arrows highlighting the transitions into a set of input and output positions do not convey any special meaning and are provided only for the visibility. The notation of the net position sets are identical as they do not take part in the method implementation. The notation of input and output positions and the transitions of the nets N1 and N2 are different by the presence of their net identifier in the notation. Figure 2 shows the compound transition T3 in pink.

The composition of the above given nets is performed with the help of the following algorithm:

1. For each net N1 and N2 one builds the matrices of the input function D_i^- for the input positions and output function D_o^+ for the output positions.

    For the net N1 they have the form:

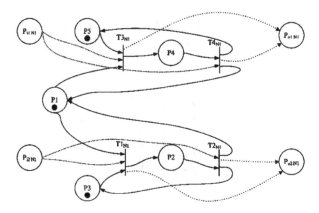

Fig. 1. Modified nets N1.

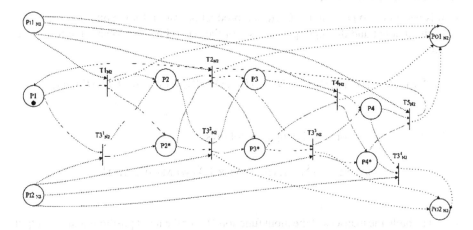

Fig. 2. Modified nets N2.

$$D_i^- = \begin{array}{c} \\ P_{i1N1} \\ P_{i2N1} \end{array} \begin{vmatrix} T1_{N1} & T2_{N1} & T3_{N1} & T4_{N1} \\ 0 & 0 & 1 & 1 \\ 1 & 1 & 0 & 0 \end{vmatrix}$$

$$D_o^+ = \begin{array}{c} T1_{N1} \\ T2_{N1} \\ T3_{N1} \\ T4_{N1} \end{array} \begin{vmatrix} P_{o1N1} & P_{o2N1} \\ 0 & 1 \\ 0 & 1 \\ 1 & 0 \\ 1 & 0 \end{vmatrix}$$

For the net N2 they have the form:

$$D_i^- = \begin{array}{c} \\ P_{i1\,N2} \\ P_{i2\,N2} \end{array} \begin{vmatrix} T1_{N2} & T2_{N2} & T3_{N2} & T4_{N2} & T5_{N2} \\ 1 & 1 & 0 & 1 & 1 \\ 0 & 0 & 1 & 0 & 0 \end{vmatrix}$$

$$D_o^+ = \begin{array}{c} T1_{N2} \\ T2_{N2} \\ T3_{N2} \\ T4_{N2} \\ T5_{N2} \end{array} \begin{vmatrix} P_{o1\,N2} & P_{o2\,N2} \\ 1 & 0 \\ 1 & 0 \\ 0 & 1 \\ 1 & 0 \\ 1 & 0 \end{vmatrix}$$

2. Let us determine the composition type (consecutive or parallel).

2a. Consecutive composition:

- It is necessary to perform the Cartesian product operation for the output position set of the first net and the input position set of the second net. We obtain the following set:

$$S_{oN1\,iN2} = \{(P_{o1\,N1}P_{i1\,N2});\ (P_{o1\,N1}P_{i2\,N2});\ (P_{o2\,N1}P_{i1\,N2});\ (P_{o2\,N1}P_{i2\,N2})\}$$

- Let us rename the obtained couple as follows:

$$S_{oN1\,iN2} = \{S_{1\,oN1\,iN2};\ S_{2\,oN1\,iN2};\ S_{3\,oN1\,iN2};\ S_{4\,oN1\,iN2}\}$$

- Let us build the matrices of the input function D_i^- for the input positions and the output function D_o^+ for the net output position but with the account of the obtained set S_{oN1iN2}. It should be mentioned that the positions from the set S_{oN1iN2} will be included in those input I and output O functions of the transitions including the positions of the generating sets $P_{o\,N1}$ and $P_{i\,N1}$, for instance: $S_{oN1iN2} = (P_{o1\,N1}P_{i1\,N2})$; the position P_{o1N1} was included into the following transition output functions: $O(T3_{N1})$, $O(T4_{N1})$; the position P_{i1N2} was included into the following transition input functions: $I(T1_{N2})$, $I(T2_{N2})$, $I(T4_{N2})$, $I(T5_{N2})$. Therefore, the position S_{oN1iN2} is included in the composition of the following transition input and output functions: $I(T1_{N2})$, $I(T2_{N2})$, $I(T4_{N2})$, $I(T5_{N2})$, $O(T3_{N1})$, $O(T4_{N1})$.

We obtain the following matrices of the input function D_i^- of input positions and the output function D_o^+ of the output positions.

For the net N1 they have the form:

$$D_i^- = \begin{array}{c|cccc} & T1_{N1} & T2_{N1} & T3_{N1} & T4_{N1} \\ \hline P_{i1\,N1} & 0 & 0 & 1 & 1 \\ P_{i2\,N1} & 1 & 1 & 0 & 0 \end{array}$$

$$D_o^+ = \begin{array}{c|cccc} & S_{1\,oN1\,iN2} & S_{2\,oN1\,iN2} & S_{3\,oN1\,iN2} & S_{4\,oN1\,iN2} \\ \hline T1_{N1} & 0 & 0 & 1 & 1 \\ T2_{N1} & 0 & 0 & 1 & 1 \\ T3_{N1} & 1 & 1 & 0 & 0 \\ T4_{N1} & 1 & 1 & 0 & 0 \end{array}$$

For the net N2 they have the form:

$$D_i^- = \begin{array}{c|ccccc} & T1_{N2} & T2_{N2} & T3_{N2} & T1_{N1} & T1_{N1} \\ \hline S_{1\,oN1\,iN2} & 1 & 1 & 0 & 1 & 1 \\ S_{2\,oN1\,iN2} & 1 & 1 & 0 & 1 & 1 \\ S_{3\,oN1\,iN2} & 0 & 0 & 1 & 0 & 0 \\ S_{4\,oN1\,iN2} & 0 & 0 & 1 & 0 & 0 \end{array}$$

$$D_o^+ = \begin{array}{c|cc} & P_{o1\,N2} & P_{o2\,N2} \\ \hline T1_{N2} & 1 & 0 \\ T2_{N2} & 1 & 0 \\ T3_{N2} & 0 & 1 \\ T4_{N2} & 1 & 0 \\ T5_{N2} & 1 & 0 \end{array}$$

On the basis of the obtained matrices we build a net which is a composition of consecutive nets N1 and N2 (Fig. 3). The nets in Fig. 3 are presented in the form of blocks for better visibility and understanding.

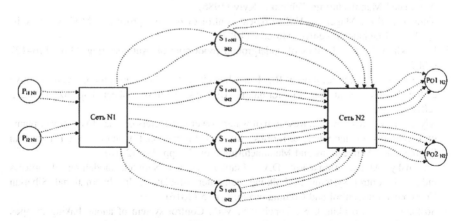

Fig. 3. Net - composition of the nets N1 and N2.

2b. Parallel composition:
If there is a necessity to provide for the parallel functioning of the CTP operations, the operations realizing the net composition are identical to the consecutive composition operations except for the first one: it is necessary to conduct the Cartesian product operation for the sets of input positions of the first and second nets as well as for the output position sets of the first and second nets, correspondingly.

These operations result in the common sets of input and output positions for the nets being united. All other operations for the building of the input function D_i^- of input positions and the net are performed on the basis of the Cartesian product result.

3 Conclusion

The proposed method for the composition of multistructural Petri nets to get equivalent control models of complicated technological processes can be used at the uniting of two, three and more nets regardless of their necessary mutual functioning type (consecutive and/or parallel). The equivalence of the proposed method and the control models obtained on its basis is confirmed by the modelling results in PIPEv4.3.0.

Acknowledgements. The article was made with the financial support of the Russian Foundation for Basic Research within the scientific project No. 16-38-00551.

References

1. Zaitsev, D., Jurjens, J.: Programming in the Sleptsov net language for systems control. Adv. Mech. Eng. **8**(4), 1–11 (2016)
2. Zaitsev, D., Petri, A.: Nets and system modelling: method instructions. Odessa National Academy of Telecommunications, Odessa (2006)
3. Konyukh, V.L., Ramazanov, R.A.: Control of underground load-haul-dump machines from the surface. Phys. Tech. Prob. Resour. Dev. **4**, 61–66 (2004)
4. Sleptsov, A.I., Yurasov, A.A.: Automation of the Designing Control Systems of Flexible Automated Manufacturing. Tekhnika, Kyiv (1986)
5. Yuditskiy, S.A., Magergut, V.Z.: Logical control of discrete processes. Models, analysis, synthesis. Machinostroyeniye, Moscow (1987)
6. Konyukh, V.L.: Peculiarities of underground robot control. Autometering **43**(6), 116–127 (2007)
7. Amelin, K.S., Andriyevskiy, B.R., Tomashevich, S.I.: Data transmission with adaptive coding between quadcopters in formation. Big Syst. Control Collect. Pap. **62**, 188–213 (2016)
8. Klyuev, R.V., Bosikov, I.I.: Research of water-power parameters of small hydropower plants in conditions of mountain territories. In: 2nd International Conference on Industrial Engineering, Applications and Manufacturing, vol. 53, pp. 1–5 (2016)
9. Braginsky, M.Ya., Tarakanov, D.V., Tsapko, S.G.: E-Network modelling of process industrial control systems in building computer simulators. In: International Siberian Conference on Control and Communications, p. 5 (2016)
10. Soshkin, S.V., Soshkin, G.S., Topchayev, V.P.: Control system of anode baking in open annular furnace. Non-ferr. Met. **9**(873), 62–67 (2015)

11. Minca, E., Dragomir, O.E., Dragomir, F.: Producer-consumer distributed energy production systems modelling with a new approach of recurrent synchronized fuzzy petri nets. In: 8th World Congress on Intelligent Control and Automation, pp. 1668–1673 (2010)
12. Dedegkaev, A.G., Maslakov, M.P.: Modelling of glasswork technological process with modified Petri Nets by the example of Irsteklo, LLC. Sustain. Dev. Mountainous Territories **4**, 35–40 (2012)
13. Dedegkaev, A.G., Maslakov, M.P., Maslakov, D.P.: Application of modified Petri Nets at composition of models of parallel technological operations. In: Collection of Papers of All-Russian Scientific Conference of Young Scientists, Post-graduates and Students, Publishing House of Southern Federal University, Gelendzhik, November 2012. Automation Problems. Regional Control. Communication and Automation, pp. 36–40 (2012)
14. Maslakov, M.P., Dedegkaev, A.G., Antipov, K.V., et al.: Application of modified petri nets as control models of the technological process of glass batch preparation. Glass Ceram. **72**(7–8), 324 (2015)
15. Maslakov, M.P., Dedegkaev, A.G., Antipov, K.V.: The activity count of transitions Petri networks of technological processes. In: 7th International Conference Science and Technology, pp. 20–25 (2016)
16. Maslakov, M.P., Maslakov, D.P.: Operations on Petri Nets. In: Materials of International Extra-Mural Research-To-Practice Conference, Publishing House Siberian Association of Consultants, Novosibirsk, June 2012. Physical and Mathematical Sciences and Information Technologies: Relevant Problems, pp. 12–16 (2012)
17. Maslakov, M.P., Dedegkaev, A.G.: Method of modification of Petri Nets to build operating models of complicated technological processes. Sci. Prospects **3**(78), 39 (2016)
18. Maslakov, M.P.: The automation of design technological preparation process of preparation furnace charge. The thesis for a degree of candidate of technical sciences. North Caucasian Institute of Mining and Metallurgy, Vladikavkaz (2013)
19. Maslakov, M.P., Dedegkaev, A.G.: Method for modification of Petri Nets to build control models of complicated technological processes. Sci. Prospects **3**(78), 39–45 (2016)
20. Maslakov, M.P., Antipov, K.V., Dobaev, A.Z.: Method of activity transition graphs conversion into modified Petri Nets of technological processes. In: International Conference on Industrial Engineering, Applications and Manufacturing, p. 4 (2017)
21. Maslakov, M.P., Kulakova, S.V.: Development of method of consecutive and parallel composition of modified Petri Nets to obtain integral and equivalent model to control CTP operations. Sci. Bus. Ways Dev. **10**(76), 83–87 (2017)

Control System of Tension Device in Winding Fabrics from Composite Materials

A. Mikitinskiy[1](✉), B. Lobov[1], and P. Kolpachshan[2]

[1] Platov South-Russian State Polytechnic University, 132 Prosveshcheniya Street, Novocherkassk 346428, Russian Federation
mialexp@mail.ru
[2] Rostov State Transport University, 2 Rostovskogo Strelkovogo Polka Narodnogo Opolcheniya Sq., Rostov-on-Don 344038, Russian Federation

Abstract. Products made by winding are widely used in many industries. The winding technology is improved and control systems must adopt to the changing operating modes. It may be possible only with the use of modern systems of winding control. There are special requirements for synthesis of tension control systems regulators. The synthesis must take into account the specific features of winding. Known synthesis methods do not take into account the peculiarities of the "wet" composite tape. This article deals with the mathematical description and linearization of the elastic tape, a structural diagram of a linearized control object (elastic tape) and a structural diagram of the tensioner mechanical part. The received description was linearized. The Bode plots of initial and simplified models were graphed. The initial model substitution for a simplified one does not significantly affect the description of the control object but significantly simplifies the control system synthesis. The use of synchronous motors with permanent magnets as an electromechanical converter is most appropriate. The structural diagram of tensioner position system is developed. The simulation of this system was realized. The developed control system meets the requirements of the composite materials winding.

Keywords: "Wet" composite tape · Mathematical model · Control system · Compensating device

1 Introduction

Fabrics made from composite materials by winding method have recently found their application in many branches: in spacecraft and space rockets, aircraft, chemical industry owing to their chemical-mechanical features [1]. There are two methods of winding fabrics: "dry" and "wet". Both methods are being widely used. Up to the present time the control systems used in winding were created using direct-current drives with thyristor voltage regulators. Control systems had analogue control circuits, which complicated its production, debugging, limited functional drive capabilities. In connection with the winding technology upgrading control systems were imposed stricter requirements concerning drives operation speed, possibility of their adaptation to the changing operating conditions in the process of winding. Due to the above-mentioned facts control system of "wet" composite tape tension using synchronous engine with vector control system of brand-name Siemens in the capacity of electrical actuating motor was developed at SRSPU (NPI) South Russian State Polytechnic University (Novocherkassk Polytechnic Institute).

Let's consider the process of "wet" winding of fabrics. The track of winding is shown schematically in Fig. 1.

Bobbins with material 2 are installed on the bobbin-carrier 1. The quantity of bobbins depends on the technological process and can vary from 6 to 1200 pieces. Still dry material 3 passes through tension devices 4 and is formed into the roving on the special guide bar 5. Roving passes through the tub 6 with binder. The quantity of the binder is controlled by the special sensor 7.

Binder excess is removed by wringer rolls 8, clearance between which is changed by the electrical drive 9. Tension device 10 consists of the stationary roller 11 and the roller 12, moving along the tension device by means of the electrical drive 13. Thus, spanning angle of stationary rollers 11 and 12 with "wet" tape is changed. Band tension is measured by the tester 14. Roller 15 lays the tape on the fabric 16 along the special trajectory. For this purpose, it is equipped with the laying devices which are not shown in the figure for the simplicity. Fabric 16 is set into rotation by the drive 17. It should be noted that the winding process is controlled by several interconnected systems.

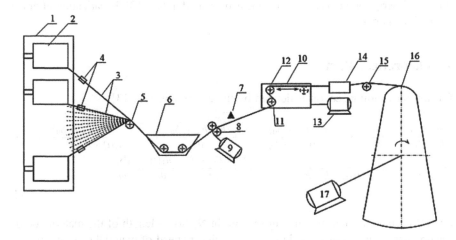

Fig. 1. Fabrics winding by the "wet" method.

Tension on the winding section is supported on the given level by changing moving roller position 12, and therefore, spanning angle of stationary rollers 11 and 12 by "wet" tape.

2 Problem Statement

It is necessary to develop modern control system for qualitative production of composite fabrics by the method of "wet" winding. Performed analysis showed, that such system should have three control circuits (internal circuit-drive moments, middle circuit - roller positions of tension device, external circuit - tension of the winding tape). Presented work deals with creation of control system by tension device position. To create and investigate such system it is necessary to have mathematical description of the tape-winding track taking into account the main peculiarities of elastic winding material. Such description should make it possible to carry out control system synthesis, perform analysis of static, dynamic and power system performance.

3 A Short Review of References

A great quantity of works concerning technological processes in ferrous metallurgy, pulp and paper, textile, chemical and electromechanical industries were devoted to the mathematical description of elastic tape. Differential equations deduced by a number of authors [2–9] describe textile materials, metallic tape, cables behavior in winding and rewinding. Mathematical description of elastic "dry" tape is given in [10–13]. Mathematical description of elastic "wet" tape is presented in [12, 13], linearization of drawn expressions is made.

4 Theoretical Part

Tension on winding section is defined by the expression [12, 13]:

$$\frac{dS_1}{dt} = \frac{1}{l_1(t)} \cdot \frac{dl_1(t)}{dt} \cdot S_1 - \frac{(E \cdot F_S - S_0)^2}{l_1(t) \cdot E \cdot F_S} \cdot v_1 + \frac{E \cdot F_S - S_0}{l_1(t)} \cdot v_2 - \frac{2 \cdot (E \cdot F_S - S_0)}{l_1(t) \cdot E \cdot F_S} \cdot v_1 \cdot S_1$$
$$+ \frac{1}{l_1(t)} \cdot v_2 \cdot S_1 - \frac{1}{l_1(t) \cdot E \cdot F_S} \cdot v_1 \cdot S_1^2 + \frac{E \cdot F_S - S_0}{l_1(t)} \cdot \frac{dl_1(t)}{dt} + \frac{dS_0}{dt}, \tag{1}$$

where S_1 – tape tension on winding section, in N; $l_1(t)$ – length of the material deformation zone at time moment t, in m; v_1, v_2 – linear speed of material entering into the winding area and linear speed material winding on the fabric respectively, in m/s, S_0 – tape tension on the previous section, in N; E, F_S – elasticity module and tape cross-section area respectively, in N/m², m².

This expression is nonlinear and it is difficult to use it for control system analysis and synthesis. Linearization of this expression is done in [12, 13] linearized equation is the following.

$$\Delta\dot{S}(t) + \left(\frac{\partial F}{\partial S_1}\right)^0 \cdot \Delta S_1(t) + \left(\frac{\partial F}{\partial v_2}\right)^0 \cdot \Delta v_2(t) + \left(\frac{\partial F}{\partial \alpha}\right)^0 \cdot \Delta\alpha(t) + \left(\frac{\partial F}{\partial S_0}\right)^0 \cdot \Delta S_0(t) + \left(\frac{\partial F}{\partial \dot{\alpha}}\right)^0 \cdot \Delta\dot{\alpha}(t) - \Delta\dot{S}_0 = 0, \tag{2}$$

It should be taken into consideration that tension device with synchronous engine is used in actuating unit of the control system.

We find dependence of the spanning angle of the tension roller α by "wet" tape on the movement of the movable roller on the axis (the second roller is stationary). Simplified picture of the tension device is shown in Fig. 2.

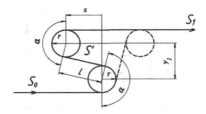

Fig. 2. Simplified picture of the tension device.

The distance between the centers of movable and stationary rollers is marked by y_1, and rollers radii by r. After transformation we get:

$$\alpha(x) = \begin{bmatrix} \arccos\{-\dfrac{2 \cdot r \cdot y_1}{x^2 + y_1^2}(1 - \sqrt{1 - \dfrac{(x^2 + y_1^2) \cdot (4 \cdot r^2 - x^2)}{4 \cdot r^2 \cdot y_1^2}}\}, x \leq 0; \\ \arccos\{-\dfrac{2 \cdot r \cdot y_1}{x^2 + y_1^2}(1 + \sqrt{1 - \dfrac{(x^2 + y_1^2) \cdot (4 \cdot r^2 - x^2)}{4 \cdot r^2 \cdot y_1^2}}\}, x > 0, \end{bmatrix} \quad (3)$$

We linearize the deduced expression:

$$\Delta\alpha(t) = k_5 \cdot \Delta x(t), \quad (4)$$

where $k_5 = \left(\dfrac{\partial \alpha}{\partial x^0}\right)^0$, x^0 - position value of the movable roller, in which linearization is done.

Structural scheme of the generalized mathematical model of the tension device mechanical part is given in Fig. 3.

Here is: ΔM_M – moment developed by the motor, $N \cdot m$; $\Delta\Omega_M$ – rotation frequency of the motor shaft, rad/s; J_M – reduced moment to the motor shaft-second moment of the mechanism, $kg \cdot m^2$; k_7 – contact ratio of the reducing unit.

Coefficients are calculated according to the formulae (5):

$$k_1 = \dfrac{\left(\frac{\partial F}{\partial \alpha}\right)^0}{\left(\frac{\partial F}{\partial S_1}\right)^0}; k_2 = \dfrac{\left(\frac{\partial F}{\partial v_2}\right)^0}{\left(\frac{\partial F}{\partial S_1}\right)^0}; k_3 = \dfrac{\left(\frac{\partial F}{\partial S_0}\right)^0}{\left(\frac{\partial F}{\partial S_1}\right)^0}; k_4 = \dfrac{\left(\frac{\partial F}{\partial \alpha}\right)^0}{\left(\frac{\partial F}{\partial S_1}\right)^0}; T_1 = \dfrac{1}{\left(\frac{\partial F}{\partial S_1}\right)^0} \Bigg\}, \quad (5)$$

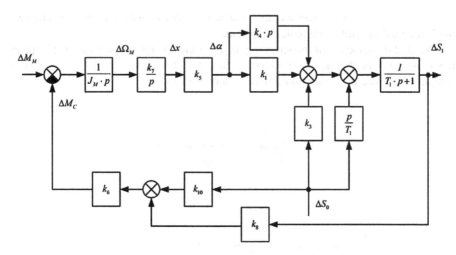

Fig. 3. Structure chart of mathematical model of tension device mechanical part with elastic "wet" tape.

Possibility of simplified expression application should be evaluated. For this purpose, we plot logarithmic amplitude-frequency characteristic (LAFC) and phase-frequency characteristic (PFC) of initial and simplified models of tension device which are shown in Fig. 4.

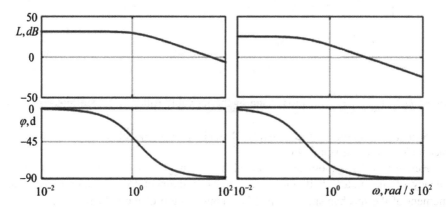

Fig. 4. LAFC and PFC of initial and simplified model of tension device.

From drawn diagrams it is seen that substitution of initial model for simplified one doesn't influence significantly the control object description. That's why a simplified expression should be used in control system synthesis.

In our opinion, it is mostly reasonable to use ac electronic motors in the capacity of actuating units [14] or synchronous motor with permanent magnets (SMPM) [15] in electric drives of tension devices.

Analysis done in works [15, 16] showed that to realize structure synthesis tasks and control device parameters under vector control of SMPM moments supporting magnetic linkage at the rated transmission level it is acceptable to use structure chart shown in Fig. 5.

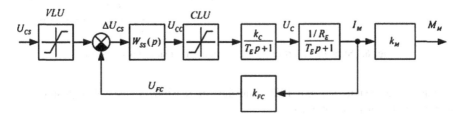

Fig. 5. Structure chart of electrical part of electric drive with SMPM.

The following symbols are used in Fig. 5: U_{CS} – stator current control signal to SMPM; VLU – converter voltage limiting unit; U_{FC} – feedback current signal of SMPM; ΔU_{CS} – control signal to current controller, $\Delta U_{CS} = U_{CS} - U_{FC}$; $W_{CC}(p)$ – transfer function of electric motor current controller; U_{CC} – control signal to converter; CLU – current limiting unit of SMPM moment; U_C – output converter voltage; k_C – converter voltage transfer ratio, $k_C = U_C/U_{CC}$; I_M – electric motor stator current; $R_E = R_1$ – equivalent stator resistance; $T_E = L\sigma_1/R_1$ – electromagnetic stator time constant; R_1, $L\sigma_1$ - ohmic resistance and leakage inductance of stator winding; k_{FC} – feedback current transfer ratio of SMPM, $k_{FC} = U_{FC}/I_M$; M_M – electromagnetic moment; k_M – transfer ratio for SMPM, $k_M = M_M/I_M$.

It should be noted that practically all modern electric drives have built in SMPM moment controller. As shown in [16] in analysis and synthesis of such systems the structure chart of the electric part of electric drive can be substituted by aperiodic link of the kind:

$$W_M = \frac{k_{M1}}{T_M S + 1}, \qquad (6)$$

where k_{M1} – amplification factor of moment circuit, Nm/V; T_M – circuit time constant, s.

Having added the description of the electrical part of electric drive with SMPM to the structure chart of tension device mechanical part we will obtain generalized structure chart of winding machine, comprising two circuits: internal circuit of motor current and external circuit of tension roller position control (Fig. 6).

The following symbols are used in Fig. 6: $W_{S\alpha}(p)$ – the transfer function of position control tension pulley; α_{CS} – the reference signal at the device position; U_C – device position feedback signal.

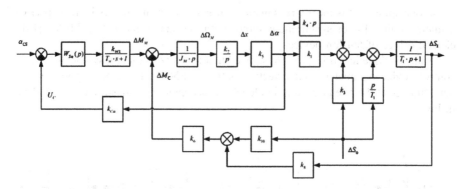

Fig. 6. Structure chart of control system of tension device position

Suggested system has got PID controller of tension device roller position and internal circuit of motor control.

Simulation results of developed system are shown in Fig. 7. Thus the following values of tape parameters and winding track are taken: $EF_S = 1000$ N, $r = 0,025$ m, $y_1 = 0,095$ m, $S_0 = 100$ N, $v_2 = 0,3$ m/s, $f = 0,14$, $k_7 = 1/40$. Simulation shows that the developed control system transfers tension device roller in 0,5 s by 4 rad, providing given tension level in winding.

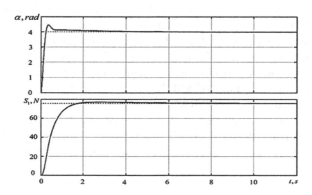

Fig. 7. Simulation results of the developed control system.

5 Conclusion

1. Linearization of the initial nonlinear mathematical description of elastic tape doesn't contribute noticeable error to the dynamic processes, taking place in winding. That's why in further calculations it is reasonable to present mathematical description of tape-winding track of the winding machine in the form of linear differential equations.

2. Synthesized control system of tension device roller position has got a good response speed and makes it possible to change winding material tension efficiently.

References

1. Rosato, D.V., Grove, K.S.: Glass fabric winding. Mechanical engineering, Moscow (1969)
2. Ivanov, G.M., Levin, G.M., Khutoretsky, V.M.: Automated multimotor DC electric drive. Power, Moscow (1978)
3. Druzhynin, N.N.: Continuous rolling mill as an object of automation. Metallurgy, Moscow (1975)
4. Basharin, A.V., Novikov, V.A., Sokolovsky, G.G.: Control of electric drives. Energoizdat, Leningrad (1982)
5. Krysin, N.N., Krysin, M.V.: Technological processes of structures forming, winding and gluing. Mechanical engineering, Moscow (1989)
6. Bondarev, N.I.: Electromechanical systems of automatic monitoring and control tape materials tension. Power, Moscow (1980)
7. Ilina, S.T.: Development and research of the automatic control system of tension basis on looms, dissertation, Moscow (1973)
8. Fainberg, U.M.: Automatic control in cold rolling. Metallurgiya, Kharkov (1960)
9. Kholodenko, N.O.: Problem solving of mathematical description of elastic tape as the control object in winding objects of complex geometric form. In: Economy, Science and Education in the 21 Century. III Regional Theoretical and Practical Conference of Scientists, Students and Postgraduates, pp. 376–380. AR-Consalt, Moscow (2011)
10. Mikitinskiy, A.P., Bekin, A.B., Altunyan, L.L.: Electric drives of winding devices of fabrics from composite materials. In: Proceedings of YII International Conference on Automatic Electric Drive, pp. 516–520 (2014)
11. Kravchenko, O.A., Mikitinskiy, A.P.: Tension control system in winding fabrics from composite materials. In: Theory and Practical Work Integration as the Mechanism of Home Technology and Production Development, pp. 130–132 (2015)
12. Mikitinskiy, A.P.: Mathematical model of tape-drive machine track for winding fabrics from composite materials. Electromech. News 1, 62–66 (2016)
13. Mikitinskiy, A.P.: Mathematical model of machine track tape handler for "wet" winding of fabrics from composite materials, In: Transactions IX International Conference on Automatic Electric Drive, pp. 447–450 (2016)
14. Lobov, B., Mikitinskiy, A., Kolpachshan, P.: The control of dosing pumps by modern electric drive. In: 2nd International Conference on Industrial Engineering, Applications and Manufacturing (2017). https://doi.org/10.1109/icieam.2016.7911499
15. Vinogradov, A.B.: Vector Control of AC Electric Drives. Ivanovo State Power University, Ivanovo (2008)
16. Kolpachshan, P., Zarifian, A., Andruschenko, A.: Systems approach to the analysis of electromechanical processes in the asynchronous. In: Traction Drive of an Electric Locomotive. Rail Transport-Systems Approach. Studies in Systems, Decision and Control, vol. 87, pp. 67–134 (2017)

Analysis of Drive of Mechatron Control System of Rotary Trencher

M. E. Shoshiashvili[✉], I. S. Shoshiashvili, and T. P. Kartashova

Platov South-Russian State Polytechnic University, 132 Prosveshcheniya Street,
Novocherkassk 346428, Russian Federation
shosh61@yandex.ru

Abstract. The problems of kinematic and force analysis of the mechanical excavating part of the (actuator) drive in the mechatronic control system of the trencher rotor position are considered. Analytical expressions that represent a solution for the inverse and direct problems of the lever mechanism kinematics of the rotor drive are obtained. The inverse kinematic problem makes it possible to control the position of the excavating part by an indirect parameter that is the position of the control cylinder rod as well as to evaluate the correctness of the kinematic parameters selection for the mechanical drive and their compatibility with the design parameters of actuator hydraulic cylinders. Based on the direct kinematics problem solution, the analysis of the kinematic error of the trencher rotor control system is conducted. The obtained value of the load on the control cylinder rod can be used for a mathematical description of the entire system operation, as well as for the evaluation and calculation of drive hydraulic parameters.

Keywords: Rotary trencher · Automatic control system · Rotor drive kinematics · Regulation of the rotor position · Kinematic error · Power analysis · Given force

1 Introduction

Among the main tasks of construction and road machines automating, the most urgent one is to control the actuator's spatial position. This is due to the fact, that in many technologies the quality of planning work is determined by how accurately the actuator moves along a given trajectory [1, 2].

For the multibucket rotary excavator (hereinafter trencher), the task of controlling the actuator (rotor) position in automatic mode can be solved by creating an automatic control system (ACS) for the actuator position [5, 9], the functional diagram of which is shown in Fig. 1.

A signal U_h^{task} corresponding to the preset penetration depth of the actuator comes to the subsystem input of automatic position control (APC) of the trencher's rotor from the block for setting the parameters of the trencher automatic control system. Taking into account the difficulties, arising in the direct measurement of the rotor penetration depth in regard to the trench surface h_{tr}, this parameter should be determined indirectly by the size of the control hydraulic cylinder rod (HC) L_{rot} extension, and the control

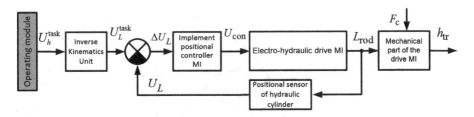

Fig. 1. Functional diagram of the mechatronic CS of the trencher's rotor position.

loop should be formed with feedback according to the position of HC rod of the rotor drive. Thus, the relation between the control parameter U_L^{task} corresponding to the preset value of the HC rod extension and the parameter U_h^{task} is calculated in the inverse kinematics unit (IKU).

The control loop itself includes an actuator position controller, which receives an error signal to the input $\Delta U_L = U_L^{task} - U_L$, where U_L is the signal from the HC rod position sensor, the electrohydraulic actuator drive, whose output parameter is the displacement of the hydraulic cylinder rod L_{rod} [3], and the mechanical part of the drive with the lever system and rotor. The adjustable parameter is trench penetration h_{tr} depth.

2 Kinematical Analysis of the Trencher's Actuator Drive Mechanism

The mechanical part of the actuator's drive of one of the types of trenchers is a lever mechanism with rotating rotor, shown in Fig. 2, which responds to the resistance of the soil cut off [4].

We perform a kinematic analysis of the mechanical part of the trencher's actuator drive (see Fig. 2). To do this, we establish relation between movement of the HC rod L_{rod} and the value of the rotor's penetration h_{tr}.

The penetration value is defined as:

$$h_{tr} = H_0 + H_1 - R_r, \qquad (1)$$

where H_0 – the distance between the point O_2 of attachment of the rotor lever mechanism and the surface through which the vehicle moves; H_1 – the vertical distance between the point O_2 point B on the axis of the rotor rotation; R_r – radius of the rotor. The negative value of the quantity h_{tr} in expression (1) indicates the penetration of the rotor. The variable component of the expression (1) H_1 is determined from the triangles $O_1 O_2 A$ and $O_2 A_1 B$:

$$H_1 = L_3 \sin \beta,$$

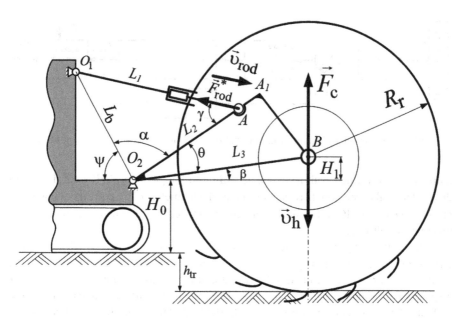

Fig. 2. Kinematics of the mechanical part of trencher's actuator drive.

where $\beta = 180° - \psi - \alpha - \theta$ and $\alpha = \arccos \frac{L_b^2 + L_2^2 - L_1^2}{2L_b L_2}$, L_b – base designed distance between points O_1 and O_2; L_2 – distance between points O_2 and A_1 of rigid frame; L_3 – distance between points O_2 and B of rigid frame of the mechanism; L_1 – distance between points O_1 and A, determined as:

$$L_1 = L_0 + L_{rod}, \qquad (2)$$

where L_0 – constant (minimum) designed distance between points O_1 and A.

Eventually, considering some transformations, we obtain the dependence $h_{tr} = f(L_{rod})$, relation representing the solution of the direct kinematics problem in the form:

$$h_{tr} = H_0 - R_p + L_3 \sin\left(\psi + \theta + \arccos \frac{L_b^2 + L_2^2 - (L_0 + L_{rod})^2}{2L_b L_2}\right), \qquad (3)$$

Expression (3) can be adopted for a mathematical description of the mechanical part of the actuator drive (see Fig. 1).

The solution of the inverse problem of kinematics $L_{rod} = f(h_{tr})$ is based on the following expressions:

$$L_{rod} = L_1 - L_0;$$

$$L_1 = \sqrt{L_b^2 + L_2^2 + 2L_b L_2 \cos \alpha};$$

$$\alpha = 180° - (\alpha_0 + \beta + \theta);$$

$$\beta = \arcsin(H_1/L_3);$$

$$H_1 = h_{tr} + R_r - H_0.$$

Having combined these equations, we obtain the IKU solution for the mechanical part of the trencher actuator drive:

$$L_{rod} = \sqrt{L_b^2 + L_2^2 + 2L_bL_2 \cos\left(\alpha_0 + \theta + \arcsin\frac{h_{tr} + R_r - H_0}{L_3}\right)} - L_0, \quad (4)$$

The graphic interpretation of expression (4) is shown in Fig. 3. When calculating the following parameters were accepted: $L_b = 2$ m; $L_2 = 0{,}7$ m; $L_3 = 2{,}5$ m; $H_0 = 0{,}5$ m; $\alpha_0 = 60°$; $\theta = 55°$; $R_p = 1{,}5$ m.

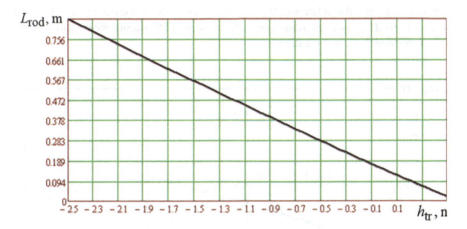

Fig. 3. The graph of dependence of control HC rod displacement on the penetration depth of the trencher actuator.

The graphical dependence shown in Fig. 3 allows evaluating the correctness of the choice of the kinematic parameters for mechanical drive, which is to correspond to condition [10]

$$L_{max} \leq (0{,}5 \div 0{,}7) L_1^{min}, \quad (5)$$

where $L_{max} = L_{rod}^{max}\left(h_{tr}^{min}\right) - L_{rod}^{min}\left(h_{tr}^{max}\right)$ – piston stroke of the control cylinder, h_{tr}^{max} corresponds to uplifted actuator (transport position) and h_{tr}^{min} – actuator's position at maximum penetration. Condition (5) allows selecting or calculating the control HC of actuator drive, the piston stroke of which cannot structurally exceed a certain value $(0{,}5 \div 0{,}7)L_1^{min}$, where L_1^{min} – overall dimensions of HC at fully retracted rod. For the accepted parameters of the mechanism we obtain: $L_{max} = 0.83$ m; $L_1^{min} = 1.422$ m, and expression (5) is performed at $L_{max}/L_1^{min} = 0.583$.

To estimate the kinematic error of the actuator position by penetration depth h_{tr}, we perform the expansion of the function (3) in the Taylor series [6, 7]:

$$\Delta h_{tr} = \left|\frac{\partial h_{tr}}{\partial L_{rod}}\right| \Delta L_{rod}.$$

After some transformations, we obtain expression:

$$\Delta h_{tr} = \left| L_3 \cos\left(\psi + \theta + \arccos\frac{L_b^2 + L_2^2 - (L_0 + L_{rod})^2}{2L_b L_2}\right) \right.$$
$$\left. \cdot 2(L_0 + L_{rod})\sqrt{\frac{2L_b L_2}{(L_0 + L_{rod})^2 - (L_b - L_2)^2}} \right| \Delta L_{rod}, \qquad (6)$$

Figure 4 shows the relations $\Delta h_{tr} = f(h_{tr})$ formed for different values ΔL_{rod}. In expression (6) ΔL_{rod} is a rod motion error of the control HC during the operation of the APC of the trencher rotor position.

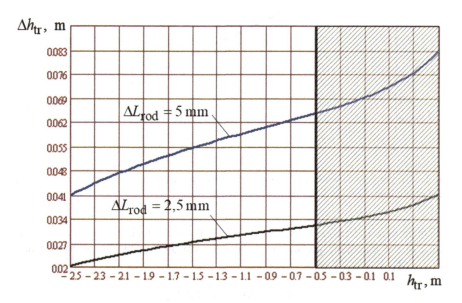

Fig. 4. Graphs of kinematic error of actuator penetration of trencher $\Delta h_{tr} = f(h_{tr})$.

Taking into account that the actuator motion control in automatic mode starts, as a rule, from the penetration depths of $h_{tr} = -0.5$ m and below, the maximum theoretical error of the rotor penetration will not exceed:

$$\Delta h_{tr} = \begin{cases} 0,032 \text{ mm} & at \ \Delta L_{rod} = 2,5 \text{ mm}; \\ 0,065 \text{ mm} & at \ \Delta L_{rod} = 5 \text{ mm}. \end{cases}$$

3 Evaluation of the Strength of the Actuator on the Actuator's Hydraulic Drive

To evaluate the strength of the actuator on the hydraulic drive, we calculate the given force F_{rod}^* acting on the control hydro cylinder rod. From the power balance equation developed by forces and moments of forces applied at different points of the links and performing a progressive or complex plane motion [7], we get:

$$F_{rot}^* = \begin{cases} F_c \frac{v_h}{v_{rod}}, & if \ \angle(\vec{F}_c; \vec{v}_h) = 0 \ and \ \angle(\vec{F}_{rod}; \vec{v}_{rod}) = 0 \\ or \ \angle(\vec{F}_c; \vec{v}_h) = 180° \ and \ \angle(\vec{F}_{rod}; \vec{v}_{rod}) = 180°; \\ -F_c \frac{v_h}{v_{rod}}, & if \ \angle(\vec{F}_c; \vec{v}_h) = 0 \ and \ \angle(\vec{F}_{rod}; \vec{v}_{rod}) = 180° \\ or \ \angle(\vec{F}_c; \vec{v}_h) = 180° \ and \ \angle(\vec{F}_{rod}; \vec{v}_{rod}) = 0, \end{cases} \quad (7)$$

where Fc is the total force acting on point B of the actuator's traveling mechanism (see Fig. 2). When recording power balance, only the forces acting on the rotor were taken into account: gravity of the rotor with loaded buckets and response from the soil being cut off. The gravity forces and moments of forces of the rigid structure O2A1B (see Fig. 2) were neglected in this case.

To calculate the relation v_h/v_{rod}, including in expressions (7), we take derivative $v_h = dh_{tr}/dt$ as follows:

$$v_h = \frac{dh_{tr}}{dt} = \frac{dh_{tr}}{dL_{rod}} \cdot \frac{dL_{rod}}{dt} = \frac{dh_{tr}}{dL_{rod}} v_{rod} \quad (8)$$

from where according to (6)

$$\frac{v_h}{v_{rod}} = \frac{dh_{tr}}{dL_{rod}} = L_3 \cos\left(\psi + \theta + \arccos\frac{L_b^2 + L_2^2 - (L_0 + L_{rod})^2}{2L_b L_2}\right) \cdot 2(L_0 + L_{rod}) \sqrt{\frac{2L_b L_2}{(L_0 + L_{rod})^2 - (L_b - L_2)^2}}.$$

4 Conclusion

The analysis of the kinematic parameters of one of the structures of the actuator's drive mechanism of the trencher carried out showed that the drive mechanism kinematics should be taken into account when designing APC for an actuator. In order to minimize the error in regulating the actuator penetration, it may be necessary to optimize the dimensions of the lever mechanism and location of the attachment points of the structure to the caterpillar. The analytical dependencies obtained can form the basis of such procedure.

Analysis of expression (8) (see Fig. 4) shows that at constant motion speed of the control HC rod the vertical movement speed of the actuator in the operation range of penetrations will vary approximately 1.5 times, which may require the introduction of speed control loop into the control system. On the other hand, the power analysis showed that the force F_{rod}^* applied to the HC rod will change in a similar way, which will also affect the drive dynamics. In addition, expressions (7) and (8) can be used in the mathematical description of automated trencher operation.

References

1. Tikhonov, A., Demidov, S., Drozdov, A.: Automation of Construction and Road Machines: Manual. MGRS, Moscow (1986)
2. Buchholz, V., Snagin, V.: Automatic Rotary Excavators. Nedra, Moscow (1986)
3. Anisimov, A., Kondrashev, V., Lichoded, K., et al.: Dynamics of Hydraulic Systems. SRSTU, Novosherkassk (2012)
4. Karnaukhov, N.N., Merdanov, Sh.M., Shefer, V.V., et al.: Operation of Handling, Construction and Road Machinery. Construction Machines. TyumGNGU, Tyumen (2012)
5. Kartashova, T.P., Lazaridi, K.M., Shoshiashvili, M.E.: To the issue of automation of rotary excavators. In: Innovations in Science - Innovations in Education, Novocherkassk, pp. 80–87 (2016)
6. Korn, G., Korn, T.: Handbook of Mathematics. For Scientists and Engineers. Science, Moscow (1974)
7. Timofeev, G.A.: Theory of mechanisms and machines: a course of lectures. Higher Education, Moscow (2009)
8. Chmil, V.P.: Theory of mechanisms and machines: manual. Lan, St. Petersburg (2012)
9. Shoshiashvili, M., Shoshiashvili, I., Kartashova, T., et al.: Mobile robotic complex based on rotary trencher. In: Proceedings of the Theoretical and Applied Problems of Science and Education, Tambov (2016)
10. Shoshiashvili, M.E., Lazaridi, K.M., Evkhuta, O.M., et al.: Design of a mechatronic module with rotational kinematic pair and electrohydraulic drive of translational action. North-Caucasian region. Tech. Sci. **1**, 65–70 (2014)
11. Shoshiashvili, M.E., Shoshiashvili, I.S.: Principles of construction of control devices for mechatronic pipe-lay complexes. In: 2nd International Conference on Industrial Engineering, Applications and Manufacturing (2016)
12. Shoshiashvili, M.E., Shoshiashvili, I.S., Kartashova, T.P.: Automatic control system of rotary trencher. In: International Conference on Industrial Engineering, Applications and Manufacturing (2017)
13. Tsikerman, L.: Automation of production processes in road construction. Transport, Moscow (1992)
14. Beletskiy, B., Bulgakova, I.: Construction machines and equipment. Lan, Moscow (2012)
15. Dotsenko, A., Dronov, V.: Construction Machines. INFRA-M, Moscow (2014)
16. Shoshiashvili, M.E., Slutsky, V.P., Zagorodnuk, E.V.: Methodological aspects of building control devices for mobile robotic system. In: Proceedings of the Higher Educational Institutions. North-Caucasus region. Engineering, vol. 3, pp. 21–23 (1997)

17. Shoshiashvili, M.E., Shoshiashvili, I.S.: Mechatronics complexes for building and repair of the main pipelines. Algorithmization and modelling. In: Proceedings of the Higher Educational Institutions. North-Caucasus region. Engineering, vol. 4, pp. 3–8 (2014)
18. Shoshiashvili, M.E.: Algorithmization control insulation-laying column. In: Proceedings of the Higher Educational Institutions. Electromechanics, vol. 4, pp. 93–96 (1999)
19. Shoshiashvili, M., Shoshiashvili, I.: Mechanics of controlled machines. SRSTU (NPI), Novosherkassk (2012)
20. Erofeev, A.: Theory of automatic control. Politechnika, Moscow (2008)

Digital Platform for Non-destructive Testing

A. Kosach, I. Moskvicheva, and E. Kovshov(✉)

Joint Stock Company «NIKIMT - Atomstroy», 43/2 Altufevskoe Highway,
Moscow 127410, Russian Federation
ekl77@bk.ru

Abstract. Considers the possibility of applying the principles of informatics in non-destructive testing according to industrial enterprise paradigm-based Industry 4.0. It is discussed the need for automation of non-destructive testing based on the development of a digital platform, including the electronic workflow of technological maps and the management of low-level technological equipment for control monitoring. It is pointed out that many innovative enterprises are moving to the digital format of product manufacturing, automating workshops and individual sites, uniting them into a single information space. In this case, the emergence of universal automated systems for non-destructive testing is a natural development of the production complex, including the methodology Industrial Internet of Things (IIoT). As an illustration of the practical application of the survey results, a study on the direction of automation of non-destructive testing in the development and testing of an intelligent digital platform for data collection and processing by an automated mass spectrometric leak detection system is presented. The estimation of economic efficiency of development and introduction of new software, which is the foundation for the information system for development of technological maps for various types of non-destructive testing, objects and elements of industry has been made. Based on expert opinions and calculations, conclusions on the economic feasibility and effectiveness of new software at an industrial laboratory for non-destructive testing were drawn.

Keywords: Non-destructive testing · Industrial Internet of Things · Technological map · Electronic document workflow · Digital platform

1 Introduction

Non-destructive testing (NDT) of products is an integral part of the production cycle [1]. Automation of control allows to significantly reduce the influence of the human factor on the results, to increase the reliability of the system and the accuracy of data analysis, and also to shorten the time for monitoring and evaluation of results, especially on the production lines of digital production.

Many innovative enterprises are moving to the digital format of product manufacturing, automating workshops and individual sites, uniting them into a single information space. In this case, the emergence of universal automated systems for non-destructive testing is a natural development of the production complex [2], including the methodology Industrial Internet of Things (IIoT).

In the manufacture and operation of products, one of the time-consuming processes is the development and coordination of technological documentation [3–5]. There is no single information database now that unites the data of all participants in the development, production and operation of the facility in real time.

Most methods of non-destructive testing - visual and measuring, radiographic, ultrasonic, leakage detection, and their technological preparation have only partial automation [6–9]. The engineer conducts a visual assessment on the output data and uses a manuscript technological map. Today, the market offers a wide range of software for analyzing unstructured and weakly structured data, including Big Data. There are tools for processing digital results of certain types of non-destructive testing, usually supplied as part of a technological complex for the control of products. However, at the same time, the presented solutions are based on different software architectures, incompatible computer equipment, and do not cover the whole spectrum of types of non-destructive testing. So the most urgent task is the creation of automated systems for technological preparation for control - the development of technological maps and the processing of data obtained during monitoring [10].

2 Workflow Automation

The technological control map is an obligatory part of the technical documentation at any enterprise carrying out high-tech production. By automating the main technological and production processes, it is possible to achieve more flexible controllability of the enterprise and to ensure the required quality of the products.

Engineering-technological workflow is a set of actions for the formation, publication and use of technological maps produced by a multitude of participants over a set of documents. Many actions when creating a technological map with manual input of information represent filling each line of the map. Many actions when creating a technological map with automated filling: filling block-points.

There are two ways of forming the process of document circulation. With the manual version, there are 15 consecutive actions, and with the automated version, 6, with five final actions common to both scenarios (Fig. 1).

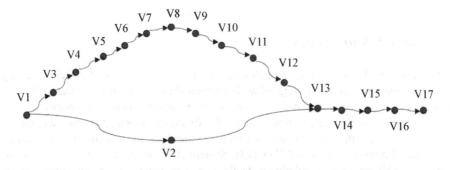

Fig. 1. Graph with two scenarios for the formation of a technological map.

Thus, with the weights of the graph edges specified in Table 1, the length of the shortest path is 14 min. From the last relation it is clear that to the vertex V17 it is necessary to go through the vertex V2.

Table 1. Name and weight of edge of graph.

Name of edge	Weight of edge, minutes	Name of edge	Weight of edge, minutes
K1	6	K10	6
K2	1	K11	4
K3	3	K12	15
K4	3	K13	1
K5	11	K14	1
K6	2	K15	2
K7	2	K16	2
K8	3	K17	2
K9	4		

Taking into account the automation of the formation of technological maps of visual and measurement control, the time of technological preparation is reduced on average from 61 min to 14 min with a full registration of one technological map.

During the researches, software modules for automated compilation of technological maps for visual and measurement control of welded joints have been developed. The developed software, even at the initial level of its application, significantly shortens the duration of the development of the technological map on the basis of formalizing the stage of preparation and coding of the initial design-technological, methodological and normative-technical information. The software is the foundation for an information system for the development of a technological map for various types of control, industrial objects, control elements (welded joints, surfacing, basic material, etc.). This information system based on the use of the universal exchange XML-format [11] and can be integrated into the electronic design and technological workflow. It allows optimizing the data processing, to reduce the labor costs for individual technological operations.

3 Control Automation

One of the results within the research carried out to automate non-destructive testing was the development and testing of an intelligent digital platform for data collection and processing by an automated mass-spectrometry leakage detection system.

To implement the management system, the Enterprise Open ESB bus was chosen, services are implemented in the form of Web-services [12], the functional part of which is executed mainly in Java and C++ [13]. System elements exchange messages based on the SOAP protocol. A relational database management system with open source code has been used to store data and results of calculations. Using these technologies allows analyzing and filtering in the main software tools Microsoft Office (for example, MS Excel). Since the data used in the system is predominantly XML-formatted, the system can be included in a single enterprise information space and exchange data with MES/ERP subsystems. Communication between the system parts and OPC servers (OPC UA, OPC DA) [14] also uses Modbus RTU and Modbus TCP communication protocols (Fig. 2).

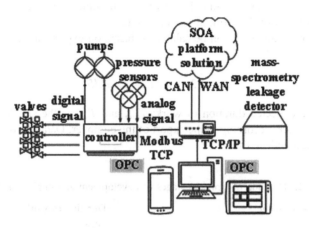

Fig. 2. Scheme of automated system of leakage detection.

The applied tools are cross-platform and belong to the Rapid Application Development category; they are constantly improving, expanding their functionality, while ensuring high ergonomic requirements and consumer properties on the part of the developer [15].

4 Economic Efficiency

The development and implementation of software allows to reduce the labor costs of the process engineer associated with routine work, and to arrange a more logical and functional connection between the interacting departments at the enterprise, which increases the economic performance of the units and the enterprise as a whole.

The specific nature of the manifestation of the economic effect requires special methods for determining it. In general, we can distinguish three main groups of methods that allow determining the effect of implementation: financial (quantitative), qualitative and probabilistic. In the case of the research project, it is sufficient to use the financial method to determine the effectiveness, since the amount of money spent and the period of software development and implementation are small.

According to the methodology proposed in [16], an indicator of the effectiveness of designing and implementing new software is the expected economic effect, determined by the formula:

$$E = E_A - N \cdot K_p, \quad (1)$$

where E_A is the annual cost savings; N – normative coefficient ($N = 0.15$), it shows the size of the minimum acceptable efficiency. In calculating the annual economic effect, the normative coefficient of economic efficiency of capital investments is 0.15, since automation measures relate to new technology; K_P – the capital costs for design and implementation, including the initial cost of the program, it is calculated taking into account the duration of the work at a certain stage.

The duration of the work can be calculated based on expert estimates according to the formula:

$$T_o = (3 \cdot T_{\min} + 2 \cdot T_{\max})/5, \quad (2)$$

where T_0 is the expected duration of work; T_{min} and T_{max} – the smallest and largest duration of the work, according to the expert of the engineer. Data of expected duration of works are resulted in Table 2.

Table 2. Duration of works at stages of development and implementation.

Work name	Duration of work, days		
	Min	Max	Expected
Development of terms of reference	1	2	2
Technical assignment analysis	2	3	3
Registration and approval	1	2	2
Development of the algorithm	5	10	7
Software development	10	20	14
Software debugging	10	20	14
Software testing by related organizations	10	20	14
Software implementation in the enterprise	20	30	24
Training of technologists	1	2	2

As a result, from (2) we have that development and implementation are 82 days or 4 months. Capital costs during the projection phase of the K_K are calculated by the formula:

$$K_K = C + Z_P + M_P + H, \quad (3)$$

where C is the initial cost of the software product - the cost of the FastReport software package (cost approx. $200), which is the most attractive for its functional and consumer properties report generator for various environments and development platforms (.Net, VCL, FMX, etc.); Z_P – wages of specialists involved at all stages of development and implementation; M_P – the cost of using the computer at the design and implementation stage - taking into account the implementation at a large enterprise, accept for zero; N – overhead costs at the design and implementation stage - taking into account the implementation at a large operating enterprise, accept for zero.

If the user saves ΔT_i when doing i-type work with the application of the program, then the increase in labor productivity P_i is determined by the formula:

$$P_i = \left(\frac{\Delta T_i}{F_i - \Delta T_i}\right) \cdot 100, \qquad (4)$$

where F_i is the time that was planned by the user to perform of the i-type work before the implementation of the program (hours).

Table 3. The work of engineer-technologist.

Type of work	Before automation (F_i) depending on the complexity of the (T_C), minutes	Time saving (T) depending on the complexity of the (T_C), minutes	Increase of labor productivity (P_i) depending on the complexity of (T_C),%
Data entry	20 ÷ 40	15 ÷ 35	300 ÷ 700
Analysis and sampling of data	5 ÷ 10	4 ÷ 9	400 ÷ 900
Preparation and printing of reports	5 ÷ 20	4 ÷ 19	400 ÷ 1900

Table 3 shows the work algorithm of a process engineer in the development of a technological map before and after the application of software, as well as the increase in labor productivity calculated with the help of (4).

The savings associated with increasing the productivity of the engineer are determined by the formula:

$$\Delta P = Z_P \sum_i \frac{P_i}{100}, \qquad (5)$$

where Z_P – the average annual wages of the engineer-technologist.

The economic effect of the development and implementation of new software, in accordance with (1), will amount to \$ 22,500 per year. To analyze the effectiveness of the created software, expert evaluations of specialists in non-destructive testing of the profile production laboratory, competent in the development of technological maps, were collected and processed. The results of the analysis are shown on Fig. 3.

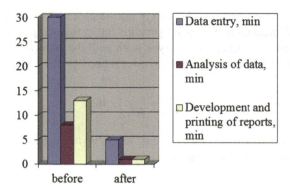

Fig. 3. Estimation of resource costs for the generation of a technological map before and after application of the developed software.

5 Conclusion

As part of the implementation and operation of the pilot project of the intelligent digital platform, the following tasks were successfully accomplished: improving the efficiency of the specialist's work due to the rapid formation of technological maps; the acceleration of the processes of coordination and issuance of technological maps; the increase of efficiency of work with the normative and technical documentation due to creation of a convenient replenished electronic library; prompt provision of current technical documentation of production facilities.

The intelligent digital platform for automation of non-destructive testing is built based on a modular architecture, which makes it flexible and easily scalable, allows the use of virtual resources for various engineering calculations and information storage. The use of this platform in the automation of non-destructive testing allows you to save a large amount of computing resources and time, reduce the complexity of calculations.

Undoubtedly, the indicated approaches will significantly reduce the material and time costs for the process of conducting industrial safety expertise, the exploitation of a variety of process equipment and hazardous facilities in various industries.

Summarizing the previously mentioned earlier in this article, only an integrated approach with a wide application of the digital intelligent platform, automation of non-destructive testing and electronic technological document workflow would be able to form an objective assessment of the technical condition, including the possibility for further operation of various devices at hazardous production facilities.

Acknowledgments. The authors express their deep gratitude to the leadership of the research Institute of design and construction technology (NIKIMT) for financial support throughout the entire R &D cycle.

Special thanks to the Expert Centre (NIKIMT) for providing statistical information on NDT methods and the possibility of experimental operation of hardware and software solutions for the automated development of process maps types of NDT and practical testing in the expanded debug mode intelligent digital platform as an aggregator handler of data flows physical experiments performed during NDT.

We also thank the Department of information technology (NIKIMT) for the creation of modern software and hardware infrastructure and all essential spectrum for software development and numerous computational experiments in the course of the conducted scientific research.

References

1. Mix, P.E.: Introduction to Nondestructive Testing: A Training Guide. Wiley, New York (2005)
2. Rathod, V.R., Anand, R.S., Alaknanda, A.: Comparative analysis of NDE techniques with image processing. Nondestr. Test. Eval. **27**(4), 305–326 (2012)
3. Moskvicheva, I.S., Kovshov, E.E.: Preconditions for the development of electronic technological maps of non-destructive testing in industry. Econ. Soc. **16**(3–3), 81–86 (2015)
4. Kovshov, E.E., Moskvicheva, I.S.: Information and software of the automated subsystem for the development of technological maps of nondestructive testing of welded joints in industry. Mod. High Technol. **9**(1), 51–56 (2016)
5. Kovshov, E.E., Moskvicheva, I.S.: Automation of the development of technological maps of non-destructive testing as a way to increase the economic efficiency of production. Mod. High Technol. **5**(3), 454–458 (2016)
6. Hameed, W.A., Mayali, Y., Picton, P.: Segmentation of radiographic images of weld defect. J. Glob. Res. Comput. Sci. **4**(7), 1–4 (2013)
7. Yazid, H., Arof, H.: Automated thresholding in radiographic image for welded joints. Nondestr. Test. Eval. **27**(1), 69–80 (2012)
8. Moghaddami, A.A., Rangarajani, L.: A method for detection welding defects in radiographic images. Int. J. Mach. Intell. **3**(4), 307–309 (2011)
9. Hartley, R.I., Zisserman, A.: Multiple View Geometry in Computer Vision. Cambridge University Press, Cambridge (2004)
10. Kosach, A.A., Kovshov, E.E.: Software and hardware for industrial automation in the management of remote leakage detection control. Contemp. Eng. Sci. **10**(8), 367–374 (2017)
11. Pylkin, A.N., Tishkin, R.V., Trukhanov, S.V.: Tasks of DATA MINING and their decision in modern relational DBMS. Herald Ryazan State Radio Eng. Univ. **4**(38), 60–65 (2011)
12. Alonso, G., Casati, F., Kuno, H., et al.: Web services – Concepts. Architectures and Applications. Springer, Heidelberg (2004)
13. Kovshov, E., Kosach, A.: Automate remote computer monitoring of environmentally hazardous products. In: Conference Proceedings Application of Contemporary Non-destructive Testing in Engineering, pp. 95–102 (2017)
14. Papazoglou, M., Heuvel, V.: Service-oriented design and development methodology. Int. J. Web Eng. Technol. **2**, 414–442 (2006)
15. Kovshov, E.E., Volkov, A.E., Charaev, G.G.: Reduction of the costs for development and operation of the application information system based on open software solutions. Herald MSTU Stankin **27**(4), 136–140 (2013)
16. Poddubny, A.: Calculation of the economic effect from the introduction of the automation system. In: Expert Opinions (2018).http://antegra.ru/news/experts/_det-experts/4. Accessed 28 Sept 2018

Intelligent Support System of Batch Selection for Electric Arc Furnace: Consolidation of Empirical and Expert Information

O. S. Logunova[✉] and N. S. Sibileva

Nosov Magnitogorsk State Technical University, 38 Lenina Avenue,
Magnitogorsk 455000, Russian Federation
logunova66@mail.ru

Abstract. The paper analyzes the traditional technique for preparing a technological letter on the formation of the structure of charge materials for electrical arc furnace. The necessity of introducing an intelligent decision support system with adaptation functions based on accumulated empirical and expert information in order to expand the capabilities of traditional technology is shown. The structure of a new system of intellectual support for the selection of charge materials for an electric arc furnace, the introduction of which will eliminate the shortcomings of a traditional technology is presented. The developed system consists of two subsystems. The first subsystem is necessary for gathering information to form the structure of charge materials. The second subsystem provides information for solving the problem of choosing the structure of charge materials. The new system uses two types of information: expert and empirical, which consolidation makes it possible to formalize the problem taking into account the accumulated empirical data and the knowledge of experts. The decomposition diagram of the consolidated expert and empirical information in the intelligent support system is presented on the basis of the main characteristics: the way of presenting information, the purpose of the information, the information belonging to a specific module in the system, the solution of the functional problem, the result of using the information.

Keywords: Intellectual support system · Formation of charge materials · Arc steel smelting furnace · Expert information · Empirical information · Information consolidation · Information decomposition

1 Introduction

Modern production complexes are complex systems, the functioning of which occurs under conditions that are difficult for an analytical description. The quality and effectiveness of decisions taken depends on the experience of experts, which is based on the information accumulated in the production system and depends on the expert's skill level. In this regard, it is necessary to expand the capabilities of traditional technologies through the introduction of intellectual decision support systems, since such systems are built on the basis of expert and empirical information.

The use of intelligent decision support systems is widely used in various fields. Thus, the authors [1–3] have developed intellectual systems that are actively used in medicine. In particular, for the diagnosis of blood vessels, for decision-making when choosing medicines and for ultrasound diagnosis. The authors of papers [4–6] suggest the use of intelligent systems for solving problems in the field of earth sciences. They calculate the processing indices of agricultural productive lands taking into account unique features of their configuration [4]; predict the zone of the expected earthquake [5] based on the analysis of experimental data; demonstrate the technique of digital processing of radargrams of ice cover [6]. The authors of works [7–11] use the intelligent systems in the field of image processing in various areas. Work [7] considers the approach to allocation of significant objects on images in various areas; in [8, 9] the application of the intellectual system was considered from the point of view of managing the production of continuous cast billets; in [10] a mathematical apparatus for processing images of the microstructure of cast irons having in the structure randomly distributed inclusions of graphite is proposed; work [11] offers an effective algorithm for preprocessing digital images for further detection of traffic signs in real time.

Most of the works, such as [12, 13], consider the use of intelligent algorithms for managing complex projects. In work [12] the basic principles of system development are presented under condition of management of a large amount of resources.

The authors of [13] investigate the applicability of the use of temporary logics in the organization of intelligent support for complex systems. A large number of studies are aimed at developing solutions for the intelligent management of production [14–16].

The management aspect is presented in [14, 15]. Work [16] shows the intellectual support of the automated technological complex with the purpose of reception of comprehensible schedules.

Thus, at present, there are a large number of intelligent systems, each of which has unique features. The authors analyzed the existing theoretical and practical work and find out that at present there are no modules for intellectual support of the process of selection the batch structure for an electric arc furnace (EAF) with adaptation functions based on accumulated empirical and expert information.

2 Structure of the Intellectual Support System for the Process of Batch Selection for EAF

In order to increase the efficiency of making decisions on the composition of the batch for the electric arc furnace, the authors conducted an analysis of the traditional technique for preparing a technological letter on the structure of batch for an electric arc furnace (EAF). The obtained results of the analysis made it possible to develop the structure of a new system for collecting and preparing information for selection of the batch structure [17]. The structure of the new system for functioning under the new conditions of intellectual support is shown in Fig. 1.

Figure 1 represents the following notations: Q_E – request for experimental data (query experiment); Q_{RR} – request for resources (query resources); A_E – answer on experimental data (answer experiment); A_{RR} – answer on the availability of resources

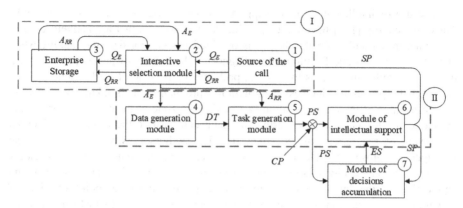

Fig. 1. Structure of a new system for collecting and preparing information for selection of the batch structure.

(answer resources); *DT* – a table of experimental data (data table); *PS* – statement of the problem to solve; *SP* – solution of a specific problem; *ES* – existing solutions; *CP* is the checking point.

The problem that led to the investigation of information flows in the new intellectual support system: the lack of consolidation between the accumulated data represented as an array of empirical information, and the knowledge of experts in the form of a human resource concentrated in heuristic decision algorithms based on unique professional experience.

The new system of intellectual support for the batch selection contains two subsystems: subsystem I is a subsystem for collecting information for the formation of the batch structure; subsystem II prepares information for solving the problem of selection the batch structure.

Block 2 is the link between the accumulated information of the Enterprise Storage (3) and the Source of the call (1) for solving the problem. The source of the call is the receipt of an order for the release of finished products from the consumer or a request about the reasons for the presence of a product failures.

Interactive selection module (2) makes it possible to consolidate the experience, customer requirements and the fact that adequate resources to fulfill the order.

The data generation module (4) provides the principle of empiricality for obtaining regularities describing the change in the parameters of steelmaking.

The task generation module (5) makes it possible to set in the interactive mode target indicators (minimum electric power consumption, minimum values of residual elements, etc.), restrictions on the structure of charge materials (content of scrap metal, liquid iron, alternative materials, etc.).

As a result, a formal statement of the problem is performed with objective functions and constraints in the form and terms of industry-specific production features. An example of the formation of the problem of stabilizing residual elements in steel using alternative materials in the charge is given in [18].

To implement the principles of a priori and intellectuality, the module of intellectual support (6) is included in the system, which implements the solution of the generated task in the form of ready-made recommendations. Before executing the solution in the CP, it is checked that there is a solution to this problem in the archive of the local database. If there is an equivalent or similar task, the solution is extracted from the archive and offered for use in the intelligent support module. The existing solution is used in full or as the basis for a new solution with correction of constraints or objective function. A new solution of the generated task is transferred to the source of the call and, implementing the principle of adaptability, is transferred to the module for accumulating decisions.

3   Consolidation of Empirical and Expert Information

A new system of intellectual support is built using two types of information: empirical and expert. Table 1 provides a description of the characteristics of each type of information.

The diagram of the decomposition of using the consolidated empirical and expert information is shown in Fig. 2.

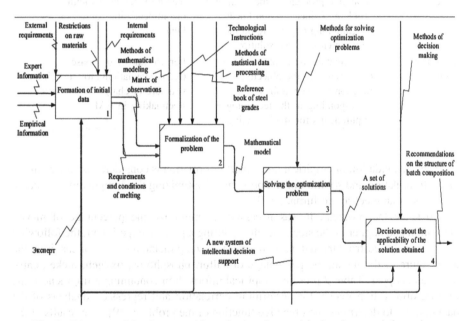

Fig. 2. Diagram of the decomposition of using the empirical and expert information.

Table 1. Description of each type of information.

Characteristic	Empirical information	Expert information
Method of representation	The results of a passive experiment that includes a matrix of observations derived from the selection of data from the Enterprise Storage, in accordance with the research task - Q_E	Expert knowledge of the technology of steel smelting and physical and chemical processes
Purpose of information	Formation of an observation matrix for subsequent statistical processing, including the choice of the objective function and the construction of a system of constraints in the form of interdependent equations and inequalities [19]	Determination of acceptable ranges for future decisions, selection of a set of input data, interpretation of the results obtained
Belonging to the module	The data generation module (4), the task generation module (5)	The data generation module (4), the task generation module (5), the intelligent support module (6)
Functional problem	Formalization of the technical and economic problem in the form of a problem of linear or non-linear programming depending on the requirements for the solution	Defining logical rules for decision-making when constructing recommendations based on fuzzy logic
Result of using	Mathematical description of the task, including the objective function and constraints corresponding to the technological requirements for steel smelting	Formed knowledge base (recommendations) for choosing the structure of charge materials for steelmaking in EAF

The consolidation of empirical and expert information is considered on the example of the formulation and solution of the problem of optimizing the composition of charge materials in an electric arc furnace [19].

To solve the problem, it was necessary to minimize the percentage of nickel, chromium and copper in the steel, therefore, during a passive experiment, the following data were selected from the database: the mass of scrap metal, the mass of cast iron, the temperature of the metal, the percentage of sulfur, phosphorus, oxygen, nickel, chromium and copper. Table 2 presents empirical information containing a fragment of the melting observation matrix for performing correlation and regression analysis of the data in order to determine the objective function of the problem [19]. The results of the correlation analysis showed the presence of intra-correlation and multicollinearity between the indicators selected for the study and made it possible to identify causal relationships between them.

Table 2. A fragment of the initial data of the task

Under current	Scrap mass	Cast iron mass	Outlet temperature	FeO	O2	Chemistry				
						Cr	Ni	Cu	S	P
36	136,9	50	1604	32,5	41,6	0,018	0,078	0,074	0,016	0,003
37	137,7	50	1622	27,5	47,2	0,028	0,075	0,071	0,019	0,003
36	131,1	50	1653	29,5	38,1	0,027	0,068	0,12	0,023	0,006
33	136,3	50	1634	34,5	41,2	0,042	0,066	0,125	0,025	0,004
35	134	50	1614	28,5	56,8	0,029	0,052	0,09	0,025	0,004
34	155,2	50	1610	30,3	41,1	0,066	0,087	0,175	0,035	0,006
36	130,9	50	1606	34,4	60,2	0,04	0,082	0,16	0,034	0,003
34	134,2	50	1640	27,7	49,1	0,023	0,06	0,096	0,015	0,002
36	135,2	50	1649	33,9	50,1	0,016	0,06	0,065	0,012	0,002
34	142,1	50	1648	29,2	49,6	0,017	0,048	0,057	0,013	0,002
37	138,8	50	1634	37,6	56,4	0,027	0,056	0,156	0,006	0,028
33	123,5	65	1617	33,1	44,3	0,048	0,085	0,18	0,006	0,039
34	159,4	60	1609	34	50,3	0,07	0,09	0,187	0,004	0,038
40	102,9	60	1631	32,5	52,6	0,049	0,091	0,192	0,005	0,045
38	168,7	61	1627	28,1	61,7	0,077	0,087	0,179	0,005	0,041
37	105,9	60	1655	28,7	53,3	0,038	0,091	0,194	0,005	0,038
37	137,6	60	1625	30,4	51,6	0,023	0,079	0,138	0,007	0,021

The expert determines the permissible ranges for the solution according to the requirements of the technological letter [17]. Consolidated information is further processed in the modules of task generation (5), data generation (4) and intellectual support (6). The solution of the problem is carried out using three methods of multi-criteria optimization, that is why the possibility of the applicability of each of the three methods to the problem in question is also examined from the point of view of compliance of the obtained solutions with technological limitations. These methods are based on the principle of converting a multi-criteria optimization problem to a single-criteria one.

When solving a problem by the method of successive concessions by deviating from optimal solutions according to more important criteria, a certain quasi-optimal solution is calculated. In the method of constraints, objective functions are brought into normalized form and then a compromise solution is calculated.

The criterion folding method involves transforming the set of available particular criteria into one global criterion. The obtained solutions are analyzed and systematized, and used as expert information to select the structure and composition of charge materials for an electric arc furnace.

4 Conclusion

1. Investigation of the information structure of the technological letter on the choice of charge materials for melting in the EAF made it possible to build the structure of a new system for collecting and preparing information for the formation of the structure of charge materials.
2. A distinctive feature of the new system is the consolidation of empirical and expert information, makes it possible to perform the task formalization taking into account the accumulated empirical data and expert knowledge.
3. The study of the characteristics of empirical and expert information makes it possible to form a scheme for decomposition of the use of consolidated information flows to obtain recommendations on the choice of the structure of charge materials for chipboard in the conditions of the functioning of a new system of intellectual decision support.

References

1. Scherrer, A., Jakobsson, S., Kufer, K.: On the advancement and software support of decision-making in focused ultrasound therapy. J. Multi-Criteria Decis. Anal. **23**(5–6), 174–182 (2016)
2. Butenko, D., Butenko, L., Bolshakov, A.: Decision support when choosing medications based on the hierarchy analysis method. Softw. Syst. **3**, 96–100 (2016)
3. Kychkin, A.: Intelligent information and diagnostic system for examining blood vessels. J. Comput. Syst. Sci. Int. **52**(3), 439–448 (2013)
4. Raevich, K., Zenkov, I.: Intelligent system to support managerial decision-making in the problems of agricultural land assessment. Proc. Irkutsk State Tech. Univ. **112**(5), 95–104 (2016)
5. Pashayev, A., Alizada, A., Aliev, T., et al.: Intelligent seismic-acoustic system for identifying the location of the focus of an expected earthquake. Mechatron. Autom. Manag. **16**(3), 147–158 (2015)
6. Labunets, L.: Automatic intellectual signal processing in subsurface radar systems. J. Commun. Technol. Electron. **60**(4), 362–374 (2015)
7. Kiy, K.: An automatic real-time system for detecting objects and landmarks in images based on processing color images. Mech. Manag. Inform. **6**, 268–276 (2011)
8. Logunova, O., Matsko, I., Posohov, I., et al.: Automatic system for intelligent support of continuous cast billet production control processes. Int. J. Adv. Manuf. Technol. **74**(9–12), 1407–1418 (2014)
9. Logunova, O., Matsko, I., Safonov, D.: Simulation of the thermal state of the infinite body with dynamically changing boundary conditions of the third kind. Bull. South Ural State Univ. Math. Model. Program. **27**, 74–85 (2012)
10. Chichko, A., Likhouzov, S., Sobolev, W., et al.: Computer processing of images of microstructures of pig-iron for the decision of the problem of classification casting on the workability. Herald Polotsk State Univ. Appl. Sci. **3**, 103–109 (2011)
11. Yakimov, P.: Preprocessing of digital images in systems of location and recognition of road signs. Comput. Opt. **37**(3), 401–405 (2013)

12. Rizvanov, D., Yusupova, N.: Intelligent decision support for resource management of complex systems based on multi-agent approach. Ontol. Des., 297–312 (2015)
13. Stoyanova, O., Ivanova, I., Baguzova, O.: Intellectual substitution of solutions for managing complex projects. Bull. Educ. Dev. Sci. Russ. Acad. Sci. **1**, 88–90 (2012)
14. Sazykina, O., Kudryakov, A., Sazykin, V.: Intellectual support of production management solutions. Path Sci. **1**(9), 91–93 (2014)
15. Chernyahovskaya, L., Fedorova, N., Nizamutdinova, R.: Intellectual substitution of work in the operational management of business processes. Bull. Ufa State Aviat. Technol. Univ. **15**(2), 172–176 (2011)
16. Bubnov, D.: Intellectual support of decision-making in the management of automated technological complexes. Technol. Mech. Eng. **7**, 64–66 (2011)
17. Logunova, O., Sibileva, N., Pavlov, V.: Intellectual support in the structuring of batch within an arc furnace. Steel Transl. **46**(10), 733–738 (2016)
18. Logunova, O., Filippov, E., Pavlov, I., et al.: Multicriterial optimization of the batch composition for steel-smelting arc furnaces. Steel Transl. **43**(1), 34–38 (2013)
19. Logunova, O., Sibileva, N., Pavlov, V.: The resulting comparative analysis of the solution of the multicriteria optimization problem for the development of the structure of the charge materials of the arc steelmaking furnace. Math. Softw. Syst. Ind. Soc. Soc. Spheres **2**(5), 54–64 (2014)

Automatic Control of Emulsifier Consumption in Oil Treatment Unit

K. V. Arzhanov[1(✉)] and A. V. Arzhanova[2]

[1] Tomsk State University of Control Systems and Radioelectronics,
40 Lenina Avenue, Tomsk 634050, Russian Federation
otdellltomsk@yandex.ru
[2] Tomsk State Polytechnic University, 30 Lenina Avenue,
Tomsk 634050, Russian Federation

Abstract. The paper examines the problem of oil-well product processing on an oil treatment unit, particularly a matter of water-oil emulsion breaking with the use of demulsification. The oil treatment unit is considered in general terms, the water-oil emulsion breaking process has been analyzed, and an automatic control system has been design as the result of the work. Modal predictive control has been selected as a control algorithm. The developed automatic control system allows users to maintain an optimal process control, to reduce costs, and to improve the quality of treated oil. In the research we face the problem of automatic control to supply demulsifier in complex oil preparation in real time depending on temperature and output amount with the use of PID and MPC controllers. Models of automatic control system of demulsifier feed with different types of regulators are constructed, results are compared and conclusions are drawn about the appropriateness of the use of a particular type of regulator.

Keywords: Automation · Oil treatment · Demulsifier · Proportional integral derivative controller · Model predictive control

1 Introduction

Oil and gas industry is one of the crucial branches of economic development for Siberian regions of the Russian Federation. In Western Siberian region, there are 670 oil fields which average depletion varies from 37 to 60%. During production of crude oil, a working agent (it is water in 80% of cases) is injected to the developed reservoir to maintain the reservoir pressure and the water cut gradually increases, as a result. The average water cut in Russian oil fields is 50–90%, and the oil-water separation increases the oil production costs [1, 12].

The crude from oil deposits is a mixture of oil, oil-associated gas and solids [2]. The untreated product cannot be transported through main pipelines as the multi-component multiphase flow causes the loss of transportation rate due to the friction force, and the oilfield brine and solids attack the equipment enhancing its abrasion. So, the crude has to be refined at the production stage.

When extracting from a layer, moving along the tubing in the wellbore, and also along the oil and water production pipelines, a water-oil emulsion is formed. It's a mechanical mixture of insoluble in each other and finely dispersed liquids.

There are two types of emulsions: "oil in water" and "water in oil". The type of emulsion depends on the ratio of phase volumes, as well as on temperature, surface tension at the oil-water boundary, etc.

In Fig. 1 are presented as:

- A. Two immiscible liquids, not yet emulsified
- B. An emulsion of Phase II (water) dispersed in Phase I (oil)
- C. The unstable emulsion progressively separates
- D. The surfactant (outline around particles) positions itself on the interfaces between Phase II and Phase I, stabilizing the emulsion

To separate water from oil, the surface active agents such as demulsifiers are used, which contribute to the water particle coalescence, heating and separation. The demulsifier consumption is the most cost-intensive part of the oil refining [3, 13]. Coalescence is the process of merging the droplets into larger drops up to the film formation, which leads to the separation of the oil and water phases.

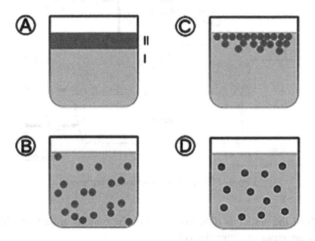

Fig. 1. Types of emulsions.

2 Purpose

The aim of this research is to develop the automated dosage of demulsifiers in oil treatment units in real time depending on the temperature and the amount of treated oil.

3 Design of Control System

When the oil flows through the pipes in an oil treatment unit (OTU), water and oil are continuously mixed which results in water-oil emulsion, the breaking of which is the time- and cost-consuming process.

The emulsion stability is a property no to break and to be separated into a dispersed phase and a dispersed medium over a period of time [2, 3, 16].

The examples of the water-oil emulsion breaking are the processes of settling, filtering, demulsification, thermal treatment, electric influence, magnetic field interference [2, 3].

Gravitational cold separation is used if water content in the formation fluid is high. Settling is carried out in periodic and continuous settlers.

Raw tanks that similar to oil storage tanks are commonly used as periodic settlers. After filling these tanks with crude oil, water is plate out in their lower part.

In continuous settlers, water is separated by continuous passage of the treated mixture through a settler. Schematic scheme of a continuous settler is shown in Fig. 2.

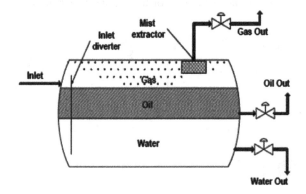

Fig. 2. Three phase horizontal separator.

The length of the settler is determined from the condition that droplets of a given size needed to be separate from the oil.

The essence of the method of intratubular demulsification is that a special substance is added to the mixture of oil and water - a demulsifier in the amount of 15–20 g per ton of emulsion. The demulsifier destroys the armor shell on the surface of the water droplets and by that provides the conditions for their fusion in collisions. Subsequently, these enlarged droplets are separated easily in the settlers due to the difference in the phase densities.

The thermal effect is that the oil subjected to dehydration is heated before settling. When heated, on the one hand, the strength of the armor shells on the droplet surface decreases, and, therefore, their fusion is facilitates, on the other hand, the viscosity of the oil in which the droplets settle decreases, and this increases the rate of separation of the emulsion.

The emulsion in tanks, heat exchangers and tubular furnaces is heated to a temperature of 45–80 °C.

The thermochemical method consists of a combination of thermal effect and in-tube demulsification.

The electrical effect on the emulsion is carried out in apparatuses called electric dehydrators. Under the action of the electric field at opposite ends of the drops of water electric charges appear. As a result, droplets attract to each other and merge. Then they settle on the bottom of the tank.

Filtration is applied to destroy unstable emulsions. As a filter material, substances that are not wetted by water but wetted by oil are used. Therefore, oil penetrates through the filter, and water does not.

The separation in the field of centrifugal forces is performed in centrifuges, which are a rotor with a large number of rotations. An emulsion is supplied to the rotor along the hollow shaft. Here it is divides by the forces of inertia, since the drops of water and oil have different densities.

When dehydrating, the water content of the oil is reduced to 1–2%.

The kinetic process of destabilization can be rather long – up to several months, or even years for some products.

Temperature affects not only the viscosity but also the surface tension in the case of non-ionic surfactants or, on a broader scope, interactions of forces inside the system.

There are several factors which impact the emulsion stability: the surface tension, the dispersed medium viscosity, temperature, the particle fineness, etc. However, the main factor of the emulsion stability is the size of emulsified droplets. When the droplet size is reduced, the gravity impact decreases and is prevailed by the forces which hold the droplets in suspension [4].

The main purpose of a demulsifier is to displace water droplets from the emulsion surface.

There are a number of demulsifiers available which effectiveness is determined by the breaking rate as a result of reducing the liquid interface tension. The amount of consumption is defined individually in each field according to the oil quality and a type of demulsifier. The uncontrolled demulsifier consumption during the oil treatment results in increased economic costs.

The general view of an oil treatment unit is presented in Fig. 3, where the following streams are indicated: I – oil in place; II – associated petroleum gaz; III – oil after the second separation stage; IV – formation water; V – oil trapped from water settling drum; VI – solids, sludge; VII – treated formation water cleared from oil and solids; VIII – oil for processing facility; IX –flared gas; X– injection station water.

The following equipment is used in oil treatment: 1 – the first stage separator; 2 – knock-out drum; 3 – the demulsifier dosing skid; 4 – tubular furnace; 5 – three-phase separator; 6 – oil dehydration tank; 7 – the formation water settler; 8 – degasser; 9, 10 – pumps; 11 – the associated gaz meter; 12 – the incoming oil meter; 13 – the formation water meter.

After adding the demulsifier, the oil product is delivered to the furnace for heating, which increases the difference of water and oil density and reduces the oil viscosity, as a result. So, the breaking is accelerated.

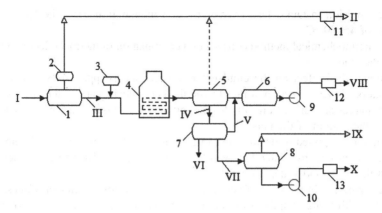

Fig. 3. Oil treatment unit.

After heating, the oil product, which consists of oil-water and gas, is delivered to the three-phase separator. Due to the lower density, oil rises to the surface of the water phase. Simultaneously, the associated gas is separated and directed to the gas section and then, to the associated gas meter.

We focus on three-phase gravity separators as they form the main processes in the upstream petroleum industry, and have a significant economic impact on the produced oil quality.

Degassing of oil is carried out in order to separate gas from oil. The apparatus in which this occurs is called the separator.

Considering the water particle settling in water-oil emulsion, we know the relationship between the settling velocity and the droplet diameter, temperature and density, which implies [5] that the settling velocity is proportional to the square of the water droplet diameter and the square of temperature.

If the Reynolds number [5] $Re = 0..1$:

$$U = \frac{1}{18}\frac{D^2(\rho_\text{в}-\rho_\text{н})}{\mu_\text{вяз}} = \frac{1}{0{,}003294}D^2(\rho_\text{в}-\rho_\text{н})\cdot(1+0{,}0337T+0{,}000221T^2), \qquad (1)$$

if $Re = 1..700$:

$$U = (\frac{4}{39})^{2/3}\frac{D(\rho_\text{в}-\rho_\text{н})^{2/3}(1+0{,}0337T+0{,}000221T^2)^{1/3}}{(0{,}000183\rho_\text{н})^{1/3}}, \qquad (2)$$

where Re is the Reynolds number, $\text{Re} = \dfrac{DV\rho_\text{н}}{\mu_\text{вяз}}$; U – the velocity of the water droplet settling, m/s; V – droplet velocity, m/s; D – droplet radius, m³; $\rho_\text{в}, \rho_\text{н}$ – water and oil densities, kg/m³; $\mu_\text{вяз}$ – oil dynamic viscosity, Pa × s; T – temperature, K;

As the equations show, depending on the Reynolds number, the settling velocity depends on the droplet diameter and the temperature of emulsion. According to the oil treatments specifications, the preheat temperature of emulsion is usually constant.

As the Reynold number in the separator is no more than 1, only the first formula (1) is used when selecting the oil treatment mode at OTU.

To maintain the most efficient separation of the water-oil emulsion, the near-optimal relationship of the demulsifier consumption and the furnace temperature to be provided in real-time. This relationship can be monitored by measuring the emulsion temperature and the OTU performance through the measuring the product oil [5, 17, 18].

The automated control loop diagram is presented in Fig. 4, where the transformation coefficients of the demulsifier dosage and the tubular furnace are presented as the coefficients according to the Eq. (1), and the separator is presented as an integral unit with transport delay.

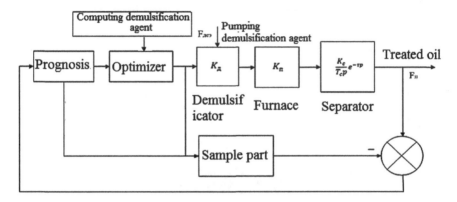

Fig. 4. Block diagram.

To develop the control system, the following control algorithms were considered: FF/FB, Cascade, Override, Split-range control, Model Predictive Control (MPC) [6–8, 11, 14].

All those algorithms are efficient when all the parameters of the control system components are available and just slightly modified during the OTU operation. The MPC algorithm is referred to the algorithms which optimize control according to the simulation which predicts changes of the differential state vector [9, 10, 13]. The MPC algorithms solves the optimization problems at each stage, which provides the optimal, or near-to-optimal, system parameters in real time.

The application of the MPC and PID controller in the system was studied.

4 Experimental Results

To study the process, the simulation of the demulsification with a PID controller was designed. The object parameters were used from [5, 6, 15]. The numerical model of the oil demulsification is presented in Fig. 5, the simulation result is shown in Fig. 6. The simulation was performed with Matlab in the Simulink application.

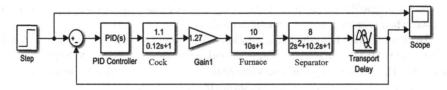

Fig. 5. Numerical model of the oil demulsification.

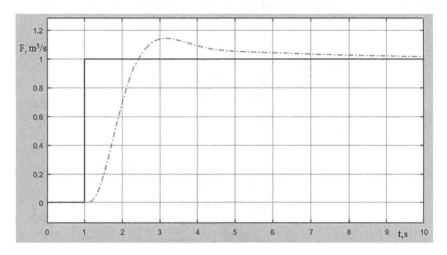

Fig. 6. The result of simulation using the PID control.

The oil demulsification model with the use of the MPC algorithm is presented in Fig. 7. The simulation result is presented in Fig. 8.

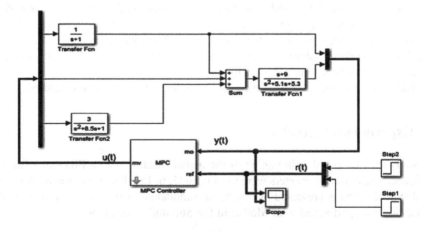

Fig. 7. Numerical model of the oil demulsification using the MPC.

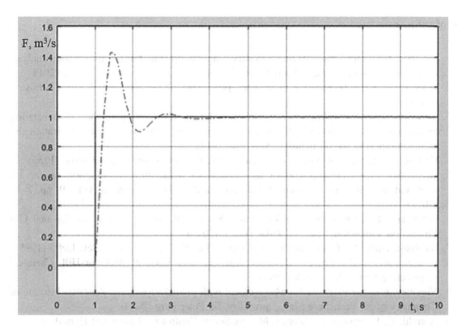

Fig. 8. The result of simulation using the model predictive control.

According to the results of simulation, PID controllers as well as MPC provide the specified quality of control and can be efficiently used in the automated control of demulsifier consumption in the oil treatment units.

5 Conclusion

During the current research, the literature survey was carried out. It was revealed that the oil treatment units are not provided with the automation control of the demulsifier consumption and the furnace temperature, which results in lower efficiency of the oil treatment.

Because of the continuous oil water cut, the demulsifier consumption control according to the proposed algorithm is considered to be worthwhile. The real time control referred to the temperature and the amount of the treated oil provides the relevant consumption of demulsifier in the amount which is required for the effective treatment of oil subject to the transportation through the main pipelines.

In simulation of the oil dehydration, the MPD did not demonstrate any distinct advantage over the PID controller. So, it is recommended to use the PID controller in a real control system as the sophisticated MPC model is hard to be realized.

References

1. All about the mineral resource sector in Russia and worldwide. Mineral information analysis center. http://www.mineral.ru/Facts/russia/131/288/index.html. Accessed 25 Sept 2018
2. Manovyan, A.K.: Technology of primary processing of oil and natural gas. Chemistry, Moscow (2001)
3. Verevkin, A.P., Kiryushin, O.V., Urazmetov, Sh.F.: Management of the oil preparation process on the aggregative stability of water-oil emulsion automation, telemechanization and communication in the oil industry, Moscow (2012)
4. Clayton, V.: Emulsions. Their theory and technical applications. Izinl, Moscow (1950)
5. Arzhanova, A.V.: Optimization of the process of demulsifier consumption on the oil treatment unit. Strategy of sustainable development of Russian regions, vol. 39, pp. 7–15 (2017)
6. Gromakov, E.I., Liepinsh, A.V.: Designing Automated Control Systems for Oil and Gas Production. Publishing house of TSU, Tomsk (2016)
7. Stephanopoulus, G.: Chemical Process Control. Prentice Hall of India, New Delhi (1995)
8. Coughanowr, D.R.: Process Systems Analysis and Control. McGraw-Hill Chemical Engineering Series, New York (1991)
9. Veremey, E.I., Sotnikov, M.V.: Management with predictive models. Research Institute of Chemistry of St. Petersburg State University, St. Petersburg (2014)
10. Borrelli, F., Bemporad, A., Morari, M.: Predictive Control for Linear and Hybrid Systems. Cambridge University Press, Cambridge (2015)
11. Pal, R.: Techniques for measuring composition (oil, water content) of emulsions. Colloids Surf. **84**, 141–193 (1994)
12. GOST R 51858-2002 Oil. Standartinform, Moscow (2002)
13. Mullaev, B.T.: Designing and optimization of technological processes in oil production (with algorithms for solving fishing tasks), vol. 1, pp. 247–285 (2003)
14. Pozdnyshev, G.N.: Stabilization and Destruction of Emulsions. Nedra, Moscow (1982)
15. Veremey, E.I., Eremeev, V.V., Sotnikov, M.V.: Model predictive control toolbox (2018). http://matlab.exponenta.ru/modelpredict/book1/index.php. Accessed 12 Oct 2018
16. Molchanov, G.V., Molchanov, A.G.: Machines and Equipment for Oil and Gas Extraction: a Textbook for Universities. Nedra, Moscow (1984)
17. Kireev, V.A.: A short course of physical chemistry. Chemistry, Moscow (1978)
18. Filimonova, E.I.: Fundamentals of Oil Refining Technology. Publishing House of the Yakut State University, Yaroslavl (2010)

Comparative Analysis of Automatic Methods for Measuring Surface of Threads of Oil and Gas Pipes

D. S. Lavrinov[✉], A. I. Khorkin, and E. A. Privalova

Ural Federal University, 19 Mira Street, Yekaterinburg 620002,
Russian Federation
dslavrinov@gmail.com

Abstract. In this paper a comparative analysis of methods for measuring complex profile surfaces such as threaded surfaces were carried out. The oil and gas pipe and couplings threads were used as an example of a complex surfaces. A theoretical and experimental evaluation of the reliability, performance and accuracy of the geometric measuring systems in shop floor environment was made in cooperation with executive and ordinary staff. Coordinate measurement machine with stylus, profilometer, conoscopic, confocal, interferometric and laser triangulation methods were involved to solve thread measurement problem. The optimal method of thread geometry measurement for a shop floor was tested and identified as a result of the analysis of pipes and couplings threads. The laser 2D triangulation scanning method was verified as the only suitable for thread measuring problem solution. The results of measurement using this method show a possibility to achieve technical requirements for the premium oil and gas pipes thread.

Keywords: Laser triangulation · Scanner · Contact · Non-contact · Measurement · Thread · Conoscopic · Interferometric · Characteristic of accuracy

1 Introduction

Manual geometry measurement means are one of the biggest current problems of an oil and gas pipe manufacturers. A calipers and other special gauges often require a surface recovery and verification. This increase the amount of non-quality production and maintenance costs. Human factor problems, performance and accuracy requirements leads to the development of new coordinate-measuring machines (CMM).

A wide range of problems in measuring geometry determines the variety of coordinate-measuring technologies and approaches. In the course of creating systems for measuring geometry, the task for developers is to select the optimal method and equipment. Therefore, specialization of technology for various operating conditions, surface features and performance requirements take place. In fact, for each task the existing methods and technologies are modified or new ones are created.

Measurement of premium oil and gas pipe thread is one of the complex problems on automated production lines of metallurgical plants (shown in Fig. 1).

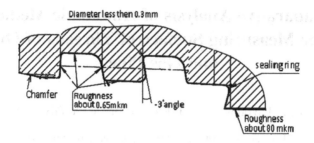

Fig. 1. Drawing of premium thread example.

At the moment all manufacturers of pipes and couplings use two-level geometry inspection. The first level is the total control of the thread by manual gauges and calibers, the second is the selective control of the geometry using casts [1]. The first level takes about 1 min per one part, ensuring that the connection is checked for screwing, but not giving information about the geometry of the teeth and the roughness of the thread.

The second level provides information on the geometry of the teeth and the roughness of the thread, but takes too much time - about 20 min per part. With the advent of premium couplings with sealing rings, grooves and threads of special geometry, contact methods have ceased to provide control of required parameters.

According to the standards the accuracy of measurement should be 5 microns with performance 100 000 points per second. It should also be taken into account that most of the products have a diameter of more than 70 mm.

To solve this problem now operators of production lines use manual thread gauges or casts form the thread surface. All these methods are not integrated in SCADA and depends on human factor. Therefore, pipe plants sometimes supply defective production. It causes accidents.

As you can see at the Fig. 1, the thread has some features, such as the negative angles of the sides of the teeth, the sealing ring, various roughnesses from 0.65 to 80 microns, and small diameters. Reference element for basing of thread is chamfer cone, which axis coincides with axis of thread. Measuring of 4 sections of thread is enough according to API standard [1]. Therefore, precision positioning system is needed to measure reference element and four pipe or coupling sections.

According to required parameters, there are list of existing measuring methods and systems to be considered [2]:

- Contact method with portal CMM or profilometer
- Conoscopic method with portal CMM
- Confocal sensing method with portal CMM
- Interference method with portal CMM
- Laser triangulation method with portal CMM

2 Contact Methods of Measurement

Theoretical and experimental analysis of existing methods should be held to find or create suitable solution.

Historically, the first devices used to control the geometry of the object were the contact probes. The probe method of measuring the geometry of the surface is a wide spreaded on the production plants (shown in Fig. 2).

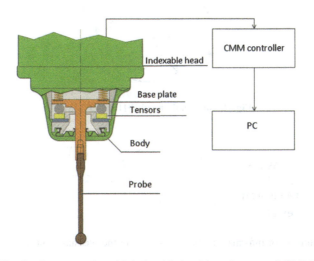

Fig. 2. Contact probe with indexable head based on portal CMM.

As probes are used thin rods with tips made of metal, ceramics, ruby or sharpened needles of high hardness. The tips on the rods are often balls of different diameters (minimum 0.3 mm). The probes are driven along a certain trajectory relative to the surface. In case of probe mount on CMM with indexable head the angle of probe could be changed from −105 to +105 at A axis and from −180 to +180 at B axis [3] (shown in Fig. 3).

The mechanical vibrations of the probe are converted into electrical vibrations and enter the processing unit (CMM controller), where the measured profile of the object is formed.

The use of digital processing of the signal from the contact probe allowed to reduce the electronic noise to a level equivalent to the rms surface roughness of 1–2 nm with a needle load of 0.5–2 mg. Modern CMMs have a high sensitivity (about 4 nm in space) [4]. But due to the finite size of the touching surface of the probe, this type of measuring probes reproduces the microrelief only in the case of far-away microroughnesses or slightly-wavy surfaces, as can be seen in Fig. 4.

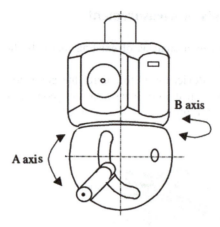

Fig. 3. Indexable head PH10 on CMM.

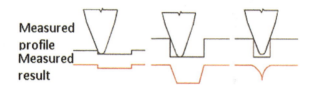

Fig. 4. Distortion of information about geometry in the measurement of a complex part.

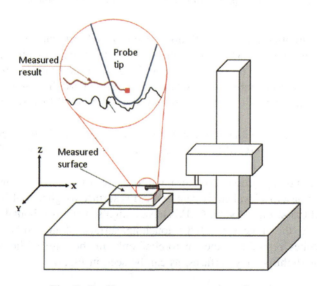

Fig. 5. Profilometer measures rough surface.

Comparative Analysis of Automatic Methods for Measuring Surface of Threads 87

In the case of closely located microroughness, the probe, sliding along their vertices, transmits only the general contour of the relief. The resolving power of the CMM depends on the radius of the rounding of the probe and the nature of the topography of the surface. In addition, the probes wear out and due to their fragility they often break. When the probe is in the groove, it is possible to break it.

Measurement of diameters less than 0.3 mm using probe with spherical tip is impossible, because minimum diameter of sphere is 0.3 mm. Therefore, only needles should be used.

Needles are usually used in profilometers, devices for measuring the profile of an object in a given section (shown in Fig. 5). Experiment was held using Mitutoyo profilometer for measurement of premium casing coupling 168 mm diameter (shown in Fig. 6).

Result of experiment shows impossibility to measure small diameters of internal thread using profilometer, because probe angle about 45°–55° is required to reach these places. It is impossible to lead probe inside with 125 mm diameter coupling. Required angle could be reached in case of external thread. However, during experiment two tips of probes were broken, because of grooves and roughnesses (shown in Fig. 7).

It can be stated, that contact method doesn't suitable for measurement of complex part with small diameters, because needles are required to measure them. In turn, needles are very fragile and not suitable to grooved surface.

Fig. 6. Experimental measurement of casing coupling using profilometer Mitutoyo and broken probe.

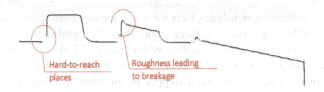

Fig. 7. Point cloud of the coupling thread section collected using profilometer.

Also, measuring time for contact methods is about 20 min per pipe/coupling, or 500 points per second. This is not suitable for automatic production lines, because one CNC produces one pipe or coupling per minute.

3 Conoscopic Method of Measurement

A relatively new method of measuring the geometry of a surface is conoscopic holography.

In classical holography, a hologram is created by recording an interference pattern formed between an object beam and a reference beam using a coherent light source [5].

One of the problems of optical methods of geometry inspection is the measurement of the surface with sharp differences in altitude. The larger the field of view of the optical system, the lower its numerical aperture (shown in Fig. 8). The light that hits the surface through the interferometer lens must be collected again to focus on the digital matrix in order to process the information and create the desired 3D surface map.

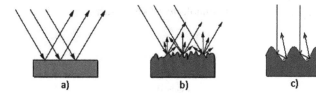

Fig. 8. Examples of reliefs of measured surfaces and reflection of rays: (a) a smooth surface reflects incident light with an angle equal to the angle of incidence; (b) the rough surface reflects a part of the light at an angle equal to the angle of incidence, and part dissipates in different directions; (c) the threaded surface also reflects and diffuses the light as in case b, but in addition creates interference noise from the adjacent tooth.

The light reflected from the surfaces of a larger angle, than lens can collect, either does not hit on the digital matrix, or hit in a distorted form, which makes it impossible to accurately measure (shown in Fig. 9).

In the paper [6], an experiment is described to assess the error of a measuring system based on CMM and ConoProbe Mark3.0 cone sensor, manufactured by Optimet, a leader in the market of conoscopic sensors. Turbine blade was measured with error about 0.030 mm.

Taking into account the submicron accuracy of the sensor according to the documents [5], this indicates a significant increase in the measurement error when scanning polished, well-reflecting surfaces at large (more than 45°) angles of incidence.

In this paper, thread of tubing coupling with diameter 73 mm. was measured using a ConoProbe Mark3.0 sensor and a special sensor displacement system, which is a carriage driven by a ball screw and a stepper motor (shown in Fig. 10). The sensor

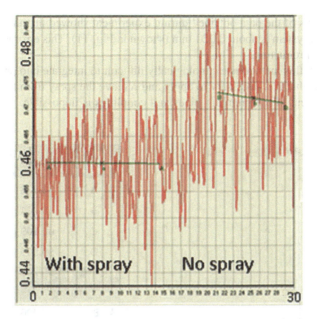

Fig. 9. Accuracy of measurement by a conoscopic sensor of a mirror surface with application of a spray and without.

Fig. 10. Experimental set up for coupling measurement using conoscopic probe.

error is about 0.0002 mm. The error of the displacement system is about 0.001 mm. The scanning speed of the sensor is no more than 9000 points per second.

During experiment it was found, that without applying a special spray to the thread surface the measurement error is 0.040 mm, which does not satisfy the requirements normative documents [1]. Spray removes the mirror effect from the surface and reduce interference from the adjacent tooth. The structure of the spray is a granules of about 0.010 mm. in diameter. Therefore, the measurement error after application became about 0.020 mm (shown in Fig. 9).

The analysis of the data obtained during experiment showed that the measurement error with this system is about 0.020 mm (shown in Fig. 11). This error is sufficient for non-premium production.

As a result, this method is technologically difficult to integrate into production line, due to the fact that it is necessary to apply and remove the spray from the surface, and also measurement of the premium pipes and couplings is impossible.

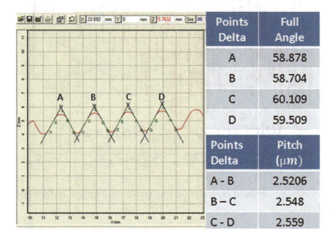

Fig. 11. Accuracy of measurement by a conoscopic sensor of a mirror surface with application of a spray and without.

4 Confocal Method of Measurement

Confocal sensing is one of fast and precise methods of geometry measurement. Light is focused using a lens on a measured object, reflects from it and returns into the sensor. In a monochromatic confocal sensor a single-color light is used and the sensor and object have to be mechanically moved with respect to each other to keep the object at the focal point of the lens. This makes the technique very slow and unsuitable for shop floor measurement of pipes and couplings.

In contrast, white-light confocal sensors use light composed of a many colors. The lens focuses each color at a slightly different location, and by measuring the returned light's exact color we can evaluate the distance with up to nanometric precision.

An example of confocal chromatic principle of measurement is CHRocodile CLS sensor made by Precitec company [7] (shown in Fig. 12).

A white light source is used to illuminate the surface of a part. The light travels via fiber from the CHRocodile control unit to an optical probe which then spreads the focal length over a discrete number of points creating a full spectrum of light as shown in the graph. Sensor contains 192 channels of measurement, which provide 2D line (5 mm length) scanning of an object. Based on the wavelength of the reflected light, a very precise (lateral resolution: less than 1 μm) distance measurement can be taken up to 2 000 times per second (or 384 000 points per second). Also sensor has high acceptance angle, up to 45°, which provides possibility to measure slope surfaces, such as thread teeth. The optical probe determines the measuring range (up to 35 mm), or focal depth of the spectrum (shown in Fig. 13). Because of the high numerical aperture of the probes and dynamic range of the sensor, it is possible to measure on nearly all materials.

However, these sensors are very cumbersome. They cannot be mount on indexable head. It causes to problem not only with scanning of sealing elements of premium production, but also with scanning of internal surfaces of holes diameter less than 100 mm. Therefore, they are suitable only for non-premium production and only for outside thread.

Fig. 12. Precitec CHRocodile CLS confocal chromatic sensor.

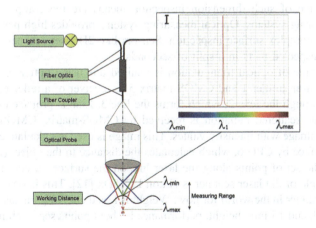

Fig. 13. Scheme of the confocal chromatic sensor.

5 Interferometric Method of Measurement

An example of modern interferometric sensor is Hexagon hp-o (shown in Fig. 14), which is based on frequency-modulated interferometric optical distance measurement method.

Diameter of this sensor is about 3 mm, weight is less then 190 g, measuring range is up to 10 mm., performance is about 1000 points per second, resolution is 0.9 nm. This sensor is easy mount on indexable head. The big disadvantage of this sensor is a very small acceptance angle (for 10 mm. range, angle is about 0.3°). Therefore, it cannot be used to measure sharply sloped thread surface.

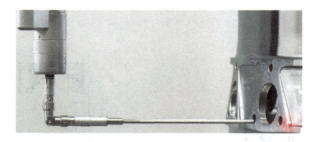

Fig. 14. Hexagon hp-o interferometric sensor.

Of course, pitch of thread could be measured using interferometer sensor [9], but it just one from set of parameters.

6 Laser Triangulation Method of Measurement

Non-contact geometry measuring methods are show the biggest efficiency when used in high performance production lines [10].

An example of such dimension inspection means are the computer vision and optical measuring systems. Optical measuring systems provides high performance and precision for complex surface inspection. Sensors like 2D laser scanners can measure areas of an inspected parts in a split of secconds.

The principle of optical triangulation is realised due using opticzl scheme, which includes the laser emitter 1 and CMOS-matrix 5 as receiver of a reflected light. Laser beam goes through the lens 2, which forms the line 3. Line is projected on the object surface 7. The scattered radiation is received by CMOS-matrix. CMOS matrix generates digital image with intensity values. This image is used to calculate contour of the measured surface by CPU 6, which calculates the distance to the object (Z coordinate) for each of the set of points along the laser line on the surface (X coordinate) [11].

An example of 2D laser scanner is Nikon LC15Dx [12]. This is one of the precise and fastest scanner in the world for now. Weight is less then 370 g., measuring range is 18 mm. width and 15 mm. height, performance is about points/sec, probing error is 1.9

micron, which is very close to contact method precision (shown in Fig. 16). This sensor is easy mount on indexable head (Fig. 15).

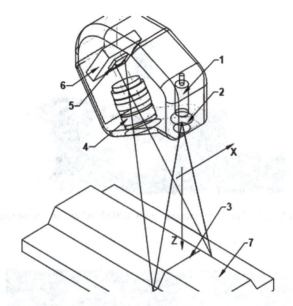

Fig. 15. Scheme of the 2D laser scanner.

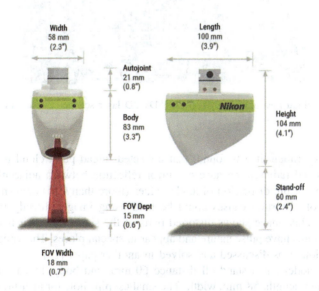

Fig. 16. Nikon LC15Dx 2D laser scanner.

In this paper, external premium pipe thread diameter 245 mm. was measured using LC15Dx 2D laser scanner with CMM Altera (shown in Fig. 17).

Fig. 17. Nikon LC15Dx 2D laser scanner with CMM Altera in measurement of external premium pipe thread.

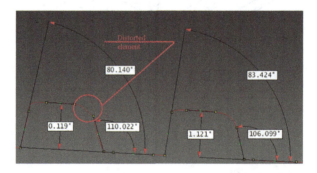

Fig. 18. Point cloud measured by Nikon LC15Dx 2D laser scanner and analysed in CAMIO software.

During experiment it was found, that collected thread point cloud is distorted in places with small radiuses, because of mirror reflection between adjacent teeth. This was the reason of big dispersion of angle values (more than 0.5°) (shown in Fig. 18).

Analysis of distortion causes could be held using images directly from scanner video matrix. This image shows distorted parts of the signal and their features.

These features have pulse nature and appear in special places. The same problem of signal distortions was discussed and solved in another paper [13].

LC15Dx model has a stand-off distance 60 mm. and body dimensions 104 mm. height, 100 mm. length, 58 mm. width. The smallest pipe hole for measurement by this scanner is about 170 mm, which is not satisfy to the requirements of pipe producers. But it is possible to create special 2D scanner for diameters less then 70 mm (for example LS2D made by LLC Geomera).

If problem of filtration of these distortions is solvable, then laser triangulation method satisfies technical requirements (shown in Fig. 19).

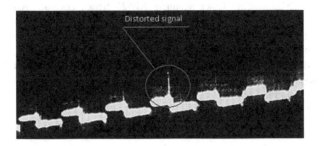

Fig. 19. Point cloud measured by Nikon LC15Dx 2D laser scanner and analysed in CAMIO software.

Acknowledgments. As a result of the work comparative analysis of automatic methods for measuring the surface of threads of oil and gas pipes was held. Analysis of technical characteristics and experimental thread measurements revealed options to achieve the required parameters of accuracy and productivity. According to experiments the most possible and universal solution for internal and external thread measurement is laser 2D triangulation in combination with special filtration algorithms.

References

1. American Petroleum Institute: API Specification 5CT, Specification for Casing and Tubing. API Publishing Services, Washington (2005)
2. Halevy, A.: Non-contact distance measurement technologies (2015). https://www.optimet.com/blog/non-contact-distance-measurement-technologies. Accessed 27 Sept 2018
3. Zexiao, X., Chengguo, Z., Qiumei, Z.: A simplified method for the extrinsic calibration of structured-light sensors using a single-ball target. Int. J. Mach. Tools Manuf. **44**(11), 1197–1203 (2004)
4. Hexagon, A.B.: LEITZ PMM-C LINE ultra high accuracy coordinate measuring machines (2013). http://apps.hexagon.se/downloads123/hxmt/Leitz/general/brochures/Leitz%20PMM-C%20Line_brochure_en.pdf. Accessed 27 Sept 2018
5. Optical Metrology Ltd.: An introduction to optimet conoscopic holography sensors family (2015). http://www.optimet.com/our_technology.php. Accessed 27 Sept 2018
6. Cheng, X., Li, Z.W., Shi, Y.: A high-reflective surface measurement method based on conoscopic holography technology. In: Proceedings of the SPIE, Optical Metrology and Inspection for Industrial Applications III, vol. 9276, p. 927608 (2014)
7. Dupraz, D.: CHRocodile CLS: capteur chromatique ligne dedie a la metrologie haute cadence et haute resolution. In: Proceedings of 17th International Congress of Metrology, p. 13011 (2015)
8. Renishaw plc: How do interferometric systems work? (2012). http://www.renishaw.com/en/how-do-interferometric-systems-work-38612. Accessed 27 Sept 2018
9. Rajeswara, R.: Interferometer and screw thread, gear measurement. In: Proceedings of Mechanical Measurements and Metrology 10ME42B (1959)

10. Zhang, Zh., Feng, Q., Gao, Z., et al.: A new laser displacement sensor based on triangulation for gauge real-time measurement. Opt. Laser Technol. **1**(40), 252–255 (2008)
11. RIFTEK Ltd.: Laser Scanners RF625 Series User's Manual (2017). https://riftek.com/media/documents/rf625/manual/2D_Laser_Scanners_RF625_Series_eng.pdf. Accessed 27 Sept 2018
12. Nikon Metrology NV: LC15Dx - closing the gap with tactile probe accuracy (2017). https://www.nikonmetrology.com/en-gb/product/lc15dx. Accessed 27 Sept 2018
13. Izquierdo, M., Sanchez, M., Ibanez, A., et al.: Sub-pixel measurement of 3D surfaces by laser scanning. Sens. Actuators **76**, 1–8 (1999)

Influence of Induction Heating Modes on Thermal Stresses Within Billets

Yu. E. Pleshivtseva and E. A. Yakubovich[✉]

Samara State Technical University, 244 Molodogvardeyskaya Street,
Samara 443100, Russian Federation
eyakubovich@mail.ru

Abstract. The paper investigates the influence of heating modes on the thermal stresses within the billets appearing during the induction heating prior hot forming operations. The temperature distribution within the billet is considered as an output controlled function depending on time and spatial coordinates. Applicable control actions are a coil voltage or a coil current. The technological restrictions on the maximum temperature, the maximum level of thermal stresses and the maximum temperature gradient between the hottest and the coldest points within steel cylindrical billet at any time moment of the process are substantiated and formulated. Therefore, the aim of the study is to choose a control input that minimizes time of heating and assures the required final temperature distribution to compliance with all selected restrictions. The special optimization approach based on the unique alternance method of optimal control theory is applied to minimize the time of heating taking into consideration the main technological restraints. The computational results are shown for the time-optimal heating of steel billets utilizing the software package ANSYS. Comparative analysis of time-optimal heating mode with conventional mode of heating under lower constant voltage allows to argue that time of heating can be substantially decreased that corresponds increasing of heating installation productivity up to more than 15.0%. At the same time, the time-optimal heating mode with consideration of restraints provides more homogeneous temperature distribution within the heated billet in comparison with typical heating modes.

Keywords: Induction heating · Billets · Thermal stress · Optimal control · Hot forming

1 Introduction

The induction heating of workpieces (billets, bars, slabs, etc.) prior to hot working is widely used in modern industrial production. This is due to a number of practical and technological advantages that this electrothermal process offers in comparison with alternative heating methods. In addition to increasing equipment efficiency and flexibility, a topic of increasing interest in the field of induction heating is the control of the stress and strain within workpieces are subjected to during heating.

The paper investigates the influence of heating modes on the thermal stresses within the billets appearing during the induction heating prior hot forming operations.

Technological processes of metal heating prior a treatment by pressure should satisfy some requirements imposed on the behavior of the temperature field during the heating stage.

The most typical requirement is caused by the fact that that a maximum temperature within workpiece during the heating stage should not exceed a specified admissible level. If this requirement is not provided, then undesirable irreversible adverse changes could take place in material microstructure and even metal melting can appear. The similar requirement is imposed for steady-state mode of continuous heating regarding maximum temperature within all cross-sections of a heated workpiece.

Not only maximum temperature but also a difference between maximum and minimum temperatures within the volume of a heated workpiece should be constrained during heating process providing that the maximum tensile thermal stresses (caused by temperature gradient) would be below or equal to the admissible value that is prescribed according to ultimate stress limit of the material heated. At the same time, the thermal stresses depend (in a rather complicated way) on temperature distribution within the heated workpiece.

Violation of these requirements often leads to irreversible spoilage of product, i.e. crack development.

It is very important when choosing the proper process parameters for induction heating to take into account technological requirements influencing the formation of thermal stresses within the billets. Technological constraints complicate the solution of the optimization problem and appropriate computational procedures [1, 2].

In the presented researches the optimization technique based on alternance method of optimal control theory [3] is applied to the solution of time-optimal control problem of steel billet heating with additional restraints utilizing 2D ANSYS model.

2 Problem of Thermal Stresses Formation

The heating of metals prior to hot forming reduces materials' inherent resistance to plastic deformation. Induction heating processes for subsequent hot working involve a number of important process parameters – some of the most critical are: maximum admissible temperature during heating, maximum temperature gradient within the transverse cross-section, and necessary heating time.

The maximum admissible temperature $T_{max}(t)$ depends on the chemical composition of the metal alloy and the characteristics of the forging process [4]. The target forging temperature, the temperature of the workpiece after induction heating (and transportation/loading time if significant), must be sufficiently high to reduce deformation resistance, provide safe and productive equipment operation, and facilitate tooling longevity. Exceedingly high forging temperatures, however, can result in unacceptable structural changes (grain coarsening, decarburization, intergranular oxidation, insipient melting, and steel burning), excessive scale formation, and unnecessarily high energy consumption. Determining an optimal forging temperature is especially important for complex alloys, which often exert a high resistance to plastic deformation during hot working [5].

The degree of cross-sectional temperature uniformity, which can be quantified by the maximum allowable temperature difference, $\Delta T_{max}(t)$, is also important because excessive temperature gradients over the cross-section of the workpiece (after heating) cause uneven deformation. Temperature non-uniformity during heating can also be a concern – particularly when working with metals that have relatively high ductile-brittle transition temperatures (e.g. high-carbon steels and bearing steels). Significant thermal stresses $\sigma(t)$ can arise from a temperature difference $\Delta T(t)$. Due to their tendency to promote crack formation and propagation, the most dangerous stresses that can appear are the tensile stresses, which must be considered when estimating the permissible temperature difference $\Delta T_{max}(t)$ [6].

Thermal stresses resulting from temperature gradients are particularly important to consider when heating using electromagnetic induction. The electromagnetic skin effect phoenomenon results in inherently non-uniform induced heat generation in the heated workpiece. The skin effect is very highly pronounced when heating magnetic steels (below Curie temperature) and when heating low-resistivity materials such as low-alloy aluminium and copper grades. This importance of controlling the transverse temperature gradient in an induction heated workpiece is amplified by the high heating intensities that induction heating can provide.

The probability of cracking a workpiece during heating can increase with the presence of residual thermal stresses within its volume. These stresses arise if the workpiece was previously subjected to heating and cooling (e.g. in previous rolling, casting, or piercing processes). During the cooling stage, the cooler outer layers of the metal reach the transition temperature from the plastic state to the elastic state earlier. While cooling, tensile stresses appear in the inner layers and they do not disappear as a result of the low plasticity of the outer material. If the workpiece is heated again, the thermal stresses that arise within it can add to these residual tensile stresses, exacerbating the risk of cracking. Residual stresses related to previous phase transformations can also be present prior to heating and should also be considered if heating previously heat-treated materials (i.e. hardened steels).

Cracking of the workpiece is caused by the appearance of stresses and deformations that exceed critical values [7]. The main cause of stress and strain development during heating is non-uniform thermal expansion (and resistance to expansion) resulting from induced temperature gradients; however metallurgical phase and structural changes also contribute to the stress state of the workpiece. Heating steels for the purpose of hot working provides the formation of an austenitic structure and the dissolution of carbon and alloying elements within the workpiece cross-section. The heating process involves phase transformations and changes in the specific volume of different phases and structural morphologies. Changes in stress distributions and magnitudes inherently occur due to the difference in the mechanical properties of the individual phases.

While structural and phase changes do contribute to the stress-deformed state of the heated workpiece, the main factor in the development of the stresses is the induced temperature gradient. Due to the nature of induction heating (i.e. electromagnetic skin effect) and the fact that faster heating times can be achieved by rapid initial heating of the wokrpiece, the transverse temperature gradient is typically the largest during the initial stage of heating [8].

When heating magnetic materials with relatively poor thermal conductivity (i.e. carbon steels and non-austenitic stainless steels), surface-to-core temperature differences can approach or exceed 700 °C during this stage of heating. If a higher-than-optimal frequency is selected, the temperature difference can be substantially higher.

When the sub-surface region of a workpiece is at a substantially lower temperature than the surface region, localized tensile stresses appear in order to compensate for the inherent compressive stresses developed in the surface region (due to thermal expansion). As the temperature gradient grows these sub-surface thermal tensile stresses (in combination with potential tensile residual stresses from previous operations) can result in failure – particularly if the relatively low temperature of the subsurface material still exhibits brittle mechanical behavior. Before the elastic-to-plastic transition, the development of stresses is determined by the shear modulus of the material; once the material was transferred to plastic state the main characteristic is the yield strength.

It is advisable to evaluate the stressed-deformed state of a workpiece in the process of induction heating on the basis of the rheological model of metal behavior within the framework of the idealized theory of plasticity of a hardening body [9], using fracture criteria that follow from semiempirical theories of the limiting state [10]. For example: $KR = \sigma_i/\sigma_p$, where σ_i is the stress intensity and σ_p is the destructive stress. It is assumed that the cracks are formed most often in the case of $KR = 1$ in the regions of metal stretching. This approach allows one to predict the formation of zones of defects, for example, mesh ones, which are often observed in practice and represent a grid of mutually intersecting meandering cracks penetrating to a depth of up to 25 mm.

3 Optimization of Induction Heating Modes

It is very important when choosing the proper process parameters for induction heating to take into account all mentioned above physical properties and requirements influencing the formation of thermal stresses within the billets. A significant effect can be provided by optimization of the operational modes of induction heaters taking into consideration main technological restraints on maximum admissible temperature during heating, maximum temperature gradient within the transverse cross-section, and thermal stresses resulting from temperature gradients.

The induction heating process is considered here as a dynamic system under control. Then a temperature distribution within the billet, $T(r, l, t)$, is considered as an output controlled function depending on time t and spatial coordinates $l \in [0; R]$, $y \in [0; L]$, R is a billets' radius, and L is a billets' length.

Control inputs influence the temperature distribution. Applicable control actions are a coil voltage $U(t)$ or a coil current. A density of electromagnetic heat sources power represents a typical example of time-dependent control actions that are used typically to control the process of heating by induction. It is reasonable to consider the formulation of optimal control problem when it is required that the control inputs vary within prescribed range, i.e. the following constraint is taken into account:

$$0 \leq U(t) \leq U_{\max}, \ t \in [0, t^0] \qquad (1)$$

where a maximum allowable value of inductor voltage $U(t)$ is restricted by power supply limitations.

Temperature greatly affects the formability of metals. The most technologies of metal heating prior to hot forming must provide a desired temperature T* within the heated workpiece with a tolerance depending on the technological application specifics. In most typical processes, the maximum admissible value ε of absolute deviation of temperature distribution T(r, l, to) from the required temperature T* at the end of heating $t = to$ is prescribed as follows:

$$\max_{x \in \Omega} |T(r, l, t^0) - T^*| \leq \varepsilon \qquad (2)$$

A proper choice of certain controls leads to necessity of determination of a goal function that provides estimation of the economic and technical efficiency of an induction heating system under consideration.

Let us consider the problem of minimizing total heating time, t^0 representing a cost function in the case when maximum productivity is required.

According the explanation above the following technological requirements are taken into account. The following requirement provides that maximum temperature T_{max} within a heated billet does not exceed a specified admissible value T_{adm} during the whole process:

$$T_{\max}(t) = \max_{l \in [0; R]; y \in [0; L]} T(r, l, t) \leq T_{adm}, \ 0 \leq t \leq t^0 \qquad (3)$$

The next requirement provides that the value $\sigma_{\max}$ of maximum tensile thermal stresses would be below the specified admissible value, σ_{adm}, at any moment of the process. The value σ_{adm} should correspond to ultimate stress limit of the material to be heated. This requirement can be written in the form of the following constraint that should be satisfied:

$$\sigma_{\max}(t) = \max_{l \in [0; R]; y \in [0; L]} \sigma(r, l, t) \leq \sigma_{adm}, \ 0 \leq t \leq t^0 \qquad (4)$$

The third requirement demands that the maximum value $\Delta T_{max}(t)$ of the temperature gradient between the hottest $T_{max}(t)$ and the coldest $T_{min}(t)$ points within a heated billet volume should be below a certain admissible value ΔT_{adm} at any time moment of the process. This constraint can be written in the following form:

$$\Delta T_{\max}(t) = \max_{l \in [0; R]; y \in [0; L]} (T_{\max}(t) - T_{\min}(t)) \leq \Delta T_{adm}, \ 0 \leq t \leq t^0 \qquad (5)$$

If the value ΔT_{adm} will be exceeded, then irreversible adverse changes appear in material structural properties and even irreversible damage to a product could take place.

Therefore, we are interested in choosing a control input $U_{opt}(t)$ restricted by (1) that minimizes time of heating and assures the required final temperature distribution according to (2) under conditions (3), (4) and (5).

The formulated problem can be solved applying the theory and techniques of optimal control by systems with distributed parameters described detailed in [11–14]. The general algorithm for problem solution consists of following stages:

- Procedure of parameterization of control action
- Reduction of initial problem to the problem of semi-infinite optimization
- Application of alternance method of optimal control theory to solve the problem of semi-infinite optimization

4 Computational Results

Let us consider the results of solution of the formulated problem for the following initial data: length of inductor is 1.4 m; length of the billet is 0.8 m; radius of billet is 0.04 m; required temperature is 1200 °C; number of turns is 37; frequency of the current is 1000 Hz [15].

The computational procedure has been performed according to the previously described optimization technique for the time-optimal heating of steel billets utilizing software package ANSYS®. It is considered that electromagnetic, temperature and thermal stresses fields are strongly interrelated. The computational procedure starts with an electromagnetic field simulation that provides the distribution of induced power within the billet as well as the electrical parameters of the heating system. Then the thermal analysis is carried out with consideration of obtained induced power distribution, and heat losses from the surface; the temperatures field within the workpiece is calculated until the end of the current time step. Heat transfer due to radiation and thermal convection of air are taken into account as thermal boundary conditions. Thus, on the next step it is possible to perform electromagnetic analysis and update the properties of material according to the previous thermal analysis. As a result of simulation, the temperature distribution within the billet is provided as well as the values of electrical parameters are obtained [16].

The structural static analysis can be performed using the results of electromagnetic and thermal simulation that leads to determination of thermal stresses distribution within the billet. Then the next iteration repeats the coupled field analysis.

All materials properties are temperature dependent and ensure the appropriate simulation accuracy during entire heating cycle. Material properties are evaluated for each time step of the transient thermal analysis within volume of the heated billet. Advantage will be taken to billet's symmetry, which allows treating it as rotationally symmetric workpiece. The field analysis is performed in the spatial domain within the whole longitudinal cross-section including the coil, the heated workpiece, the refractory and the surrounding air [1].

Figures 1, 2 and 3 show the results of optimal induction heating that provides a holding of temperature $T_{max}(t)$ at the admissible value of $T_{max} = 1250$ °C, the maximum value, σ_{max}, of tensile thermal stresses is kept at the admissible value $\sigma_{adm} =$

300 MPa, and the maximum value $\Delta T_{max}(t)$ of the temperature drop is held at the level of $\Delta T_{adm} = 220$ °C.

As it is shown in [15], the time-optimal control algorithm $U_{opt}(t)$ by a coil voltage can be divided into 6 control intervals: I – an accelerated heating under maximum voltage $U_{opt}(t) = U_{max}$ during 20 s; II – the interval of holding the maximum tensile thermal stress σ_{max} at admissible level σ_{adm} with the duration of 30 s; III – the interval of holding the maximum temperature drop ΔT_{max} at admissible level ΔT_{adm} with the duration of 250 s; IV – heating under maximum voltage $U_{opt}(t) = U_{max}$ during 120 s; V – the interval of holding the maximum temperature T_{max} at admissible level T_{adm} with the duration of 70 s; VI – the interval of temperature soaking under $U_{opt}(t) = 0$ with the duration of 13 s.

Figure 1 demonstrates how the maximum, $T_{max}(t)$, and the minimum, $T_{min}(t)$, temperatures within a heated billet volume vary during the time-optimal heating process.

The maximum tensile thermal stress, $\sigma_{max}(t)$, within the heated billet during the time-optimal heating process is shown in Fig. 2.

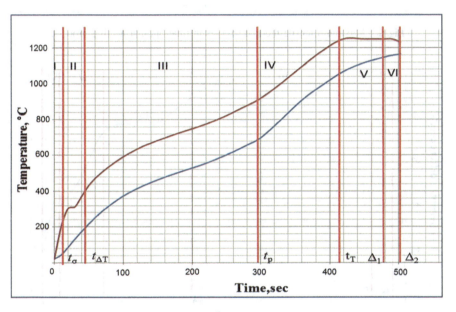

Fig. 1. Maximum (red line) and minimum (blue line) temperatures in time-optimal heating process.

Figure 3 represents the temperature drop between the hottest $Tmax(t)$ and the coldest $Tmin(t)$ points within a heated billet volume in time-optimal heating process.

The maximum value ε of absolute deviation of temperature $T(r, l, t^o)$ within the heated volume from the required temperature T^* at the end of heating $t =$ to is equal to 33 °C. Minimal possible heating time to is equal to sec. Comparative analysis of time-optimal heating mode with conventional mode of heating under lower constant voltage

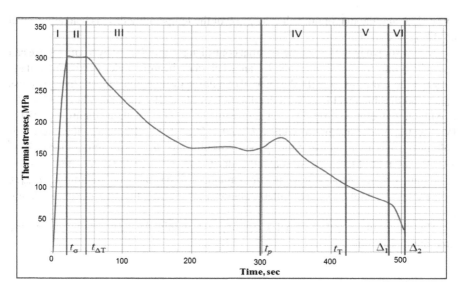

Fig. 2. Maximum tensile thermal stresses in time-optimal heating process.

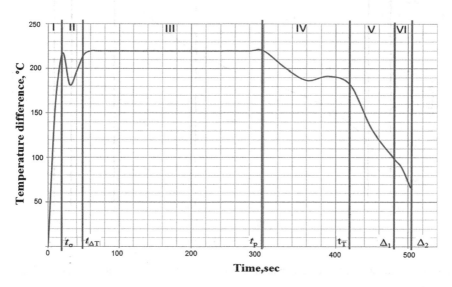

Fig. 3. Maximum temperature gradient in time-optimal heating process.

$U(t) = const = 380B$ allows to make a conclusion that time of heating can be decreased from 598 s to 503 s that corresponds increasing of heating installation productivity up to 15.9%.

At the same time, the time-optimal heating mode with consideration of restraints (3)–(5) provides more homogeneous temperature distribution within the heated billet in comparison with typical heating modes.

For example, at the end of heating under constant voltage $U(t) = const = 380B$ the value ε of absolute deviation of temperature $T(r, l, t^o)$ within the heated volume from the required temperature T^* is equal to 45 °C. Therefore, in this case the optimization of heating mode leads to increasing the heating accuracy by 26.7% [15].

5 Conclusion

The problems of control of temperature patterns and thermal stresses within steel cylindrical billets in time-optimal induction heating process are investigated and solved based on unique alternance optimization method. The optimization approaches may be used to develop the effective induction heating technological regimes that reduce heating time and thermal stresses during the process.

Acknowledgment. The researches were partly funded by the Russian Foundation of Basic Researches (Projects No. 16-08-00945, 17-08-00593).

References

1. Rapoport, E.Ya., Pleshivtseva, Yu.E.: Optimal control of temperature modes of induction heating. Science, Moscow (2012)
2. Rapoport, E.Ya., Pleshivtseva, Yu.E.: Optimal Control of Induction Heating Processes. Taylor Francis Group, Boca Raton (2007)
3. Rapoport, E.Ya.: Alternance method in applied optimization problem. Science, Moscow (2000)
4. Van Tyne, C.J., Walters, J.: Warm and hot working applications. In: Induction Heating and Heat Treatment, vol. 4C, pp. 293–301 (2014)
5. Srivastava, S.K.: Forging of heat-resistant alloys. In: Forming and Forging, vol. 14, pp. 231–236 (1988)
6. Brown, D., Rudnev, V., Dickson, P.: Induction heating of billets, rods, and bars. In: Induction Heating and Heat Treatment, vol. 4C, pp. 330–345 (2014)
7. Kachanov, L.V.: Fundamentals of fracture mechanics. Science, Moscow (1974)
8. Rudnev, V., Loveless, D., Cook, R., et al.: Handbook of Induction Heating. CRC Press, New York (2003)
9. Kolmogorov, V.L.: Stress, strain, destruction. Metallurgy, Moscow (1970)
10. Nadai, A.: Plasticity and fracture of solids. Mir, Moscow (1969)
11. Rapoport, E.Ya., Pleshivtseva, Yu.E.: Algorithmically precise method of parametric optimization in boundary-value optimal control problems for distributed parameters systems. Optoelectron. Instrum. Data Process. **45**(5), 103–112 (2009)
12. Rapoport, E.Ya., Pleshivtseva, Yu.E.: The successive parametrization method of control actions in boundary value optimal control problems for distributed parameter systems. J. Comput. Syst. Sci. Int. **48**(3), 351–362 (2009)
13. Rapoport, E.Ya., Pleshivtseva, Yu.E.: Optimal control of induction heating of metals prior to warm and hot forming. In: Induction Heating and Heat Treatment, vol. 4C, pp. 366–401 (2014)
14. Rapoport, E.Ya., Pleshivtseva, Yu.E.: Optimal control of nonlinear objects of engineering thermophysics. Optoelectron. Instrum. Data Process. **48**(5), 429–437 (2014)

15. Pleshivtseva, Yu.E.: Modeling and optimization of metal induction heating before hot forming and heat treatment. In: Proceedings of the 28th ASM Heat Treating Society Conference, pp. 670–675 (2015)
16. Pleshivtseva, Yu.E., Korshikov, S., Wipprecht, S., et al.: Simulation of primary heating stage in resource efficient forging chain. Heat Process. **13**(1), 85–90 (2015)

Propellant Consumption Optimal Adaptive Terminal Control of Launch Vehicle

A. F. Shorikov[1(✉)] and V. I. Kalev[2]

[1] Ural Federal University, 19 Mira Street, Ekaterinburg 620002
Russian Federation
afshorikov@mail.ru
[2] S&D Association of Automatics, 145 Mamina-Sibiryaka Street,
Ekaterinburg 620075, Russian Federation

Abstract. This paper provides the solution technique of propellant consumption optimal adaptive terminal control problem for launch vehicles. The initial nonlinear continuous-time model of plant is approximated by linear discrete-time dynamical system using linearization along reference trajectory and subsequent discretization. It is assumed that state vector and control vector are constrained by convex, closed and bounded polyhedral sets with finite number of vertices in corresponding finite-dimensional vector spaces. The problems of optimal open-loop and adaptive (closed-loop) control problems are formulated for approximated system. The solution of the main problem is reduced to sequential solving of auxiliary problems. An optimal adaptive terminal control algorithm is developed based on computation and analysis of forward and backward reachable sets of approximating system. The computation of polyhedral reachable sets is implemented by general recurrent algebraic method. The performance of proposed algorithms is shown at the numerical example of optimal adaptive terminal control for launch vehicle's third stage.

Keywords: Optimal adaptive control · Terminal control · Propellant consumption · Launch vehicle · Reachable sets

1 Introduction

Most of liquid-propellant launch vehicles have four basic on-board control systems that used for guidance, navigation, stabilization and propellant consumption control. The propellant consumption control problem can be classified as a terminal control problem. Its main objective is to completely and simultaneously consume oxidizer and fuel for a given time. The quality criterion of this propellant consumption terminal control problem is the deviation of state vector at final time instant (engines cutoff) from its desired (reference) value.

Over the past 40 years, propellant consumption terminal control problem has received much attention and the obtained results of these investigations are widely presented in literature. The main work in this line of research is the monograph of Petrov [1]. This work shows the way to solve the propellant consumption control problem of launch vehicle as the stochastically optimal control problem.

Currently [2], optimal terminal control problems of dynamical systems with known probabilistic characteristics of a-priori unknown parameters of the system are well studied. However, in actual conditions of launch vehicles production it is often very expensive or even impossible to realize a large number of tests that let us obtain the probabilistic characteristics of uncertain parameters [2]. Thus, the information about the probabilistic characteristic is either lacking or unreliable [2–4]. Therefore, we motivated in considering of parameters uncertainty as the set which bounds all the possible values of the parameters.

The purpose of this work is to solve the optimal adaptive propellant consumption terminal control problem. To achieve this, we use the nonlinear continuous-time plant model described in [5]. We approximate this model of plant by linear discrete-time controllable system and we assume that the sets of state and control constraints are convex, closed and bounded polyhedra with finite number of vertices in corresponding finite-dimensional vector spaces (hereinafter referred to as polytopes, having in mind all these properties). To solve the main problem, it is necessary to solve an auxiliary optimal open-loop terminal control problem similarly as in [2, 3]. It is expected that found optimal terminal control values in approximated system will be close to the same in initial nonlinear continuous-time system in terms of given quality criterion. Thus, the solution of optimal adaptive terminal control problem can be reduced to the recurrent algorithm [4] which is based on sequential (at each control step) solving of corresponding optimal open-loop terminal control problem.

2 Problem Statement

The nonlinear continuous-time dynamical model [5] is considered at the time interval $[t_0, t_1]$ and reproduces the operation mode of launch vehicle propulsion system. The scalar control action $u(\tau)$ represents the change of throttle position in the fuel line, which allows to change the oxidizer and fuel mass rates simultaneously.

The values of propellant mass rates can be calculated from the following nonlinear algebraic expression:

$$
\begin{aligned}
m_o(\tau) &= \frac{(P + c_1 U(\tau)^2 + c_2 U(\tau))(K + U(\tau))}{(I + c_3 U(\tau)^2 + c_4 U(\tau))(1 + K + U(\tau))}; \\
m_f(\tau) &= \frac{P + c_1 U(\tau)^2 + c_2 U(\tau)}{(I + c_3 U(\tau)^2 + c_4 U(\tau))(1 + K + U(\tau))}
\end{aligned}
\qquad (1)
$$

where $\tau \in [t_0, t_1]$; P, I, K are the nominal values of thrust, specific impulse and propellant consumption ratio coefficient, respectively; $U(\tau) = c_5 u(\tau) + \Delta K$, where $u(\tau)$ is the scalar control action; $c_1, c_2, \ldots, c_5$ are some coefficients; ΔK is an error of throttle initial setting.

The values of propellant masses in the tanks depend on the propellant mass rates (1) and can be found from:

$$\begin{aligned} M_o(\tau) &= M_o^0 + \Delta M_o - \int m_o(\tau)d\tau; \\ M_f(\tau) &= M_f^0 + \Delta M_f - \int m_f(\tau)d\tau \end{aligned} \quad (2)$$

where M_o^0, M_f^0 are required initial masses of oxidizer and fuel in propellant tanks; ΔM_o, ΔM_f are propellant tanks filling errors.

The initial nonlinear continuous-time model described by (1), (2) is linearized along the reference trajectory:

$$\begin{cases} m_o^{ref}(\tau) = \frac{P \cdot K}{I + I \cdot K}; \\ m_f^{ref}(\tau) = \frac{P}{I + I \cdot K}; \end{cases} \begin{cases} M_o^{ref}(\tau) = M_o^0 - \frac{P \cdot K}{I + I \cdot K}\tau; \\ M_f^{ref}(\tau) = M_f^0 - \frac{P}{I + I \cdot K}\tau \end{cases} \quad (3)$$

Then, linearized along reference trajectory (3) model is discretized according to piecewise constant nature of control actions. Step-by-step instruction of obtaining approximated system has been provided in [4].

Now let us consider the approximated system at the given integer-valued time interval $\overline{0,T} = \{0, 1, \ldots, T\}$. The values of propellant mass rates can be calculated by

$$\begin{aligned} m_o(t+1) &= m_o(t) + \alpha u(t), m_o(0) = m_o^{nom} + \alpha \Delta K; \\ m_f(t+1) &= m_f(t) + \beta u(t), m_f(0) = m_f^{nom} + \beta \Delta K \end{aligned} \quad (4)$$

where $t \in \overline{0, T-1}$; α, β are the coefficients obtained by linearizing of initial model; $u(t)$ is a scalar control; m_o^{nom}, m_f^{nom} are the nominal values of propellant mass rates.

Recurrence equations of propellant masses in the tanks are

$$\begin{aligned} M_o(t+1) &= M_o(t) - \Delta T(t)m_o(t), M_o(0) = M_o^0 + \Delta M_o; \\ M_f(t+1) &= M_f(t) - \Delta T(t)m_f(t), M_f(0) = M_f^0 + \Delta M_f \end{aligned} \quad (5)$$

where $t \in \overline{0, T-1}$; $\Delta T(t)$ is an estimated time between two neighboring controls.

Thus, Eqs. (4) and (5) can be written in recursive vector-matrix form:

$$x(t+1) = A(t)x(t) + B(t)u(t), x(0) = x_0, \quad (6)$$

where $t \in \overline{0, T-1}$; $x(t)$ is a state vector, $x(t) == \{m_o(t), M_o(t), m_f(t), M_f(t)\} \in \mathbb{R}^4$; $x(0) = x_0$ is a specified initial state vector; $u(t)$ is a control vector, $u(t) \in \mathbb{R}^1$; $A(t)$ is a state matrix, $A(t) \in \mathbb{R}^{4\times 4}$; it is assumed that $\forall t \in \overline{0, T-1}$ inverse matrix $A^{-1}(t) \in \mathbb{R}^{4\times 4}$ exists; $B(t)$ is a control matrix, $B(t) \in \mathbb{R}^{4\times 1}$.

For the next reasoning, we need the following assumptions.

Assumption 1. State vector of the system (6) is bounded by convex polytope with finite number of vertices:

$$x(t) \in X_1(t) \subset \mathbb{R}^4, t \in \overline{0,T}, \quad (7)$$

Assumption 2. Control vector of the system (6) is bounded by convex polytope with finite number of vertices:

$$u(t) \in U_1(t) \subset \mathbb{R}^1, t \in \overline{0, T-1}, \tag{8}$$

It should be noted that time instant T in linear discrete-time system (6) corresponds to time instant t_1 in initial nonlinear continuous-time system (1), (2) and number of control actions $u(\tau)$ in initial system and $u(t)$ in approximated system is the same.

For fixed time interval $\overline{\tau, T} \subseteq \overline{0, T}$ ($\tau < T$) and using the constraint (8) let us define the set $U(\overline{\tau, T}) \in \text{comp}(\mathbb{R}^{1 \times (T-\tau)})$ of feasible open-loop controls $u(\cdot) = \{u(t)\}_{t \in \overline{\tau, T-1}}$ as follows

$$U(\overline{\tau, T}) = \left\{ u(\cdot) : u(\cdot) \in \mathbb{R}^{1 \times (T-\tau)}, \forall t \in \overline{\tau, T-1}, u(t) \in U_1(t) \right\}.$$

Let us call the set $w(\tau) = \{\tau, x(\tau)\} \in \overline{0, T} \times X_1(\tau)$ ($w(0) == w_0 = \{0, x_0\}$) a τ-position of discrete-time dynamical system (6)–(8) and $W(\tau) = \{\tau\} \times X_1(\tau)$ ($W(0) = W_0 = \{w(0) = w_0 : w_0 = \{0, x_0\} \in 0 \times X_0\}$) a set of all its feasible τ-positions.

Then, to estimate the control process quality in dynamical system (6)–(8) at the time interval $\overline{\tau, T} \subseteq \overline{0, T}$ we define the convex functional $\gamma_{\overline{\tau, T}} : W(\tau) \times U(\overline{\tau, T}) \to \mathbb{R}^1$ so that for realization of the tuple $(w(\tau), u(\cdot)) \in W(\tau) \times U(\overline{\tau, T})$ its value is defined by

$$\gamma_{\overline{\tau, T}}(w(\tau), u(\cdot)) = \|x(T) - x_d\|_4 = \Phi(x(T)), \tag{9}$$

where $x(T) = \bar{x}(T; \overline{\tau, T}, x(\tau), \{u(t)\}_{t \in \overline{\tau, T-1}})$ is a final state of system motion (trajectory); x_d is a desired final state vector, $x_d \in \mathbb{R}^4$; $\|\cdot\|_4$ is an Euclidean norm in $\mathbb{R}^4$.

At the time interval $\overline{\tau, T} \subseteq \overline{0, T}$ it is necessary to obtain such result of control process using feasible open-loop controls $u(\cdot) \in U(\overline{\tau, T})$, that functional $\gamma_{\overline{\tau, T}}$ defined by (9) takes the minimum possible value.

This goal is achieved by solving the following auxiliary problem of optimal open-loop terminal control for approximating system (6)–(9).

Problem 1. Given fixed time interval $\overline{\tau, T} \subseteq \overline{0, T}$ ($\tau < T$) and realization of τ-position $w(\tau) = \{\tau, x(\tau)\} \in W(\tau)$ ($w(0) = w_0$) in dynamical system (6)–(9) it is necessary to find such set $U_\gamma^{(e)}(\overline{\tau, T}, w(\tau)) \subset U(\overline{\tau, T})$ of optimal open-loop controls $u(\cdot) = \{u(t)\}_{t \in \overline{\tau, T-1}}$ defined by

$$U_\gamma^{(e)}(\overline{\tau, T}, w(\tau)) = \{u^{(e)} : u^{(e)}(\cdot) \in U(\overline{\tau, T});$$
$$\gamma_{\overline{\tau, T}}(w(\tau), u^{(e)}(\cdot)) = \min_{u(\cdot) \in U(\overline{\tau, T})} \gamma_{\overline{\tau, T}}(w(\tau), u(\cdot)) = c_\gamma^{(e)}(\overline{\tau, T}, w(\tau))\}, \tag{10}$$

using finite sequence of only one-step operations. The functional $\gamma_{\overline{\tau, T}}$ is defined here by (9). The number $c_\gamma^{(e)}(\overline{\tau, T}, w(\tau))$ we call the optimal value of open-loop control process result at time interval $\overline{\tau, T}$ for discrete-time dynamical system (6)–(9) relative to

τ-position $w(\tau)$ and functional $\gamma_{\overline{\tau,T}}$. It should be noted that the solution of above problem (10) exists and later we will provide the constructive algorithms for finding of this solution.

Taking into account the stated above we can formulate the goal in optimal adaptive (closed-loop) control problem for approximating system (6)–(9) as follows.

At the time interval $\overline{0,T}$ it is necessary to choose the control $u(\cdot) = \{u(t)\}_{t \in \overline{0,T-1}}$ ($\forall t \in \overline{0,T-1} : u(t) \in U_1(t)$) of the plant (6) as adaptive (closed-loop) control based on the information about t-position $w(t) = \{t, x(t)\} \in W(t)$ at every time instant $t \in \overline{0,T-1}$ so that at the finish of control process realization the functional $\gamma_{\overline{0,T}}$ defined by (9) takes the minimum possible value. Therefore, using above reasoning and similarly to [3] we can formalize the achievement of this goal as follows.

The feasible adaptive control policy U_a in discrete-time dynamical system (6)–(9) at time interval $\overline{0,T}$ we call the mapping which for every time instant $\tau \in \overline{0,T-1}$ and feasible realization of τ-position $w(\tau) = \{\tau, x(\tau)\} \in W(\tau)$ ($w(0) = w_0$) prescribes the set $U_a(w(\tau)) \subseteq U_1(\tau)$ of controls $u(\tau) \in U_1(\tau)$. Let us denote the set of all feasible adaptive control policies for considered process as U_a^*.

The bundle of trajectories of system (6)–(9) is the set

$$\begin{aligned} X(\cdot; \overline{0,T}, w_0, U_a) = \{x^*(\cdot) &: x^*(\cdot) \in \mathbb{R}^{4 \times (T-1)}; \\ \exists u^*(\cdot) \in U(\overline{0,T}), \forall t \in \overline{0,T}, x^*(t) &= \bar{x}(t; \overline{0,T}, x_0, u^*(\cdot)); \\ w^*(t) = \{t, x^*(t)\} &\in W(t), w^*(0) = w_0; \\ u_t^*(\cdot) = \{u^*(\tau)\}_{\tau \in \overline{0,t-1}}, \forall t \in \overline{0,T-1}, u^*(t) &\in U_a(w^*(t))\} \end{aligned} \quad (11)$$

that corresponds to initial position $w(0) = w_0 \in W_0$ and feasible policy $U_a = U_a(w^*(\tau)) \in U_a^*$, $t \in \overline{0,T-1}$, $w^*(t) = \{t, x^*(t)\} \in W(t)$ at time interval $\overline{0,T}$. Thus, we can formulate the nonlinear multi-step optimal adaptive control problem for dynamical system (6)–(9).

Problem 2. Given time interval $\overline{0,T}$ and initial position $w_0 = \{0, x_0\} \in W_0$ in discrete-time dynamical system (6)–(9) it is necessary to compute the optimal adaptive control policy $U_a^{(e)} = U_a^{(e)}(w(\tau)) \in U_a^*$, $w(t) = \{t, x(t)\} \in W(t)$, $t \in \overline{0,T-1}$, ($w(0) = w_0$), which satisfies to

$$\begin{aligned} \gamma_{\overline{0,T}}(w_0, U_a^{(e)}) &= \min_{x(T) \in \bar{x}(T; \overline{0,T}, w_0, U_a^{(e)})} \Phi(x(T)) = \min_{U_a \in U_a^*} \min_{x(T) \in \bar{x}(T; \overline{0,T}, w_0, U_a)}, \\ \Phi(x(T)) &= \min_{U_a \in U_a^*} \gamma_{\overline{0,T}}(w_0, U_a) = c_{a,\gamma}^{(e)}(\overline{0,T}, w_0) \end{aligned} \quad (12)$$

using finite sequence of only one-step operations. The functional $\gamma_{\overline{0,T}}$ is defined here by (9). The number $c_{a,\gamma}^{(e)}(\overline{0,T}, w_0)$ we call the optimal value of adaptive control result at time interval $\overline{0,T}$ for discrete-time dynamical system (6)–(9) relative to initial position w_0 and functional $\gamma_{\overline{0,T}}$. Note that the solution of above problem (12) exists. Therefore, the next section is devoted to providing of the algorithm for finding of this solution.

3 Solutions of the Problems 1 and 2

In this section, we first introduce some significant definitions and propositions, and then we present the algorithms of solving the Problems 1 and 2.

3.1 Preliminaries

At the present time, the using of reachable set construction approaches is widespread in theoretical and practical control problems [2–6, 9]. Let us introduce the following definitions.

Definition 1. The forward reachable set of linear discrete-time dynamical system (6) states at the time instant $\vartheta \in \overline{\tau+1,T}$ corresponding to the tuple $(\tau, X(\tau)) \in \overline{0, T-1} \times 2^{\mathbb{R}^4}$ (here and below, for any set Y a symbol 2^Y denotes the set of all subsets of the set Y) is the set defined by

$$G_+(\tau, X(\tau); \vartheta) = \{x(\vartheta)|x(\vartheta) \in \mathbb{R}^4;$$
$$x(t+1) = A(t)x(t) + B(t)u(t) \in X_1(t+1);$$
$$t \in \overline{\tau, \vartheta - 1}, x(\tau) \in X(\tau), u(t) \in U_1(t)\}.$$

Definition 2. The backward reachable set of system (6) states at the time instant $\tau \in \overline{0, \vartheta - 1}$ corresponding to the tuple $(\vartheta, X(\vartheta)) \in \overline{1,T} \times 2^{\mathbb{R}^4}$ is the set defined by

$$G_-(\vartheta, X(\vartheta); \tau) = \{x(\tau)|x(\tau) \in \mathbb{R}^4;$$
$$x(t) = A^{-1}(t)[x(t+1) - B(t)u(t)] \in X_1(t);$$
$$t \in \{\vartheta, \vartheta - 1, \ldots, \tau + 1, \tau\}, x(\vartheta) \in X(\vartheta), u(t) \in U_1(t)\}.$$

The detailed description of reachable set computation modified (using [8]) algorithm for linear discrete-time dynamical systems implemented in this work one can find in [6, 7].

The necessary and sufficient condition for solution of Problem 1 follows from the previous reasoning and results of work [4], that is following proposition holds.

Proposition 1. Given fixed time interval $\overline{\tau, T} \subseteq \overline{0, T}(\tau < T)$, initial position $w_0 \in W_0$ in dynamical system (6)–(9), state vector $x(\tau) \in G_+(0, \{x_0\}; \tau)$ of the plant (6) that defines the τ-position $w(\tau) = \{\tau, x(\tau)\} \in W(\tau)$ of considered system and the set $X^{(e)}_{\gamma_{\tau,T}}(T) \in 2^{\mathbb{R}^4}$ which is approximated with specified tolerance by polytope (i.e. the set of optimal in terms of (9) terminal states of the system (6)), let the set $U^{(e)}(\overline{\tau,T}, \{x(\tau)\}, X^{(e)}_{\gamma_{\tau,T}}(T)) \subseteq U(\overline{\tau,T})$ be already generated. Thus, the following equations hold:

$$U_\gamma^{(e)}(\overline{\tau,T},w(\tau)) = U^{(e)}(\overline{\tau,T},\{x(\tau)\},X_{\gamma_{\tau,T}}^{(e)}(T));$$
$$c_\gamma^{(e)}(\overline{\tau,T},w(\tau)) = \Phi_{\overline{\tau,T}}^{(e)}.$$

Remark 1. The solution of Problem 1 consists in the computation of the optimal open-loop controls set $U^{(e)}(\overline{\tau,T},\{x(\tau)\},X_{\gamma_{\tau,T}}^{(e)}(T)) = U_\gamma^{(e)}(\overline{\tau,T},w(\tau))$ and the optimal value of open-loop control result $\Phi_{\overline{\tau,T}}^{(e)} = c_\gamma^{(e)}(\overline{\tau,T},w(\tau))$ at time interval $\overline{\tau,T}$ by recursive solving of multiple linear and convex programs.

Using the solution of Problem 1 for all $\tau \in \overline{0,T-1}$ and all τ-positions $w^{(e)}(\tau) = \{\tau, x^{(e)}(\tau)\} \in W(\tau)$ $(w^{(e)}(0) = w_0)$, where $x^{(e)}(\tau) = \bar{x}(\tau; \overline{0,T}, x_0, u^{(e)}(\cdot)), u^{(e)}(\cdot) \in U^{(e)}(\overline{0,T},\{x_0\},X_{\gamma_{0,T}}^{(e)}(T))$, we can build the following set:

$$\widetilde{U}_*^{(e)}(w^{(e)}(\tau)) = \{\tilde{u}^{(e)}(\tau) : \tilde{u}^{(e)}(\tau) \in U_1(\tau), \tilde{u}^{(e)}(\tau) = u^{(e)}(\tau); \\ u^{(e)}(\cdot) \in U^{(e)}(\overline{\tau,T},\{x^{(e)}(\tau)\},X_{\gamma_{\tau,T}}^{(e)}(T))\}, \tau \in \overline{0,T-1} \quad (13)$$

Then, let us define the control policy $\widetilde{U}_a^{(e)} == \widetilde{U}_a^{(e)}(w^{(e)}(\tau)) \in \widetilde{U}_a^*$ ($\tau \in \overline{0,T-1}, w(\tau) \in W(\tau), w(0) = w_0$) for the adaptive (closed-loop) control process in approximating system (6)–(9) at the time interval $\overline{0,T}$ from the class of feasible control policies U_a^*. This control policy is formally described by following relations:

(1) for all $\tau \in \overline{0,T-1}$ and for all τ-positions $w^{(e)}(\tau) == \{\tau, x^{(e)}(\tau)\} \in W(\tau)$ $(w^{(e)}(0) = w_0)$ let

$$\widetilde{U}_a^{(e)}(w^{(e)}(\tau)) = \widetilde{U}_*^{(e)}(w^{(e)}(\tau)) \subseteq U_1(\tau), \quad (14)$$

(2) for all $\tau \in \overline{0,T-1}$ and for all τ-positions $w^{(e)}(\tau) == \{\tau, x^{(e)}(\tau)\} \notin W(\tau)$ $(w^{(e)}(0) \neq w_0)$ let

$$\widetilde{U}_a^{(e)}(w^{(e)}(\tau)) = U_1(\tau), \quad (15)$$

where $x^{(e)}(\tau) = \bar{x}(\tau; \overline{0,\tau}, x_0, u_\tau^{(e)}(\cdot)), u_\tau^{(e)}(\cdot) = \{u^{(e)}(t)\}_{t\in\overline{0,\tau-1}}, u^{(e)}(\cdot) \in U^{(e)}(\overline{0,T},\{x_0\},X_{\gamma_{\tau,T}}^{(e)}(T))$.

Let $\tilde{u}^{(e)}(\cdot) = \{\tilde{u}^{(e)}(t)\}_{t\in\overline{0,T-1}} \in U(\overline{0,T})$ be the realization of control at time interval $\overline{0,T}$, which is computed using policy $\widetilde{U}_a^{(e)} \in U_a^*$ at this time interval and $\tilde{u}^{(e)}(T-1)$ satisfies to (10) for $\tau = T-1$. Thus, we can compute the following:

$$\tilde{c}_{a,\gamma}^{(e)}(\overline{0,T},w_0) = \gamma_{\overline{0,T}}(w_0,\tilde{u}^{(e)}(\cdot)), \quad (16)$$

The following proposition results from the Proposition 1 and formulae (13)–(16).

Proposition 2. Given initial position $w(0) = w_0 = \{0, x_0\} \in W_0$ in discrete-time dynamical system (6)–(9), the control policy $\widetilde{U}_a^{(e)} \in U_a^*$ at time interval $\overline{0,T}$ defined by (14), (15) is the optimal adaptive control policy for Problem 2, that is $U_a^{(e)} = \widetilde{U}_a^{(e)} \in U_a^*$ and the number $\widetilde{c}_{a,\gamma}^{(e)}(\overline{0,T}, w_0)$ is the optimal result of Problem 2, that is $c_{a,\gamma}^{(e)}(\overline{0,T}, w_0) = \widetilde{c}_{a,\gamma}^{(e)}(\overline{0,T}, w_0)$ corresponding to implementing of this policy at time interval $\overline{0,T}$ to considered control process.

Thus, to solve the optimal adaptive (closed-loop) control problem for discrete-time dynamical system (6)–(9) in the class of feasible adaptive control policies, we propose the recursive algorithm, which reduces the initial multi-step problem to solving of the finite sequence of optimal open-loop control Problems 1. Each Problem 1 reduces to implementation of finite sequence of one-step operations and solving the linear and convex programs.

3.2 Algorithm of Optimal Open-Loop Terminal Control

Let the τ-position $w^{(e)}(\tau) = \{\tau, x^{(e)}(\tau)\} \in W(\tau)$ ($w^{(e)}(0) == w_0$) be already generated in discrete-time dynamical system (6)–(9). Then, for the computation of the set of optimal open-loop controls $U^{(e)}(\overline{\tau,T}, \{x(\tau)\}, X_{\gamma_{\tau,T}}^{(e)}(T)) \subset U(\overline{\tau,T})$ the following algorithm is developed [3, 6].

(1) The sequential computation of forward reachable sets $G_+(\tau, x^{(e)}(\tau); T), \forall t \in \overline{\tau,T}$.
(2) Optimization of the functional $\gamma_{\overline{\tau,T}}(w(\tau), u(\cdot))$ defined by (9) on the set $G_+(\tau, x^{(e)}(\tau); T)$, i.e. finding the set $X_{\gamma_{\tau,T}}^{(e)}(T) \subset \mathbb{R}^4$ of final states of the system (6)–(9) from the solving of convex program

$$X_{\gamma_{\tau,T}}^{(e)}(T) = \{x^{(e)}(T) | x^{(e)}(T) \in G_+(\tau, x^{(e)}(\tau); T);$$
$$c_\gamma^{(e)}(\overline{\tau,T}, w(\tau)) = \Phi_{\overline{\tau,T}}^{(e)} = \min_{x(T) \in G_+(\tau, x^{(e)}(\tau); T)} \|x^{(e)}(T) - x_d\|_4\}.$$

(3) The sequential computation of backward reachable sets $G(T, X_{\gamma_{\tau,T}}^{(e)}(T); \tau)$, $\forall t \in \{T-1, \ldots, \tau+2, \tau+1\}$.
(4) The sequential computation of following sets (intersections of forward and backward reachable sets)

$$G(\tau, x(\tau); t) = G_+(\tau, x(\tau); t) \cap G_-(T, X_{\gamma_{\tau,T}}^{(e)}(T); t), \forall t \in \overline{\tau, T-1};$$
$$G(\tau, x(\tau); \tau) = \{x(\tau)\}; G(\tau, x(\tau); T) = X_{\gamma_{\tau,T}}^{(e)}(T).$$

(5) Solving the boundary-value problems at interval $\overline{\tau,T}$ for dynamical system (6)–(9) with the boundary conditions $x(\tau)$ and $x(T) = x^{(e)}(T) \in \mathbf{X}_{\gamma_{\tau,T}}^{(e)}(T)$, that is finding the following set of optimal open-loop controls

$$U^{(e)}(\overline{\tau,T},\{x(\tau)\},X^{(e)}_{\gamma\frac{}{\tau,T}}(T)) = \{u^{(e)}(\cdot)|u^{(e)}(t) \in U_1(t), \forall t \in \overline{\tau,T-1};$$
$$x^{(e)}(t+1) = A(t)x^{(e)}(t) + B(t)u^{(e)}(t) \in G(\tau,x(\tau);t+1);$$
$$x^{(e)}(\tau) \in X_1(\tau)\}.$$

3.3 Algorithm of Optimal Adaptive Terminal Control

Let the τ-position $w^{(e)}(\tau) = \{\tau, x^{(e)}(\tau)\} \in W(\tau)$ ($w^{(e)}(0) = w_o$) be already generated in discrete-time dynamical system (6)–(9). Then, for $\tau \in \overline{0,T-1}$, the optimal adaptive control policy $U_a^{(e)} = \widetilde{U}_a^{(e)} \in U_a^*$ is realized according to the following algorithm.

(1) Solution of Problem 1, i.e. computation of the set $U_\gamma^{(e)}(\overline{\tau,T}, w(\tau)) = U^{(e)}(\overline{\tau,T},\{x(\tau)\},X^{(e)}_{\gamma\frac{}{\tau,T}}(T)) \in U(\overline{\tau,T})$ and the number $c_\gamma^{(e)}(\overline{\tau,T}, w(\tau)) = \Phi^{(e)}_{\frac{}{\tau,T}}$ at time interval $\overline{\tau,T}$;

(2) Computation of the set of controls $U_a^{(e)}(w^{(e)}(\tau)) = \widetilde{U}_a^{(e)}(w^{(e)}(\tau)) \subseteq U_1(\tau)$;

(3) Selecting of any control $u^{(e)}(\tau) \in U_a^{(e)}(w^{(e)}(\tau))$;

(4) Computation of $\tau + 1$-position $w^{(e)}(\tau+1) = \{\tau+1, x^{(e)}(\tau+1)\}$, $x^{(e)}(\tau+1) = \tilde{x}(\tau+1;\overline{\tau,\tau+1}, x^{(e)}(\tau), u^{(e)}(\tau))$ based on nonlinear plant (1), (2);

(5) If $(\tau+1) \leqslant T-1$ go to stage 1, otherwise go to stage 6;

(6) Computation of number $c_{a,\gamma}^{(e)}(\overline{0,T}, w_0) = \gamma_{\overline{0,T}}(w_0, u^{(e)}(\cdot))$ (optimal result of main problem) based on the realization of control $u^{(e)}(\cdot) = \{u^{(e)}(t)\}_{t \in \overline{0,T-1}} \in U(\overline{0,T})$ generated by optimal adaptive control policy $U_a^{(e)} = \widetilde{U}_a^{(e)} \in U_a^*$.

In [3] it has been shown that for the class of linear discrete-time dynamical systems with corresponding properties the introduced adaptive control algorithm is an optimal adaptive control algorithm according to cost function (9). Thus [2], it should be noted that the whole algorithm of optimal adaptive control reduces to solving a finite number of linear and convex programming (optimization) problems.

4 Numerical Example

Let us show the efficiency of proposed approach on numerical example, where we simulate the solution of propellant consumption optimal adaptive terminal control problem for launch vehicle [5]. Initial nonlinear continuous-time system is described at time interval $[0, t_1]$ by the Eqs. (1), (2) with the parameters given in Table 1.

The initial nonlinear continuous-time system is associated with the simplified linear discrete-time system obtained from linearization along reference trajectory (3) and discretization. Thus, the recursive vector-matrix system corresponds to the Eqs. (4), (5) and has the form

$$x(t+1) = A(t)x(t) + B(t)u(t), t \in \overline{0,15}$$

where $x(t) = \{x_1(t), x_2(t), x_3(t), x_4(t)\} \in \mathbb{R}^4$; $x_1(t)$ is an oxidizer mass rate; $x_2(t)$ is a mass of oxidizer in the tank; $x_3(t)$ is a fuel mass rate; $x_4(t)$ is a mass of fuel in the tank; $u(t) \in \mathbb{R}^1$ is a scalar feasible control action; $x(0) = x_0 = (1\,180.097,\ 119\,600,\ 460.975,\ 45\,300)^T$; matrices $A(t)$ and $B(t)$ are time-invariant and have the values

$$A(t) = \begin{pmatrix} 1 & 0 & 0 & 0 \\ -6.25 & 1 & 0 & 0 \\ 0 & 0 & 1 & 0 \\ 0 & 0 & -6.25 & 1 \end{pmatrix}, \quad B(t) = \begin{pmatrix} 17 \\ 0 \\ -12 \\ 0 \end{pmatrix}, \quad \forall t \in \overline{0,15}$$

The state vector coordinates are constrained for all $t \in \overline{0,16}$:

$$x_1(t) \in [1\,176; 1\,224], \quad x_3(t) \in [432; 468]$$

The performance of control process is estimated by the value of cost function at time $T = 16$ corresponding to time instant $t_I = 100$ s for initial nonlinear continuous-time system

$$\Phi(x(T)) = \sqrt{(1\,200 - x_1(T))^2 + x_2(T)^2 + \frac{64}{9}\left((450 - x_3(T))^2 + x_4(T)^2\right)}$$

where the first and the third terms under the square root denote the deflection of propellant mass rates from the required values; the second and the fourth terms denote the amount of residual propellant in the tanks.

The solution of optimal open-loop control problem for linear discrete-time system is the set of optimal controls $U^{(e)}(\cdot) = \{u^{(ej)}(\cdot)\}_{j \in \overline{1,k}}, \forall j \in \overline{1,k} : u^{(ej)}(\cdot) = \{u^{(ej)}(t)\}_{t \in \overline{0,15}}$. The set $U^{(e)}(\cdot)$ generates the optimal trajectories $x^{(ej)}(\cdot) = \bar{x}(\cdot; \overline{0,16}, x_0, u^{(ej)}(\cdot)), j \in \overline{1,k}$, the final state of which $x^{(ej)}(16) = \bar{x}(16; \overline{0,16}, x_0, u^{(ej)}(\cdot))$ is for every $j \in \overline{1,k}$ the same state vector $x^{(ej)}(16) = (1200.75,\ 40.64,\ 450.2,\ 10.95)$ that gives to the functional Φ the minimum value $\Phi^{(e)} = \Phi(x^{(e)}(16)) = 50.0622$. Then, we select two optimal open-loop control $u^{(ej)}, j = \overline{1,2}$ and substitute them in initial nonlinear continuous-time system (1), (2) to compare this control with found adaptive control policies.

Table 1. Parameters of nonlinear continuous-time system.

I, s	P, kg	K	M_o^0, kg	M_f^0, kg	ΔM_o, kg	ΔM_f, kg	ΔK	c_1	c_2	c_3	c_4	c_5	t_I, s
320	528 000	8/3	120 000	45 000	−500	300		8/75	8/75	−1 675	2 250	−15	−12	100

The solution of main problem is the sets $U_a^{(e)}(\cdot) = \{u_a^{(e,j)}(\cdot)\}_{j \in \overline{1,d}}, \forall j \in \overline{1,d}$: $u_a^{(e,j)}(\cdot) = \{u_a^{(e,j)}(x(t))\}_{t \in \overline{0,15}}$. For the comparison, we select two optimal adaptive terminal control policies $u_a^{(e,j)}(x(t)), j \in \overline{1,2}$, i.e. for $d = 2$ and substitute it, step by step, to initial nonlinear continuous-time model (1), (2). The projections of trajectories of nonlinear system with optimal open-loop terminal control and optimal adaptive terminal control are shown on Figs. 1–2.

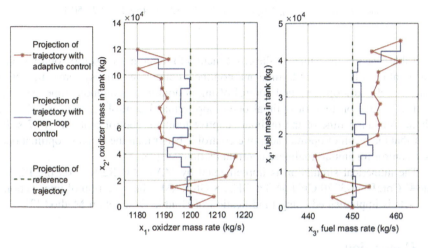

Fig. 1. The projections of phase trajectories with $u^{(e,1)}$ and $u_a^{(e,1)}$.

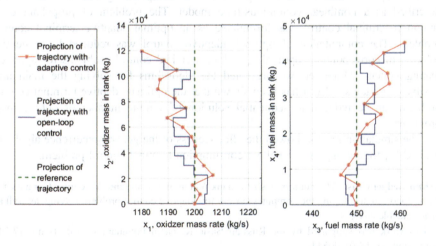

Fig. 2. The projections of phase trajectories with $u^{(e,2)}$ and $u_a^{(e,2)}$.

Table 2. The results of solving the problem.

	Control	$x_1(T)$	$x_2(T)$	$x_3(T)$	$x_4(T)$	$\Phi(x(T))$
Reference trajectory	$u(t) \equiv 0$	1 200	0	450	0	0
Nonlinear system	$u^{(e,1)}(t)$	1 200.6	55	449.7	53.9	153.8
	$u_a^{(e,1)}(x(t))$	1 200	101.3	450	27.3	124.8
	$u^{(e,2)}(t)$	1 200.6	55.6	449.7	55.3	157.7
	$u_a^{(e,2)}(x(t))$	1 200.1	86.9	449.9	23.3	106.7
Linear system	$u^{(e,j)}(t)$	1 200.7	−22.9	450.2	−7.7	50.1

Table 2 shows the values of the cost function for reference trajectory, optimal open-loop control of linear discrete-time and nonlinear continuous-time systems and optimal adaptive (closed-loop) terminal control of initial nonlinear continuous-time system.

The simulation time of optimal open-loop terminal control problem is 0.4 s. The whole simulation time of the optimal adaptive terminal control problem is 27 s. It is clear that the result of optimal adaptive control is better than the result of optimal open-loop control for initial nonlinear continuous-time system.

The numerical simulation is implemented by MATLAB R2014a on the PC with Intel© Core™ i7-3770 CPU @ 3.4 GHz, 8 Gb RAM. The operation of intersection of two sets is implemented using modification of Double Description Method [3].

5 Conclusion

In this paper, a launch vehicles propellant consumption dynamical system was described as a nonlinear continuous-time model. The problem of propellant consumption terminal control was formulated as an optimal adaptive terminal control problem. The computation of optimal adaptive control was reduced to sequential solution of corresponding optimal open-loop terminal control problems for approximating system. Moreover, we presented the algorithms for solving the main and auxiliary problems. Also it has been shown that the optimal adaptive terminal control problem can be reduced to sequential solution of a finite number of optimization problems.

The simulation results showed the effectiveness of the general recurrence algebraic method [3] for solving the propellant consumption terminal control problem.

Acknowledgements. The authors wish to thank Vladimir Bulaev and Alexander Goranov for their support in developing the computer system for adaptive control problem solving, as well as their valuable advices.

This work was supported by the Russian Basic Research Foundation, projects no. 17-01-00315 and no. 18-01-00544.

References

1. Petrov, B.N., Portnov-Sokolov, Yu.P., Andrienko, A.Y., et al.: On-board Terminal Control Systems. Mashinostroenie, Moscow (1983)
2. Shorikov, A.F., Goranov, A.Yu.: About one adaptive correction of optimal open-loop control for spacecraft rendezvous. In: International Conference on Industrial Engineering, Applications and Manufacturing, pp. 1–6 (2018)
3. Shorikov, A.F.: Minimax Estimation and Control in Discrete-Time Dynamical Systems. University Publ., Ural (1997)
4. Goranov, A.Yu., Shorikov, A.F.: Modified general recursion algebraic method of the linear control systems reachable sets computation. In: CEUR Workshop Proceedings, vol. 1825, pp. 95–102 (2017)
5. Kalev, V.I., Shorikov, A.F.: Fuel consumption terminal control problem statement for liquid-propellant rockets. Russ. Phys. J. **59**(8–2), 45–48 (2016)
6. Bulaev, V.V., Goranov, A.Yu., Kalev, V.I.: The optimal open-loop control construction of complex mechanical plants motion. Bull. South Ural State Univ. Ser. Comput. Technol. Autom. Control Radio Electron. **17**, 20–28 (2017)
7. Shorikov, A.F., Bulaev, V.V.: A modification of the generalized recursion method of linear control systems reachable sets computation. In: CEUR Workshop Proceedings, vol. 1825, pp. 88–94 (2017)
8. Fukuda, K., Prodon, A.: Double Description Method Revisited. Lecture Notes in Computer Science, vol. 1120, pp. 91–111 (1996)
9. Kurzhanskiy, A.A., Varaiya, P.: Reach set computation and control synthesis for discrete-time dynamical systems with disturbances. Automatika **47**, 1414–1426 (2011)

Analysis of Adaptive Machines with Two Process-Connected Working Movements

M. A. Lemeshko(✉), M. D. Molev, and A. G. Iliev

Don State Technical University (Branch) in Shakhty,
147 Shevchenko Street, Shakhty 346500, Russian Federation
lem-mikhail@ya.ru

Abstract. The article is devoted to the study of an adaptive hydraulic drive of technological machines in which two working movements are performed. By the examples of tunnel and drilling machines the analysis of known circuit solutions of such hydraulic drive is carried out. The structural schemes of a hydraulic drive of machines in which the self-regulation for optimum operating conditions is carried out are investigated. The description of the principle of adaptive control of two working movements of the executive bodies of this machine is made by the example of a mountain loading machine with a hydraulic drive. Their technological connection and possibilities of the adaptive hydraulic drive are shown. The possibility of controlling the machine with feedback through kinematic links is shown. By increasing the load on the loading body, automatically decreases the speed of the body of the machine which leads to a decrease in the load on the loading body and stabilize the machine. The examples of other adaptive machines are given. These and other adaptive machines with hydraulic differentials have identical structure of the automated hydraulic drive. The generalized structure of these machines is presented and analyzed and the technique of their research is offered.

Keywords: Automated control · Adaptive hydraulic drive · Technological machine · Analysis of circuit design · Block diagram · Drill machine · Method of stabilization

1 Introduction

Known production machines, in which the workflow is provided by two working-class movements. For example, rotary type drilling machine [1]. The drive of this machine performs two related movements: rotation and translational movement in the direction of drilling. For this purpose, it is advisable to use the methods of hydraulic automation, and in particular to use an automated hydraulic drive, which uses hydraulic differentials [2]. A feature of such schemes is the implementation of feedback force via kinematic couplings [3]. In this automated drive, it is not necessary to use sensors and measuring instruments, comparison schemes of controllers and control algorithms typical for automated control systems. Experience in the development, research and application of this type of drives has shown the ability to provide high performance and stability of these machines [4]. This kind of drive it is advisable to call adaptive, and method of controlling the working process of the machine –adaptive control [3–5]. On the basis of generalization of a number of known and investigated adaptive machines it was succeeded to develop the generalized structure of these machines which is applicable for the analysis and modeling of the automated drive executed on the basis of hydraulic differentials.

2 Topicality

For the rational mode of operation of such machines, it is necessary to regulate and maintain the necessary ratio of each working movement. In the case of a drilling machine, this is the speed of rotation and the translational speed of its feed into the hole or the moment of resistance to the rotation of the drill and its feed force. In this case, the variables are the drilling depth, the strength of the drill rock, the blunting of the cutting end and the possible deformation of the drill and the machine.

The most practical way of such control is the modification of the machine, which uses the principle of adaptation to drilling conditions.

Structurally, the drive of the adaptive machine is designed in such a way that the machine automatically changes the feed force and rotation speed of the drill with increasing strength of the drill rock, or rotation speed and feed speed. As an estimate of the strength of the drilled rock, the value of the moment of resistance to rotation is used. If the moment of resistance to rotation decreases, the feed force or feed speed increases, the rotation speed decreases. With this control, the drilling speed is stabilized in modes close to rational.

Actual is the question of coordination of characteristics of the drive and required laws of regulation. To do this, it is necessary to develop a method of such coordination, perform an analysis of the adaptive control scheme of one of the machines, for further use of this analysis on other similar machines with two technologically related working movements.

There are known adaptive machines [5, 6] in which their operation modes are optimized under changing operating conditions, and such optimization is performed by the machine itself due to its structure, in particular, by means of double-differential connections between the drives performing two main movements.

Let us analyze structural diagrams of some of such adaptive machines. Figure 1 gives the diagram of the hydraulic loading machine with adaptive properties [7]. To analyze adaptive properties, the machine hydraulic drive can be presented in the form of a closed-loop hydraulic drive with allocation of differential groups. Figure 2 shows a hydraulic drive of the given loading machine identical with the diagram shown in the Fig. 1. In the loading machine, the hydraulic drive is not actually closed.

If we connect the hydraulic lines14 at the inlet to the hydraulic pump 3 with the ejection lines15 through the chokes 9 and 10, we obtain a structural scheme of the hydraulic drive with three contours (Fig. 2).

Two hydraulic differentials can be distinguished on the obtained scheme.

The first differential is the flow divider from the outlet of the hydraulic pump 3. The cost of the three circuit elements: 3, 4, 11 associated expression: $Q_3 = Q_{11} + Q_4$.

This is an important differential, because it connects the load of the actuator 1 (A.D.1), and the load of the actuator group 5(A.D.2), 6 (A.D.3).

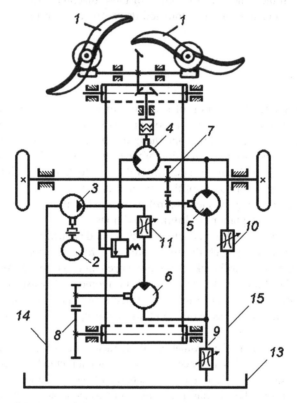

Fig. 1. Diagram of the hydraulic loading machine: 1 – actuating device, 2 – hydraulic motor of the actuating device, 3 – feeding hydraulic motor, 4 – hydraulic pump, 5 – loading conveyor.

The second differential is the summator of pressures and costs through the reversing hydraulic motor 5.

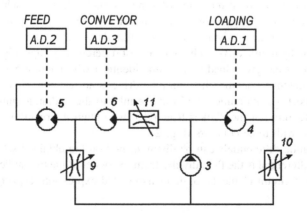

Fig. 2. Transformed structural diagram of the loading machine hydraulic motor.

If the load on the drive 4 increases, then the consumption of the drive 5,6 increases. Technologically, this means an increase in the speed of movement of conveyors and reducing the feed – the speed of moving the machine to the material being loaded.

This loading machine works as follows. Hydraulic fluid from the hydraulic pump 3 is fed to the hydraulic motor 4 of the loading mechanism-Executive device A.D.1 (hydraulic motor of the loading body), while the second hydraulic circuit, the hydraulic fluid flows from the pump through the throttle 11 to the hydraulic motor 6 of the conveyor drive loading machine and then to the entrance of the reversible feed motor 5 driving the loading machine (actuator device A.D.2).

The other input of this hydraulic motor is connected to the hydraulic line from the hydraulic motor output 4 which is also connected to the drain manifold through the throttle 10. The operation condition of the hydraulic circuit is maintained by the main flow of the hydraulic fluid through the hydraulic motor 4, the hydraulic motor 5 and the throttle 9.

In this case, the rotational speed of the reversing motor decreases, the speed of the loading machine towards the mine blade also decreases, while the rotational speed of the conveyor drive increases.

This protects the conveyor from overloading. At a lower conveyor speed, it would be overloaded. As the load on the hydraulic motor 4 decreases, the speed of the loading machine (engine speed 5) in the direction of its movement automatically becomes higher. The considered loading machine performs continuous movement with variable speed.

The speed of this machine is adjusted depending on the moment of resistance to rotation of the drive of the raking paws. Thus, the self-regulation of the loading machine is performed.

In the considered scheme of control of the loading machine, adaptability consists in self-regulation, self-adjustment to changing working conditions. This self-regulation or adaptability is provided by two hydraulic differentials [8].

3 Practical Relevance

As a rule, metal cutting, rock cutting, destruction of construction materials by means of actuating devices with bits, teeth, cutters, and other cutting tools is achieved through the adjustment of the ratio between two speeds in the direction of the cutting tool movement and in the direction perpendicular to the cutting vector—feed. At that, the optimality of operation modes is guaranteed by the adjustment of the combination of forces applied to the above mentioned directions.

Rotary drilling machines also perform two related movements: rotating and feeding it forward.

The problem of adaptive control of cutting is automatic maintenance of optimal [9] the ratios of these efforts, within the limitations of [10]. First of all, technological limitations due to strength properties, wear resistance, dynamic phenomena, etc.

Within the limits of strength and power constraints to maintain drilling performance by increasing the strength of the drill rock or by increasing the blunt of the cutting head, it is necessary to increase the feed force. In a number of scientific works, for example in work [11] it is shown why and as with increase of strength of a drilled material, the moment of resistance to rotational movement decreases.

The well-known study of drilling machine with properties of adaptation [12, 13]. The target function of adaptation in these machines is automatic adjustment to rational modes with protection of the Executive body and its cutting part from torque overloads.

For rotary drilling machines, the adaptation is ensured by the implementation of the coupling: the dependence of the feed force on the torque resistance. The greater the moment of resistance to rotation, the less force feeding. In the developed variants of adaptive drilling machines provides a decrease in the force of feeding the rod into the face with an increase in the torque of the rotation resistance, and in the case of jamming the rod into the face, automatically reverses its movement in the direction of feeding, i.e. the rod is removed from the face to the state when the torque is not restored.

One of the first drilling machines of such a type to successfully pass the industrial tests is the self-adjusting drilling machine BMVA–1 developed by the group of authors headed by the professor Vodyanik [13].

The principle of the hydraulic drive in other machines is similar to the above given description of the loading machine work. In Fig. 4 one can see some examples of drives of the machines and other devices with the adaptive structure: – the hydraulic drive of the method of the cutting process intensification.

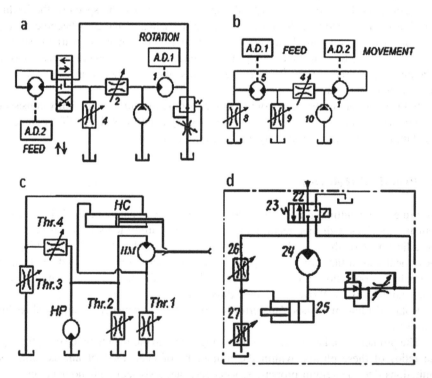

Fig. 3. (a) The hydraulic diagram explaining the method for cutting intensification (i.c. No. 929331); (b) the hydraulic diagram of the device to stabilize the thickness of the sheared layer at mechanical treatment of cammed surfaces (i.c. No. 483224) (c) the hydraulic diagram of the production terminal module of an industrial robot (i.c. 1318392), (d) the hydraulic diagram of the adaptive drilling machine (The patent of the Russian Federation No. 2473767).

Figure 3(a) [14], the hydraulic drive of the device to stabilize the thickness of the sheared layer at mechanical treatment of cammed surfaces – Fig. 3(b) [15], – the hydraulic diagram of the production terminal module of an industrial robot – Fig. 3(c) [16], – the hydraulic diagram of the adaptive drilling machine – Fig. 3(d) [12].

Let us note that structural diagrams of these machines with various intended uses are identical assuming the open-loop circuits are closed-loop, i.e., the drain lines are connected.

Obviously, the given diagrams of the drives of different machines have an identical structure which can be represented in the form of the bridge connection of drives and throttles (Fig. 4).

A detailed description of the method of the structural diagrams representation in the form of bridge connections is given in the patent description [17].

Representation of the structural diagram of the adaptive drilling machine in the form of a bridge connection is suggested with the purpose of use of the electrohydraulic analogy in calculation tasks and for simulation of the process of work of hydraulic machines with similar structures.

The Fig. 4 shows the example of such representation used in the works [17, 18], [19].

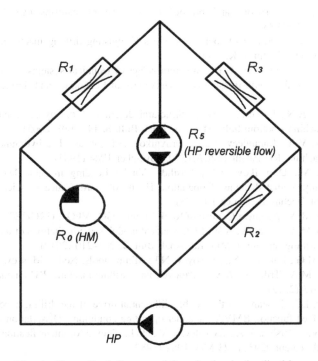

Fig. 4. Generalized diagram of the adaptive hydraulic drive.

4 Conclusions

- Adaptive properties of the hydraulic drive of technological machines can be provided by the use of hydraulic differentials
- The conducted analysis of adaptive machines with two process-connected working movements demonstrated the identity of the structural diagrams of these machines
- The studied adaptive machines can be represented as a bridge connection that is easily analyzed and calculated by the methods of electrohydraulic analogies.

References

1. Vodyanik, G.M., Ryleev, E.V., Drovnikov, A.N.: Research and development of the drilling machine with automatic adjustment of the feed speed for drilling of vertical blastholes up to 2.5. m. Report on the Ec. Agr. Work 1340, p. 111 (1968)
2. Vodyanik, M., Travnikov, A.N., Streltsov, I.P.: Adaptive drive of mining machines on the basis of the machine, hydromachine and mechanical differentials. Min. Mach. Electromech. **10**, 54–57 (2007)
3. Drovnikov, A.N.: Theory and practice of adaptive mechanisms use. VINITI RAN, Novocherkassk (1983)
4. Vodyanik, G.M.: Drilling of rocks with the self-adjusting drilling machine. Mining rock destruction tool. Technics, Kiev (1970)
5. Travnikov, A.N., Dibrova, G.D.: Structural-adaptive technical systems. Mines. Education and Science of the South-Russian State University of Economics and Service, Rostov-on-Don (2011)
6. Drovnikov, A.N., Lemeshko, M.A.: Structural diagram of the dynamic model adaptive drilling machine - bottom hole. Min. Inf. Anal. Bull. **8**, 147–149 (2003)
7. Boltovskiy, V.A., Karandeev, Yu.E., Drovnikov, A.N., et al.: The hydraulic drive of the loading machine. RU Patent 1126701, 30 November 1984 (1983)
8. Vodyanik, M., Drovnikov, A.N., Vasiliev, Yu.A.: Loading machine lateral grip with automatically regulated mode of operation. Herald of North-Caucasus scientific center of high school. Technical science 1 (1973)
9. Lemeshko, M.A.: Adaptive Control of Rock Cutting. GOUVPO YURGUES, Shakty (2010)
10. Lemeshko, M.A., Trifonov, A.V.: Mathematical model of the limitations of adaptive control by rotary drilling machines. Min. Inf. Anal. Bull. **2**, 207–210 (2012)
11. Krapivin, M.G., Rakov, L.Ya., Sysoyev, N.I.: Mining tools. Nedra, Moscow (1990)
12. Lemeshko, M.A., Trifonov, A.V.: Adaptive rotary drilling machine. RU Patent 2473767, 27 January 2013 (2013)
13. Kuznetsov, I.C.: Research of the double-differential drive of the drilling machine with the automatic feed function BMVA - 1 under operating conditions. Dissertation (1976)
14. Tolubets, V.I., Pershin, V.A., Gvozdev, V.V., et al.: Method for intensification of the cutting process. RU Patent 929331, 23 May 1982 (1982)

15. Drovnikov, A.N., Vodyanik, G.M., Pershin, V.A.: Device for stabilizing of the sheared layer thickness at mechanical treatment of cammed surfaces. RU Patent 483224, 5 September 1975 (1975)
16. Drovnikov, A.N., Darda, I.V.: Terminal robot module, RU Patent 1318392, 23 June 1987 (1987)
17. Lemeshko, M.A., Volkov, R.Yu.: Use of the electrohydraulic analogy method to simulate the work of the adaptive drilling machine. Serv. Tech. Prod. Probl. 3(29), 62–65 (2014)
18. Lemeshko, M.A., Vasin, M.A., Saj, D.E.: Mathematical model of optimal control rotary drilling machines. Eastern Eur. Sci. J. **3**, 193–197 (2014)

Approach to Building an Autonomous Cross-platform Automation Controller Based on the Synthesis of Separate Modules

G. M. Martinov, I. A. Kovalev[✉], and A. S. Grigoriev

Moscow State Technological University "STANKIN",
1 Vadkovsky Lane, Moscow 127055, Russian Federation
ilkovalev@mail.ru

Abstract. The paper is dedicated to building a programmable automation controller to execute simple technological tasks (packaging, press, etc.). The proposed approach allows to select the minimum sufficient configuration of the software and hardware platform and application software, depending on the technical task for the controlled object, taking into account the computing capabilities of the platform. It is shown that the cross-platform nature of the proposed solution allows to transfer the kernel of the control system to various platforms without significant alterations in the architecture of the solution, reducing the financial and time costs for designing the final solution. The proposed solution allows the user, on the one hand, to make a choice from the whole variety of technical solutions of one that fully meets the requirements of the task, and on the other hand, the options for eliminating emergency situations associated with the inability of the software and hardware platform to perform the tasks. A practical example of finding a synthesized solution to the control of hydraulic cylinders is also considered.

Keywords: Automation · PAC · PLC · CNC · Industry · Synthesis

1 Introduction

There are various field where the use of powerful high-cost systems is, if not excessive, then at least leads to large financial costs of the enterprise.

Simple technological operations such as product packaging, diagnostic information collection, machine electroautomatic do not require CNC systems, but at the same time with the development of the level of informatization, it is necessary to provide the possibilities for flexible adjustment and modernization of systems, to maintain the integration in smart manufactories. Most PLC and PAC do not have such capabilities because of the peculiarities of their building architectures. For such cases, it is proposed to use an autonomous solution of the automation controller of technological processes with the synthesis of the necessary modules, the requirements of which correspond to the technical task for the control object [1–3].

It is worth noting that the choice of a hardware execution platform is also important. Cross-platform must provide functioning both on standard PC platforms (x86) with CISC, and on ARM-RISC platforms. The first ones have proven themselves as reliable proven solutions, the development that has been going on for a long time and there is a large number of supported solutions. The second - a new trend in the field of mobile technology, these processors are low power, with low power consumption, and for certain tasks, their capacity is more than enough [4, 5].

2 Building Structural Model of the Components for Synthesis Automation Controller

The problem of building such control systems is that there is currently no approach that will allows selecting the optimal software and hardware execution platform for the automation controller's mathematical kernel. Currently such platforms are either redundant in terms of their functionality, or their performance may not be sufficient to fulfill all the terms of technical task, which will lead to the project being rolled back to the initial stages of development [6].

The final version of the functioning of the system depends on its configuration, and, therefore, on the set of modules used and their interrelationships.

To ensure that the control system meets all the requirements of the technical task for the control object, it is necessary to formalize the separate modules, identify the control object itself, its main technological functions, on basis of which the control program will be created in the future [7].

When working with complex technological equipment, and the tendencies go to the constant complication and expansion of its functionality, the foreseeability of the entire technical task is lost. It is necessary to draw up a map of technical task, which displays: the control object, the purpose of the control object and the requirements for the control system, which will contain basic information about the management system and the required supported functionality [8, 9].

A structural model of the components of the cross-platform automation controller has been developed, which made it possible to identify groups of mandatory and optional modules, as well as to describe the interaction options for individual modules of the system, depending on the technical task for the control object [10, 11].

Mandatory modules are always present in the control system, the use of optional modules depends on the task of controlling specific technological equipment. On the left (Fig. 1) are terminal communication clients with the automation controller kernel, through which the management programs, device configurations and system diagnostics are loaded.

Mandatory modules include communication modules with terminal clients, self-test of the system and a master module responsible for the operability of the final solution. The modules to be selected include: the module for working with the machine electroautomatics, the module for diagnosing the technological system and its individual components, the module for verifying the model and devices, the module for working with motion, and others. It is also worth noting that the proposed set of modules is extensible.

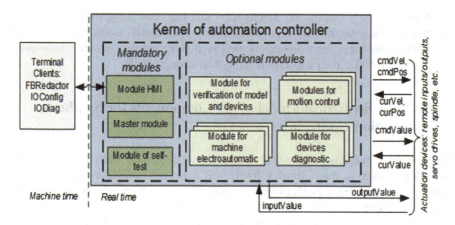

Fig. 1. Structural model of the components of the automation controller.

3 Architectural Model of Cross-platform Controller

The proposed approach allows you to select only those modules that are necessary to perform specific tasks based on a technical assignment. In this case, after selecting the necessary modules, the hardware and software execution platform is tested for compliance with the terms of the technical task with the capabilities of the target computing platform [12]. If any of the specification (CPU load, RAM, etc.) does not meet the requirements, then the solution can be offered either to replace the computing platform, or change its specification, if possible (increase RAM, for example). To ensure this, the core of the control system must have the cross-platform property. To implement the cross-platform functions, the following architectural model is presented, shown in Fig. 2 (levels 1, 2 and 3) and Fig. 3 (levels 4, 5, 6, 7).

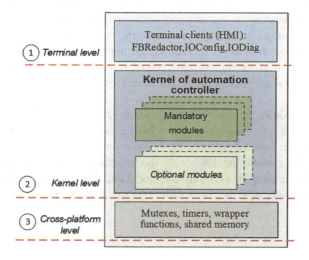

Fig. 2. Levels 1–3 architectural model for cross-platform functions.

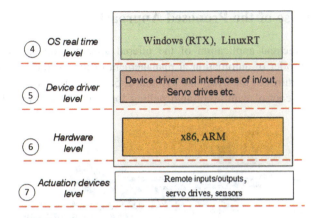

Fig. 3. Levels 4–7 architectural model for cross-platform functions.

At the highest level (level 1) of the architectural model, terminal clients are located to implement the operator interface [13].

At the kernel level (level 2) of the control system is the mathematical kernel of the synthesized control system for the standalone solution and for the solution embedded in the CNC system, which includes mandatory and optional modules.

The embedded cross-platform (level 3) includes both standard PC-based platforms with CISC architecture (x86) for more complex management tasks, and on ARM-based platforms with RISC architecture for simple tasks and control objects. At the same time porting the kernel of the synthesized automation controller from one platform to another occurs with minimal changes due to the use of wrapper functions, platform independent implementation of timers and mutexes [14].

The real-time OS level (level 4) includes MS Windows with RTX real-time extension for PC architecture and LinuxRT for PC and ARM architectures, support for other RTOS, for example, MS DOS, Solaris, QNX, VxWorks, etc., can be added if necessary [15].

The level of drivers (level 5) and interfaces provides data exchange in real time.

The hardware level (level 6) includes x86 or ARM architectures. It is necessary to orientate these two directions when building the architecture of the created automation controller of technological processes. The proposed approach avoids the redundancy of the solution and binds the cost of the synthesized automation controller to the complexity of the managed object, depending on the technical task for it.

The level of the actuation devices (level 7) consists of the remote input/output modules, field buses and sensors connected to them, drives, etc., which are connected to the controller.

The proposed approach to building a cross-platform architecture allows, on the one hand, to expand the scope of the product being developed, on the other hand, it provides the potential customer with the choice of a hardware/software platform, depending on the specific needs of production, technology and the specifics of the tasks and resources of the enterprise [16].

4 Practical Use of the Proposed Approach

Let's consider practical application of the offered approach on an example of the independent controller of automation at management of the hydraulic cylinder.

The design features of the hydraulic cylinder L-500 (OAO «ENIMS») are aimed at implementing the control with a nominal pulling force of 500 kN (Fig. 4) [20].

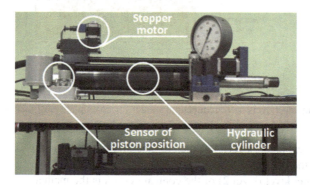

Fig. 4. Experimental stand for hydraulic cylinder L-500.

To control the hydraulic cylinder does not need a sophisticated control system as CNC. At the same time, the use of the PLC as a control system is inadequate, because it is necessary to support various interfaces and be able to configure the system for motion control [17].

The proposed approach includes a method comprising the following steps:

1. Selection of modules based on the technical task;
2. Determination of computing capabilities of the target software and hardware platform;
3. Synthesis of the controller through the alignment of logical relationships between optional and mandatory modules;
4. Verification of the operability of the resulting solution.

In this project, two execution platforms were involved. In the priority was the Russian development Tion Pro 28, as an auxiliary version was Intel Dual CPU E2200. Based on the technical task analysis, a map of technical task was drawn up (Fig. 5).

In the first step of the method, optional and mandatory modules were defined: communication with the terminal, self-test, master module (responsible for building logical relationships), data verification, work with traffic stepper motors), machine electroautomatics (control of a hydroelectric station, sensors).

At the second step, the initial testing of the target execution platforms was performed (Table 1). The column "Without controller kernel" means that on the hardware platform the code of only the operating system with utilities is running without the code of the developed mathematical kernel of the controller [18]. "Empty modules" are OS the kernel code of the controller with modules, but without the components of the control program and the building logical relationships.

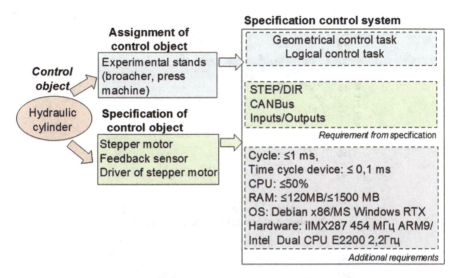

Fig. 5. Experimental stand for hydraulic cylinder L-500.

Table 1. Initial platforms load.

	Without controller kernel (Tion/PC)	Empty modules (Tion/PC)
Load CPU, %	2/10	18/23
Load RAM, MB	50/800	170/260
Cycle tact, ms	–	0,3/01

At the initial stage it was revealed that the Tion Pro does not have the necessary specification (128 RAM versus 220 required), so further work using this platform is not advisable.

Depending on the incoming technical task for the control object and the developed control program, the number of optional modules may vary, but the variant of changing the number of components of the control program is more likely, and the load specification of the hardware and software platform will also change.

The target PC platform has been tested to take score the increase in the number of components of the control program. Tests were conducted on the basis of the developed specialized load testing tools. Figure 6 shows the results for CPU and RAM loads (also tests were conducted for the parameters of the response time of devices and the clock of the control program). In the third step of the method, the mandatory master module assembles the optional modules and their submodules into a kernel of the automation controller based on the generated configuration file. In addition, this master module at this step builds up the interrelations between the individual modules of the final solution.

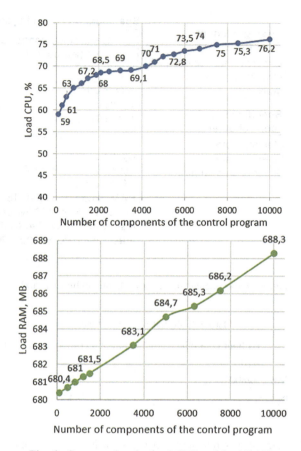

Fig. 6. Stress testing for load CPU and load RAM.

In the fourth step of the method, a synthesized controller with boundary conditions for the components of the control program is obtained. The boundary conditions are obtained based on the results of the second and third steps as a function of the constraint.

In this example, the following restriction functions and the recommended mean maximum values of the control program components were determined based on stress tests.

Equation (1) is restriction function for load CPU.

$$f(x) = (3,42\ln(x) + 12, 18), \qquad (1)$$

The recommended maximum average number of components of the control program for load CPU is 3407 components.

Equation (2) is restriction function for load RAM.

$$f(x) = (0,0016x + 1060, 9), \qquad (2)$$

The recommended maximum average number of components of the control program for load RAM is 274436 components. This value also includes RAM, which is occupied by the OS and utility utilities.

Equation (3) is restriction function for cycle time.

$$f(x) = (0,0001x + 0, 16), \qquad (3)$$

The recommended maximum average number of components of the control program for cycle time is 18400 components.

Equation (4) is restriction function for time cycle devices.

$$f(x) = (1,63\ln(x) + 87, 04), \qquad (4)$$

The recommended maximum average number of components of the control program for cycle time devices is 2836 components.

When designing the control program for autonomous solution, it was necessary to focus on the number of control program components to characterize the response time of devices and CPU utilization, since these values are the smallest ones obtained for other specification: $2836 < 3407 < 18400 < 274436$.

Verification of the operability of a synthesized automation controller used as an autonomous solution, compliance with the items of additional technical task requirements was identified. A stock of more than 20% of the recommended average maximum number of components allows you to extend the management program if necessary.

5 Conclusion

The proposed approach allowed at the initial stages of project development to determine the optimal implementation platform and to avoid the time and financial costs that could arise in the later stages of the project.

Acknowledgements. This research was supported by the Ministry of Education and Science of the Russian Federation as a public program in the sphere of scientific activity (N 2.1237.2017/4.6).

References

1. Nikishechkin, P.A., Kovalev, I.A., Nikich, A.N.: An approach to building a cross-platform system for the collection and processing of diagnostic information about working technological equipment for industrial enterprises. In: MATEC Web of Conferences. ICMTMTE, vol. 129, pp. 1–4 (2017)
2. Neugebauer, R., Denkena, B., Wegener, K.: Mechatronic systems for machine tools. CIRP Ann. Manuf. Technol. **56**(2), 657–686 (2007)

3. Martinov, G., Kozak, N., Nezhmetdinov, R., et al.: Method of decomposition and synthesis of the custom CNC systems. Autom. Remote Control **78**(3), 525–536 (2017)
4. Grigoriev, S.N., Martinov, G.M.: An ARM-based multi-channel CNC solution for multi-tasking turning and milling machines. Procedia CIRP **46**, 525–528 (2016)
5. Vaishak, N.L., Chandra, C.R.: Embedded robot control system based on an embedded operating system, the combination of advanced RISC microprocessor (ARM), DSP and ARM Linux. Int. J. Eng. Innovative Technol. **2**(6), 143–147 (2012)
6. Mehrdad, S., Nassehi, A., Newman, S.T.: A novel methodology for crosstechnology interoperability in CNC machining. Robot. Comput. Integr. Manuf. **29**(3), 79–87 (2013)
7. Geetha, A.: Modelling a computer numerical control machine-2 axis. Middle-East J. Sci. Res. **20**(1), 62–64 (2014)
8. Bushuev, V.V., Evstafieva, S.V., Molodtsov, V.V.: Control loops of a supply servo drive. Russ. Eng. Res. **36**(9), 774–780 (2016)
9. Martinov, G.M., Lyubimov, A.B., Bondarenko, A.I.: An approach to building a multiprotocol CNC system. Autom. Remote Control **76**(1), 172–178 (2015)
10. Kovalev, I.A., Nikishechkin, P.A., Grigoriev, A.S.: Approach to programmable controller building by its main modules synthesizing based on requirements specification for industrial automation. In: International Conference on Industrial Engineering, Applications and Manufacturing, pp. 1–4 (2017)
11. Cloutier, M.F., Paradis, C., Weaver, V.M.: Design and analysis of a 32-bit embedded high-performance cluster optimized for energy and performance. In: Hardware-Software Co-Design for High Performance Computing, IEEE, pp. 1–8 (2014)
12. Martinov, G.M., Kozak, N.V., Nezhmetdinov, R.A.: Implementation of control for peripheral machine equipment based on the external soft PLC integrated with CNC. In: International Conference on Industrial Engineering, Applications and Manufacturing, pp. 1–4 (2017)
13. Martinova, L.I., Sokolov, S.S., Nikishechkin, P.A.: Tools for monitoring and parameter visualization in computer control systems of industrial robots. In: Proceedings of Advances in Swarm and Computational Intelligence, Beijing, pp. 200–207 (2015)
14. Ma, X., Han, Z., Wang, Y., et al.: Development of a PC-based open architecture software-CNC system. Chin. J. Aeronaut. **20**(3), 272–281 (2007)
15. Mori, M., Fujishima, M., Komatsu, M., et al.: Development of remote monitoring and maintenance system for machine tools. CIRP Ann. Manuf. Technol. **57**(1), 433–436 (2008)
16. Chen, J.Y., Tai, K.C., Chen, G.C.: Application of programmable logic controller to build-up an intelligent industry 4.0 platform. Procedia CIRP **63**, 150–155 (2017)
17. Martinova, L.I., Kozak, N.V., Nezhmetdinov, R.A., et al.: The Russian multi-functional CNC system AxiOMA control: practical aspects of application. Autom. Remote Control **76**(1), 179–186 (2015)
18. Martinova, L.I., Pushkov, R.L., Kozak, N.V., et al.: Solution to the problems of axle synchronization and exact positioning in a numerical control system. Autom. Remote Control **75**(1), 129–138 (2014)

The Controlled Proportional Electromagnet-Driven Devices for Axial Rotor Loading in Test Rigs for Studying Fluid-Film Bearings

A. V. Sytin[1], V. V. Preys[2], and D. V. Shutin[1(✉)]

[1] Orel State University, 95 Komsomolskaya Street,
Orel 302026, Russian Federation
rover.ru@gmail.com
[2] Tula State University, 92 Lenina Avenue, Tula 300012, Russian Federation

Abstract. The paper describes the design of the device for axial rotor loading in the experimental rig for studying fluid-film bearings. The device allows applying the controllable force to the rotor and is able to create arbitrary load patterns unlike the passive loading devices that are limited by only one or several types of the created load. The paper shows the structure of the test ring in general and the way of how the designed load device is integrated into it. The controlled force in the device is created by the proportional electromagnetic drive. The required parameters of the drive are based on the parameters of the test rig and bearings being tested on it, such as film thickness, load capacity, rotation speed. The appropriate electromagnetic drive has been chosen taking into account its force-stroke characteristic. The resulting lever-drive system provides covering the whole range of the required loads in the test rig. The paper also described how the drive is integrated into the informational and measurement system of the test rig.

Keywords: Test rig · Fluid-film bearings · Loading device · Proportional electromagnetic drive · Informational and measurement system

1 Introduction

Bearings are the highly loaded and demanding parts of rotor systems and mostly define their reliability, operating life and economy. In the most responsible applications the requirements to bearings are even higher. They must keep operability at deformations caused by high temperatures, have high damping properties and high limit speed [1].

Experimental studies of fluid-film bearings are often implemented using original test rigs that contain an electric drive and a rotor mounted on the bearings being studied. The very important units in such rigs are rotor loading systems. They allow studying bearings in the wide range of external forces acting on them. The typical loading scheme for the most of test rigs is fixed step-by-step loading using a set of weights. Changing the loading weight requires stopping rotation of the system. It is also necessary to start recording a new set of experimental data each time through the operator's command. So, additional time is required for experimental studies implemented at such systems. Moreover, in such cases it is often impossible to get data and analyze transition loading processes.

The present paper describes a test rig for studying fluid-film bearings with the loading device free from the problems mentioned above.

2 Designed Test Rig

Designing of the new test rig was based on the following requirements:

- The automated control of the test rig and data acquisition process
- The ability to operatively and automatically change of the load in order to obtain the physical parameters of the studied bearings and analyze
- Adequacy of the primary converters used for acquisition data about the controlled objects and their parameters
- The software must be adopted to the algorithms of data acquisition and processing
- Safety during the experiment (all rotating elements must be covered with housings, all electric units must be grounded)

The mechanical part of the test rig is a massive basement with the rotor-bearing unit and the electric drive mounted on it (Fig. 1). The electric drive is an asynchronous motor AIR80A2C2 with vertical rotor (power N = 1.5 kW, rotation frequency n = 2850 rpm) controlled with the variable-frequency drive Lenze ESMD (Fig. 1) connected to the information and measurement system (Fig. 2) [2].

The variable-frequency drive allows:

- Choosing the type of the acceleration curve: linear, U-type, S-type, minimal, optimal
- Setting the rotation frequency within the range from 0 to 12000 rpm
- Implementing a run-out or stopping using direct current or a braking resistor
- Imitating transitional processes in some aspects, in particular – set oscillations of the rotor rotation frequency
- Limiting the electric drive power

The test rig (Fig. 1) consists of the shaft 1 that rotates in the two bearings: the radial bearing 2 and the radial-thrust bearing 3. The bearing 3 contains the changeable sleeve 4 that is a pivot of the studied thrust fluid-film bearing. Also there are openings in the bearings 2 and 3 for mounting the displacement sensors 5 tracking the rotor movements in the radial and the axial directions. The bearings are fixed in the steel housing 6 that also has threaded openings for mounting the elements of the lubricant (water) supply system 7 and the pressure sensors 8. The steel housing is mounted on the frame 9 mounted on the massive bed 10. The lower part of the frame 9 is a sealed tank with the system of the lubricant discharge 11 for studying the "submerged" bearings. The walls of the tank are made of transparent material for observation of experiments.

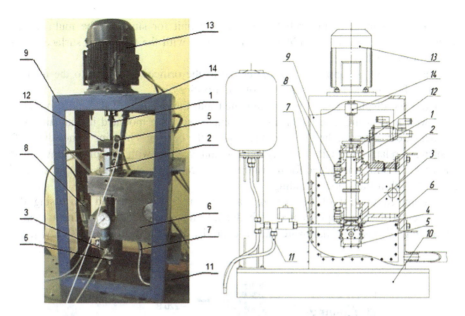

Fig. 1. The test rig.

The original design of the loading device allows solving the mentioned above tasks of the research. The device contains an electromagnetic actuator and a mechanical lever converter for generating axial load at any rotation frequency of the shaft. The device also allows controlling the generated force and the displacement of the output unit manually or according to the pre-developed program.

The arrangement of the radial-thrust bearing unit for studying the tapered land thrust hydrodynamic bearing (THDB) includes a shaft with the thrust end surface, a sleeve of the radial fluid-film bearing and a combined sleeve of the THDB. A sleeve with the pivot with inclined bearing surfaces attached with screws is mounted in the housing of the bearing unit.

At this arrangement it is possible to make experimental studies of the THDB at various directions of the lubricant supply: (1) from the side of the larger radius of the THDB; (2) from the side of the smaller radius of the THDB; (3) without the forced lubricant supply (submerged type).

When lubricant supply is from the side of the larger radius of the THDB, lubricant travels from the collector through the orifices to the radial gap, and then to the cavity and to discharge.

Lubricant supply is from the side of the smaller radius of the THDB is implemented by creating an excess pressure to the cavity. Submerged bearing is obtained by placing the whole bearing unit in the lubricant.

Design of the radial/thrust bearing node for the study of the THDB includes a thrust rotor disk and the tapered land thrust bearing.

The arrangement of the radial-thrust bearing unit for studying the multi-wedge thrust hydrostatic bearing (THDB) includes a shaft with the thrust end surface and a sleeve of the THDB.

The lubricant is supplied from the choke to the orifices, and then to the feeding chamber through the collector and distribution orifices.

The axial displacement sensor is mounted into the opening of the end cap and is fixed with a lock-nut. The radial displacement sensors are mounted in mutually perpendicular directions in the threaded openings of the bearing's housing. Since the shaft diameter is relatively small (12 mm) and is commensurate with a displacement sensor diameter (18 mm), the sleeve for measuring displacement is put on the output zone of the shaft with a transitional landing.

Automated control of the test rig, as well as acquisition and processing the experimental data is implemented with the informational and measurement system (IMS) (Fig. 2) based on hardware and software by National Instruments (USA) [3].

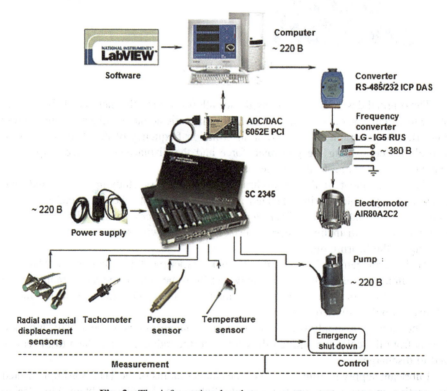

Fig. 2. The informational and measurement system.

The base of the IMS is a multi-functional board NI SC-2345 that contains multiple analog and digital inputs and outputs, counters and timers. The board has a calibration certificate and is mounted into a PCI slot of a PC. The board is used for providing input and output of control and measurement signals used in the rig. NI-DAQ software is used for providing operation of the board in Windows 2000/NT/XP/Me/9x. LabView software [4, 5] is used for implementation of algorithms, processing the experimental data.

The primal converters used as sensors in the test rig are: displacement sensor Pepperl&Fuchs IA5-18GMI3 [6]; pressure sensor KRT-S; temperature sensor PT-S. The proximity sensor is used for measuring the frequency of the shaft's rotation [7]. The pump control and emergency shutdown of the test rig is implemented with relays SC-RLY01.

The control devices are connected to the ADC via an adapter board SC-2345. The board NI6052E and the module SC-2345 are connected with a cable SH 68-68-EP. The adapter board SC-2345 and the modules SCC and SC-RLY01 are mounted in a joint housing.

The software for adjusting the IMS checks operability of all its devices, if the sensors are mounted in a right way according to the measurement schemes.

The software also calibrates the sensors and shows the measurement results in real-time mode. The operation of the test rig can be implemented both in manual or in automatic mode according to the previously set algorithm. Such algorithms take into account particular features of different types of thrust fluid-film bearings. So, studying the run-out of the THDB requires that the time for acquiring and recording the data from the sensors would be greater than the time of operation of the electric drive. The result of operation of the software during the experiment is a file containing the recorded data. The software also can use various filters for pre-processing the data.

The similar experiments have been held in the [8].

Design of the rig allows loading the rotating shaft with either static force, or with dynamic force generated by the controllable electromagnetic actuator. Such load regimes allow testing the stiffness of the fluid film and checking operability of the damper [9].

The calculation of the forces acting in the system has been made during the designing in order to choose the most relevant electromagnetic drive. The main equation used in the calculation was [10]:

$$P = Gü \frac{\mu v L B^2}{h_0^2} \qquad (1)$$

where $Gü = 0.0066$ is the Gumbed number at the maximal bevel height of 100 μm, $\mu = 1 \cdot 10^{-3}$ Pa·s is the lubricant viscosity, v is the relative speed of the surfaces motion, L and B are the length and the width of the inclined surface, h is the minimal thickness of the lubricant film (see Fig. 3 that shows the calculated values of the axial load at different h and rotation speed of the shaft).

The calculated range of the loads allows choosing the appropriate electromagnetic drive, that may be a proportional electromagnetic drive PEM 10 [11, 12]. The chosen drive being used together with the lever in the test rig covers the whole range of the required loads in the system. The force-stroke characteristic of the electromagnetic drive PEM 10 in a graphical form is shown in the Fig. 4.

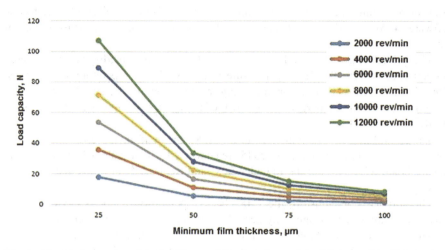

Fig. 3. The load capacity of the bearing at different values of the lubricant film thickness and rotation speed of the shaft.

The electromagnetic actuator used for generating the load force is a proportional actuator EM 3.4-48 [13]. This device provides axial load on the shaft at any rotation frequency within the frequency range of the electric motor. The force and the displacement of the output shaft of the electromagnetic actuator are controllable and can be smoothly adjusted manually or according to the previously created algorithm.

The Fig. 5 shows a general view of the controlled loading device used in the test rig. The device includes: the proportional electromagnetic actuator 1 with the feedback sensor 2, jointed with the corner 3 to the basement 4. A force sensor 5 that is a part of the IMS is fixed with the corner 6 to the lever 8 rotating at the axis 7. On the other side of the lever there are rollers 9 fixed with the axis 10. The rollers push the disk 11 mounted at the shaft 12. The whole mechatronic device is assembled with screws 14, 15, 17 and nuts 16.

The way the loading device operates is smooth change of the value of the axial load with the proportional electromagnet and tracking the value of the loading force and the displacement of the electromagnetic actuator's output shaft.

After switching on the electromagnetic actuator starts pushing through the lever the disc mounted at the shaft. The built-in sensors track the load acting on the rotor and its axial displacement.

The system operates according to the preset mathematical law, the value of the load can be also changed in real-time mode by electronic controllers.

The presented loading device has some advantages that other mechanisms mentioned above also have, such as simplicity of construction, simple integration into the construction of existing test rigs. It can be easily integrated into the IMSs of the rigs and expand the functionality of the loading devices. The designed device is also free from a number of shortcomings of the mentioned analogues.

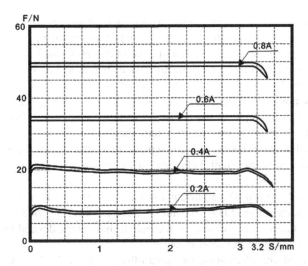

Fig. 4. The force-stroke characteristic of the electromagnetic drive PEM 10.

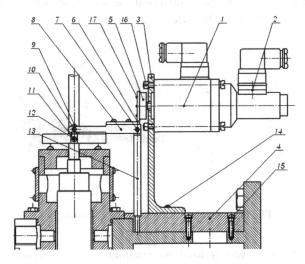

Fig. 5. The general view of the mechatronic device for controlled axial loading.

EM 3.4–48 is used for controlling safety and pressure reducing valves, chokes, flow regulators, pump regulators, direct and electro-hydraulic proportional control valves [14]. Unlike the discrete electromagnet, the output shaft of a proportional electromagnet makes the force proportional to current in the coil. The feedback sensor is used for obtaining information about the position of the output shaft.

Proportional electromagnetic actuator in the loading device is able to provide the dynamic load applied to the rotor of the test rig. Since the relations between the control signal sent to the electromagnet and the force generated by the actuator are well-known and rather simple, mathematical modeling of such processes can be easily implemented. This greatly expands the possibilities for studying the various types of axial fluid-film bearings, especially those that operate in complex conditions with loads that cannot be reproduced with the mechanical devices mentioned above.

There are several types of controlling the proportional electromagnet, by force (Fig. 6a) or by position of the output shaft (Fig. 6b) [15]. During the controlling, the voltage is given to the input of the actuator and to the amplifier 3 through the potentiometer 1. The voltage is transformed into the load current. The current passing through the coil 5 generates an electromagnetic field that causes the lengthwise displacement of the moveable shaft 3 (Fig. 6c) with the force proportional to the current value.

The systems have a positive current feedback – the current value is added to the setpoint value in the adder 2 (Fig. 6a and b). Such feedback provides maintaining the current in the coil, and thus, the force at the output shaft, at the desired level even when the external resistance on the shaft changes.

The feature of the electromagnets controlled by force is that they provide the constant force proportional to the input signal at the full range of the output shaft stroke. Returning of the shaft to the initial position after removing the control signal is implemented by the spring 4 (Fig. 6c).

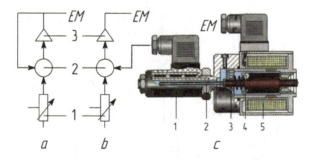

Fig. 6. Functional diagram of the electromagnet.

Current feedback is widely used for controlling electromagnets, but the most accurate position control in an electromagnetic actuator can be implemented using direct position feedback. Such controllers are used in proportional electromagnetic actuators, including the one used in the described test rig (Fig. 6). The position of the armature 3 in the actuator is determined by the value of current in the coil despite the load applied to the actuator. The MC-3300-RV controller is also used in the test rig for connecting the actuator to the LabVIEW software [16].

The ADC used in the test rig's IMS is for digitalizing the signals from the sensors. Particularly, MEGATRON Elektronik AG&Co (Germany) is used as the force sensor [17]. It is connected to the ADC/DAC board NI SC-2345 that is a multifunctional data input and output board. A high-speed chassis PXI connects the board with the PC.

Some signals from the sensors need to be converted from the current signal to the voltage signal before analog-digital conversion, modules SCC-2345 are used for this together with the 2-channel current matching devices SCC-CI20. An interface converter ADAM-4521 is used for connecting the force sensors to the NI SC-2345 board [18].

The Fig. 7 shows the structure of the loading control system based on the National Instruments hardware and LabVIEW software [19]. Automated control of measurement and data acquisition, operation frequency of various elements and required sampling frequency makes it practically impossible to use traditional methods of data acquisition. This leads to necessity to use high speed acquisition devices that must meet the following requirements: multfunctionality, programmability, flexibility, possibility for additional modification and improvement, possibility to operate with a wide range of input and output channels. It is also very important that these devices are accurate enough and could self-calibrate.

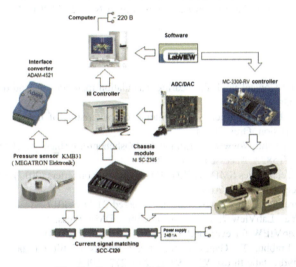

Fig. 7. Structure of the control system.

In order to control a system, it is necessary to constantly acquire data on the change of its parameters, in the present case that is rotor's displacement. For that purpose a proximity sensor is used, analog data 0–20 mA from this sensor is sent to the ADC/DAC NI SC-2345 [20]. Here a zero position corresponds to the initial position of the rotor. Then the value of the signal increases proportionally to the distance between the rotor and the sensor.

LabVIEW software allows processing of the data to control the process. A developed program analyses the data from the NI SC-2345 board and compares it to the set parameters, i.e. desired amplitude of bending and desired speed of the lateral motion. If no deviations are identified, the loading device continues operation; however, if some of the parameters deviate from the corresponding set values, the device generates the necessary adjustments of the operation. LabVIEW allows transmitting the generated signal to the corresponding output channel [21].

The main advantages of the developed control system are the following: high speed operation by means of application of hardware by National Instruments and means of

USB connection, high accuracy of parameter adjustment, absence of need of constant control and observation by an operator, compactness, reliability and simplicity of use.

3 Conclusion

The presented automated test rig with axial loading device based on a proportional electromagnet allows decreasing of time consumption of the experiments, features a flexible and dynamic loading module, which corresponds to real processes in rotor machines, and a wide range of automatically controlled parameters of the experiment.

Acknowledgments. The present research has been implemented as a part of the Ministry of Education and Science of the Russian Federation project No 9.2952.2017/4.6.

References

1. Korneev, A.Y.: Conical fluid-film bearing: calculation methodology and dynamic analysis. Dissertation, Orel (2016)
2. Alehin, A.V.: Bearing capacity and dynamic characteristics of thrust bearings of liquid friction. Dissertation, Orel (2005)
3. AIR80A2U2. http://www.elagr.ru/katalog/obshchepromyshlennye_elektrodvigateli_po_standartu_gost/15822. Accessed 21 Dec 2017
4. Lenze, A.C.: Tech Series SMD (TML/TMD) frequency converters (2017). http://www.lenze.su/SMD. Accessed 21 Dec 2017
5. National Instruments. http://www.ni.com. Accessed 21 Dec 2017
6. Suranov, A.Ya.: LabView 7: a guide to functions. DMK Press, Moscow (2005)
7. Travis, D.: LabVIEW for everyone. DMK Press, Moscow (2004)
8. Jayachandra Prabhu, T., Ganesan, N., et al.: Effects of tilt on the characteristics of multirecess hydrostatic thrust. Wear **92**(2), 269–277 (1983)
9. Prabhu, T.: Analysis of multi-recess conical hydrostatic thrust bearings under rotation. Wear **89**(1), 2940 (1983)
10. Gerasimov, S.A.: Influence of damping and parameters of axial aligned supports on dynamics of rotors. Dissertation, Oryol (2011)
11. PEM 10. http://www.elmagnit.ru/katalog/elektromagnity/spetsialnye-elektromagnity/pem-10. Accessed 21 Dec 2017
12. Electromagnets of type PEM 10. http://www.atrium-tr.ru/products/privody-elektromagnity/proporcionalnye-elektromagnity//elektromagnity-tipa-pem-10. Accessed 21 Dec 2017
13. NGO: Measuring equipment. http://www.npoit.ru. Accessed 21 Dec 2017
14. Schmitt, A., et al.: Hydraulic Training Course: proportional engineering and valve technology 2. Mannesmann Rexroth GmbH (1986)
15. Pepperl + Fuchs – Russia. http://www.pepperl-fuchs.ru. Accessed 21 Dec 2017
16. New Scale Technologies. https://www.newscaletech.com. Accessed 21 Dec 2017
17. MEGATRON. https://www.megatron.de. Accessed 21 Dec 2017
18. Advantech. http://www.advantech.ru. Accessed 21 Dec 2017
19. Blum, P.: LabVIEW. The style of programming. DMK Press, Moscow (2008)
20. Izerman, R.: Digital Control Systems. Mir, Moscow (1984)
21. Fedosov, V.P., Nesterenko, A.K.: Digital signal processing in LabVIEW. DMK Press, Moscow (2007)

Research of Process of Automatic Opening of Air-Penetrating Flexible Containers for Free-Flowing Products

A. M. Makarov(✉), O. V. Mushkin, M. A. Lapikov, and Yu. P. Serdobintsev

Volgograd State Technical University, 28 Lenina Avenue, Volgograd 400005, Russian Federation
amm34@mail.ru

Abstract. This paper presents studies of vacuum grippers when working with breathable materials. Such devices can be used in packaging equipment in the food, chemical industry, construction and agriculture. It will create conditions for the comprehensive automation of the packaging process. Consequently, productivity and quality of packaging will increase, while material and labor costs will decrease. The research results are necessary for the development of reliable vacuum grippers for automatic packaging systems for polypropylene bags and can be used in structural-parametric synthesis, tuning and optimization of automatic packaging lines for packaging various materials in flexible containers. This will allow us moving to a more productive and environmentally friendly production. In the framework of the study, mathematical dependences were obtained for calculations of pass-through of vacuum channels of the gripping devices for flexible breathable matter. An experimental setup has been developed and constructed to study the dependence of pressure in vacuum grippers of various diameters, the number and diameter of vacuum channels on the reliability and quality of holding of flexible breathable containers. The results of the studies confirmed the possibility of using of vacuum gripping devices for handling of breathable and flexible materials.

Keywords: Vacuum gripping · Flexible container · Air-penetrating · Sorting · Free-flowing products · Mathematic modeling

1 Introduction

In a modern automated production robotics and mechatronic systems are used. Issues related to improving the efficiency of mechatronic systems, including in the field of packaging of bulk materials are considered in [1, 2]. It is shown that there are many problems that arise in technical processes of handling of production objects. Industrial handling systems are used in various operations and work with components that vary in strength, mass, sizes etc. An important task is handling of breathable flexible containers (FC) in filling its' of free-flowing products (FP). The task includes a several difficult actions: gripping of FC from stack; opening and holding of FC during filling with FP, transfer of filled FC to conveyer belt for its' transportation to the area of sealing (stitching, gluing etc.) [3, 4]. One of the most important stages is the process of individual separation and opening of flexible container [5–7].

Vacuum gripping devices VGD [8–12] can be used to solve this problem. It's explained by the fact that VGDs have multiple advantages compared to other GDs: equal distribution of the load on the surface; simplicity of constructions; possibility of working with large nomenclature part.

In the devices [13, 14], developed by the authors, VGDs can also be used for to perform a full cycle of packing bulk materials in flexible containers, including closure and sealing. Figure 1 shows one of the schematic options of such devise.

Its' differentiating feature is the ability to conduct a full cycle of operations of handling flexible container in sorting FP in automatic mode. This is achieved by using two units of VGDs 4, 5 and 16. Unit of VGDs 4, 5 performs separation of FC from the stack and it's feeding to filling position, unit of VGDs 16 performs opening and closing of FC when it is filled with free-flowing products.

The suggested devise has a number of advantages, compared to alternatives: simplicity of construction and kinematic diagram, which allows to use automated drive for moving of the grippers; full cycle of handling of FC during sorting is provided, including separation of FC from the stack, its' gripping, opening and feeding under the loading nozzle, dozing of FP, stitching of FC and its' transportation to the place of storing.

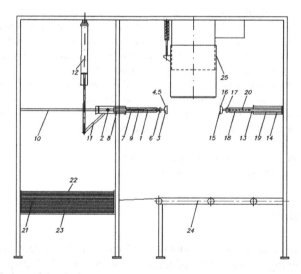

Fig. 1. General view of the device: 1 – gripping mechanism, 2 – pneumatic cylinder, 3 – VGD, 4, 5 – vacuum suctions, 6 – bar, 7 – lever, 8 – supporting element, 9 – spring, 10 – frame, 11 – rod, 12 – turning pneumatic cylinder, 13 – opening mechanism, 14 – pneumatic cylinder, 15 – VGD, 16 – vacuum suctions, 17 – bar, 18 – lever, 19 – supporting element, 20 – spring, 21 – stack of FCs, 22 – FC, 23 – lifting table, 24 – conveyor belt, 25 – loading nozzle.

However, while developing such systems, it is necessary to take into consideration features of the process of manipulating FC, material of which it is made, and also the properties of the material of FPs used to fill FCs. For analysis and evaluation purposes of the processes taking place with VGD in interacting with the material of FC, we'll describe mathematical model of VGD [15].

2 Mathematical Modelling of VGD

Let's imagine that VGD is an opening through which the air flows. In this case, an important factor of VGD functioning would be the pass-through of an opening.

By the opening is meant pipeline in which the length is significantly smaller than diameter [16]. Let's assume that the opening is in the wall separating two infinitely large objects (Fig. 2a).

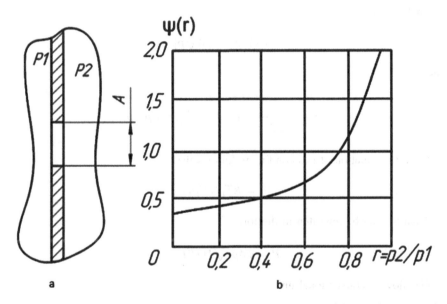

Fig. 2. Data for calculating pass-through of the opening: (a) diagram of the opening, connecting infinite volumes; (b) a graph to define function ψ.

In low vacuum and viscose mode of gas flow, the law of preserving energy for adiabatic outflow of gas can be documented in the form of equation of increasing of kinetic energy of gas and changes of its' enthalpy:

$$G\omega_{r2}^2/2 = G(I_1 - I_2), \qquad (1)$$

where G – gas flow; ω_{r2} – speed of gas at the exit of the opening; I_1 и I_2 – enthalpy of gas before and after going through the opening. Using the fact that $I = c_p T$, let's rewrite the Eq. (1) in the following way:

$$\omega_{r2}^2/2 = c_p T(1 - T_2/T_1). \qquad (2)$$

It is known from technical thermodynamics, that the following expressions are true for adiabatic process:

$$v_1/v_2 = (p_2/p_1)^{1/\gamma}, \gamma = c_p/c_v, c_p - c_v = R/M.$$

Let's transform the equation of preserving energy (2) to:

$$\omega_{r2} = \sqrt{\frac{2\gamma}{\gamma-1} p_1 v_1 \left[1 - \left(\frac{p_2}{p_1}\right)^{(\gamma-1)/\gamma}\right]}. \qquad (3)$$

The flow of gas through an opening (kg/s) taking into consideration written for expression (3):

$$P = \omega_{r2} A / v_2 = \psi A \sqrt{p_1/v_1}, \qquad (4)$$

where

$$\psi = r^{1/\gamma} \sqrt{\frac{2}{\gamma-1} [1 - r^{(\gamma-1)/\gamma}]}, r = p_2/p_1;$$

From the equation of gas conditions follows that:

$$v_1 = RT_1/(Mp_1)$$

Then (4) can be rewritten in the form:

$$P = \psi A p_1 \sqrt{M/RT_1}, \qquad (5)$$

Gas flow in conventional units:

$$Q = \frac{PRT_1}{M} = \psi A p_1 \sqrt{\frac{RT_1}{M}}. \qquad (6)$$

Reduction of the relations of gases $r = p_2/p_1 < 1$ leads to the volume of gas flowing through diaphragm and the final speed of flow in area p_2 is increasing until the relation of p_2/p_1 reaches the critical value corresponding to the speed of sound. If the process of expiration is adiabatic, then the critical value:

$$r_k = \left(\frac{2}{\gamma+1}\right)^{\frac{\gamma}{\gamma-1}}. \qquad (7)$$

In the relation of pressures $p_2/p_1 < r_k$ the volume of gas flow remains constant. In the area of relations $p_2/p_1 > r_k$ pass-through of the opening can be defined by expression:

$$U_{o\theta} = \frac{Q}{p_1 - p_2} = \psi \frac{A}{1-r} \sqrt{\frac{RT_1}{M}}. \tag{8}$$

For air and other diatomic gases at $\gamma = 1.4$ from (4) we'll get:

$$\psi = 2.65 r^{0.714} \left(1 - r^{0.236}\right)^{1/2}; \tag{9}$$

$$U_{o\theta} = \psi \frac{91A}{1-r} \sqrt{\frac{T_1}{M}}, \tag{10}$$

where M – molecular mass, kg/mol; T_1 – absolute temperature, K; A – area of the opening, м²; r – relation of pressures; p_2/p_1; ψ – function of r, calculated according to (9) and shown in Fig. 2b.

At room temperature ($T = 293$ K) for air ($M = 29$ kg/mol) Eq. (10) can be simplified:

$$U_{o\theta} = \psi \frac{289A}{1-r} \text{ at } 1 > r \geq 0.528. \tag{11}$$

In closed flow mode $\psi = 0.69$. Then the pass-through of opening:

$$U_{o\theta} = \frac{200A}{1-r} \text{ at } 1 > r \geq 0.528. \tag{12}$$

Usually, when calculating, pass-through relation of pressures r is not known beforehand. Calculations are done using the method of successive approximations. For the first approximation it can be assumed that pass-through is equal to the minimum value of $U_{ov} = 200\,A$ m³/s and does not depend on r. For round openings this value is $U_{ov} = 160\,d^2$ m³/s.

Therefore, the pass-through of the opening is connected to the holding force and directly depends on diameter of channels, through which the air outflows, and the degree of vacuum, and, consequently, the holding force of flexible container is as high, as the difference of diameter of air channels and vacuum suction cups is as small. The increase in the quantity of vacuum suction cups allows to increase holding force only in the case where they are connected to different channels of vacuum machine and it allows to provide for necessary consumption [17].

3 Materials and Methods of Experimental Research

An experimental setup (Figs. 3 and 4) has been developed and constructed to evaluate the adequacy of the mathematical model and conduct experiments in automatic mode. An experiment was performed to determine the necessary vacuum pressure for reliable retention of an empty FC [18].

Automatic control system of the experimental setup is based on Mitsubishi Electric's FX3U programmable controller and Camozzi pneumatic equipment. Microcontroller Arduino UNO is used to receive and process information from sensors, which is connected to the PLK using an circuit board with transistor logic, which converts the signals from the outputs of the 5 V microcontroller to 24 V signals. These signals control the electric air distributors, displacing pneumatic cylinders and vacuum gripping devices and switching vacuum on different VGDs (Fig. 5). An ultrasonic sensor is installed opposite the neck of the bag allows to determine the success of its capture and disclosure, measuring the distance to the flexible container. The vacuum pressure in the system is measured by a sensor in the microcontroller and a vacuum gauge to display and verify the accuracy of the readings. A servomotor that is connected to the microcontroller can adjust the degree of vacuum depression by opening the air channel shutter by a certain percentage. Information about current pressure of vacuum, total number of conducted and successful experiments is shown on the display. Information from the microcontroller is transmitted to personal computer by a USB port (virtual COM-port).

The microcontroller is programmed to conduct 100 tests with different pressures in the vacuum system in automatic mode.

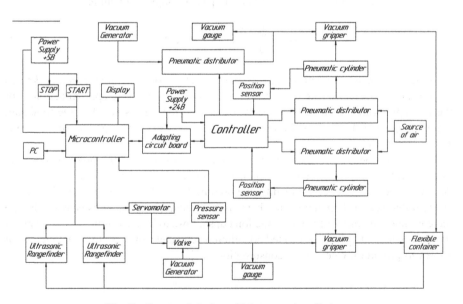

Fig. 3. Structural design of laboratory installation.

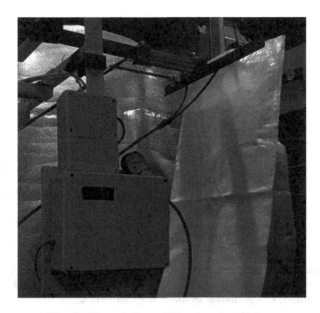

Fig. 4. General view of laboratory installation.

After starting the program by pressing the "Start" button, the microcontroller sends a signal to the servomotor and PLC. The servomotor opens the air duct shutter by a minimum percentage for the vacuum to enter the gripper. The screen displays the current vacuum pressure.

With defined intervals, microcontroller is sending two signals, which through adapting circuit board, PLK and two pneumatic distributors controls reciprocating movement of pneumatic cylinder rods with VGDs attached to them. The fact of successful gripping and opening of the flexible container is registered with the ultrasonic sensor and transferred to the display. Then, one of the VGDs is cut off from the source of vacuum with the help of pneumatic distributor, and flexible container is returned to the initial position. Then the cycle repeats. After each series of experiments, it is repeated with a greater degree of vacuum in the gripping devices.

If after a series of experiments, the capture probability is 100% (all captures are successful), another one series is carried out and the test is completed. The test results are transferred to a computer for further processing in Microsoft Excel.

Tests have been conducted for vacuum gripping devices, containing 2, 3, 4 and 5 vacuum grippers with diameters of 30 mm, 40 mm and 50 mm.

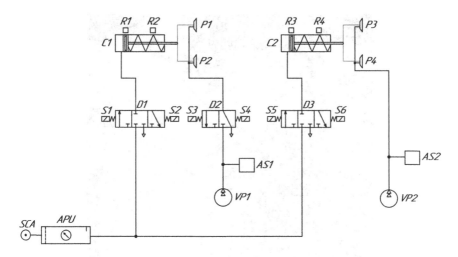

Fig. 5. Pneumatic diagram of experimental installation: SCA – source of compressed air, APU – air preparation unit, VP – vacuum pump, AS – air sensor, P – pneumatic distributor, S – solenoid, P – pneumatic gripper, C – pneumatic cylinder, R – reed switch.

4 Results and Discussion

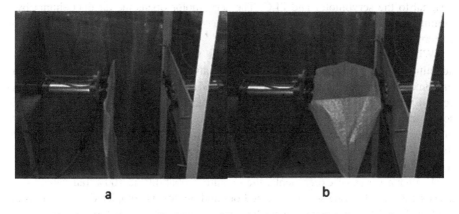

Fig. 6. Experiments of holding and opening FC: a – holding, b – opening.

Figure 6 shows photos of experimental installation in two positions with open and closed FC.

Based on analyzed data collected during tests, graphs showing probability of gripping and opening FCs depending on the degree of vacuum created in vacuum gripping devices taking into consideration the quantity and diameter of vacuum grippers. Results of the test of gripping device for gripping and holding FC with 2, 3, 4 and 5 GD with a diameter of 50 mm are shown on Fig. 7. Reliable gripping of FC is achieved at the degree of underpressure of −24 kPa (with two vacuum grippers),

−22 kPa (with three vacuum grippers), −16 kPa (with four vacuum grippers), −22 kPa (with five vacuum grippers).

Figure 8 shows test results for opening FC with two vacuum grippers with diameters of 40 mm and 50 mm. Reliable gripping of FC for them was achieved at the degree of underpressure of −20 kPa (with the diameter of vacuum grippers of 40 mm) and −22 kPa (50 mm).

Therefore, probability of successful gripping and holding of air-penetrating polypropylene flexible containers increases with the increase in a number of vacuum grippers, while gripping and holding is achieved at lower degree of underpressure. The same tendency would be observed during opening of FC, besides, the probability of opening increases with the decrease in diameter of vacuum grippers, which correlates with increase in speed of lowering of carrying capacity of VGD caused by leakage of air through air-penetrating material of FC and faster under pressurizing.

Experimental test results coincide with results of mathematical modeling and confirm that the force of holding flexible container is higher when the difference in diameter of air channels and vacuum suction cups is lower. At the same time, increase in the number of vacuum suction cups allows to increase holding force only in the case where they are connected to different channels of vacuum machine and it allows for required consumption.

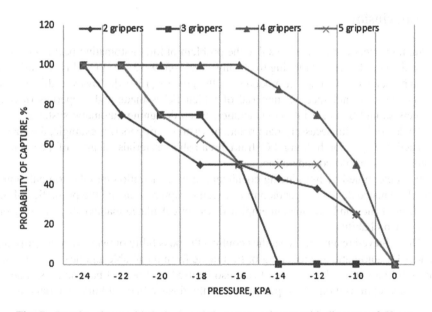

Fig. 7. Results of test with 2, 3, 4, and 5 vacuum grippers with diameter of 50 mm.

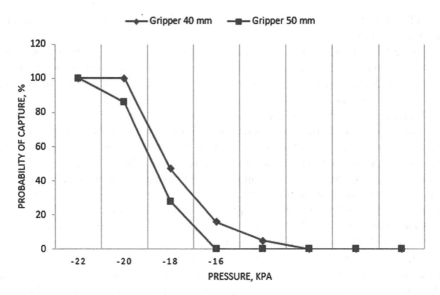

Fig. 8. Results of test with 2 vacuum grippers with diameter of 40 mm and 50 mm.

5 Conclusion

Vacuum grippers can be used to solve the problem of fully automating packing of bulk materials into FCs, from grabbing to closing FCs [1–5, 9–12, 16, 17]. This is confirmed by mathematical and computer modeling of the process [13–15]. However, this process is very complex and specific, material of which FC is made and properties of bulk products should be taken into consideration when designing automatic systems.

To determine the necessary parameters of vacuum grippers (for example, the size of the suction cups) for holding FC from breathable materials, degree of penetration should be taken into consideration.

It is determined by introducing coefficient of air penetration into the computational formula. This coefficient depends on the vacuum capture area and the properties of the FC material and can be defined in empirical way or calculated mathematically using the theory of probability.

The achieved results of the studies confirm the possibility of using vacuum gripping devices for automatic manipulating of FCs made from breathable and flexible materials during filling with bulk products. It can also be used in structural-parametric synthesis, tuning and optimization of the parameters of the dosage lines of bulk materials in FC.

Acknowledgements. The work has been performed with the financial support through the Grant of the President of the Russian Federation for young Russian scientists – candidates of science MK-2619.2017.8.

References

1. Suzdaltsev, V.A., Suzdaltsev, I.V., Chermoshentsev, S.F., et al.: Formation of expert system decision explanation in the object classification process. In: Proceedings of XIX IEEE International Conference on Soft Computing and Measurement, pp. 319–320 (2016)
2. Suzdaltsev, V.A., Suzdaltsev, I.V., Bogula, N.Y.: Fuzzy rules formation for the construction of the predictive diagnostics expert system. In: Proceedings of XX IEEE International Conference on Soft Computing and Measurements, pp. 481–482 (2017)
3. Bogachev, Y.A.: The mechanism for capturing and disclosure of sacks to the packing machines. RU Patent 2028257, 9 October 1995 (1995)
4. Buler, A.G.: Method and device for automatic suspension bags. RU Patent 2174486, 10 October 2001 (2001)
5. Keppler, C.: Better productivity: it's in the bag. Text. Rental **92**(3), 88–89 (2008)
6. Kreymborg, M.: Packaging for bulk materials, especially bag made of plastic film. DE Patent 2007/115538, 12 April 2006 (2006)
7. Grigorev, S.N., Gribkov, A.A.: Precision of dosing-system components. Russ. Eng. Res. **30**(8), 779–780 (2010)
8. Kozyrev, Y.G.: Gripping device and tools of industrial robots. KNORUS, Moscow (2011)
9. Sysoev, S.N., Bakuto, A.V., Privedenets, I.A.: Construction of high-speeding vacuum gripper. In: STIN, vol. 12, pp. 15–19 (2007)
10. Makarov, A.M., Serdobintsev, Yu.P., Mushkin, O.V., et al.: Modelling and research of automated devices for handling of flexible containers during filling with free-flowing products. In: XX IEEE International Conference on Soft Computing and Measurements, pp. 281–283 (2017)
11. Makarov, A.M., Serdobintsev, Y.P., Mushkin, O.V., et al.: Research of the process of automatic gripping of air-penetrating flexible containers with vacuum during sorting of free-flowing products. In: Industrial Engineering, Applications and Manufacturing, p. 5 (2017)
12. Makarov, A.M., Rabinovich, L.A., Serdobintsev, Yu.P.: Apparatus for automatically opening, closing and retaining the bags. RU Patent 2469928, 20 December 2012 (2012)
13. Kovalev, A.A., Makarov, A.M., Kristal, M.G., et al.: Development of a device for automatic opening, withholding and closure of flexible containers using vacuum grippers. In: News VSTU, Ser. Innovative Technologies in Machine-Building, vol. 11, no. 173, pp. 53–56 (2015)
14. Makarov, A.M., Kristal, M.G., Kovalev, A.A., et al.: The device for the automatic opening, holding and closing the flexible containers. RU Patent 155000, 20 September 2015 (2015)
15. Bontem, R., Mpinga, V.: Developments in the field of sealing bags. Feeds **7**, 32–33 (2013)
16. Rozanov, L.N.: Vacuum Equipment: Textbook for Colleges Specializing in "Vacuum Equipment". Higher Education, Moscow (1990)
17. Makarov, A.M., Serdobintsev, Yu.P., Krylov, E.G.: Automated filling of flexible containers with free-flowing products. Russ. Eng. Res. **34**(11), 734–736 (2014)
18. Shang, R., Li, S.: The study on CAD technology of reconfigurable and module for vertical bag packaging machine. Adv. Mater. Res. **299–300**, 1124–1127 (2011)

Automatic Control of the Oil Production Equipment Performance Based on Diagnostic Data

K. F. Tagirova and A. R. Ramazanov[✉]

Ufa State Aviation Technical University, 12 Karla Marksa Street, Ufa 450077,
Russian Federation
ramazanov.anvar@gmail.com

Abstract. In this paper, we are considering the possibility of automatic control of oil production equipment performance based on diagnostic data. It was analyzed of existing approaches to the well production rate estimation equipped with sucker rod pump unit. Then we were examined the data acquisition from detectors of oil well with sucker rod pumps. An estimating production rate based on wattmeterograms and dynamograms are being considered, and compared their effectiveness. During the testing of the calculation algorithms, an estimate of the relative error in determining the production rate from wattmetering data was obtained. Refining the experimental conditions, improving the procedure for identifying unknown parameters of the model and telemetry parameters will improve the accuracy of the estimate. The coupling coefficient between the calculated and measured values of the flow rate shows the consistency of the approach to the determination of the flow rate over the area of the dynamogram or the wattmetering data, which allows estimating the flow rate of a well without a flowmeter. The algorithm for estimating the flow rate according to the wattmeterogram data makes it even more cost-effective to estimate the production rate by refusing a dynamograph. Thus, operational control of the operating modes of each well is provided by expanding the functionality of the control station based on the use of diagnostic information in addition for management purposes.

Keywords: Sucker rod pump unit · Wattmetrogramm · Dynamogramm · Oil rate · Automatic control

1 Introduction

In Russia, a lot of exploited oil wells are low-yielding, and the most common method of extracting oil from these wells is the use of sucker rod pumping units (SRPU).

The main characteristic of the oil producing well is the flow rate, i.e. the amount of oil well production received within 24 h. To increase the efficiency of the well, it is necessary to control the well operating mode by establishing and maintaining the productivity of pumping equipment with the appropriate rate of fluid flow to the bottom of the well. For this, it is required to determine the current production rate of the well in real time.

Basically, rather expensive sensors – flowmeters – are used to determine daily production rate of a well. There are methods allowing to estimate the well flow rate without a flowmeter, using the information about force dependence at the rod string suspension point from the stroke of the polished rod (dynamogram). An alternative option that makes it even more affordable to estimate the well rate by refusing a dynamograph is the algorithm for calculating the rate from the data of dependence of the electric drive power consumption of the rocking machine from the shifting of rod string suspension point (wattmeterogram) based on the analysis of the technological time series of the accumulated parameters. This algorithm has been known for a long time, but there are no examples of its practical application in the literature. In this article, we consider the possibilities of practical application of flow rate estimation algorithms based on wattmetering data in comparison with the dynamometer flow calculation algorithms and flowmeter data.

2 Channel for Primary Information Obtaining from the SRPU Sensors

Primary information from the sensors comes in real time by creating a permanent communication channel between the dynamograph, the equipment for fixing the currents consumed by the electric drive installed on the SRPU, and an intelligent control system created on the basis of a control computer.

It is suggested to use the following sequence of information transfer and transformation (Fig. 1): signals from the force sensors and dynamograph motion sensors (usually analog) are fed to the normalizing device and then to the analog-digital converter from which the digitized signal is fed to a controller connected with a computer and performing direct equipment diagnostics and monitoring of the SRPU state. Based on the results of diagnostics and equipment state monitoring, the operation of the plant is performed through actuators. Likewise, the currents consumed by the SRPU electric drive are fixed for transmission to an intelligent control system and to determine the current production rate according to the wattmetering data.

The structure of primary information obtaining and transferring to the SRPU control station is shown in Fig. 1.

To estimate the well production rate, the formula proposed by

$$Q = F_s \cdot S_{stroke} \cdot n \cdot a \qquad (1)$$

Takhautdinov is widely used [7]: where Q – flow rate of well fluid (barrels/day); F_s – the area of the plunger section of the borehole pump (square inch); S_{stroke} – effective stroke length of polished rod (inch); n – rocking machine oscillation frequency (min^{-1}); a – pump feed ratio.

Performance of the borehole pump is determined by formula [7]:

$$Q = 0.144 \cdot F_s \cdot S_{stroke} \cdot n \cdot a \cdot \beta, \qquad (2)$$

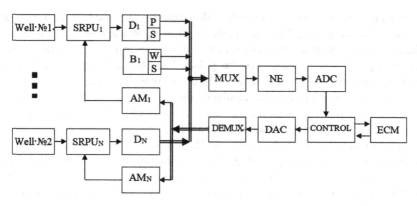

Fig. 1. Intelligent diagnostic system with automatic information input: D – dynamograph, B – fixation of the currents consumed by the electric drive (W – power, S – rod stroke), AM – actuator, NE – normalizing device, ADC – analog-to-digital converter, DAC – digital-to-analog converter, MUX – multiplexer, DEMUX – demultiplexer, CONTROL – controller.

where F_s – the area of the plunger section, sm; S_{stroke} – stroke length of polished rod, m; n – number of oscillations, 1/min; a – pump feed ratio; β – pump filling ratio.

Coefficient a depends on several quantities:

$$a = K_1 \cdot K_2 \cdot K_3 \cdot K_4 \tag{3}$$

where K_1 – coefficient of leakage in oil well tubing; K_2 – coefficient characterizing the change in the oil volume; K_3 – coefficient of leakage in the pump; K_4 – plunger stroke length to the operating rod stroke length ratio $(S_{stroke}/1)$.

Coefficient β is defined as:

$$\beta = \frac{S_{eff}}{S_{stroke} \cdot (R_H + 1)} \tag{4}$$

where R_H – the ratio of the gas volume to the volume of oil in the cylinder at the injection pressure; S_{eff} – plunger stroke length with open discharge valve, mm; S_{stroke} – stroke length of plunger, mm.

2.1 Evaluation of Well Production Rate from Dynamometer Data

At present, a method for dynamometry has become very popular for diagnosing downhole pumping equipment.

Information on the technical condition of the SRPU underground part is obtained in the form of force signals P(t) and the stroke signals S(t) from the outputs of the sensors converting the parameters of the movement of walking beam or polished rod into an electrical signal.

Researches have shown that the shape of the dynamogram, which is the dependence of the force on the displacement from the shifting of rod string suspension point, corresponds to a definite state of the SRPU.

Otherwise, using formulas (1)–(4) and measuring on the practical dynamogram (Fig. 2) the values l and corresponding to the plunger movement from the moment of opening the discharge valve (point Γ) to its closing (point A), the pump performance can be defined.

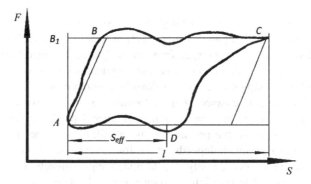

Fig. 2. Well flow rate determination by dynamogram.

Practically determined by the dynamogram, the flow rate differs from the actual production rate of the well. The effective plunger stroke Sэф, determined algorithmically, differs from the real stroke, which is due to:

- Measurement instrument error
- Type of algorithm for finding characteristic points of the dynamometer

General algorithm for processing a dynamogram for estimating the effective stroke of a rod consists of the following steps:

1. The functional dependence F(S) is analyzed to isolate the lines of the processes of load redistribution between the rod string and the oil well tubing.
2. Process junctions that are characteristic points of the dynamometer are analyzed.
3. From the found points of the dynamogram, the effective stroke of the pump plunger and the flow rate are calculated from the dynamogram.

In [2] to analyze the segments Seff (effective stroke of the rod) and l (the length of the stroke from the "bottom dead" point to the upper "dead point"), the first and second derivatives of the function F(t) are analyzed, with help of which a theoretical dynamogram is constructed and the characteristic points of a practical dynamogram are analyzed. The maximum deviation of the estimated production rate from the real was 15%.

2.2 Evaluation of Well Production Rate from Wattmeteringdatas

The experimental-calculation method for estimating the performance of SRPU using wattmeterogram leads to Krichke in his works. By Krichke [4], SRPU performance Q is proportional to the work done by drive on the wellhead. The operation is determined by the active power consumed by the drive of the pumping unit and the pressure developed by the pump at the wellhead.

$$Q = \frac{86400 \cdot 10^3 \cdot N_{avg} \cdot \rho}{p \cdot 10^6} \cdot \eta \qquad (5)$$

where 86400 – number of seconds in a day; N_{avg} – average power per whole number of oscillation cycles, kW; ρ – liquid density in surface conditions, kg/m^3; p – the average pressure incident on the pump plunger from the weight of the lifted liquid column by the electric motor and the loads on the cranks during the time of the wellhead rod stroke upwards, MPa; η – complex efficiency of downhole equipment.

To determine the performance Q from formula (5), it is necessary to know the values of p and ρ. The pressure on the pump plunger depends on the density of the liquid column in the pipes, the pressure in the manifold and other factors. The efficiency of downhole equipment depends on the frictional forces in the moving parts of the SRPU. These parameters can only be determined approximately. In connection with this, [10] it is suggested to take into account not the absolute values of pressure and power, but their relative increments. For the interdependencies detection, active power diagrams are fixed at various wellhead pressures and corresponding wellhead dynamograms:

$$Q = \frac{43200 \cdot 10^3 \cdot \Delta N}{\Delta p \cdot 10^6} \cdot \eta \qquad (6)$$

where 43200 – the number of seconds in a day, divided in half (since the formula takes into account the half stroke of the rod – the up stroke); Δp, ΔN – average difference pressure at the wellhead (Mpa) and power (kW) during the time of the wellhead stroke upwards, respectively.

To determine the beginning of the wellhead stroke up on the wattmeterogram, the algorithm proposed by Tsapko [11] can be used. In most cases, the power peak of the wattmeterogram is in the middle of the half-cycle.

Thus, to determine the starting point of the plunger up stroke, it is necessary to know the pumping period of the rocking machine (RM), the power peaks P_{max1} and P_{max2}, and the sample numbers corresponding to the maximums of the P. Further, the count corresponding to the start of the plunger stroke up is determined

$$N_{start} = N_{max1} - T/4$$

where N_{start} – a count corresponding to the start of the plunger stroke; N_{max1} – count corresponding to the first maximum of the wattmeterogram.

At the same time, the power corresponding to the count $N_{нх}$ will be the engine idling power.

Research [4, 12] allows to establish that the cyclic wattmeterograms of the SRPU drive motor contain information not only about the state of the underground part of the well, but also about the state of the ground equipment.

The power consumed by RM drive motor is determined by the magnitude of the resultant torque of forces on the crank shaft of the gearbox:

$$W_{motor} = M \frac{\pi n}{30} \frac{1}{\eta_{motor} \eta_{trans} \eta_{gear} \eta_{vbd}} = MA \qquad (7)$$

where n – electric motor speed, min^{-1}; η_{motor} – electric motor efficiency; η_{trans} – efficiency of the transmission mechanism, which takes into account the friction in the gearbox and at the V-belt drive; η_{gear} – gear ratio; η_{vbd} – transmission ratio of the V – belt transmission, equal to the ratio of the pulley diameters of the reducer and the electric motor:

$$\eta_{vbd} = \frac{d_{gear}}{d_{motor}} \qquad (8)$$

Formula (8) does not take into account the elasticity of V-belt transmission and the non-rigid characteristic of the motor.

The RM energy parameters are determined by analyzing the area of the diagram, which is proportional to the energy expended by the electric motor for lifting the liquid and the energy going to the losses in the ground and underground equipment of the well.

The area of the diagram per oscillation cycle is proportional to the energy $\int_0^t W_c(t)dt$, consumed by the motor of the RM drive. The algebraic sum of the areas located between the wattmeterogram and the line AP connecting its beginning and end is proportional to the energy $\int_0^t W_{rod}(t)dt$ consumed on the polished rod (Fig. 3).

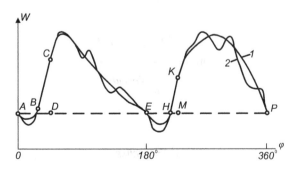

Fig. 3. Power curve of the RM motor.

The difference obtained between the areas proportional to $\int_0^t W_c(t)dt$ and $\int_0^t W_{rod}(t)dt$, characterizes the energy losses in the ground equipment of the installation, which are equal to:

$$\int_0^t W_{SE}(t)dt = \int_0^t W_c(t)dt - \int_0^t W_{rod}(t)dt \qquad (9)$$

Hydraulic (useful) energy, which goes to the lift of the liquid:

$$\int_0^t W_{hydraulic}(t)dt = \int_0^t W_{rod}(t)dt - \int_0^t W_{UE}(t)dt \qquad (10)$$

where $\int_0^t W_{UE}(t)dt$ energy losses in the underground part of equipment.

Then, the efficiency of the downhole η_{WE}, the ground η_{SE} of the pumping unit equipment and the overall efficiency η_U:

$$\eta_{WE} = \frac{\int_0^t W_{hydraulic}(t)dt}{\int_0^t W_{rod}(t)dt} \qquad (11)$$

$$\eta_{SE} = \frac{\int_0^t W_{rod}(t)dt}{\int_0^t W_C(t)dt} \qquad (12)$$

$$\eta_U = \frac{\int_0^t W_{hydraulic}(t)dt}{\int_0^t W_C(t)dt} \qquad (13)$$

or

$$\eta_U = \eta_{WE}\eta_{SE}$$

Taking into account the known relationships, as well as formulas (5) and (6), we obtain the supply of a pumping unit, determined through the spent energy of the well:

$$Q = \frac{75 \cdot 1,36 \cdot 3600 \cdot 24 \cdot \eta_{WE}}{1000Ht} \int_0^t W_{rod}(t)dt \qquad (14)$$

where η_{WE} – efficiency of downhole equipment, Q – delivery of the pumping unit, tons/day, H – liquid lifting height, t – time of full cranking, s; $\int_0^t W_{rod}(t)dt$ – the energy spent on the polished rod per oscillation cycle, kW/s.

The formulas (11)–(13) for determining the efficiency $\eta_{WE}, \eta_{SE}, \eta_U$ supply Q are independent of the RM balancing, since the algebraic sum of the areas per oscillation cycle taken with respect to the zero line, is close to zero.

2.3 Assessment of Well Production Rate on Field Data

Initial data for testing the considered algorithms for estimating the well production rate are presented by a set of normalized and time-deployed dynamograms, corresponding wattmeterograms and measurements of automated group metering stations (AGMS). Measurements were taken for four wells equipped with SRPU. The series of experiments includes 41 fixations of the dynamogram, wattmeterogram and measurement according to AGMS.

Based on the results of the preliminary calculation, the wattmeterogram is automatically divided by the periods of the complete oscillation cycle.

Based on the first 16 oscillation periods, the coupling factor for a given well is calculated for each experiment. The coupling coefficient makes it possible to identify unknown parameters of the model (14). On the remaining selected periods, the production rate and its deviation from the nearest time to the AGMS are determined.

The results of the Aliyev and Ter-Khachaturov algorithm are shown in Table 1.

Table 1. Calculations for the dataset 03112016.

Period number	Coupling factor	Period number	Calculated production rate	Deviation of calculated production rate, %	The production rate according to the data of the AGMS
0	0,818	16	5,601	9,816	5,1
1	0,835	17	4,867	4,578	
2	0,843	18	5,107	0,129	
3	0,757	19	4,921	3,510	
4	0,864	20	4,565	10,488	
5	0,851	21	4,711	7,631	
6	0,718	22	5,346	4,816	
7	0,822	23	6,058	18,792	
8	0,954	24	5,290	3,725	

(*continued*)

Table 1. (*continued*)

Period number	Coupling factor	Period number	Calculated production rate	Deviation of calculated production rate, %	The production rate according to the data of the AGMS
9	0,991	25	4,854	4,820	
10	0,753	26	4,736	7,131	
11	0,842	27	6,261	22,763	
12	1,056	28	5,308	4,078	
13	0,996	29	4,944	3,065	
14	0,672	30	5,052	0,943	
15	0,646	31	5,729	12,330	
		32	5,883	15,354	
		33	5,719	12,146	
		34	4,954	2,867	
		35	5,492	7,691	
		36	5,533	8,496	
		37	4,620	9,415	
		38	4,196	17,723	
		39	5,383	5,540	
Average value of the coupling factor	0,838	Average value of the calculated production rate	5,214	Average deviation of the calculated production rate, %	2,23

For 03112016 dataset it can be seen that the average deviation of the calculated flow rate from the measurements is 2.3%. Calculations are based on the following data:

- Flow rates measured at the AGMS by the dates of reading of wattmetergrams
- Lowering depth of the pump
- Period
- Density foil

Based on the available set of experimental data, the procedure for calculating the production rate by the algorithm for estimating areas with the identification of unknown parameters of AlievTer-Khachaturov's model is implemented.

For one of the experiments, a graph is given illustrating the relationship between evaluation of the production rate by wattmeterogram, the evaluation of the production rate by dynamogram and the readings of the AGMS. The energy calculation of the production rate according to the dynamogram is based on the algorithm proposed in the [12] (Fig. 4).

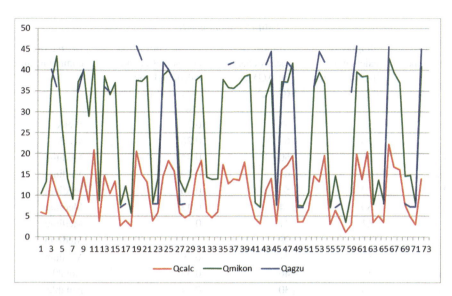

Fig. 4. Relationship between evaluation of the production rate by wattmeterogram (Q_{calc}, conventional units), the evaluation of the production rate by dynamogram (Q_{mikon}, m^3/day), and the readings of the AGMS (Q_{AGZU}, m^3/day).

The correlation coefficient between the calculated and measured values shows the consistency of this approach to the determination of the production rate in the area of the dynamogram.

Table 2 shows the relationship between the calculated production rate in the wattmeterogram and the measurement with the AGMS for a series of experiments.

Table 2. Relationship between the calculated production rate and the measurement with the AGMS.

№ of experiment	Number of calculation measurements (Q_{calc})	Number of measurements by AGMS (Q_{agzu})	Linear correlation coefficient of the calculated and measured production rate	Degree of communication
27	56	48	0,03	Weak
12	50	42	0,04	Weak
14	74	42	0,1	Weak
39	62	58	0,1	Weak
8	48	19	0,11	Weak
28	75	61	0,13	Weak
15	36	31	0,23	Weak
4	62	34	0,24	Weak

(*continued*)

Table 2. (*continued*)

№ of experiment	Number of calculation measurements (Q_{calc})	Number of measurements by AGMS (Q_{agzu})	Linear correlation coefficient of the calculated and measured production rate	Degree of communication
29	57	31	0,24	Weak
26	75	60	0,26	Weak
13	76	50	0,34	Medium
16	85	65	0,39	Medium
25	70	56	0,40	Medium
33	39	34	0,42	Medium
3	61	47	0,45	Medium
5	61	38	0,45	Medium
10	89	54	0,49	Medium
11	53	46	0,53	Notable
31	55	30	0,54	Notable
19	57	46	0,57	Notable
35	69	53	0,59	Notable
24	56	20	0,60	Notable
6	37	37	0,61	Notable
20	73	57	0,62	Notable
2	59	45	0,66	Notable
34	56	42	0,66	Notable
18	72	46	0,69	Notable
38	91	49	0,72	Strong
36	39	33	0,74	Strong
37	109	83	0,78	Strong
30	102	84	0,87	Strong

From Table 2 it can be seen that part of the experiments makes it possible to estimate the notable and strong correlation between the calculated values of the flow rate and the measurements of the AGMS.

3 Analysis of Results and Conclusions

During the testing of the calculation algorithms, an estimate of the relative error in determining the production rate from wattmetering data was obtained in the case of manual breakdown of wattmeterogramfor periods of 15–20%.

When using automatic breakdown for the oscillation periods based on the set parameters of the accompanying telemetry, the essential character of the connection between the calculation results and the measurements by the AGMS is revealed. Refining the experimental conditions, improving the procedure for identifying unknown parameters of the model and telemetry parameters will improve the accuracy of the estimate.

The coupling coefficient between the calculated and measured values of the flow rate shows the consistency of the approach to the determination of the flow rate over the area of the dynamogramor the wattmetering data, which allows estimating the flow rate of a well without a flowmeter. The algorithm for estimating the flow rate according to the wattmeterogram data makes it even more cost-effective to estimate the production rate by refusing a dynamograph. Thus, operational control of the operating modes of each well is provided by expanding the functionality of the control station based on the use of diagnostic information in addition for management purposes.

References

1. Chigvintsev, S.V., Chigvintseva, A.S.: Virtual sensors for the control system of a deep-well pumping unit. In: Electrotechnology, Electric Drive and Electrical Equipment of Enterprises. Proceedings of II All-Russian Scientific and Technical Conference, p. 194 (2009)
2. Tagirova, K.F., Vulfin, A.M., Sabitov, A.R., et al.: Dating of the diagnostics of well sucker-rod pumping units on the basis of intelligent analysis of dynamometric data. Autom. Telemechanization Commun. Oil Ind. **11**, 23–28 (2014)
3. Khakimyanov, M.I., Pachin, M.G.: Monitoring of sucker rod pump units on result of the analysis wattmeter cards. Electron. Sci. J. Oil Gas Bus. **5**, 26–36 (2011)
4. Krichke, V.O.: Measuring information system for wells equipped with pumping jack IIS-SK. Autom. Telemechanization Commun. Oil Ind. **11**, 16–18 (1976)
5. Aliev, T.M., Ter-Khachaturov, A.A.: Automatic monitoring and diagnostics of downhole sucker rod pumping units. Nedra, Moscow (1988)
6. Dregotesku, N.D.: Oil production using deep-well pumping unit. Nedra, Moscow (1966)
7. Osovskiy, S.: Neural networks for information processing. Finance and Statistics, Moscow (2004)
8. Tagirova, K.F., Vulfin, A.M., Ramazanov, A.R., et al.: Modified algorithm of determining the current DSRP operating parameters accorging to the dynamometry data. Autom. Telemechanization Commun. Oil Ind. **12**, 37–41 (2015)
9. Tagirova, K.F., Vulfin, A.M., Ramazanov, A.R., et al.: Improving the efficiency of operation of sucker-rod pumping unit. Oil Ind. **7**, 82–85 (2017)
10. Svetlakova, S.V.: Information-measuring system for dynamometry of wells equipped with sucker-rod deep pumps. Dissertation, Ufa State Oil University (2008)
11. Goldstein, E.I., Tsapko, I.V., Tsapko, S.G.: Some aspects of selecting method of balancing of well sucker-rod pumps. Autom. Telemechanization Commun. Oil Ind. **10**, 33–37 (2010)
12. Krichke, V.O.: Debitomer (flowmeter). RU Patent 2018650, 30 August 1994 (1994)
13. Vulfin, A.M., Tagirova, K.F.: Enhancement of accuracy of deep-pumping equipment based on data mining. Opt. Mem. Neural Networks **24**, 28–35 (2015)
14. Vasilyev, V.I., Ilyasov, B.G.: Intelligent control systems. Theory and practice. Training material, pp. 33–62 (2009)
15. Tagirova, K.F., Vulfin, A.M.: Neural network algorithms of information processing in the tasks of diagnosing downhole pumping equipment of oil company. Autom. Telemechanization Commun. Oil Ind. **12**, 28–32 (2013)

Technology of Building Movement Control Systems of Products in Warehouse Areas of Industrial Enterprises Using Radio Frequency Identification Methods

A. V. Astafiev[✉], A. A. Orlov, and T. O. Shardin

Vladimir State University, 87 Gorkogo Street,
Vladimir 600000, Russian Federation
alexandr.astafiev@mail.ru

Abstract. At present, in connection with the need to move to new intellectual digital production technologies and implement international quality standards, it is necessary to introduce new science-based approaches to controlling the movement of products and small-scale mechanization of warehouses. This is due to the fact that the warehouses of large industrial organizations, at the current level of hardware and software, cannot fully comply with domestic and foreign quality standards in the field of product tracking, regulated by GOST and ISO. The article is devoted to the development of a technology of building movement control systems of products in the warehouse areas of industrial enterprises using radio frequency identification methods. The main methods of development of such systems based on technical vision and radio frequency identification are considered. The main normative documents and interstate standards regulating the described processes are given. The analytical review of the Russian and foreign scientific and technical base on development of methods and algorithms of movement control systems is conducted. The comparative analysis of analog systems was carried out. The simulation model for the movement of industrial products and the structure of the hardware-software complex movement control systems are developed. The article presents the results of experimental studies in laboratory and industrial conditions.

Keywords: Positioning · Radio · Frequency identification · Racks

1 Introduction

According to the program "Development of the Digital Economy in Russia until 2035", one of the technologies that determine the transition to the digital economy is identification technology [1].

In particular, the development of AutoNet and EnergyNet markets requires the development of autonomous navigation and positioning systems [1], such as RTLS (from the English. Real-time Locating Systems - a real-time positioning system). The development of methods for constructing RTLS systems has an important role in the implementation of projects of unmanned production and "smart cities".

The main problems of the implementation of autonomous navigation systems currently include:

1. The lack of high-precision information and communication technologies with low energy consumption for the implementation of optimal management methods for vehicles and vehicles in the conditions of unoccupied production and "smart cities";
2. The lack of effective and resistant to external influences sensors and sensor networks, providing positioning of vehicles and vehicles and monitoring their movement with high accuracy;
3. Weak elaboration of cybersecurity issues of the proposed sensor networks for positioning and monitoring of vehicles.

Based on this, the aim of the work is to develop a technology for building automated systems for monitoring the movement of products in the territories of industrial enterprises based on radio frequency identification methods. To achieve this goal, you must complete the following tasks:

1. To conduct a review and analysis of domestic and foreign scientific literature on the topic of building industrial automation systems;
2. To conduct a review and comparative analysis of systems of analogues;
3. Conduct a study of the technological process and develop a simulation model of the movement of industrial products;
4. To conduct experimental research and modernization of the developed methods for constructing automated systems for monitoring the movement of products.

2 Review of the Russian and Foreign Scientific and Technical Base

The development of systems for monitoring the movement of objects includes 2 major tasks:

1. Real-time positioning;
2. Identification of relocatable products.

The issue of controlling the movement of objects in a controlled area has been dealt with by a large number of scientists around the world for many decades. One of the first solutions to this problem can be described as manually fixing the passage of an object relative to a certain point (checkpoint, checkpoint, check-in point, etc.). With this approach, there is no question of any automation of control over the movement of speech.

The second stage in the development of algorithms and methods for controlling movement and positioning can be identified as the use of stationary sensors to determine the movement of objects on the stage. Examples of this approach are permanently installed vision systems, radio frequency identification, presence sensors, etc. These technologies were quite suitable for solving everyday problems, for example, access control to an object, video recording during movement, etc.

In industry, equipment manufacturers have also actively implemented such systems for positioning objects relative to their own sensors. For example, monitoring the position of an object on a conveyor, positioning a crane truck relative to a bridge, etc. Such technologies, with proper implementation, allow you to position objects in the process of moving them with these units, but do not give a complete picture of the movement. That is, the movement of an object outside such devices remains untraceable.

Currently, the development of mobile technologies and the Internet of Things (Internet of Things) has given rise to technologies that allow more efficient organization of indoor positioning [2]. For this, technologies such as:

- GPS [3]
- GLONAS [4]
- Wi-Fi [5, 6]
- ZigBee [7]
- Z-wave
- UHF [8]
- Active RFID
- Passive RFID
- Z-Wave [9]
- UWB-IR [10]
- Bluetooth
- Inertial sensors (accelerometer, gyroscope) [11]

The use of satellite navigation systems such as GPS and GLONAS for positioning indoors in a classic form is not possible. Some enterprises (such as Navigine [12]) offer artificially created GLONASS-, GPS-systems for indoor positioning by installing expensive repeaters. Such systems can effectively track the movement of beacons in a controlled area. The main disadvantages of this technology is the small coverage area and the high cost of equipment. Control of a site measuring 20×20 m will cost at least 500,000 rubles. Therefore, such systems are mainly used in stores selling fur and jewelry.

Another class of technologies that have found wide application in positioning are: Wi-Fi (2.4 GHz), RFID (Active - 2.4 GHz; UHF - European range 863–868 MHz) and UWB (3.1–10, 6 GHz). The use of such devices in combination with geodesy methods (for example, triangulation, trilateration, etc.) allows you to organize positioning in an automated territory. The disadvantages of these technologies are the short identification range, high sensitivity to physical barriers and high cost. Projects based on such technologies are implemented mainly with the initial presence of a large number of devices of this type [13–18].

The use of these high-tech technologies allows to automate the processes of control of movement of industrial products at enterprises of various spheres of life and, ultimately, to improve the efficiency and reliability of control of transportation and warehouse accounting of manufactured products. However, they are not without drawbacks.

The use of existing software and hardware solutions is more focused on the organization of automated warehouse accounting and is less suitable for automation of traffic control. In confirmation of this, at a number of industrial enterprises of the Vladimir region and neighboring regions, developers of RFID systems have attempted to organize the traceability of products by automatic motion control based on radio frequency identification. As a result, it turned out that automatic control of the movement of industrial products is possible only in certain parts of the production process. These areas are conveyor line and transport tunnels, where transportation of products is made along the permanently installed equipment, radio frequency identification (RFID tunnels). In other production and storage areas, the implementation of automatic control of the movement of products is impossible. This is due to the lack of methods and algorithms for identifying products during their transportation on untyped routes (e.g., stackers, loaders, cranes, etc.) and the complexity of the orientation of the moved products marked side to the sensor [13–18].

A large number of uncertainties in the logistics routes of technological processes, strict requirements of standardization bodies, possible errors in the installation of equipment, as well as errors of the first and second kind of identification algorithms themselves do not allow to move from automated to automatic traffic control, therefore, additional scientific research is required to find additional data sources, methods of analysis to increase the share of successful identification of marked objects.

3 Overview of Analogues

At the moment, to implement the goal, there are several similar solutions. We will perform a comparative analysis of these analogs to identify their merits and demerits. The result of the comparative analysis is summarized in Table 1 for clarity.

- VITRONIC - automatic recognition system is used to read barcodes in various industries
- OptiCode is an industrial scanner for high-speed barcode reading
- Group of companies Lumenta (system SmartVision M3200-IP) - system of automatic identification of marking of pipes

As a result of the comparative analysis of the presented analogs, it can be concluded that these solutions are not entirely suitable for use in the automation of non-typed routes, since they are intended for conveyor lines, or they represent only the hardware implementation to which the software part needs to be developed by forces third-party organizations. In some cases, there is no support for multi-code reading, as a result of which a significant amount of funds will be required for the development, which is unprofitable.

Table 1. Advantages and disadvantages of analog systems.

Parameter\system	VITRONIC	OptiCode	Lumenta
Barcode recognition	+	+	+
Recognition of two-dimensional codes	+	−	+
Read multiple codes in one plane	+	−	−
The ability to automate non-typed routes	−	+/−	−

As a result of the comparative analysis of the presented analogs, it can be concluded that these solutions are not entirely suitable for use in the automation of non-typed routes, since they are intended for conveyor lines, or they represent only the hardware implementation to which the software part needs to be developed by forces third-party organizations. In some cases, there is no support for multi-code reading, as a result of which a significant amount of funds will be required for the development, which is unprofitable.

4 Development of a Simulation Model for the Movement of Industrial Products

The development of almost any algorithm is based on a data model. To develop an algorithm for forecasting and preventing abnormal situations, it is proposed to use the following simulation model.

The model represents a certain section of the territory of the enterprise with certain zones of goods receipt (for example, a conveyor, small mechanization means, etc.) and its shipment (for example, a conveyor belt, a loading zone for road and rail transport, etc.).

The goods receipt area is a generator of transactions (products) of the simulation model, each of the storage areas is a drive, and the shipping zone is a transaction execution point. After the transaction is generated, it can be moved to one of the drives, after which it can move randomly between other drives. In practice, this corresponds to the acceptance of goods and movement through the warehouse. At certain points in time, shipment orders arrive, in accordance with which the goods move to the shipping zone. Control of the movement of products through the warehouse is proposed to be performed using the multicode labeling approach. Thus, with each movement, the product is labeled and analyzed.

The scheme of this simulation model is shown in Fig. 1.

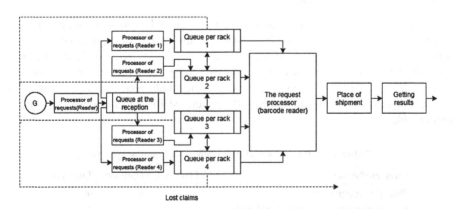

Fig. 1. Model outline.

Technology of Building Movement Control Systems of Products in Warehouse Areas   175

It should be noted that during the operation of the simulation model, the following contingencies may arise during the transportation of products, which must be taken into account:

- Supernumerary situation 1 – during the movement of the product, the same marking is read
- Supernumerary situation 2 – in the process of identification, a non-existent marking fell into the field of view of the reader
- Supernumerary situation 3 – During the transfer to the product, the marking of another object, by mechanical action (marking off) or deliberate action of personnel (intentional re-gluing of the marking) fell on the product

5 Development of the Structure of the Hardware and Software Complex MCS

At the latest industrial enterprises, the control of the movement of products is carried out by specialized means-automatic identification marking systems. In modern companies issued all sorts of types and names of products, one of these may be pipe products, which is a metal pipe long up to 12 m and a wall thickness of up to 500 mm.

To move the products, you will need a crane with a traverse capable of lifting rolled products. There are three types of traverse suitable for this task:

- Chain (pipes are fixed with chains and hooks)
- Vacuum (pipes are fixed by vacuum suckers)
- Magnetic (with a permanent magnet)

The most common and effective in most enterprises is the magnetic traverse. It allows you to capture several pipes at the same time.

The movement of products is carried out mainly by magnetic cranes. The length of the trolley crane is 32 m. On modern magnetic cranes (for example, list the various cranes) the Fixed load can be at a distance of 3 to 16 m to the nearest support and at a height of up to 8 m.

For successful reading, transmission and storage of this information requires special equipment, algorithms and methods.

The equipment must be in direct proximity to the identified product, as well as allow to determine the location to control its movement. That is why it needs to be placed directly into the magnetic traverse, which performs manipulations to capture the pipes (Fig. 2).

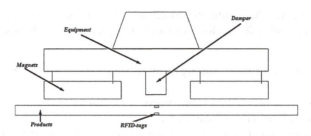

Fig. 2. The scheme of the crane with a movable work piece and RFID-mark.

The equipment itself is a kind of single device consisting of four main components connected to each other (Fig. 3).

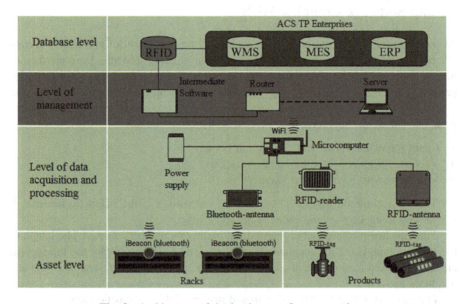

Fig. 3. Architecture of the hardware-software complex.

The architecture of the developed technology can be divided into 4 levels:

1. At the asset level, storage sites are equipped with beacons with Bluetooth Low Energy (iBeacon). Products are labeled with high-frequency RFID tags. For metal products, labels with a substrate of ferrite materials are used, which are placed on each face.
2. Level of data acquisition and processing. The client module is a stand-alone device that includes a microcomputer, RFID reader, power supply, Bluetooth adapter, WiFi and RFID antennas.

3. The level of management. This level is represented by a server that is a fixed computing device with the appropriate software and has access to the network, a router and middleware.
4. Database level. Includes a database of RFID systems with the ability to interface with the automated process control system: a warehouse management system (WMS) database, a software package that implements an ERP strategy, a production process control system (MES) database, etc.

The level of data acquisition and processing is inherently a client, which interacts with the server using WiFi technology.

Some parts of the hardware can be replaced with a similar one in order to reduce the cost of the system, increase productivity, or increase battery life.

6 Experimental Results

To test the evaluation of the reliability of the results obtained, an experimental study was carried out. The subject of the study is a system for automatic control over the movement of products. The object is a product, an enterprise or a warehouse.

Laboratory experience was conducted with a prototype system in conditions close to real production. To the moving cart was attached a model of the part of the beam, with the reading and processing device mounted on the side. Under the traverse on the cart is mounted a pallet for products in the form of pipes. View of the laboratory setup is shown in Fig. 4.

Fig. 4. View of the laboratory setup.

For the experiment was created two racks (c1, c2), each of which is labelled iBeacon-tag. The movement of products made from rack 1 to rack 2. During the movement of production readings from their RFID-tags and iBeacon-tags of racks have been processed and are presented in tables (Tables 2 and 3).

Table 2. Readings from iBeacon-tags of racks.

Time	Rack 1	Rack 2
35:22:00	0	43
35:28:00	0	26
35:34:00	0	0
35:40:00	0	0
35:46:00	3	0
35:52:00	34	0
35:58:00	37	0
36:04:00	27	0
36:10:00	27	0
36:16:00	16	0

Table 3. Readings from RFID tags of products.

Time	Product 1	Product 2	Product 3	Product 4
35:22:00	65	16	91	89
35:28:00	61	5	101	101
35:34:00	75	20	105	105
35:40:00	81	11	108	107
35:46:00	73	9	110	109
35:52:00	57	21	98	95
35:58:00	38	7	103	110
36:04:00	38	21	99	110
36:10:00	61	2	107	110
36:16:00	35	3	78	80

The duration of the experiment was 1 min. During the experiment with iBeacon-tags of rack has received 212 readings and from RFID tags of products - 2715. Graphic interpretation of the data presented in Figs. 5 and 6.

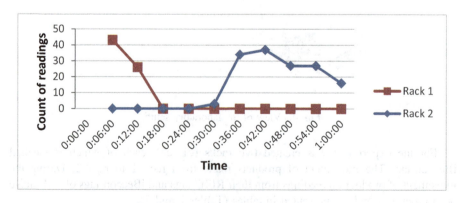

Fig. 5. Readings from iBeacon-tags of racks.

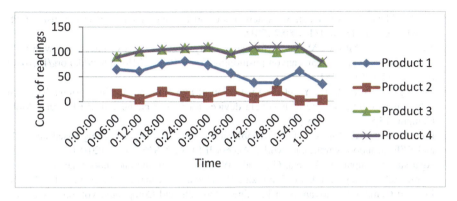

Fig. 6. Readings from RFID tags of products.

From the graph in Fig. 4 shows that at time 00:18:00 readings from iBeacon-tag of the first rack were no longer received by the reader. At time 0:24:00 began to receive readings from iBeacon-tag of the rack number 2. This fact indicates that there was a movement of the transport device from the area of the rack 1 to the area of the rack 2.

Figure 5 shows the readings from the RFID-tags of products, time-spaced with an interval of 6 s. During the experiment, 4 items were moved. According to the graph it is clear that a stable signal came from all 4 product tags. This allows us to say that during the movement of the conveying device the products moved along with it.

7 Conclusion

In the course of the work, an analytical review of the Russian and foreign scientific and technical base on development of methods and algorithms of MCS. Comparative analysis of analog systems was carried out. Simulation model for the movement of industrial products and the structure of the hardware-software complex MCS is develop.

In the course of the work, more than 150 experiments were conducted in the laboratory. Also, the installation was tested in an industrial plant. The results of the experiment showed the reliability of the movement identifications in the amount of 97.3%. As a result, information was also collected that allowed adjustments to the system to improve its efficiency.

References

1. Development of the digital economy in Russia program until 2035. http://innclub.info/wp-content/uploads/2017/05/strategy.pdf. Accessed 15 Sept 2018
2. Internet of Things in 2020 roadmap to future. https://mafiadoc.com/internet-of-things-in-2020-roadmap-for-the-future-eposs_5a17c5b41723ddb59566366a.html. Accessed 18 Aug 2018

3. Xu, R., Chen, W., Xu, Y., Ji, S.: A new indoor positioning system architecture using GPS signals. Sensors **15**, 10074–10087 (2015)
4. Tabatabaei, A., Mosavi, M.-R.: Performance analysis of GLONASS integration with GPS vectorised receiver in urban canyon positioning. Surv. Rev. (2018). https://doi.org/10.1080/00396265.2018.1481181
5. Zhuang, Y., Syed, Z., Li, Y., et al.: Evaluation of two WiFi positioning systems based on autonomous crowdsourcing of handheld devices for indoor navigation. IEEE Trans. Mob. Comput. **15**(8), 1982–1995 (2016)
6. Luo, J., Zhang, Z., Liu, C., et al.: Reliable and cooperative target tracking based on WSN and WiFi in indoor wireless networks. Access IEEE **6**, 24846–24855 (2018)
7. Agarwal, P., Gupta, A., Verma, G., et al.: Wireless monitoring and indoor navigation of a mobile robot using RFID. In: Panigrahi, B., Hoda, M., Sharma, V., et al. (eds.) Nature Inspired Computing. Advances in Intelligent Systems and Computing, vol. 652. Springer, Singapore (2018)
8. Gunther, E.J.M., Sliker, L.J.: A UHF RFID positioning system for use in warehouse navigation by employees with cognitive disability. Disabil. Rehabil. Assist. Technol. **12**(8), 832–842 (2017). https://doi.org/10.1080/17483107.2016.1274342
9. Ying, J., Pahlavan, K., Li, X.: Precision of RSS-based indoor geolocation in IoT applications. In: IEEE 28th Annual International Symposium on Personal Indoor and Mobile Radio Communications, pp. 1–5 (2017)
10. Vashistha, A., Gupta, A., Law, C.L.: Self calibration of the anchor nodes for UWB-IR TDOA based indoor positioning system. In: IEEE 4th World Forum on Internet of Things (2018). https://doi.org/10.1109/wf-iot.2018.8355163
11. Grinyak, V.M., Grinyak, T.M., Tsybanov, P.A.: Indoor positioning using bluetooth devices. Territory New Opportunities **2**(41) (2018)
12. Navigine — navigation and tracking of movements inside buildings. https://vc.ru/tribuna/11212-navigine. Accessed 18 Aug 2018
13. Astafiev, A.V., Orlov, A.A., Popov, D.P.: Development the algorithm of positioning industrial wares in-plant based on radio frequency identification for the products tracking systems. In: International Conference Information Technology and Nanotechnology. Session Image Processing, Geoinformation Technology and Information Security, vol. 1901, pp. 23–27 (2017)
14. Astafiev, A.V., Orlov, A.A., Privezencev, D.G.: Method of controlling the movement of large metal products with the use of algorithms for localization and recognition of bar code markings. In: Dynamics of Systems, Mechanisms and Machines, Dynamics (2016)
15. Orlov, A.A., Provotorov, A.V., Astafiev, A.V.: Methods and algorithms of automated two-stage visual recognition of metal-rolling billets. Autom. Remote Control 1099–1105 (2016). https://doi.org/10.1134/S000511791606014X
16. Zhiznyakov, A.L., Privezentsev, D.G., Zakharov, A.G.: Using fractal features of digital images for the detection of surface defects. Pattern Recogn. Image Anal. **25**(1), 122–131 (2015)
17. Nhat, D.D.: Researches and application of RFID technology. Int. Res. J. **5**, 34–37 (2015)
18. Kamozin, D.Y.: Comparison of the effectiveness of bar-code technology and RFID technology's application in logistics processes. Bull. Baikal State Univ. **3**, 71–75 (2013)

Approach to Development of Specialized Terminals for Equipment Control on the Basis of Shared Memory Mechanism

P. A. Nikishechkin[✉], N. Yu. Chervonnova, and A. N. Nikich

Moscow State Technological University "STANKIN", 1 Vladkovskii Lane, Moscow 127055, Russian Federation
pnikishechkin@gmail.com

Abstract. The development of modern production technologies and industrial equipment leads to an increase in the amount of data that must be monitored when managing technological processes. Often, standard terminal tools are not enough to fully control and manage complex processes, especially when the equipment has complex electrical automation, which requires visualization and control of a large number of parameters. Thus, the task of creating mechanisms for building additional portable terminals with a convenient and flexible user interface designed to solve the problems of control and management of technological equipment is relevant. The article is devoted to the study of the features of building modern means of collecting, processing and transmitting information at the workplace, as well as the creation of a mechanism for the rapid development of human-machine interfaces for controlling technological equipment. An architectural model for constructing a control system and an algorithm for creating terminal solutions are presented. A practical example of the formation of a simple terminal solution for monitoring the work of a training and demonstration stand is considered.

Keywords: Automation · PLC · CNC · Human-machine interface · Modular approach · Control terminal

1 Introduction

At modern industrial enterprises, there is a tendency to move from the automation of individual machines and processes to a new concept that provides a digital representation of all physical assets with subsequent integration into a digital system. In addition to production processes, any enterprise cannot work without solving tasks such as feasibility planning, scheduling and operational management. Thus, the process of managing modern production can be implemented in the form of a multi-level hierarchical system. It is customary to designate the modern automation information system of a large industrial enterprise as a pyramid, based on automated process control systems, in the middle part - MES-systems (Manufacturing Execution System), then ERP-systems (Enterprise Resource Planning). The pyramid of information systems in production symbolizes the flow of information - from the lowest production level, to systems of a higher level of management.

For effective management of complex systems, which include modern industrial enterprises, it is necessary to coordinate the work of all levels of the enterprise. It is advisable to carry out this coordination by integrating the information, mathematical, organizational and technical support of the individual subsystems within a single system, based on a holistic view of the production, financial and organizational activities of the enterprise or holding. One of such important tasks is the realization of the ability to visualize the operation of control objects from a lower production level, and transfer the control interface to levels higher [1–3].

In case of logic control systems, that along with the hardware also have the program that implements a control algorithm, there is the task of joining the control system, hardware components, control algorithm and user interface to a single integrated control system with an accessible user interface. In order to get that done, design and programming environment of logic control programs should have the possibility of engineering and software maintenance of user interface and monitoring means for the implementing control system. For control systems of increased complexity (for example, CNC machines) operating in real time, the development of an operator interface for control objects is impossible due to the complexity of the software implementation.

Man-machine interaction with such equipment is implemented using standard terminal solutions, the set of which is predefined, has limited functionality and does not allow changing it. However, in logical control systems that are responsible for the coordinated operation of aggregates and mechanisms of simple control objects (conveyors, storage mechanisms, packaging machines, etc.), accelerated creation of an operator interface based on a set of graphic objects is an urgent task today. In this case, graphic objects for creating an operator interface must be connected with a logical control program and receive relevant data [4–6].

One of the analogues of the proposed mechanism is SCADA, designed to ensure the operation of systems for collecting, processing and displaying relevant information about the control object. SCADA is a software package that implements data collection and analysis for engineers at higher levels of management. This is an important and necessary product for a modern enterprise, but it has certain disadvantages: high price; low level of openness for the user, which does not allow creating their own information processing algorithms. To do this, it is advisable to develop your own system for implementing the operator interface based on a set of your own graphic objects and the mechanism of their binding to real objects. Thus, we can conclude that it is necessary to develop a mechanism for the rapid development of flexible operator interfaces for controlling various equipment [7–9].

2 Development of a Mechanism for Creating Simple Terminal Solutions

The computing and control platform created at MSUT "STANKIN" allows to solve many different automation tasks. The main product of this platform is the AxiOMA Control CNC system for controlling various complex machine tools. The developed CNC system also includes a software-implemented controller that implements the solution to the problem of controlling electro-automatic machines. The CNC system is made on a modular basis, which increases the flexibility and openness of the system. This principle made it possible to implement on the basis of this platform an independent solution in the form of a system for solving logical control problems, as well as simple motion control problems [10, 11].

Dual computer system is used in the working out system, as well as in «AxiOMA Control» CNC system, it includes the control system core, functioning in real time, and the terminal part (Fig. 1) [12].

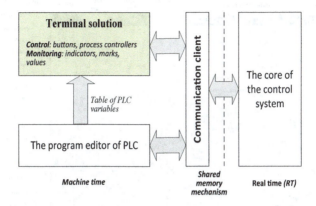

Fig. 1. Architectural model of the work of specialized terminal solutions.

External/remote terminal is connected to the core of the control system (machine, robot) and provides its real-time control. As a remote terminal, specialized solutions can be used the panel for operator, a machine panel, or, for the implementation of simple operations, a classic PC [13].

Data transfer between the system core and the terminal is carried out using a special TCP/IP communication client: by an Ethernet connection or WiFi. For the interaction of the control core with the terminal, a shared memory mechanism is used, which is a separate memory area with predefined cell addresses. This allows express settings of the terminal components for controlling and debugging the system [14, 15].

The main steps of working out simple terminal clients for processing equipment control could be pointed out:

- PLC program development
- Parameterization of the hardware - configuration of input/output modules and servo drives
- Systematization of the main variables of inputs/outputs, for control and monitoring implementation of basic parameters
- Development of the terminal solution with the help of special interface builder
- Binding of terminal components to PLC variables of the program

Development of control algorithms is carried out in the PLC-developed editor, with the help of which programs are created in the language of function blocks diagram, as well as configuration of the hardware (I/O, servodrives). In the developed PLC program, a number of basic variables are integrated, and it is of necessity to visually control them or their values, using a special screen of PLC variables.

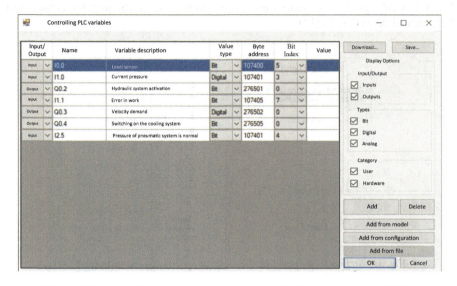

Fig. 2. The screen interface for creating and configuring of variables of PLC.

As you can see from the Fig. 2, each variable has its own set of parameters: description, type of value (bit, float, integer), and address. Variables can contain values from different areas of system memory: inputs, outputs, or PLC variable area. Each memory area has its own identifier [16–19].

The visual part of the terminal solution is created using the interface builder. This builder allows to create and place the graphic components and configure them. The set of visual components includes elements for controlling the parameters of the ongoing process and components for managing these processes (buttons, knobs, input fields). Figure 3 shows an example of creating a terminal solution with the help of a builder, for technological equipment monitoring and controlling [20].

The user interface designer has a component menu, in which there is a list of objects that the operator can add to monitor and control processes; the main area in which the terminal solution is created; parameter area, with the help of which each graphic component is configured and linked to memory areas for updating in real time [21, 22].

The designer terminal interfaces contains a panel component, which contains the core objects, which can be used by the user to create the interface; the terminal decision in which to create the operator interface; the settings panel terminal components, which is used to set up each of the components, including the tuning peg to the kernel memory.

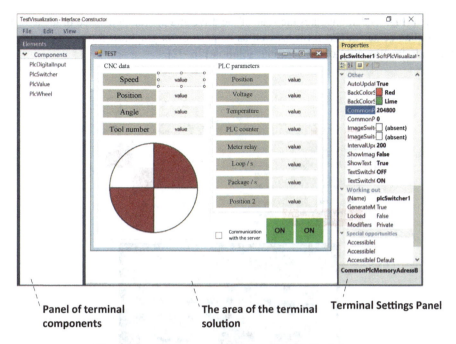

Fig. 3. Development environment of terminal solutions.

In this example, PlcValue objects are used to display the state of bit variables (true/false) or variables with numerical values. Also, the example uses a component PlcSwitcher for signaling to the controller outputs, and component PlcWheel for visual display of the position and velocity of any object rotation [23–26].

Binding of graphical components to PLC variables in the interface designer can be done manually, by specifying the address of the shared memory, or automatically, using the binding to a variable from the corresponding table in the PLC editor [25].

3 An Example of Developing a Simple Terminal Solution for Technological Equipment Control

As a practical example of using described mechanisms is represented the creation of a terminal solution that allows to visualize the state of the work of the training and demonstration stand that imitates the work of the relay ladder logic system of the milling processing center. The stand consists of fields, visualizing the operation of the basic units of machine tools: power, emergency button status of the work, the tool changing mechanism, safety guards, control coolant, augers for removing shavings, as well as control of the exit axis outside working area.

To create the terminal described above, graphic components were used, tied to the memory areas of the PLC variables of the program for managing this stand. These variables are responsible for the state of all nodes of the stand. Also, the terminal solution (Fig. 4) demonstrates the state of the drive and the rotation speed of the axis of the milling center in text form, as well as its position in graphical form, which shows the wide functionality of the terminal solution [26–32].

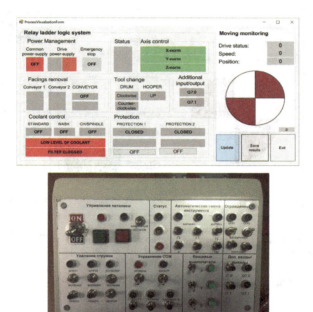

Fig. 4. (a) Developed terminal solution; (b) training and demonstration stand.

The developed terminal solution allows to visualize the state of operation of all units of the demonstration stand, as well as to transmit information to higher levels, which is important for the construction of modern production facilities that meet the concept of building a smart production "Industry 4.0" [22, 33].

4 Conclusion

The mechanism for developing simple terminal represented in this article offers the possibility to create efficiently a HMI for heterogeneous technological equipment control. The distinguishing characteristic of the presented mechanism is the solution for connection of terminal components with the core of the system and directly with the object of control by using the shared memory mechanism, which allows to receive information from control objects and provide its management without interfering into the core system operation. The developed terminal components make it possible to implement a large number of logical control tasks, while allowing both control over the operation of the equipment and implement its direct management.

The developed mechanism simplifies the process of startup and commissioning of technological equipment and allows you to increase control over the entire process by continuously monitoring it using graphical interfaces. This allows you to increase the level of informatization of the enterprise and provides the ability to transfer data to higher levels of the enterprise.

Acknowledgements. This research was supported by the Ministry of Education and Science of the Russian Federation as a public program in the sphere of scientific activity (N 2.1237.2017/4.6).

References

1. Martinov, G.M., Martinova, L.I.: Trends in the numerical control of machine-tool systems. Russ. Eng. Res. **30**(10), 1041–1045 (2010)
2. Martinova, L.I., Grigoryev, A.S., Sokolov, S.V.: Diagnostics and forecasting of cutting tool wear at CNC machines. Autom. Remote Control **73**(4), 742–749 (2012)
3. Grigoriev, S.N., Martinov, G.M.: Scalable open cross-platform kernel of PCNC system for multi-axis machine tool. Procedia CIRP **1**, 238–243 (2012)
4. Martinova, L.I., Pushkov, R.L., Kozak, N.V., et al.: Solution to the problems of axle synchronization and exact positioning in a numerical control system. Autom. Remote Control **75**(1), 129–138 (2014)
5. Nezhmetdinov, R.A., Sokolov, S.V., Obukhov, A.I., et al.: Extending the functional capabilities of NC systems for control over mechano-laser processing. Autom. Remote Control **75**(5), 945–952 (2014)
6. Martinov, G.M., Grigorev, A.S.: Diagnostics of cutting tools and prediction of their life in numerically controlled systems. Russ. Eng. Res. **33**(7), 433–437 (2013)
7. Martinov, G.M., Lyubimov, A.B., Grigoriev, A.S., et al.: Multifunction numerical control solution for hybrid mechanic and laser machine tool. Procedia CIRP **1**, 277–281 (2012)
8. Martinova, L.I., Kozak, N.V., Nezhmetdinov, R.A.: The Russian multi-functional CNC system AxiOMA control: practical aspects of application. Autom. Remote Control **76**(1), 179–186 (2015)
9. Martinova, L.I., Sokolov, S.V., Nikishechkin, P.A.: Tools for monitoring and parameter visualization in computer control systems of industrial robots. In: Tan, Y., Shi Y., Buarque, F., Gelbukh, A., Das, S., Engelbrecht, A. (eds.) Proceedings of Advances in Swarm and Computational Intelligence, pp. 200–207. Springer, Cham (2015)
10. Martinov, G.M., Kozak, N.V.: Numerical control of large precision machining centers by the AxiOMA control system. Russ. Eng. Res. **35**(7), 534–538 (2015)
11. Martinov, G.M., Grigoryev, A.S., Nikishechkin, P.A.: Real-time diagnosis and forecasting algorithms of the tool wear in the CNC systems. In: Tan, Y., Shi, Y., Buarque, F., Gelbukh, A., Das, S., Engelbrecht, A. (eds.) Advances in Swarm and Computational Intelligence, vol. 9142, pp. 115–126. Springer, Cham (2015)
12. Correa, J.E., Toombs, N., Ferreira, P.M.: A modular-architecture controller for CNC systems based on open-source electronics. J. Manuf. Syst. **44**(2), 317–323 (2017)
13. Martinov, G.M., Nezhmetdinov, R.A.: Modular design of specialized numerical control systems for inclined machining centers. Russ. Eng. Res. **35**(5), 389–393 (2015)
14. Martinov, G.M., Lyubimov, A.B., Bondarenko, A.I., et al.: An approach to building a multiprotocol CNC system. Autom. Remote Control **76**(1), 172–178 (2015)
15. Martinov, G.M., Obuhov, A.I., Martinova, L.I.: An approach to building specialized CNC systems for non-traditional processes. Procedia CIRP **14**, 511–516 (2014)

16. Grigoriev, S.N., Martinov, G.M.: Research and development of a cross-platform CNC Kernel for multi-axis machine tool. Procedia CIRP **14**, 517–522 (2014)
17. Martinov, G.M., Nezhmetdinov, R.A., Kuliev, A.U.: Approach to implementing hardware-independent automatic control systems of lathes and lathe-milling CNC machines. Russ. Aeronaut. **2**, 128–131 (2016)
18. Grigoriev, S.N., Martinov, G.M.: An ARM-based multi-channel CNC solution for multitasking turning and milling machines. Procedia CIRP **46**, 525–528 (2016)
19. Martinov, G.M., Kozak, N.V.: Specialized numerical control system for five axis planning and milling center. Russ. Eng. Res. **36**(3), 218–222 (2016)
20. Martinov, G.M., Obuhov, A.I., Martinova, L.I., et al.: An approach to building a specialized CNC system for laser engraving machining. Procedia CIRP **41**, 998–1003 (2016)
21. Grigoriev, S.N., Martinov, G.M.: The control platform for decomposition and synthesis of specialized CNC systems. Procedia CIRP **41**, 858–863 (2016)
22. Martinov, G.M., Kozak, N.V., Nezhmetdinov, R.A., et al.: Method of decomposition and synthesis of the custom CNC systems. Autom. Remote Control **78**(3), 525–536 (2017)
23. Martinov, G.M., Kozak, N.V., Nezhmetdinov, R.A.: Implementation of control for peripheral machine equipment based on the external soft PLC integrated with CNC. In: International Conference on Industrial Engineering, Applications and Manufacturing, pp. 1–4 (2017)
24. Martinov, G.M., Sokolov, S.V., Martinova, L.I., et al.: Approach to the diagnosis and configuration of servo drives in heterogeneous machine control systems. In: 8th International Conference ICSI II, pp. 586–594 (2017)
25. Kovalev, I.A., Nikishechkin, P.A., Grigoriev, A.S.: Approach to programmable controller building by its main modules synthesizing based on requirements specification for industrial automation. In: International Conference on Industrial Engineering, Applications and Manufacturing, pp. 1–4 (2017)
26. Nikishechkin, P.A., Kovalev, I.A., Nikich, A.N.: An approach to building a cross-platform system for the collection and processing of diagnostic information about working technological equipment for industrial enterprises. In: MATEC Web of Conferences, vol. 129 (2017)
27. Kaur, A., Bansal, D.: Monitoring and controlling of continue furnace line using PLC and SCADA. In: Wireless Networks and Embedded Systems (2016)
28. Younas, U., Durrani, S., Mehmood, Y.: Designing human machine interface for vehicle's EFI engine using Siemen's PLC and SCADA system. In: Frontiers of Information Technology (2015)
29. Anestisa, S., Kleopatra, P.: Exploring of the consequences of human resources multitasking in industrial automation projects: a tool to mitigate impacts. Procedia Eng. **196**, 738–745 (2017). Creative Construction Conference
30. Querol, E., Romero, F., Estruch, A.M., et al.: Design of the architecture of a flexible machining system using IEC61499 Function blocks. Procedia Eng. **132**, 934–941 (2015). The Manufacturing Engineering Society International Conference
31. Turc, T.: Using WEB services in SCADA applications. Procedia Technol. **19**, 584–590 (2015)
32. Prahofer, H., Schatz, R., Wirth, C.: Comprehensive solution for deterministic replay debugging of SoftPLC applications. IEEE Trans. Ind. Inform. **7**(4), 641–651 (2011)
33. Zhao, R., Li, C., Tian, X.: A novel industrial multimedia: rough set-based fault diagnosis system used in CNC grinding machine. Multimedia Tools Appl. **76**(19), 19913–19926 (2017)

Control System for Gas Transfer by Gas Gathering Header in the Context of Water Accumulation

M. Yu. Prakhova[✉], E. A. Khoroshavina, and A. N. Krasnov

Ufa State Petroleum Technological University,
1 Kosmonavtov Street, Ufa 450062, Russian Federation
prakhovamarina@yandex.ru

Abstract. One of the basic facilities of the gas field is a gas gathering header. It consists of a system of flow lines used for gas transportation from wells to a gas processing plant. The operation of flow lines is complicated by raw gas containing reservoir water which may be abundant especially in brown fields located in the Extreme North. In winter, the flow lines get blocked by ice plugs. As a result, flow line friction increases and well production rates decrease. The methods aimed at the removal of accumulated fluid from gas gathering header are not fully effective, so to prevent residual water content from freezing, it is suggested to use coiled tubing installed within flow lines and connected with a methanol pipeline. The methanol-water solution serves as a heat-transfer medium. The rejected heat from air cooling unit heats the medium in the heat exchanger. The control system maintains operational parameters of heat-transfer flow providing a balanced transfer of gas. This system makes it possible to solve two problems: secure positive temperature of gas flow all the way from a well to switching valve building and recover rejected heat of the air cooling unit enhancing efficiency and reliability of gas gathering header operation.

Keywords: Gas gathering header · Ice formation · Coiled tubing · Heat tracer · Methanol-water solution · Rejected heat from air cooling unit

1 Introduction

Basic operational issues of wells and gas gathering headers, in particular in the regions of Extreme North, are hydrated gas plugs and freeze-up. These incidents are caused by fluid contained in gas flow. The fluid is reservoir water that appears in bottom-hole zones of development wells when reservoir pressure decreases and gas-water contact becomes higher. Gas flow and mass flow rates crucially decline with field exploitation to the point when gas flow no longer carries out built-up fluid resulting in its recurrent accumulation in production tubing (TBG) and gas gathering headers, or flow lines. In summer, liquid plugs do not considerably affect operation of flow lines and wells, but in winter when soils freeze, the plugs accumulated in depression areas solidify leading to significant increase in flow line friction and wellhead pressure [1, 2]. In the event of production decline as a result of natural or process reasons, it may lead to self-kill of wells and their complete shut-down, which, in its turn, causes the flow lines to freeze completely. In Urengoyskoye oil, gas and condensate field it is not uncommon for a gas gathering header to freeze for several months [3], that is why it is important to remove the accumulated water both from wells and gas gathering headers.

2 The Main Part

2.1 Research Significance

Today approximately 65% of gas gathering headers operate under conditions of accumulating liquid, which equals 1100 km (at total length of 1690 km). Experts predict that in the future process of liquid removal from wells and flow lines will worsen and by 2030, 87% of total pipeline length will be affected. For this reason, removal of accumulated water from wells and gas gathering headers is arranged throughout northern production fields.

In case water-cut is extreme, the water is removed continuously either by well pump or by reducing production tubing diameter to increase gas flow rate [4]. If fluid accumulates gradually then it is periodically removed by blowdown of wells and flow lines to the flare [5]. This is the easiest, most efficient and, in reality, basically the only infallible and all-round method of maintaining well stock and gas gathering headers in good working order [6]. It helps restore production for a long period (several days). Blowdown is performed 5–6 times a month and lasts for 15–30 min. Apart from permanent gas losses and pollution charges the main problem with this method of production maintenance remains physical impossibility to constantly treat a large number of sites.

Should there occur an increased water intrusion, the flow line will systematically switch to a reduced pressure line, which is also used for water removal [7]. These lines are created when a compressor cuts down the pressure low enough for the liquid to be carried out from wells and flow lines. On the one hand, this method helps to avoid gas losses and environmental pollution, but on the other hand, it reduces economic efficiency of natural gas production due to installation of additional equipment and increased power consumption.

Therefore, as it happens, today there is no optimal method to remove water from gas gathering headers, which proves significance of research of this issue.

2.2 Research Objective

The problem of flow line water-cut could be considered from a different perspective. Studies that researched the effect of liquid plugs on total flow line friction have shown that in flow lines with no significant elevation differences liquid plugs do not have a considerable influence on the line friction. Liquid in a gas gathering header does not pose a threat up until it turns into ice. Therefore, the main goal of this study should be rephrased: instead of a system of complete liquid removal from gas gathering headers we need to develop a system to prevent residual water content from freezing in gas gathering headers in cold periods. Therefore, a control system for gas transfer by flow lines within gas fields should ensure positive gas flow temperature along the whole length of flow lines.

Separate elements of gas gathering headers are heated to prevent hydrate formation. As a rule, heating cables are used for this purpose [8, 9]. However, this system only proves effective on some elements of gas gathering headers, such as shutoff valves or local short sections. Otherwise, heating of long gas gathering headers (average of 70–100 km per one field) would require hundreds of kilowatts of electric power to treat one flow line.

2.3 Theory

On production sites of oil and gas industry, the so-called heat tracers are used to prevent liquids from freezing in pipelines. Heat tracers are lines installed along the main pipeline (or as a main piping shell) for pumping of heat-transfer medium for heating (cooling) of material transferred by the main pipeline [10]. To use heat tracers for flow line heating, we need to solve three problems: first, we need to find a source of large quantity of heat, accessible throughout the entire winter period of gas gathering header operation, second, we need to select a heat tracer, third, we need to install the heat tracer along the gas gathering header or in-side it.

Let us analyze the suggested solution.

As a source of heat for treatment of gas gathering headers, it is reasonable to use rejected heat, for its dissipation requires considerable resources. In gas field, the rejected heat comes from air cooling units of boosting compressor station (BCS ACU), which consume the most electric energy and are expensive to maintain and operate [11]. Moreover, another advantage of use of heat from air cooling units is that it does not introduce extra heat into the process, which, otherwise, would re-quire to be released on the same air-cooling unit.

It is suggested to use coiled tubing as a heat tracer [reference book on geology] and to place it inside existing flow lines. By the pad, at tube outlet, TBG connect with existing methanol line forming a closed loop. Heat-transfer medium circulates in the loop and prevents liquid from freezing, ensuring its removal during compressor blowdown. Methanol-water solution serves as heat-transfer medium. It is delivered into the well and to certain areas of flow lines to prevent hydrate formation.

In standard operation mode of gas gathering header methanol-water solution is delivered to well pads through methanol pipeline system (Fig. 1).

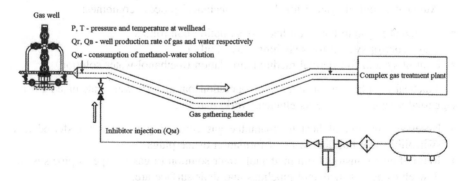

Fig. 1. Delivery of methanol-water solution to pipeline of gas gathering system.

The suggested system of flow line heating by recovery of waste heat from gas ACU required for liquid removal from wells and gas gathering headers in winter time is demonstrated by Fig. 2.

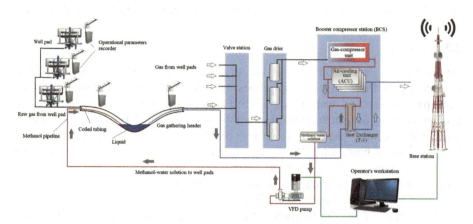

Fig. 2. The suggested system of flow line heating by recovery of waste heat from gas ACU.

This is how the system operates.

It is suggested to recover the heat from gas compressed at BCS using inter-mediate heat-transfer medium inside gas pipelines of gas gathering headers to pre-vent water from freezing inside. Methanol-water solution serves as heat-transfer medium. Its concentration will stop solution from freezing when its temperature drops to outside temperatures in the coldest period of the year.

Methanol-water solution is transferred inside gas gathering headers through installed flexible metal tube by VFD controlled pumps. Coiled tubing is suggested for use. Outer diameter 40–60 mm, pipe-in-pipe installation.

Automatic control system for this method is based on the work of industrial controllers, which analyze data from heat consumption controllers and manage pump flow rate by means of frequency regulators, maintaining required pressure and temperature throughout the system and ensuring balanced work of a gas gathering header.

Automatic control system for the entire technical process combines:

- Intelligence system based on heat consumption controllers
- Valve control system for switching valve building
- Control system for thermal medium circulation (methanol-water solution)

Practical significance: a number of operation parameters were calculated for the suggested system to ensure its efficient work, namely:

- Necessary amount of heat to guarantee gas containing water is transferred to a switching valve building without formation of ice plugs
- Ratio of heat transfers from methanol-water solution to gas in pipe-in-pipe system, flow efficiency of methanol pipelines and their surface area

- Quantity of methanol-water solution required for cooling of gas, its concentration to prevent it from freezing within subzero temperature range of winter period

When evaluating the necessary amount of heat to guarantee gas containing water is transferred to a switching valve building without formation of ice plugs it is important to calculate heat losses on gathering system pipelines. This calculation involves a certain difficulty due to uneven heat insulation of gas pipelines and their installation. Each gas gathering header pipeline may be partially buried at various depths, be installed on the ground or above ground on pipe racks. In addition, heat insulation may be partly damaged or missing. If pipeline is installed on the ground, it is not uncommon that its long segments are completely or partly under water. Significant seasonal temperature fluctuations, variations of solar radiation, changing wind speeds and snow cover depth, a vast difference between heat transfer rates of soils along the pipeline make it impossible to make an adequately accurate calculation.

You can see a typical pipeline bedding of gas gathering header of Cenomanian complex gas treatment plant in Urengoyskoye oil, gas and condensate field (Fig. 3).

Fig. 3. Typical pipeline bedding of gas gathering header of Cenomaniancomplex gas treatment plant in Urengoyskoye oil, gas and condensate field.

The soils are multiphase dispersed systems where the matter can be found in solid, liquid or gas state. Heat exchange process in this type of system is influenced by conductive, convective, radiative and mass exchange susceptibility [12]. Considering factors and conditions influencing heat transfer in soil we arrive at the system of differential equations of heat conductivity, mass exchange, convective and radiant heat transfer [13, 14]. In general, when analyzing heat exchange between soil and pipeline multiple factors are estimated to define thermal characteristics of soils. With that, the soil is viewed as a quasi-homogeneous matter, to which heat conduction equation is applied. This simplified approach proves to be adequate since many of the named factors have only slight impact on heat transfer in the soil.

For thermal calculations of a stationary pipeline, it is assumed that heat transfer rate is constant within the pipe length and in time [15]. However, observation and experiments point out that even in this case the rate of heat transfer changes within the pipeline length, which can be explained by temperature drop of transferred matter over the length.

Soils of Urengoyskoye oil, gas and condensate field are typically swampy. It is rather difficult to study heat exchange processes of pipelines installed in water-bearing soils and water. That is why when developing the suggested system we calculated soil heat transfer rates based on gas temperature difference as taken at wellhead and in switching valve building. We also included data obtained during an experiment at complex gas treatment plant-5. This direct experimental observation of thermal characteristics of soils for accurate data acquirement and its further use in thermal calculations of gas pipeline modes seems most advisable.

A simulation model was created to estimate quantity of methanol-water solution required for cooling of gas. The basic data for estimation was taken for complex gas treatment plant-5 of Urengoyskoye oil, gas and condensate field.

Optimal temperature (for the process of absorption based gas dehydration) of gas after heat exchanger is +15 °C. To ensure this temperature when methanol-water solution temperature before heat exchanger is +1 °C and gas temperature is +90 °C at the time of peak withdrawal rates it is required to deliver 100 t/h of methanol-water solution. Methanol-water solution temperature then reaches 80 °C.

Concentration of methanol-water solution should be 60% wt, which would prevent it from freezing when temperature of the solution drops to outside temperatures in the coldest period of the year (freezing temperature minus 75,7 °C) [16].

To estimate heat losses on methanol pipeline system and calculate the temperature of methanol-water solution delivered to well pads we created a hydrodynamic model of methanol pipelines. Calculation was made for buried methanol pipelines and three types of soil: dry, damp and wet sand [17]. The results are given in the table. Initial temperature of methanol-water solution was 70 °C for all cases and soil temperature was −5 °C (Table 1).

Table 1. The results of heat loss by methanol pipeline system.

Value	Soil - sand		
	Dry	Damp	Wet
Soil heat conductivity, W/mK	0.5	0.95	2.2
Total heat conductivity, W/mK	4.4	8.4	19
Methanol-water solution temp. at consumption of 60 m^3/hr, °C	26	13	2

Because methanol pipelines are installed both on the surface and under-ground crossing swampy and dry country, the actual results for longer sections will differ from calculations. It is impossible to establish heat losses of certain methanol pipelines without special research. In general, study results indicate that surface of methanol pipelines is enough for heat dissipation of methanol-water solution.

The transfer of heat from air cooling units to flow line gas through methanol-water solution also ensures a more stable temperature at gas dehydration unit inlet. It is explained by the fact that liquid cooling does not depend on outside temperature and is easier to control.

3 Conclusion

Suggested method has a great practical use for new gas and gas condensate field development. If gas gathering headers are designed and built to be used as heat-dissipating loops it would ensure complete dissipation of a large quantity of heat (up to 15–20 MW from two stages of air cooling units), which today involves 1.5 MW of electric energy at both stages of BCS air cooling units. Moreover, this heat-dissipating loop makes it possible to implement various technical solutions allowing for the cheapest electric energy production alongside natural gas production (3–6 MW) through exhaust fumes heat recovery from gas compressor units or partial heat recovery, which today is dissipated at air cooling units. Here we mean heat recovery using steam turbines with liquid vapor with low boiling temperature, for instance, freons. This method has one rather substantial drawback: at relatively low coefficient of performance (not exceeding 15–20%) these turbines require sizable water cooling towers for operation, which reduces cost-effectiveness of this implementation. However, if gas gathering headers are used as a heat-dissipating loop instead of water cooling towers, it guarantees electric energy production almost free of charges, since main investment and maintenance costs are included in the system of gas gathering header heating and gas cooling after first and second stage of BCS.

References

1. Abdel-Aal, H.K., Aggour, M., Fahim, M.A.: Petroleum and gas field processing. Dekker, Marcel (2003)
2. Kudiyarov, G., Istomin, V., Egorichev, A., et al.: Peculiarities of the in-field gas gathering systems at the latest stage of the development of cenomanian deposit Yamburgskoye gas field. Society of Petroleum Engineers (2017). https://doi.org/10.2118/187736-ru
3. Innovative technologies for gas recovery at wells examination and removal of liquid from wells and gas gathering headers. Znanie, Ufa (2012)
4. Vyakhirev, R.I., Korotaev, Yu.P., Kabanov, N.I.: Theory and practice of natural gas production. Nedra, Moscow (1998)
5. Rassokhin, G.V.: Final stage of gas and gas condensate field development. Nedra, Moscow (1977)
6. Kolovertnov, G.Yu., Krasnov, A.N., Kuznetsov, Yu.S., et al.: Automation of process of liquid removal in gas wells and flow lines. Territory Oil Gas **9**, 70–76 (2015)
7. Kolovertnov, G.Yu., Krasnov, A.N., Fedorov, S.N., et al.: Method of operating gas field in collector-beam arrangement scheme collection at final stage of deposit development. RU Patent 2597390, 10 September 2016 (2016)

8. Prakhova, M.Yu., Mymrin, I.N., Savelev, D.A.: Local automatic system of heat tracing to prevent hydrate formation on discharge line. Autom. Remote Control Eng. Commun. Oil Ind. **2**, 3–6 (2014)
9. Prakhova, M.Yu., Mymrin, I.N., Savelev, D.A.: Heaters for system of control over local heat tracing of gas condensate wellpad. Automation problems of technological processes of production, transfer and processing of oil and gas, pp. 88–92 (2013)
10. Zemenkov, Yu.D., Vasilev, D.D., Dudin, S.M.: Engineer's guidebook to exploitation of oil-and-gas and product pipelines. Infra-Inzheneria, Moscow (2006)
11. Davletov, K.M.: Improving the operation efficiency for air cooling units of gas in the fields of the Extreme North. Dissertation, Nadym (1998)
12. Reference book on geology (2017). http://www.geolib.net/tkrs/koltyubing-gnkt.html. Accessed 25 Sept 2017
13. Thermal properties and temperature conditions of soils (2016). http://biofile.ru/bio/19473.html. Accessed 12 Oct 2017
14. Bashurov, V.V., Vaganova, N.N., Filimonov, My.: Simulation of heat exchange processes in soil with account of liquid filtration. Comput. Technol. **16**(4), 3–18 (2011)
15. Thermal calculation for pipelines (2016). https://lektsia.com/2x57.html. Accessed 2 Sept 2017
16. Properties of methanol and its water solutions (2018). http://mirznanii.com/a/10128/svoystva-metanola-i-ego-vodnykh-rastvorov. Accessed 21 Sept 2017
17. Heat conductivity rates of different materials (2016). http://www.xiron.ru/content/view/58/28/. Accessed 30 Oct 2017

Active Vibration Protection System with Controlled Viscosity of Working Environment

B. A. Gordeev[1], S. N. Okhulkov[1(✉)], A. B. Darenkov[2], and D. Yu. Titov[2]

[1] Institute of Mechanical Engineering of RAS, 85 Belinsky Street,
Nizhny Novgorod 603024, Russian Federation
oxulkovs@mail.ru
[2] Nizhny Novgorod State Technical University, 24 Minima Street,
Nizhny Novgorod 603155, Russian Federation

Abstract. The article considers new approaches to optimizing the parameters of a magnetorheological transformer (MRT). The effect of a vortex magnetic field on the sedimentation of magnetic particles in a magnetorheological fluid (MRF) is considered. The creation of such a field leads not only to a change in the viscosity of the MRF, but also to the creation of a magnetic vortex. This magnetic vortex terminates the processes of sedimentation and deposition of MRF and counteracts gravitational forces. The article presents a structural diagram of a hydromount with MRT with a coaxial throttle channel controlled by a rotating magnetic field. A mathematical description of the motion of magnetorheological media in magnetic fields of a hydromount is made using the hydrodynamic equation of motion of the medium and the Maxwell equation. The article shows that with an increase in the frequency of the input vibration signal, frequency modulation of the Umov-Poynting vector occurs. Therefore, the damping process is enriched with high-frequency components, which are actively absorbed in a viscous magnetorheological medium. Hence, a rotating magnetic field in a coaxial throttle channel contributes to more efficient damping.

Keywords: Magnetorheological fluid · Viscosity · Magnetorheological transformer · Sedimentation · Magnetic field

1 Introduction

Methods regulating MRF viscosity by creating rotating magnetic field have not been widely used for active vibration protection systems such as hydromounts with MRT [1–4].

Creating such a field leads not only to MRF viscosity change but also to creating magnetic vortex [5] which stops MRF sedimentation and precipitation processes and counteracts gravity forces [6]. The works [1–8] show constant magnetic field application or the changing one with fundamental harmonic frequency of input vibration signal to control hydromount damping. MRF viscosity regulation in these cases is carried out by changing the constant magnetic field strength in choking channels and does not influence magnetic particles sedimentation.

When creating rotating magnetic field in coaxial choking channel of MRT hydromount there takes place a change of MRF flow parameters by changing its hydrodynamic resistance [9–11].

Sedimentation elimination in MRT is reached by hydromagnetic and gyroscope characteristics of the magnetic by variable magnetic field. The magnetic vortex created by the controlling field lugs away MRF particles. MRF particles having gyroscope characteristics allows to control its viscosity more effectively in MRT coaxial channel [12].

2 Rotating Magnetic Field to Control Hydromount Parameters with a Coaxial Choking Channel

It is necessary to improve the methods of controlling viscosity and managing MRF consumption at its choking in MRT [7, 13–15]. Thus it is necessary to improve the construction of the hydromount with MRT.

To create the rotating field in MRT hydromount it is advisable to use a one-phase inductor powered from one-phase AC voltage source [16, 17].

Figure 1 shows hydromount with MRT functional chart with a coaxial choking channel, controlled by rotating magnetic field.

Figure 1 shows: 1 – mounting plate; 2 – shell; 3 – the mount body; 4 – membrane; 5 – choking partition wall body; 6, 7 – working and compensating chambers; 8 – inner fixed cylinder; 9 – outer nitration cylinder; 10 – coaxial cylindrical channel with MRF; 11 – choking partition wall inlets and outlets; 12 – main choking channel; 13 – outer one-phase inductor; 14 – inner inductor; 15 – power feed cable of hydromount outer inductor.

Coaxial cylindrical choking channel 10 is formed by the outer cylinder 9 interior surface and the exterior working surface of the inner quill cylinder 8. The rotating magnetic field in MRT hydromount is created by the outer one-phase inductor windings 13, its cross-sectional cut being outlined in (Fig. 2). When the outer one-phase MRT inductor windings are placed for angle $\pi/2$ the magnetic field induction distribution in the coaxial choking channel with MRF is close to the sinusoidal one.

When creating the rotating magnetic field in a coaxial gap one of the two variants needs to be accomplished:

- Between the two windings of one-phase inductor with an equal number of winding turns $n_1 = n_2$ the phase-shifting capacitor is activated which provides the phase shift of feeding AC voltage of these windings for angle $\pi/2$
- To create rotating magnetic field by a different number of winding turns $n_1 \neq n_2$ in the inductor windings, providing the phase shift in the inductor windings also for angle $\pi/2$

The induction value in each point of coaxial cylindrical choking gap with MRF will be measured in time according to the harmonic law $B_a = B_{max}\cos\alpha$, where $\alpha = x/R$ – angular coordinate of the point on the radius circle R, expressed by в arched units (radians); x – coordinate of the point in linear units.

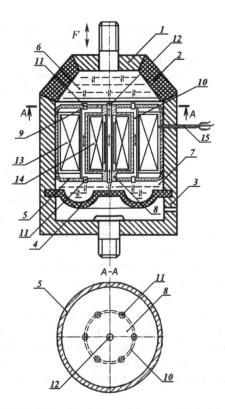

Fig. 1. Hydromount with MRT functional chart with a coaxial choking channel, controlled by rotating magnetic field.

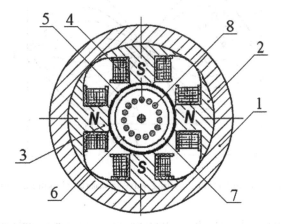

Fig. 2. One-phase inductor of MRT hydromount cross-sectional cut outline: 1 – MRT steel body; 2 – magnetic core of MRT field magnet windings; 3 – field magnets poles; 4 – field magnet windings; 5 – outer brass coaxial cylinder; 6 – inner steel coaxial cylinder; 7 – coaxial cylindrical gap with MRF; 8 – MRT choking capillary channels.

As soon as along the one-phase inductor windings 13 (Fig. 1) there runs AC current the magnetic field induction amplitude pulsates in time at the circular frequency ω of the MRT power supply source [16, 17].

The magnetic induction pulsating wave at an arbitrary point of the coaxial cylindrical choking channel 10 with MRF can be presented by the difference of two rotary waves.

$$B_p = B_{max} \cos \alpha_e \cos \omega t = \frac{B_{max}}{2} \cos(\alpha_e - \omega t) + \frac{B_{max}}{2} \cos(\alpha_e + \omega t) = B_{l+} + B_{l-}, \quad (1)$$

i.e. by the waves having half amplitude and rotating in the opposite directions - the direct one (B_{l+}) and the reverse one (B_{l-}) [16, 17].

In expression (1) $\alpha_e = z_p \cdot \alpha$ – the angular coordinate of the point on the pulsating wave period is T_a; z_p – the number of the pair of poles; $\alpha = x/R$ – the angular coordinate of the point on the radius circle R, expressed by arched units (radians); x – coordinate of the point in linear units.

The pulsating wave period T_a corresponds to the full-cycle of electromagnetic processes in time domain, so angular measure $\alpha_e = z_p \cdot \alpha$ corresponds to the phase angle angular measure.

Thus inductor MRT one-phase windings (Fig. 2) create the pulsating flux changing in time. Also along the coaxial cylindrical choking channel with MRF length there appears a pulsating wave with nodes and crests located at distance $T_a/2$ from one another (Fig. 3), where the pulsating wave length λ_a is proportional to coaxial choking gap length, performing the function of an extra choking channel.

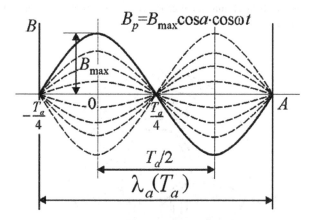

Fig. 3. The magnetic induction pulsating wave in coaxial choking channel with MRF.

Here and in expression (1) the motionless pulsating flux is replaced by the sum of identical circular fields, rotating in the opposite directions and having the same rotational frequencies: $n_{1dir} = n_{invp} = n_1$, where n_{1dir} – the direct sequence field speed, n_{1inv} – the reverse sequence field speed.

The electromagnetic moments M_{dir} and M_{inv}, created by direct and reverse magnetic fields of the outer MRT one-phase inductor, are directed in the opposite sides and its resultant moment M_{sum} is equal to the algebraic sum of the moments at one and the same magnetic field rotational frequency.

MRF sliding motion with respect to the magnetic field direct flux Φ_{dir}:

$$s_{dir} = (n_{1dir} - n_2)/n_{1dir} = (n_1 - n_2)/n_1 = 1 - n_2/n_1, \tag{2}$$

where n_2 – MRF rpm in the rotating magnetic field [16, 17].

MRF sliding motion with respect to the magnetic field reverse flux Φ_{inv}

$$s_{inv} = (n_{1inv} + n_2)/n_{1inv} = (n_1 + n_2)/n_1 = 1 + n_2/n_1, \tag{3}$$

From (2) and (3)

$$s_{inv} = 1 + n_2/n_1 = 2 - s_{dir}, \tag{4}$$

At the harmonic inlet vibration signal, alternate load affects the mounting plate 1 (Fig. 1). Assuming that in the first half-period the working chamber volume 6 decreases. There the pressure with respect to the compensating chamber 7 increases and MRF laminar flow along the main choking channel 12 starts. This channel walls have the magnetic particles speed of motion $\vec{V}_f = 0$, and at its center it is a maximum one $\vec{V}_f = $ max. At a distance r_{var} from the channel axis, $\vec{V}_f$ speed changes according to the parabolic law (Fig. 4) [18, 19]:

$$\vec{V}_f(r) = \vec{V}_{f0}\left(1 - \frac{r^2}{R^2}\right), \tag{5}$$

where R – is a pass channel radius; $\vec{V}_{f0}$ – the power fluid speed on the axis of the pass channel, determined by formula:

$$\vec{V}_{f0} = \frac{P_1 - P_2}{4\eta\, l} R^2 = \frac{\Delta p}{4\eta\, l} R^2, \tag{6}$$

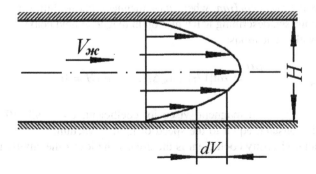

Fig. 4. Motion of fluid loose pies with different speeds at the power fluid laminar flow in the main choking channel.

Here $P_1 - P_2 = \Delta p$ – is a pressure difference in working 6 and compensating 7 chambers; l – helically-formed channel length; η – MRF dynamic viscosity.

The power fluid flow Q, i.e. the power fluid volume flowing through the main channel cross-section having the operating length l per a time unit is determined by Poiseuille formula [18, 19]:

$$Q = \frac{P_1 - P_2}{8\eta\, l} \pi R^4, \qquad (7)$$

Knowing speed $\vec{V}_{cp}$ of the fluid flow and its volume density ρ, it is possible to determine the mass-flow rate:

$$Q = S_k\, \rho\, \vec{V}_{cp}, \qquad (8)$$

where S_κ – is a flux cross-section area at the given main choking channel area 12.

From formula (8) it is possible to calculate the volume of working 6 and compensating chambers 7. In the second half-period all the process is repeated in the reverse direction.

3 Research of Magnetorheological Hydromount with Coaxial Choking Channel in the Rotating Magnetic Field

When composing the equations of magnetorheological kinds of environment motion in hydromount magnetic fields the following conditions are assumed [20, 21]. The magnetic permeability of the magnetic environment is homogeneous and isotropic in all its working capacity and does not depend on the magnetic field voltage **H**. This condition takes place at $\omega_0 \tau \ll 1$, where ω_0 – Larmor precessional frequency for the MRF polar particles, τ – free path of particles average time, electric conductivity – γ is relatively small as there are no free carriers [20, 21].

In the magnetic fields electric charges move along complex screw trajectories. Thus a particle having charge e and mass m, flying into homogeneous magnetic field of voltage H, perpendicular to the particle speed, rotates around the field line at Larmor angular frequency $\omega_0 = e\mu_0 H/m$, where μ_0 – magnetic permeability in vacuum.

The equations set, describing MRF motion in magnetic field coaxial cylindrical gap includes Maxwell's equations:

$$\frac{\partial H}{\partial t} = rot[VH] + v_m \Delta H, \qquad divH = 0,$$

where $v_m = c^2/4\pi\gamma$ – is a magnetic viscosity coefficient; $c = 2,99 \cdot 10^8$ m/s – is an electrodynamic constant, equal to the speed of light in vacuum.

The magnetic viscosity coefficient is the greater the lower the environment electric conductivity.

The environmental motion hydrodynamical equations comprise [18, 20, 21]:
Navier-Stokes generalized equation:

$$\frac{\partial V}{\partial t} + (\nabla)V = -\frac{1}{\rho}gradp - \frac{1}{4\pi\rho}[HrotH] + \frac{\eta}{\rho}\Delta V + \frac{1}{\rho}\left(\xi + \frac{\eta}{3}\right)graddivV;$$

Continuity equation:

$$\frac{\partial \rho}{\partial t} + div(\rho V) = 0;$$

Environment constituent equation:

$$p = p(\rho, T);$$

Energy conservation equation:

$$\frac{\partial}{\partial t}\left(\frac{\rho V^2}{2} + \rho u + \frac{H^2}{8\pi}\right) = -divW,$$

where ρ – environment density, V – the speed of environment motion in the gap between coaxial cylinders, u – specific internal energy, H – outer magnetic field voltage, W – energy-flux density (Umov-Poynting vector).

Dissipation of the energy density flux resultant vector W_p sharply increases, if in MRF flux the chemical equilibrium is upset. Taking into account the rheological filling parameters and the outer magnetic field ones, the resultant energy-flux density can be represented as [18, 20, 21]:

$$W_p = \rho \cdot V\left(u + \frac{p}{\rho} + \frac{V^2}{2}\right) - K\nabla T + \eta\left\{2\left[\left(\frac{\partial V_x}{\partial x}\right)^2 + \left(\frac{\partial V_y}{\partial y}\right)^2 + \left(\frac{\partial V_z}{\partial z}\right)^2\right] + \left(\frac{\partial V_x}{\partial y} + \frac{\partial V_y}{\partial x}\right)^2\right.$$
$$\left. + \left(\frac{\partial V_x}{\partial z} + \frac{\partial V_z}{\partial x}\right)^2 + \left(\frac{\partial V_y}{\partial z} + \frac{\partial V_z}{\partial y}\right)^2 - \frac{2}{3}(div\,V)^2 + \zeta(div\,V)^2\right\}V + \frac{1}{4\pi}[H\,[VH]] - \frac{v_m}{4\pi}[H\,rot\,H],$$

(9)

where K – MRF heat conduction coefficient, η and ζ – the MRF first and the second viscosity coefficients, T – Kelvin temperature.

In Eq. (9) the magnetic field voltage H is measured according to the harmonic law and is directed along the axis z:

$$H = x_0 H_0 e^{-ikx},$$

where k – is a wavenumber, then

$$rot\,H = \frac{\partial^2}{\partial x^2}H = -k^2 H_0 e^{-ikx},$$

$$[H\, rot\, H] = -k^2 H_0^2 e^{-i2kx}.$$

We shall consider MRF motion between MRT coaxial cylinders on condition $div\, V = 0$, as the working environment volume in coaxial choking gap remains constant.

Supposing the magnetic field voltage is orthogonal to the MRF velocity attitude in coaxial gap $H \perp V$ then $[VH] = v_0 H$.

Taking into account these suppositions, we rewrite Eq. (9)

$$W_{p,x=0}| = \rho \cdot v_0 (u + \frac{\Delta p}{\rho} + \frac{v_0^2}{2}) - K\nabla T + \frac{1}{4\pi} H_0^2 v_0 + \frac{v_m}{4\pi} H_0^2, \qquad (10)$$

Alfen waves absorption coefficient β (i.e. of the electromagnetic wave its propagation direction assumed as its envelope velocity direction corresponding to the field direction H) can be found from formula [20, 21]:

$$\beta = \frac{\omega^2}{2U_A^3}(\frac{\eta}{\rho} + v_m), \qquad (11)$$

where ω – is magnetic field frequency, U_A – Alfen waves propagation speed

$$U_A = \frac{H}{\sqrt{4\pi \rho}}, \qquad (12)$$

from Eqs. (11) and (12) express magnetic viscosity v_m

$$v_m = \frac{\beta 2 H^3}{(4\pi \rho)^{3/2}} - \frac{\eta}{\rho}, \qquad (13)$$

By substituting (13) in (10)

$$W_{x=0}| = \rho \cdot v_0 (u + \frac{\Delta p}{\rho} + \frac{v_0^2}{2}) - K\nabla T + \frac{1}{4\pi} H_0^2 v_0 + \frac{1}{2\pi} \beta \frac{H_0^5}{\omega^2 (4\pi \rho)^{3/2}} - \frac{\eta}{4\pi \rho} H_0^2, \qquad (14)$$

From expression (14) it follows that the energy flux density depends on H as quantic functional relation [20, 21].

Let us consider the concrete example. Assuming that:

- Mean effective vibration speed $v_0 = 0,5$ m/s
- Specific internal energy of the environment $u = 30$ J
- MRF heat-conductivity coefficient $K = 1,7$
- Kelvin temperature $0 \leq T \leq 323$ K
- Coaxial gap $2a = 0,25 \cdot 10^{-3}$ m
- Field voltage $H_0 = 2 \cdot 10^5 \sin(\omega t + \varphi(t))$ A/m
- Alfen waves absorption coefficient $\beta = 0,1$

To simplify expression (6) we consider that the fluid temperature in coaxial gap is constant (i.e. $\Delta T = 0$). We also consider that in a first approximation MRF viscosity depends neither on the magnetic field voltage H nor on the flow velocity V between coaxial cylinders (η = const). As the power fluid let us take MRF on the basis of glycerin, at 20 °C. Then $\rho = 1,26 \cdot 10^3$ kg/m^3, $\eta = 1480 \cdot 10^{-3}$ kg/m · sec.

The coaxial gap cross-section area S_κ is determined by the difference of radius squares of inner and outer coaxial surfaces gap

$$S_\kappa = \pi r_1^2 - \pi r_2^2 = \pi r_0^2,$$

where r_1 – inner surface radius of the outer cylinder; r_2 – outer surface radius of the inner cylinder coaxial gap; r_0 – gap between coaxial cylinders.
Then

$$a = (r_1 - r_2)/2$$

and

$$v_0 = \frac{a^2 \Delta p}{4\eta l}(1 - \frac{1}{3})^2 = \frac{a^2 \Delta p}{9\eta l}.$$

The pressure drop in the hydromount $\Delta p = p_1 - p_2$ (Fig. 1) $p_1 = F_0/S_1$, and $p_2 = F_0/S_\kappa$, where F_0 – is the force affecting the hydromount, $S_\kappa = 12,5 \cdot 10^{-6}$ m^2 – the coaxial gap cross-section area, $S_1 = 36,7 \cdot 10^{-3}$ m^2 – the hydromount working chamber lower section area.

Let the action force affecting the hydromount change according to the harmonic law at the frequency being equal the magnetic field frequency, so $F = F_0 \sin(\omega t)$:

$$\Delta p = \frac{F_0}{S_1}\sin(\omega t) - \frac{F_0}{S_{kah}}\sin(\omega t), \tag{15}$$

Action force $F_0 = 1000$ N.

Taking into consideration all above, let us build the dependency of energy flux density in coaxial choking channel of the hydromount with MRF $W_p(H,t)|_{x=0}$ at the magnetic field frequency $\omega = 2\pi \cdot 25$ [rad/sec] and Alfen waves absorption coefficient $\beta = 0,1$.

Graphical interpretation of energy flux density variation on a coaxial cylindrical gap 8 shown in Fig. 5.

From the diagram as shown in Fig. 5 it follows that at a time span t = 0,04 s. The pressure in a cylindrical gap between coaxial cylinders is measured within the range from zero to one MPa. The energy flux density at this changes from от 3×10^{12} up to 4×10^{12} [V/m^2] with the frequency of 625 Hz.

At the input vibration frequency increase there appears frequency modulation of Umov-Poynting vector, thus the damping process is enriched by high-frequency components being actively absorbed in viscous magnetorheological environment. In this way rotating magnetic field in a coaxial choking channel contributes to more

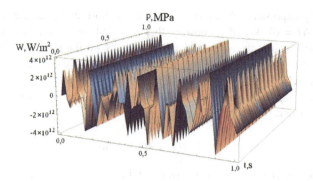

Fig. 5. The magnetic field energy flux density dependency on time and pressure at $v_0 = 0{,}5$ m/s; $f = 25$ Hz and $T = 30$ °C.

effective damping. It also follows that in the cylindrical gap there appear hydraulic pressure high-frequency pulsations in the working MRF, which destroy MRF magnetic particles sedimentation.

4 Conclusion

The MRF choking processes in coaxial cylindrical gap and the main choking channel are of different physical nature.

In the main choking channel, MRF flux control is carried out by the synchronous magnetic field inner electric magnet 14 (Fig. 1) its output vibration frequency being [22].

In the coaxial cylindrical gap, MRF flux control is carried by the rotating magnetic field of the outer one-phase inductor, independent of input vibration signal frequency.

When increasing the input vibration signal frequency there appears Umov-Poynting vector frequency modulation, thus the damping process is enriched by high-frequency components being actively absorbed in viscous magnetorheological environment. Herewith rotating magnetic field in a coaxial choking channel contributes to more effective damping. Also in the cylindrical gap there appear hydraulic pressure high-frequency pulsations in working MRF, which destroy MRF magnetic particles sedimentation.

As a result the given construction functioning safety and stability are boosted by these two processes.

Acknowledgements. The presented research results were obtained with the support of grants from the President of the Russian Federation for state support of young Russian scientists (MK-590.2018.8).

References

1. Gordeev, B.A., Erofeev, V.I., Sinev, A.V., et al.: Vibratin protection systems with the rheological media time lag and dissipation. Fizmatlit, Moscow (2004)
2. Gordeev, B.A., Erofeev, V.I., Plekhov, A.S.: Mathematical models of mobile and stationary objects adaptive vibration isolators. N. Novgorod State Technical University, Nizhniy Novgorod (2017)
3. Gordeev, B.A., Okhulkov, S.N., Plekhov, A.S., et al.: Magnetorheological fluids application in machine building. Privolzhsky Sci. J. **4**, 29–42 (2014)
4. Gordeev, B.A., Okhulkov, S.N., Plekhov, A.S., et al.: Magnetorheological fluid flux and relaxation in hydromounts choking channels. Bull. Mach. Build. **7**, 32–38 (2015)
5. Kazakov, U.B., Morozov, N.A., Nesterov, S.A.: Magnetorheological damper with piston-type magnetic system. Bull. Ivanovo State Power Univ. **6**, 1–6 (2012)
6. Gordeev, B.A., Maslov, V.G., Okhulkov, S.N., et al.: The issue of creating cylindrical magnetorheological transformer in orthogonal magnetic fields. Probl. Mach. Build. Mach. Reliab. **2**, 15–21 (2014)
7. Gordeev, B.A., Erofeev, V.I., Sinev, A.V.: Inertial electrical rheological transformers application in vibration isolation systems. Probl. Mach. Build. Mach. Reliab. **6**, 22–27 (2003)
8. Morozov, N.A., Kazakov, U.B.: Nanodispersed magnetic fluids in engineering and technology. Ivanovo State Power University, Ivanovo (2011)
9. Taketomi, S., Tikadzumi, S.: Magnetic Fluids. Mir, Moscow (1993)
10. Shulman, Z.P., Kordonsky, V.I.: Magnetorheological Effect. Science and Techniques, Minsk (1982)
11. Reicher, V.V.: Rotational viscosity of visco-elastic magnetic fluid. Colloid J. **70**, 85–92 (2008)
12. Belyaev, E.S., Ermolaev, A.I., Titov, E.U., et al.: Technology of magnetorheological fluids creation and usage for the controlled vibration isolators: monograph. N. Novgorod State Technical University, Nizhniy Novgorod (2017)
13. Neigert, K.V., Rednokov, S.N.: Magnetorheological drive of direct electromagnetic control of the upper hydraulic system valve contour flux characteristics. RU Patent 2634163, 16 March 2017 (2017)
14. Neigert, K.V., Rednikov, S.N.: Modular system of electromagnetic fluids transportation having magnetic characteristics. RU Patent 2624082, 30 June 2017 (2017)
15. Neigert, K.V., Rednikov, S.N.: Magnetorheological drive of direct electromagnetic control over the flux characteristics of the upper hydraulic system valve contour with a hydraulic bridge. RU Patent 2634166, 24 October 2017 (2017)
16. Bruskin, D.E., Zorokhovich, A.E., Khvostov, V.S.: Electrical machines and micromachines. High School, Moscow (1990)
17. Ivanov-Smolensky, A.V.: Electrical machines. Energy, Moscow (1980)
18. Loyciansky, L.G.: Fluid and gaz mechanics. Science, Moscow (1978)
19. Bashta, T.M.: Machine building hydraulics. Mashinostroenie, Moscow (1971)
20. Gordeev, B.A., Morozov, P.N., Sinev, A.V.: The outer magnetic field influence on the energy density flux in a magnetorheological transformer. Prob. Mach. Build. Mach. Reliab. **4**, 100–104 (2004)
21. Gordeev, B.A., Morozov, P.N., Sinev, A.V.: Magnetic fields influence on the fluids flux distribution speed in the channel of magnetorheological transformer square-section. Prob. Mach. Build. Mach. Reliab. **1**, 89–93 (2006)
22. Gordeev, B.A., Sinev, A.V., Kuplinova, G.S.: Hydraulic vibration mount patent of invention. RU Patent 2407029, 20 December 2010 (2010)

Increasing Precision of Absolute Position Measurement in NC Machine Tools by Means of Diminishing Inertia Loads in Measurement Unit

Ya. L. Liberman[✉] and K. Yu. Letnev

Institute of New Materials and Technologies, Ural Federal University,
19 Mira Street, Ekaterinburg 620002, Russian Federation
yakov_liberman@list.ru

Abstract. The paper describes the research and its results aimed at increasing precision of absolute position measurement in numerically controlled (NC) machine tools and preventing machining errors. The precision of that measurement unit depends on the inertia of its component parts. When abrupt braking and follow-up stopping occurs, any part of the measurement system could be subjected to torsional vibrations caused mainly by inertia of the encoder disk and transmission gear and leading to significant measurement and machining errors. The increase in precision was achieved by means of diminishing those inertia loads via developing special designs of the position encoder and the transmission between the encoder and the tool leadscrew. To decrease the inertia of the encoder, the mass of its disk was lowered by substituting the multi-track Gray-code scale with a one-track scale of one-variable group. On the other hand, a high-ratio, no-slack and low-weight gear of the wave or planetary type was proposed as such transmission which could lead to lower inertia loads there and a higher precision in general.

Keywords: Machine tools · Feed drive · Inertia loads · Position measurement · One-track position encoder · No-slack transmission gear

1 Introduction

Nowadays, a tool position encoder is often used in numerically controlled (NC) machine tools as an absolute position measurement unit [1–4]. The encoder uses transmission gear to connect with a machine tool leadscrew or rack-and-pinion gear [5–8]. Precision of such a measurement system is determined by precision of the screw, precision of the encoder and precision of the transmission gear. But, in many ways, it also depends on the inertia of component parts. In the case of abrupt braking of the feed drive with follow-up stopping, any part of the measurement system could be subjected to torsional vibrations. If the amplitude of those vibrations exceed resolution of the measurement, inaccuracies are inevitable. And those lead to significant machining errors.

The inertia loads in our measurement unit are mainly caused by the encoder disk (the encoder's main element) and transmission gear. Therefore, the most efficient way of preventing such errors and increasing the precision of position measurement would be to decrease inertia of the disk and the gear themselves.

2 The Goal of the Research and Its Tasks

The goal of the research is to increase precision of absolute position measurement in NC machine tools and prevent machining errors.

The tasks solved in the research:

- Developing such a design of the position encoder which would allow to decrease inertia loads in the measurement unit by means of diminishing the mass of the encoder disk
- Developing such a design of the transmission between the position encoder and leadscrew which would allow to decrease inertia loads in the measurement unit by means of implementing a high-ratio, no-slack and low-weight gear

3 Solution of the Research Tasks

It is known that modern position encoders use Gray-coded scale plates [9] (one such scale plate is shown in Fig. 1a). The usage is justified by the fact that in Gray code two successive values differ in only one bit (binary digit) and, thus, only one bit changes at a time, which prevents boundary readout errors. But there is one significant drawback with such a Gray-code scale – its too many tracks lead to the mass of the disk being quite large, and, therefore, the inertia of the scale becoming too high.

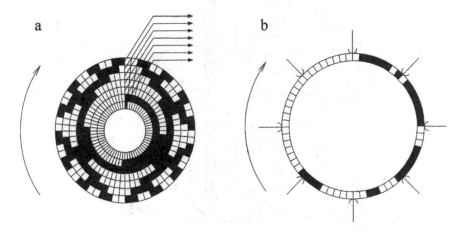

Fig. 1. Typical code scale: (a) 7-digit Gray code; (b) one-track group code.

But that mass could be diminished by substituting the Gray-code scale with a one-track scale of one-variable group code (Fig. 1b). The scale is built on the principle of calculating the number of combinations in the full set of codes, and selecting, among those combinations, such full groups which are formed by cycle-shifting some of the basic combinations [10]. Then, by choosing only one combination from each group, a code dial with a Hamming distance appropriate for us is constructed. One-variable code is characterized by the Hamming distance which is equal to one.

If – when building a code dial – we take this fact into account, we are going to get a one-track scale with a one-variable group code. The mass of such a scale would be much lesser in comparison with a multitrack Gray-coded scale, and, therefore, the inertia loads triggered by the scale – smaller.

On the basis of the stated above, a one-track scale design of the position encoder was developed [11]. The design is shown in Fig. 2, where the encoder's three-part body (its bottom 3, middle part 4, and a cover 11) contains two needle bearings 2, which house a shaft to be coded 1 and a coding scale 8 attached to the shaft. The cover 11 has an irradiating block inside, which includes a light-emitting diode (LED) lamp 13 and a reflector 12. The receiving block consists of a prism 10, which lets in the light via transparent sections and a ring 9 with holes whose axes coincide with those of photodiodes 7. The middle part of the body 4 (in between the photodiodes 7 and bottom 3)

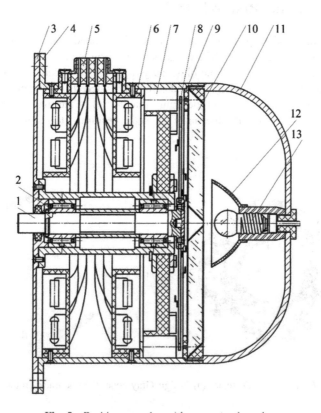

Fig. 2. Position encoder with a one-track scale.

houses two laminated plates 6 with electronic circuits of the position encoder mounted on them. The outputs of the circuitry are lead out to a socket connector 5, and the LED lamp is powered separately.

Let us consider the second source of inertial loads – transmission between the position encoder and leadscrew. This transmission gear should have a high ratio and no slack. In such measurement systems which are based on position encoders, a worm gear is mainly used; but a wave gear or a planetary gear are also utilized to function as the transmission gear. The latter two are of interest to us.

Figure 3 shows a no-slack wave gear [12]. Slack in it is dealt with as follows.

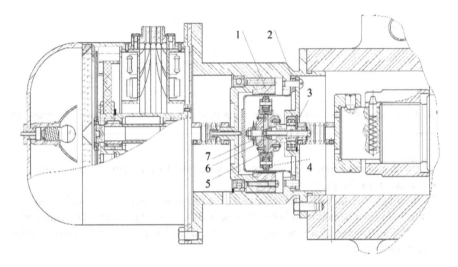

Fig. 3. Connecting the position encoder to a machine tool leadscrew by means of a wave gear.

When a shaft 3 rotates, rollers 4 move waves deforming a flexible wheel 2 and make a rigid wheel 1 rotate. If the transmission causes any wear in the toothed coupling and some slack appears, springs 7 will make the slack disappear by moving a cone 6 to the left and using pushers 5. It is going to happen automatically, without any human intervention. The transmission gear in the form of such no-slack wave gear could list low lag as its major advantage. Despite this fact, as with all wave transmissions, its flexible wheel has quite limited life-time.

To provide both a longer time and low-inertia transmission, a special design of no-slack planetary reduction gear was developed (Fig. 4).

Its body 7 has a cover 13 fastened to it, and the cover houses a planetary carrier 9 and central wheel 12 on roller bearings 10 and 11. The movement is carried from a central wheel 12 to a two-wheel satellite 5 mounted on roller bearings 6 within the planetary carriers 9 and 17. Those are tightened by a screw 14. One wheel rim rolls around the central wheel 12 using external gearing, and the second rim uses internal gearing (rigidly connected to the body 7) to roll around a central wheel 4. The planetary carrier 17 is mounted on roller bearings 16 within a cover 3 and have its internal teeth

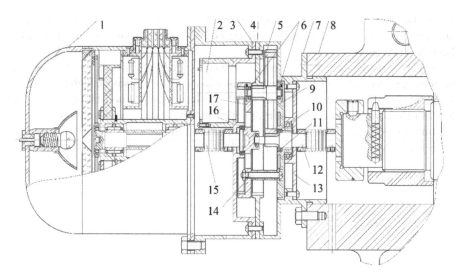

Fig. 4. Connecting the position encoder to a machine tool leadscrew by means of a planetary gear.

coupled with a torque motor 2. This motor is mounted in the cover 3 and eliminates slack appearing in the gear.

To diminish inertia loads, the wall width for the gear's elements is minimized, and a satellite 5 is implemented with holes, thus decreasing the total weight. The connection between the leadscrew of a machine tool 8 and a position encoder 1 is made using bellows couplings 15.

4 Testing of the Proposed Designs

The designed devices were tested using the methodology set by the GOST standard on positioning accuracy check (medium-inertia rundown, and its dispersion) [13, 14]. The results of the testing were processed by means of traditional approaches of mathematical statistics [15], which showed that the precision of a position measurement system based on the encoder proposed and a wave gear is 3 times higher than it is in the case of a system with a Gray-code encoder and a worm gear; similarly, it is 2 times higher when a measurement system with the same encoder and a planetary gear is compared with the one integrating a Gray-code encoder and a worm gear.

The proposed encoder along with one of the mechanisms developed to connect it to the machine tool leadscrew (a wave gear, in particular) was also tested in a high-speed lathe designed at the Department of Metal-Cutting Machines and Tools of the Ural Federal University. That lathe is intended for external machining of stepped cylinders used in precision instruments to measure machine parts. Such cylinders could be up to

15 mm in diameter and 160 mm in length. The number of steps made on the surface of a cylinder reaches up to 6 or 7 ones, and the error of their positioning should not exceed 0,005 mm. The lathe has a position-encoding NC-system based on coincidence circuits made from equivalence elements and a Sheffer stroke gate (NAND gate) (Fig. 5).

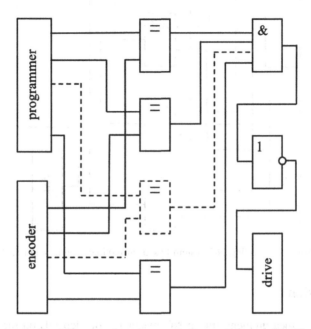

Fig. 5. Position-encoding NC-system based on equivalence elements and a Sheffer stroke gate (NAND gate).

When such a lathe is in operation, its tool-holding head can move with the speed of up to 2 m/s stopping almost without prior breaking with a precision enough to make required steps. As it was found, their end faces would deviate from programmed positions no more than the tolerance mentioned above allow, which confirms the rationality of utilizing the engineering solutions described here.

The next step in integration of the encoder with the wave gear into actual manufacturing processes has been recently made at the Department of Metal-Cutting Machines and Tools where a precision jig-boring machine with numeric adaptive control is designed. It features a boring bar as its work member, which is capable of moving along three directions: axial one and two directions perpendicular to it. All tool position encoders for those three axes have similar designs to the one in Fig. 3. Their coincidence circuits are based on modulo-2 adders and an OR gate (Fig. 6). As to adaptive elements, they are unique, based on piezoceramic detectors of vibration, which allows to eliminate almost completely vibrations of the boring bar when deep apertures are bored.

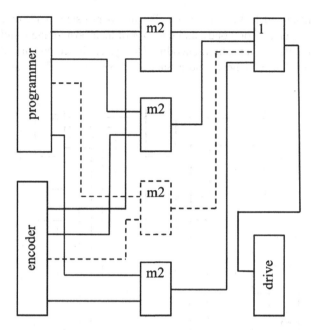

Fig. 6. Position-encoding NC-system based on modulo-2 adders and an OR gate.

5 Conclusion

Precision of the measurement unit in NC machine tools depends on the inertia of its component parts. When abrupt braking and follow-up stopping occurs, any part of the measurement system could be subjected to torsional vibrations caused mainly by inertia of the encoder disk and transmission gear. Those vibrations lead to significant measurement and machining errors and should be somehow dealt with.

It could be achieved by substituting the multi-track Gray-code scale of the encoder disk with a one-track scale of one-variable group code and, thus, diminishing its mass and inertia.

On the other hand, a high-ratio, no-slack and low-weight gear could be used as transmission between the position encoder and leadscrew, leading to lower inertia loads there. A wave gear or a planetary gear are proposed to be utilized as such. Those should have special no-slack transmission designs, with low lag being the major advantage of the former, and lesser weight and longer life-time – main features of the latter.

The designed devices were tested, showing that the precision of a position measurement system is 2 or 3 times higher when a system with the encoder proposed and a wave or planetary gear is compared with the one which includes a Gray-code encoder and a worm gear.

References

1. Liberman, Y.L., Plyuhina, E.A.: Feed drive for metal-cutting machine tools. RU Patent 96511, 29 December 2009 (2009)
2. Liberman, Y.L., Kuchin, S.G., Chepusova, E.Y.: Servo feed drive for metal-cutting machine tools. RU Patent 160849, 29 April 2015 (2015)
3. Asinovskiy, E.N., Ahmetzhanov, A.A., Gabidulin, M.A.: High-Precision Converters of Angular Displacements. Energy and Nuclear Industry Publishing, Moscow (1986)
4. Golubovskiy, Yu.M., Pivovarova, L.N., Afanasieva, Zh.K.: Photoelectric converters of linear and angular displacements. Opt.-Mech. Ind. **8**, 50–58 (1984)
5. Liberman, Y.L., Kuchin, S.G., Chepusova, E.Y.: Tool position measurement system in machines. RU Patent 154592, 20 February 2015 (2015)
6. Levashov, A.V.: Basics of Kinematic Chain Accuracy Estimation. Machine Building, Moscow (1966)
7. Margolit, R.B.: Usage and Setup of NC Machine Tools and Industrial Robots. Machine Building, Moscow (1991)
8. Pronikov, A.S.: Precision and Reliability of NC Machine Tools. Machine Building, Moscow (1982)
9. Ratmirov, V.A.: Controlling Machine Tools within Flexible Manufacturing Systems. Machine Building, Moscow (1987)
10. Sharin, Y.S., Liberman, Y.L.: Combinatorial Scales in Automation Systems. Energy, Moscow (1973)
11. Liberman, Y.L., Gordeeva, A.S.: Shaft-rotation angle-to-code converter. RU Patent 110512, 07 June 2011 (2011)
12. Liberman, Y.L., Plyuhina, E.A.: Wave gear. RU Patent 99090, 04 March 2010 (2010)
13. GOST 27843-88. Metal-cutting machine tools. Methods of positioning accuracy check
14. Reshetov, D.N., Portman, V.T.: Precision of Metal-Cutting Machine Tools. Machine Building, Moscow (1986)
15. Dunin-Barkovskiy, I.V., Kartashova, A.I.: Measurement and Analysis of Surface Roughness, Waviness and Nonroundness. Machine Building, Moscow (1978)

Calculating Lifting Capacity of Industrial Robot for Flexible Lathe System at the Design Stage

Ya. L. Liberman[✉] and K. Yu. Letnev

Ural Federal University, 19 Mira Street, Ekaterinburg 620002,
Russian Federation
yakov_liberman@list.ru

Abstract. The research was focused on the problem of increasing the efficiency of using an industrial robot within a flexible manufacturing system (FMS). It is postulated that change in the range of workpieces to be processed by that FMS negatively influences the efficiency in question. A special approach was proposed, which would allow one to select the robot at the design stage taking the capabilities of both the robot and FMS into account. A standard formula for calculating the lifting capacity of the robot was considered and altered to factor in the hollow space within a workpiece and its maximal calculated dimensions. Statistical data gathered from several general engineering companies and analyzed during the research helped to obtain the functions of diameters and lengths distribution for specific lathes and determine the fraction of workpieces in the whole manufacturing range whose dimensions do not exceed maximal calculated workpiece dimensions. Knowing the functions and the fraction, the maximal calculated workpiece dimensions and lifting capacity could be determined for any lathe. A software system developed on the basis of the method is briefly described, which suggests the lifting capacity and, thus, helps to select an appropriate robot by the catalogue or formulate a requirements specification for such a robot to be designed.

Keywords: Flexible manufacturing system · Lathe · Industrial robot · Lifting capacity · Workpiece dimensions · Hollow space · Diameters and lengths distribution

1 Introduction

Nowadays, selection of an industrial robot for a flexible manufacturing system (FMS) at the design stage usually depends on the range of workpieces to be processed by that FMS [1–3]. But when the range changes within the manufacturing capabilities of the FMS, the efficiency of using the selected robot could drastically decrease. To prevent this, it is necessary to make the capabilities of the robot and FMS agree with each other.

2 The Goal of the Research and Its Tasks

The goal of the research is to increase the efficiency of using an industrial robot within a flexible manufacturing system taking the capabilities of both the robot and FMS into account.

The tasks solved in the research:

- Formulating an approach which could held to select a more efficient and lathe-compatible robot at the design stage
- Altering a standard formula for calculating the lifting capacity of the robot to factor in the hollow space within a workpiece and its maximal calculated dimensions
- Developing a software system which could be used to calculate the lifting capacity and would help to select or design an appropriate robot

3 Solution of the Research Tasks

The easiest way to allow for the capabilities of both an industrial robot and the FMS it is supposed to be used within is to calculate the lifting capacity of the robot using the following formula:

$$Q = 0.25 \cdot \rho \cdot \pi \cdot D_{\max}^2 \cdot L_{\max}, \qquad (1)$$

where ρ is density of steel, D_{max} and L_{max} are maximal diameter and length of workpieces which could be processed by the FMS [4, 5].

But, since it is rare for any FMS to be used in its whole manufacturing range of workpiece dimensions [6, 7], and considering that (1) does not factor in the hollow space (cavities or holes) within a workpiece, it was proposed to calculate the lifting capacity using this equation:

$$Q = 0.25 \cdot \rho \cdot \pi \cdot K \cdot D_c^2 \cdot L_c, \qquad (2)$$

where K is coefficient which factors in the hollow space (coefficient of filling); D_c and L_c are maximal calculated workpiece dimensions ($D_c < D_{max}$ and $L_c < L_{max}$).

To determine dimensions and coefficient K, statistical data gathered from 10 general engineering companies were analyzed as part of the research carried out in the Ural Federal University. Two facts were taken into consideration: namely, that chuck lathes process, as a rule, workpieces with the length-to-diameter ratio $L/D < 4$, and chuck-and-center lathes – with $L/D > 0$ or an arbitrary ratio [8]. The research – which employed methods described in [9, 10] – helped to obtain the relations of $K = f(D)$ for confidence probabilities $P_K = 0.85$, 0.90 and 0.95 (Table 1 and Fig. 1) and plot histograms (bar graphs) of D and L dimensions distribution for workpieces with $L/D < 4$ and $L/D > 0$.

Table 1. Correlations between metal-filling coefficient K and diameter D.

L/D	P_K	Correlation between K and D	Correlation coefficient
<4	0,85	$K = 0,8855 - 0,0004448 \cdot D$	−0,7780
	0,90	$K = 0,9069 - 0,0003313 \cdot D$	−0,8147
	0,95	$K = 0,9270 - 0,0001908 \cdot D$	−0,9746
>0	0,85	$K = 0,8985 - 0,0004627 \cdot D$	−0,8860
	0,90	$K = 0,9185 - 0,0003463 \cdot D$	−0,9070
	0,95	$K = 0,9419 - 0,0002296 \cdot D$	−0,9184

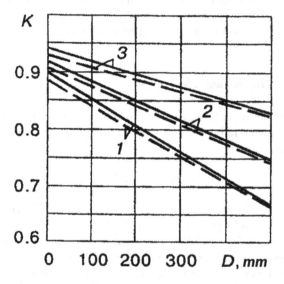

Fig. 1. Correlations between metal-filling coefficient K and diameter D for chuck (dotted line) and chuck-and-center (solid line) lathes.

The statistical analysis of those histograms gave the following functions of the diameters and lengths distribution:
for $L/D < 4$:

$$\varphi(D) = 0,138 \cdot D^{0.163} \cdot \exp(-0,0102 \cdot D);$$
$$\varphi(L) = 0,9323 \cdot L^{-0.1922} \cdot \exp(-0,00953 \cdot L), \quad (3)$$

for $L/D > 0$:

$$\varphi(D) = 0,17495 \cdot D^{0.1734} \cdot \exp(-0,0109 \cdot D),$$
$$\varphi(L) = 0,972 \cdot L^{-0.2787} \cdot \exp(-0,00578 \cdot L). \quad (4)$$

To understand the principle and procedure behind the calculation of D_c and L_c, one should consider the curves described by the functions $\varphi(D)$ and $\varphi(L)$, which are assumed to be sufficiently precise. Since those curves characterize the workpieces distribution regardless of the lathe model, it would be appropriate to assume that, in a general case, dimensions of workpieces processed by lathes of a specific model follow the laws of distribution $\varphi'(D)$ and $\varphi'(L)$ which could be derived from the functions $\varphi(D)$ and $\varphi(L)$ by truncating them [11]:

$$\varphi'(D) = \frac{\varphi(D)}{\int_{D_{min}}^{D_{max}} \varphi(D)dD}; \quad \varphi'(L) = \frac{\varphi(L)}{\int_{0}^{L_{max}} \varphi(L)dL}, \quad (5)$$

Knowing $\varphi'(D)$ and $\varphi'(L)$ for a specific lathe, it is not very difficult to determine that the portion of those workpieces in the whole manufacturing range of this lathe whose maximal dimensions do not D_c and L_c could be expressed as:

$$P_D = \int_{D_{min}}^{D_c} \varphi'(D)dD, \quad P_L = \int_{0}^{L_c} \varphi'(L)dL, \quad (6)$$

It is known [12] that $D_{min} \approx D_{max}/8.5$. Therefore, if $\varphi(D)$ and $\varphi(L)$ are specified and $P_{D(L)}$ are known, then, using (5) and (6), one could find D_c and L_c for any lathe and determine the lifting capacity of an industrial robot via (2). The value obtained should be rounded up to the next standard lifting capacity for industrial robots.

$P_{D(L)}$ are evaluated by considering the following. It is obvious that, for $D_c = D_{max}$ and $L_c = L_{max}$, the fraction of workpieces which could be in principle processed by the lathe in question and carried by a robot of the capacity calculated using (2) would be equal to P_K. If $D_c < D_{max}$ and $L_c < L_{max}$, this fraction will be no less than $P_K \cdot P_{D(L)}^2$. It follows that, if a precise fraction of workpieces carried by the robot is known, one could assume:

$$P_{D(L)} = \sqrt{\frac{P}{P_K}}, \quad (7)$$

It should be noted that, for practical purposes, special functions would be more appropriate instead of functions $\varphi(D)$ and $\varphi(L)$. Those special ones are specified in a table form (Tables 2 and 3) and ensure greater precision when calculating D_c and L_c in application to specific lathe types.

Knowing the capacity of the robot, its guiding price C might be assessed. For example, according to [13], that price for a one-arm manipulator could be calculated as

$$C = 1783 \cdot K_C \cdot Q^{0.415} \cdot N^{0.916},$$

where K_C – coefficient allowing for inflation; N – degrees of robot freedom (for a lathe, it is necessary and sufficient to have $N = 4$).

The price of a similar two-armed manipulator would be two times higher.

Table 2. Expectancy of workpieces being within a specified diameter range based on the statistics and analysis (distribution of workpiece diameters).

Range of D, mm	Range-expectancy for workpieces with		Range of D, mm	Range-expectancy for workpieces with	
	L/D < 4	L/D > 0		L/D < 4	L/D > 0
0–5	0.0105	0.0025	225–230	0.0063	0.0044
5–10	0.0183	0.0104	250–235	0.0060	0.0043
10–15	0.0228	0.0192	235–240	0.0057	0.0040
15–20	0.0257	0.0309	240–245	0.0053	0.0038
20–25	0.0282	0.0336	245–250	0.0050	0.0035
25–30	0.0297	0.0353	250–255	0.0047	0.0032
30–35	0.0305	0.0361	255–260	0.0045	0.0031
35–40	0.0312	0.0364	260–265	0.0043	0.0029
40–45	0.0313	0.0388	265–270	0.0040	0.0028
45–50	0.0315	0.0368	270–275	0.0038	0.0025
50–55	0.0312	0.0363	275–280	0.0035	0.0024
55–60	0.0305	0.0357	280–285	0.0033	0.0023
60–65	0.0293	0.0340	285–290	0.0030	0.0022
65–70	0.0283	0.0325	290–295	0.0029	0.0021
70–75	0.0272	0.0308	295–300	0.0028	0.0020
75–80	0.0262	0.0292	300–305	0.0025	0.0019
80–85	0.0250	0.0276	305–310	0.0023	0.0018
85–90	0.0243	0.0261	310–315	0.0022	0.0017
90–95	0.0233	0.0247	315–320	0.0021	0.0015
95–100	0.0225	0.0235	320–325	0.0020	0.0015
100–105	0.0215	0.0222	325–330	0.0018	0.0014
105–110	0.0207	0.0207	330–335	0.0017	0.0014
110–115	0.0200	0.0196	335–340	0.0015	0.0014
115–120	0.0193	0.0186	340–345	0.0013	0.0013
120–125	0.0183	0.0176	345–350	0.0013	0.0013
125–130	0.0178	0.0167	350–355	0.0012	0.0011
130–135	0.0170	0.0158	355–360	0.0012	0.0011
135–140	0.0162	0.0150	360–365	0.0010	0.0010
140–145	0.0155	0.0142	365–370	0.0010	0.0010
145–150	0.0148	0.0136	370–375	0.0009	0.0009
150–155	0.0142	0.0128	375–380	0.0009	0.0008
155–160	0.0137	0.0121	380–385	0.0008	0.0008
160–165	0.0130	0.0114	385–390	0.0008	0.0008
165–170	0.0125	0.0108	390–395	0.0007	0.0008
170–175	0.0120	0.0101	395–400	0.0007	0.0007
175–180	0.0113	0.0096	400–405	0.0006	0.0007
180–185	0.0107	0.0090	405–410	0.0006	0.0007

(*continued*)

Table 2. (*continued*)

Range of D, mm	Range-expectancy for workpieces with		Range of D, mm	Range-expectancy for workpieces with	
	L/D < 4	L/D > 0		L/D < 4	L/D > 0
185–190	0.0102	0.0083	410–415	0.0006	0.0007
190–195	0.0097	0.0078	415–420	0.0006	0.0007
195–200	0.0090	0.0072	420–425	0.0006	0.0006
200–205	0.0085	0.0067	425–430	0.0006	0.0006
205–210	0.0080	0.0063	430–435	0.0005	0.0006
210–215	0.0075	0.0057	435–440	0.0005	0.0006
215–220	0.0072	0.0053	440–445	0.0005	0.0006
220–225	0.0068	0.0050	445–450	0.0005	0.0006

Table 3. Expectancy of workpieces being within a specified length range based on the statistics and analysis (distribution of workpiece lengths).

Range of L, mm	Range-expectancy for workpieces with		Range of L, mm	Range-expectancy for workpieces with	
	L/D < 4	L/D > 0		L/D < 4	L/D > 0
0–10	0.1771	0.1344	500–510	0.0006	0.0020
10–20	0.1186	0.1125	510–520	0.0005	0.0019
20–30	0.0882	0.0861	520–530	0.0005	0.0018
30–40	0.0754	0.0746	530–540	0.0005	0.0017
40–50	0.0614	0.0621	540–550	0.0004	0.0016
50–60	0.0564	0.0539	550–560	0.0004	0.0016
60–70	0.0511	0.0475	560–570	0.0004	0.0015
70–80	0.0454	0.0411	570–580	0.0003	0.0014
80–90	0.0389	0.0361	580–590	0.0003	0.0014
90–100	0.0354	0.0325	590–600	0.0003	0.0013
100–110	0.0279	0.0293	600–610	0.0003	0.0013
110–120	0.0229	0.0264	610–620	0.0003	0.0013
120–130	0.0193	0.0232	620–630	0.0002	0.0012
130–140	0.0168	0.0207	630–640	0.0002	0.0012
140–150	0.0143	0.0182	640–650	0.0002	0.0011
150–160	0.0121	0.0161	650–660	0.0002	0.0011
160–170	0.0104	0.0143	660–670	0.0002	0.0010
170–180	0.0093	0.0125	670–680	0.0002	0.0010
180–190	0.0089	0.0111	680–690	0.0002	0.0009
190–200	0.0079	0.0100	690–700	0.0002	0.0009
200–210	0.0071	0.0089	700–710	0.0002	0.0009

(*continued*)

Table 3. (*continued*)

Range of L, mm	Range-expectancy for workpieces with		Range of L, mm	Range-expectancy for workpieces with	
	$L/D < 4$	$L/D > 0$		$L/D < 4$	$L/D > 0$
210–220	0.0064	0.0079	710–720	0.0002	0.0009
220–230	0.0061	0.0071	720–730	0.0001	0.0008
230–240	0.0057	0.0068	730–740	0.0001	0.0008
240–250	0.0054	0.0061	740–750	0.0001	0.0007
250–260	0.0052	0.0057	750–760	0.0001	0.0007
260–270	0.0050	0.0054	760–770	0.0001	0.0007
270–280	0.0045	0.0050	770–780	0.0001	0.0006
280–290	0.0043	0.0048	780–790	0.0001	0.0006
290–300	0.0039	0.0046	790–800	0.0001	0.0006
300–310	0.0038	0.0043	800–810	0.0001	0.0005
310–320	0.0036	0.0041	810–820	0.0001	0.0005
320–330	0.0032	0.0039	820–830	0.0001	0.0005
330–340	0.0029	0.0037	830–840	0.0001	0.0004
340–350	0.0028	0.0035	840–850	0.0001	0.0004
350–360	0.0027	0.0034	850–860	0.0001	0.0004
360–370	0.0025	0.0032	860–870	0.0001	0.0004
370–380	0.0024	0.0030	870–880	0.0001	0.0003
380–390	0.0021	0.0029	880–890	0.0001	0.0003
390–400	0.0020	0.0028	890–900	0.0001	0.0003
400–410	0.0018	0.0027	900–910	0.0001	0.0003
410–420	0.0017	0.0026	910–920	0.0001	0.0002
420–430	0.0014	0.0025	920–930	0.0001	0.0002
430–440	0.0013	0.0025	930–940	0.0001	0.0002
440–450	0.0012	0.0025	940–950	0.0001	0.0001
450–460	0.0011	0.0023	950–960	0.0001	0.0001
460–470	0.0010	0.0023	960–970	0.0001	0.0001
470–480	0.0008	0.0022	970–980	0.0001	0.0001
480–490	0.0007	0.0021	980–990	0.0001	0.0001
490–500	0.0006	0.0020	990–1000	0.0001	0.0001

On the basis of the method described here, a software system for the purposes of selecting an industrial robot and designing a flexible manufacturing system was developed in the Ural federal university [14, 15]. It allows to calculate the lifting capacity and guiding price of a robot servicing a lathe-based FMS. The system is written in Visual Basic and packed into one executable file. Through the dialogue-based graphical interface, the user should specify as its input the lathe type (chuck or chuck-and-center), maximal length and diameter of workpieces to be processed by the lathe, guaranteed fraction of the processed workpieces to be carried by the robot, and the number of robot arms.

The output contains the results of computation suggesting the robot's lifting capacity and guiding price, citing other interim calculation data for a reference. These results could be used to select an appropriate robot by catalogue or formulate a requirements specification for such a robot to be designed, and the whole module is ready to be a part of some larger PLM system.

4 Conclusion

When the range of workpieces to be processed by a flexible manufacturing system (FMS) changes, it negatively influences the efficiency of using an industrial robot within that FMS. A more rational approach would allow to select the robot at the design stage accounting for the capabilities of both the robot and FMS. The proposed alteration of a standard formula for calculating the lifting capacity of the robot factors in the hollow space within a workpiece and its maximal calculated dimensions. To determine the dimensions and coefficient of the hollow space, statistical data should be analyzed to find the functions of diameters and lengths distribution for specific lathes and determine the fraction of workpieces in the whole manufacturing range whose dimensions do not exceed maximal calculated workpiece dimensions. Knowing the functions and the fraction, the maximal calculated workpiece dimensions and lifting capacity could be determined for any lathe. For practical purposes and greater precision, special functions in a table form would be more appropriate.

A software system developed on the basis of the method should be supplied by its user with the lathe type, maximal length and diameter of workpieces to be processed by the lathe, guaranteed fraction of the processed workpieces to be carried by the robot, and the number of robot arms. The output contains the results of computation suggesting the robot's lifting capacity and guiding price. Those results could be used to select an appropriate robot or formulate a specification for such a robot to be designed, and the whole module is ready to be included into a larger PLM system.

References

1. Kozyrev, Y.G.: Industrial robots. Machine building, Moscow (1988)
2. Russian National Research Institute of Information: Technology and econonic research on machine building and robotics using electromechanical industrial robots with adaptive control, Moscow (1986)
3. Shurkov, V.N.: Basics of automated manufacturing and industrial robots. Machine building, Moscow (1989)
4. Kadyrov, Zh.N., Esenbaev, M.T.: Technological basics of robotic centers and flexible manufacturing systems. NMK, Almaty (1985)
5. Leshchenko, V.A., Kiselev, V.M., Kupriyanov, D.A.: Flexible manufacturing systems. Machine building, Moscow (1984)
6. Azbel, V.O., Egorov, V.A., Zvonitskiy, A.Yu.: Flexible automated manufacturing. Machine building, St. Petersburg (1985)

7. Timofeev, A.V.: Adaptive robotic centers. Machine building, St. Petersburg (1988)
8. Kozyrev, Y.G., Anishin, S.S., Velikovich, V.B.: Using industrial robots in machining processes. Experimental Research Institute of Machine Tools, Moscow (1976)
9. Solonin, S.I.: Using mathematical statistics in manufacturing engineering. Machine building, Moscow (1972)
10. Smirnov, N.V., Dunin-Barkovskiy, I.V.: Course on probability theory and mathematical statistics in technological applications. Science, Moscow (1969)
11. Venetskiy, G., Venetskaya, V.I.: Main mathematical and statistical terms and formulas in economic analysis. Statistic, Moscow (1974)
12. Loskutov, V.V.: Methodology of developing term projects on the course of metal-cutting machine tools. Ural Polytechnic Institute, Sverdlovsk (1964)
13. Tsarenko, V.I.: Express-method of estimating cost efficiency of industrial robots at the design stage. Prod. Mech. Autom. **9**, 29 (1985)
14. Liberman, Y.L., Zubarev, K.S.: Computer-aided system of choosing an industrial robot by its lifting capacity to match a specific lathe. Ural State Technical University, Yekaterinburg (2004)
15. Liberman, Y.L., Volkova, N.A., Kulikova, T.Y.: Computer-aided design system for robotic machining cells. Autom. Mod. Technol. **10**, 11–17 (2014)

Method to Control Quality of Assembly of Induction Motors

K. E. Kozlov[✉], V. N. Belogusev, and A. V. Egorov

Volga State University of Technology, 3 Lenina Square,
Yoshkar-Ola 424000, Russian Federation
konstantin.k-e@yandex.ru

Abstract. This paper is devoted to the method for determining the quality of assembly of induction motors on the basis of the experimental identification of their key mechanical indicators. The purpose of developing a new method is to improve the accuracy of identifying the mechanical parameters of electric motors and to create conditions for the development of methods for accelerated testing without the use of strain gauges introducing additional errors into the measurement process under dynamic operating modes and, in some cases, require significant financial investments. The method developed in this study is devoid of the above disadvantages and can be applied in a wide range of load and speed modes of operation. In the course of the study, the experimental substantiation of the method was carried out with the help of the developed hardware and software complex which demonstrated the expediency of its further application for evaluation of the technical state and the quality of manufacturing of induction motors with a high accuracy.

Keywords: Motor losses · Induction motor · Performance · Mechanical efficiency · Retardation method

1 Introduction

To improve the quality of assembling and manufacturing of asynchronous electric motors, it is first and foremost necessary to provide the manufacturing enterprises with the necessary instruments to control their technical state and to identify their key electrical and non-electrical indicators. As for electrical parameters, there are a sufficient number of methods and tools of measurement that provide the required level of accuracy. On the contrary, the existing methods and instruments to control the mechanical parameters of induction motors require their further development, since they do not provide sufficient accuracy and, as a rule, are applied for steady-state operating modes [1–5]. But the operation modes of induction motors in real operating conditions differ significantly from static ones. In this connection, it is required to develop methods and high-precision tools to control the key mechanical parameters of induction motors operating in dynamic operation modes [6–11].

The analysis of technical and scientific literature [5–7, 9] demonstrated that existing experimental methods for control of the mechanical indicators are based on a large number of assumptions (calculation methods) and averagings (dyno testing) or on a set of indirect indicators (input and output electrical parameters).

A dynamic method based on an analysis of two speed-torque characteristics of an induction motor (with a reference rotary body (disk) and without it) is presented in [11]. However, it does not take into account the change in losses in the bearings of the electric motor when a disk with a reference moment of inertia is attached to it. This affects the accuracy of the measurement. In addition, the described method requires determining the moment of inertia of the rotating parts of an electric motor (elements of the bearing assemblies and the rotor) by calculation or experimentally. The listed shortcomings cause significant errors in the tests, which limits the application of the described method to electric motors with high efficiency and operated in dynamic operation modes. Based on the literature survey it can be drawn that the widely used methods for control of the mechanical parameters of induction motors do not allow estimating the quality of their assembly with a high accuracy.

The purpose of this article is to develop an approach and a high-precision instrument to evaluate the quality of induction motors by determining their key mechanical parameters in dynamic modes of operation within a wide range of speeds.

2 Materials and Methods

The developed method is based on a comparison of two speed-torque characteristics of an induction motor when different disks of the same weight but with different known values of the inertia moments are attached. In this case, the criterion for evaluating the quality of assembly and efficiency of the tested motor is its acceleration.

2.1 The Development of a Method for the Identification of the Mechanical Parameters of an Induction Motor

Figure 1 presents a scheme for implementing the proposed method for identifying the moment of inertia of an induction motor using two disks, 5 and 6, of the same weight but with different inertia moments.

In the first stage, using a half-coupling 4 and fastening elements, we attach a disk 5 of the weight m and with the inertia moment J_1. Then we accelerate the system of rotating masses "induction motor 2, output shaft 3, half-coupling 4, disk 5" with an induction motor 2, and, with an encoder 1, we measure the angular acceleration ε_1 within the selected speed range. In this case, the torque of the induction motor 2 can be calculated as follows

$$M_{IM} = (J_1 + J_{IM}) \cdot \varepsilon_1 = (J_1 - J_2 + J_2 + J_{IM}) \cdot \varepsilon_1, \tag{1}$$

where, J_1 is the moment of inertia of the disk 5 of the weight m; J_2 is moment of inertia of the disk 6 of the weight m; J_{IM} is the moment of inertia of the rotating system "induction motor 2, output shaft 3, half-coupling 4".

In the next step, we replace the disk 5 with a dosk 6 of the same weight but with a different, smaller, moment of inertia J_2. With the help of an induction motor, the system

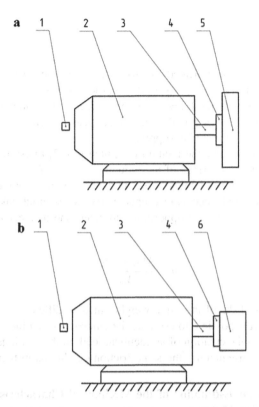

Fig. 1. Scheme for identifying the moment of inertia of an induction motor: (a) With a disk of weight m and with the inertia moment J_1; (b) With a disk of weight m and with the inertia moment J_2.

of rotating masses "induction motor 2, output shaft 3, half-coupling 4, disk 6" is accelerated, and the resultant angular acceleration ε_2 is registered with the help of an encoder 1.

In this case, the torque developed by the induction motor 2 can be calculated as follows

$$M_{IM} = (J_2 + J_{IM}) \cdot \varepsilon_2, \qquad (2)$$

Since the speed-torque characteristic of an induction motor remains relatively constant, it is therefore possible to equate the right-hand parts of Eqs. 1 and 2, and, as a consequence, determine the equation for $J_{IM} + J_2$

$$J_{IM} + J_2 = (J_1 - J_2) \cdot \frac{\varepsilon_1}{\varepsilon_2 - \varepsilon_1}, \qquad (3)$$

Substituting Eq. 3 into Eq. 2, we can calculate the torque of an induction motor as follows

$$M_{IM} = (J_1 - J_2) \cdot \frac{\varepsilon_1 \cdot \varepsilon_2}{\varepsilon_2 - \varepsilon_1}, \tag{4}$$

The torque M_{IM} is an indicated (ideal) torque, i.e., the torque that an induction motor would have in the absence of mechanical losses in it.

The use of a disk with the known different values of the moment of inertia but of the identical weight makes it possible to reduce measurement errors to errors related only to the volatility of the mains supply.

In the next stage, we use the method presented in [12] to estimate the moment of mechanical losses in the bearings of the induction motor, M_{loss}.

Knowing the level of mechanical losses in the bearing units of the induction motor, as well as its indicated mechanical characteristics (without taking into account mechanical losses), we have the opportunity to determine its mechanical efficiency as follows

$$\eta = \frac{M_{IM} - M_{loss}}{M_{IM}}, \tag{5}$$

Thus, using two disks of the same weight but with different known values of the moment of inertia, it is possible to estimate the efficiency of an induction motor with a high accuracy due to the creation of an identical load on the bearings in the process of measurement and, consequently, the same friction conditions in them.

2.2 Instruments for Estimation of the Mechanical Characteristics of an Induction Motor

To carry out the experiments and substantiate the expediency of applying the developed method, we assembled a test bench (Fig. 2), which allows registering the acceleration of a rotating system with the help of a hardware and software complex

Fig. 2. A hardware and software complex to determine the induction motor mechanical parameters.

presented in Fig. 3 and consisting of a PC with installed software, a registration unit, an incremental encoder manufactured by Autonics, and the current sensors of Honeywell. As a tested electric motor, the induction motor AIR 112MV6 (220/380 V) (4 kW, n = 940 rpm) was chosen.

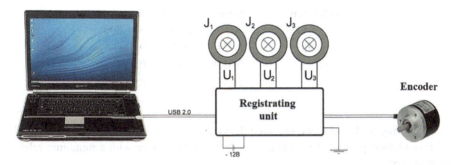

Fig. 3. A scheme of a hardware and software complex for determining the induction motor mechanical parameters.

3 Results and Discussion

In the experiment, we used two disks, 5 and 6, of the same weight and with the moment of inertia of 0.1899 kg × m² and 0.1259 kg × m², respectively.

The moment of inertia of a disk 7 was measured by two methods: by the well-known one (Fig. 4) and by the developed one (Fig. 5).

According to the existing method described in [11], we found the total moment of inertia of all rotating parts of an induction motor without the disk 7. Then we measured the total inertia of all rotating parts of an induction motor with the disk 7. The gap between the two moments of inertia was compared with the true value of the moment of inertia of the disk 7 (pre-calculated one).

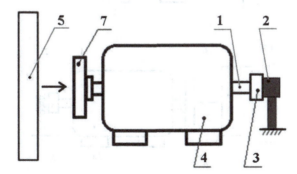

Fig. 4. Scheme for the experimental measurement of the moment of inertia of a disk by the well-known method described in [11], where 1 is an output shaft; 2 is an encoder; 3 is a coupling for attaching the encoder; 4 is an induction motor AIR 112MV6 (220/380 V); 5 is a disk with a reference inertia (0.1899 kg × m2) weighing 9.5 kg; 7 is a disk, the moment of inertia of which is determined during the experiment.

The results are shown in Table 1.

Table 1. The accuracy of determining the moment of inertia of a disk 7 with the known value of the moment of inertia.

Calculation parameter	Rotation speed, rev/min		
	200–400	400–600	600–800
J_{true}, kg · m^2	0.0247	0.0247	0.0247
J_{exist}, kg · m^2	0.0238	0.0232	0.0259
Accuracy, %	96.4	93.9	95.1

According to Table 1, we obtained the convergence of the results obtained by the existing method presented in [11] and by the calculating one with a minimum relative error of 3.6%.

Applying the method developed in this article, we estimated the moment of inertia of the disk 7 according to the scheme shown in Fig. 5. At first, we determined the total moment of inertia of all rotating parts of an induction motor without the disk 7.

Then we determined the total moment of inertia of all rotating parts of the induction motor with the disk 7. The difference module of the results obtained was compared with the true value of the moment of inertia of the disk 7. Table 2 shows the results obtained.

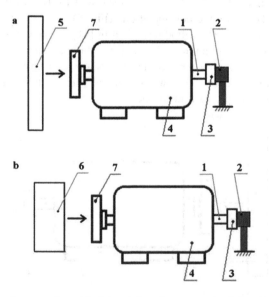

Fig. 5. Scheme for the experimental determining the moment of inertia of a disk by the developed method: (a) with a disk 5 of 9.5 kg and with the inertia of 0.1899 kg × m^2; (b) with a disk 6 of 9.5 kg and with the inertia of 0.1259 kg × m^2.

The results are shown in Table 2.

Table 2. The accuracy of determining the moment of inertia of a disk 7 with the known value of the moment of inertia.

Calculation parameter	Rotation speed, rev/min		
	200–400	400–600	600–800
J_{true}, kg · m^2	0.0247	0.0247	0.0247
J_{new}, kg · m^2	0.0249	0.0243	0.0245
Accuracy, %	98.9	98.7	99.3

According to Table 2, we obtained the convergence of the results obtained by the developed method and by the calculating one with a minimum relative error of 0.7% and a maximum relative error of 1.3%.

On the basis of the obtained results, it can be drawn that the accuracy of measuring the moment of inertia and the efficiency of an induction motor when using the method presented in this paper has increased approximately by 4 times compared to the accuracy obtained by applying the existing method presented in [11].

4 Conclusion

On the basis of the results obtained in this paper, it can be concluded that the method and instrument developed allow evaluating the quality of assembly and efficiency of induction motors with increased accuracy in comparison with existing control methods.

The proposed method can be used to estimate the technical state of induction motors at manufacturing plants, as well as at enterprises operating the research object. With the help of this method it is possible to determine the key mechanical indicators of induction motors without the use of tensometry, which allows significantly reducing the financial costs of testing. Having determined the permissible level of losses in the friction units, it becomes possible to shorten the test time.

In the future, it is planned to apply this method for asynchronous electric motors with a power of over 100 kW, as well as to develop recommendations for the application of the developed method for accelerated tests selecting the acceleration and retardation time as an evaluation criterion.

References

1. Kotelnets, N.F., Akimova, N.A., Antonov, M.V.: Tests, operation and maintenance of electric motors. Science, Moscow (2003)
2. Morris, A.S., Langari, R.: Mass, force, and torque measurement. In: Measurement and Instrumentation: Theory and Application, pp. 477–496 (2012)
3. Akimov, G.V.: Equipment for dyno tests of induction motors. Science, Moscow (1986)

4. Boldea, I., Nasar, S.A.: Unified treatment of core losses and saturation in the orthogonal. Axis Model Electric Mach. **134**(6), 355–363 (1987)
5. Kashirskikh, V.G., Zavyalov, V.M.: Estimation of parameters and condition of an induction motor at dynamical loads. Science, Moscow (2002)
6. Moon, C., DeB, Moor: Parameter identification of induction motor drivers. Automatic **32**(8), 1137–1147 (1995)
7. Myers, D.R., Chan, M.W., Vigevani, G.: Torque measurements of an automotive halfshaft using micro double-ended tuning fork strain gauges. Sens. Actuators A Phys. **204**, 79–87 (2013)
8. Wegener, G., Andrae, J.: Measurement uncertainty of torque measurements with rotating torque transducers in power test stands. Measurement **40**(7), 803–810 (2006)
9. Yu, J., Zhang, T., Qian, J.: Modern control methods for the induction motor, electrical motor products. In: Electrical Motor Products: International Energy-Efficiency Standards and Testing Methods, pp. 147–172 (2011)
10. Doppelbauer, M.: Accuracy of the determination of losses and energy efficiency of induction motors by the indirect test procedure. EEMODS, Washington (2011)
11. Egorov, A.V., Kozlov, K.E., Belogusev, V.N.: The method and instruments for induction motor mechanical parameters identification. Int. J. Appl. Eng. Res. **10**(17), 37685–37691 (2015)
12. Egorov, A.V., Kozlov, K.E., Belogusev, V.N.: Experimental identification of bearing mechanical losses with the use of additional inertia. Procedia Eng. **150**, 674–682 (2016)

Calculation and Analytical Module "Risk" for Selective Diagnostics and Repair of Main Gas Pipelines with Account of Technogenic Risks

Y. A. Bondin[1(✉)], N. A. Spirin[1], and S. V. Bausov[2]

[1] Ural Federal University, 19 Mira Street,
Ekaterinburg 620002, Russian Federation
bond_a007@list.ru
[2] Gazprom Transgaz Ekaterinburg, 14 Klary Tsetkin,
Ekatirinburg 620075, Russian Federation

Abstract. The operation of gas pipelines connected with economic and environmental risks and the risk of death. Ensuring the system reliability of gas transportation is one of the main strategic goals of PJSC "Gazprom". The allocation of areas and sections of pipeline, where there are regular accidents, maintenance and repair will increase the service life 70% of gas pipeline systems up to 45–50 years. In the framework of the implementation of the control System of technical condition and integrity of gas pipeline system in OOO "Gazprom transgaz Ekaterinburg" developed and practically used analytical module "Risk" for selective diagnostics and repair of gas pipelines with account of technogenic risks. Calculation and analytical module "Risk" has the following functional capabilities: field mapping of effected zones in certain chosen point of main gas pipeline segment; individual risk calculation in certain chosen point near main gas pipeline segment; calculation of mean cost and technogenic risk on certain chosen main gas pipeline segment.

Keywords: Gas pipelines · Maintenance and repair · Technological accidents · Economic impact · A quantitative risk assessment · Information system

1 Introduction

The whole length of major pipeline transport of hydrocarbons all over the world is about 2.3 million km. What is more, the length of pipe line infrastructure (58% of which is held by the USA and 17% by Russia) is almost twice longer than the length of pipelines in liquid hydrocarbon's circulation [1]. It can be explained by the fact that in contrast to oil and its products there is almost no alternative ways to deliver natural gas.

More over main gas pipeline is one of the most ecologically clean ways of hydrocarbons' transportation on condition that tough requirements to engineering, construction and operation are observed [2–4].

However linear part of main gas pipeline, located between sequenced line valve stations, presents the system of fire and explosion dangerous vessels. The breakdown of linear part of main gas pipeline can occur not only due to economic damage (downtime, gas leakage, accident abandonment costs) and widescale ecological losses (because of mechanical and thermal damage of natural landscape) [5] but also due to people's death. For the period of 2000–2012 there were 525 serious accidents on the sites of pipeline transportation of hydrocarbons in the USA. As a result 200 people died, 747 people were insured, material loss was 539 million dollars [6].

In 2015 the whole length of high-pressure main gas pipelines in Russia was 178 000 km [7]. All main gas pipelines of our country are the part of the unified gas supply system (UGSS) [8], that is controlled by Gazprom PJSC. Because of centralized control, great branching and presence of parallel routes for transportation, UGSS guarantees uninterrupted gas supplies even in case of peak seasonal load. Maintenance of system reliability is one of the main strategic aim of Gazprom PJSC [9].

2 Accident Rate on Russia's Main Gas Pipelines

In Russia as well as in the USA all the data about accident rates on pipelines are monitored on state level in accordance with national legislation. The Federal Service for Ecological, Technological and Nuclear Supervision (Rostekhnadzor) is the regulating authority in this sphere.

According to Rostekhnadzor [10] in 2016 there happened 11 accidents (the greatest part was fixed in Ural Federal District – 5 accidents) on the hazardous production facilities of major pipeline transport. In 2016 total damage after accidents combined 262,6 million rubles (compare with 2015 with 488,2 million rubles), among them: direct losses – 64,3 million rubles (in 2015 - 284,9 million rubles), costs for localization and elimination of accident consequences – 177,1 million rubles (in 2015 - 191 million rubles), ecological damage – 8 million rubles (in 2015 - 12 million rubles), third party damage – 13,2 million rubles (in 2015 - 300 000 rubles), accidents resulting in death and group accidents were not fixed (in 2015 – one accident resulting in death). Analyzing the results of technical investigation of the accident because it becomes clear that the key factor of accident initiation in 10 cases (90%) was the influence of internal hazards connected with depreciation, metal corrosion and pipe body cracking under pressure.

Over the last three years Rostekhnadzor fixed one serious accident on linear part of main gas pipeline in Gazprom Transgaz Yekaterinburg [11]. The accident happened on 18th February 2016.

Pipe spool breakdown (diameter 1020 mm) with leakage and gas inflammation took place on the underwater line crossing River Salmysch (57th km of main gas pipeline "Orenburg GPP-Sovkhoznoe UGSF, date of commissioning – 1977).

Technical reasons of the accident are as follows: growth of metal pipe flaw in the form of crack along the alloyage of longitudinal plant weld with main pipe metal and further on developing of the flaw on basis of stress-crack corrosion (SCC). There were no injuries. Economic damage was 36 433 thousand rubles.

One of the accident reasons on linear part of main gas pipeline is that the greatest part of unified gas supply system was built in 70–80th years of the past century and by now depreciation on linear part of main gas pipeline is more than 50% [12, 13].

The average age of main gas pipeline is about 25 years, and 15% of gas pipelines has worked out their service life (33 years) [14].

However, it is found out [14] that the age of main gas pipeline is not always the key factor influencing the frequency of accidents. In most cases accidents happen on the sections of linear part of main gas pipeline that are situated in geodynamical active regions, areas with stress condition of minerals, in fault zones and in active emanation of corrosive deep gas zones. Stress-corrosion formation and breakdown of linear part of main gas pipeline are greatly influenced by magnetic, electrical and heat anomalies and also by the preparation level of gas transportation from technical point of view.

Herein the accidents happen once during 3–4 years on some sections of linear part of main gas pipeline, once during 10–12 years on the others and once during 15–20 years on the third. But even after standard service life expiration (33 years) some sections of the pipes are almost unaffected by corrosion and breakdowns. The identification of zones and sections where accidents happen regularly, strict control of these zones' condition, performance of diagnostic operations and technical servicing and repair on the most dangerous sections will allow to increase standard service life for 70% of gas transmission systems up to 45 and in some cases even up to 50 years. Implementation of these rules will help to save thousand millions rubles [14].

3 Control System of Technical Condition and Integrity of Gas Pipeline System

Regulatory and procedural documents worked out by Rostekhnadzor [15–17] together with Gazprom PJSC [18–21] take into account the best world experience [22–25] in industrial safety maintenance, prevention of accidents and industrial injuries. These documents serve as the basis for implementation of new innovative methods in performing diagnostic operations, technical servicing and repair on linear part of main gas pipeline.

In compliance with regulatory and procedural documents [18] long-term programs in performing diagnostic operations, technical servicing and repair on linear part of main gas pipeline must be based not only on quantitative evaluation of reliability indicator [19] but also on technogenic risk on sections of main gas pipeline [20].

According to regulatory and procedural documents [20] quantitative risk analysis includes expected accident frequency evaluation, which consists of indication of potential accident frequency for each source of danger on hazardous industrial facility.

Regulatory and procedural documents [21] are recommended to be primarily applied for evaluation (forecasting) of expected accident frequency on unspecified sections of main gas pipeline in accordance with regulatory and procedural documents [20] as well.

According to above mentioned regulatory and procedural documents control system of technical condition and integrity of gas pipeline system [26] is continuously improved in Gazprom PJSC since 2009.

Unified model of executive decision-making in long-term planning of diagnostic operations, technical servicing and repair as a part of expense decrease for the whole unified gas supply system is the basis for control system of technical condition and integrity of gas pipeline system. The estimates of reliability indicators and technogenic risk on the whole range of facilities on linear part of main gas pipeline (inter-valve section) for 19 subsidiary gas transmission companies which use about 160 thousand km of main gas pipeline belonging to Gazprom PJSC are applied as source data. Expense limit is defined for each current year of planning and looks like restriction for optimized number of objects on main gas pipeline. The role of internal restrictions is fulfilled by requirements to reliability and risk indicators according to which the decisions in gas transmission companies are taken in accordance with different scenarios of diagnostic operations, technical servicing and repair.

Control system of technical condition and integrity is based on two-level structure of technical condition and integrity management of unified gas supply system (levels of gas transmission companies and Gazprom PJSC).

Control action within control system of technical condition and integrity is actually characterized by repair size: random repair of really critical pipe spools at a segment or full repair of the whole section length.

In the first case total expenses for random repair are nominally equally spread within great time period – within planning calculation horizon. Herewith the efficiency of the segment is restored up to required value only for limited operating period – until next random repair, and safety requirements can possibly not be observed in full measure.

In the second case all combined expenses for full repair are spread within one certain year. The resources after full repair are restored during specified time horizon and maximum possible reduction of technogenic risk is guaranteed thanks to repair methods.

The major components of technogenic risk within control system of technical condition and integrity of linear part in unified gas supply system are: expected direct damage caused by one accident and accident frequency on the pipeline sections [27]. To perform calculations of all components and indicators of technogenic risk on long segments of main gas pipelines it is necessary to collect and digest a lot of information, the greatest compound part of which consists of data about objects of social, economic and industrial and natural pipelines' surrounding. Solution of this problem requires application of modern computer technologies which use the functionality of geographic information system.

4 Quantitative Evaluation of Technogenic Risk for Main Gas Pipelines in Geographic Information System Arcgis Desktop

Modernization of Control system of technical condition and integrity on the objects of linear part of main gas pipelines takes place in order to implement complex approach towards formation of diagnostic operations, technical servicing and repair at Gazprom Transgaz Yekaterinburg.

- Special attention is paid to development of specialized information analysis system intended to support taking decisions when planning diagnostic operations, technical servicing and repair of the pipelines influenced by stress-crack corrosion [28]

- Along with this calculation and analytical module "Risk" was developed in geographic information system ArcGis Desktop. This package allows to define damage areas and size of technogenic risk in case of accidents on linear part of main gas pipelines [29]
- General algorithm of work for calculation and analytical module "Risk" suggests the following action sequence
- Forecasting of expected accident frequency on linear part of main gas pipelines in accordance with Part 7 of regulatory and procedural documents [21]
- Calculation of accident scenario group (see Table 1) on linear part of main gas pipelines in accordance with point 5.5.4 of regulatory and procedural documents [20]

Table 1. Set of estimated accident scenarios on linear part of main gas pipelines.

Name of estimated accident scenarios	Damage effects
C_1 Fire in foundation pit	Fragment dispersion, air wave of compression, heat radiation from the flame
C_2 Jet flames	Fragment dispersion, air wave of compression, direct influence of the flame, heat radiation from the flame
C_3 Diffusion of low-speed gas plume	Fragment dispersion, air wave of compression

In accordance with point 5.6.2 of regulatory and procedural documents [20] determination of conditional probabilities of firing/non-firing in case of main gas pipeline blow-out, and also determination of gas escape character is based on statistic data about frequency ratio depending on nominal diameter of main gas pipeline.

As an example, Fig. 1 shows conditional probabilities calculated for main gas pipeline with diameter $Dy = 1400$ mm in accordance with Pic.5.7 of regulatory and procedural documents [20].

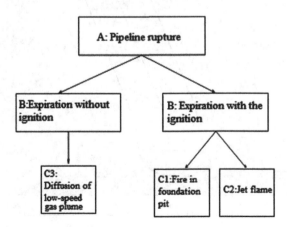

Fig. 1. Event tree for identification of accidents on main gas pipeline with diameter $Dy = 1400$ mm.

Possible amount of property and natural components destroyed or damaged as a result of accident on separate potentially hazardous area is determined for each chosen scenario.

Characteristic criterion of destructive effect is used to estimate negative impact on objects. This criterion is put by indication of threshold values for main physical affecting factor characteristics relevant to object destruction. Threshold values of affecting factors in accordance with regulatory and procedural documents [20] are shown in Table 2.

Table 2. Set of estimated accident scenarios on linear part of main gas pipelines.

Name of estimated accident scenarios	Threshold values of prevailing affecting factors
C_1 Fire in foundation pit	7 kW/m^2 for external installations, forest range, buildings and structures
C_2 Jet flames	35 kW/m^2 for underground utility systems
C_3 Diffusion of low-speed gas plume	7 MPa for external installations

When the object gets into any of these zones it becomes affected and cost of its reconstruction is added to the total material damage amount on the selected potentially hazardous area.

According to algorithms in point 5.12 of regulatory and procedural documents [20] computation mesh in the nodes of which potential risk values are determined (as shown on Fig. 2) is built on the territory adjacent to potentially hazardous area.

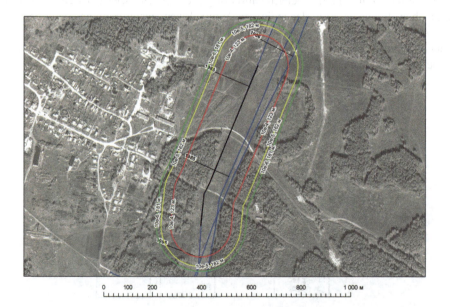

Fig. 2. Potential risk distribution.

Interface of calculation and analytical module "Risk" is represented by panel built-in in the interface of geographic information system ArcGis Desktop, which consists of set of instructions and instruments, as it is shown on Fig. 3.

Calculation and analytical module "Risk" has the following functional capabilities: field mapping of effected zones in certain chosen point of main gas pipeline segment; individual risk calculation in certain chosen point near main gas pipeline segment; calculation of mean cost and technogenic risk on certain chosen main gas pipeline segment.

As source data for calculation in calculation and analytical module "Risk" the following is taken: reports about pig inspection; data gathered by specialists of Engineering and Technical Center of Gazprom Transgaz Yekaterinburg according to form specified in table 5 of regulatory and procedural documents [21], and also spatial data received on basis of surface contour map, topographic-geodesic work results and aerospace monitoring.

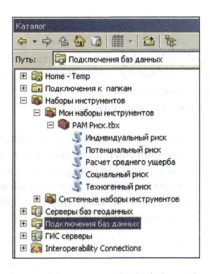

Fig. 3. Set of instructions and instruments of calculation estimating package "Risk".

At the first stage digitalization (vectorization) of surrounding objects and objects of linear part of main gas pipeline is made. Afterwards digitalization (vectorization) data is published on geo-portal ArcGis Desktop for filling in attribute data. On basis of filled in object characteristics the final cost is defined. Further on the information about technical state of linear part of main gas pipeline segment is added. And at the last stage risk calculation is performed.

Digitalization and management of special data is carried out in geographic information system ArcGis Desktop, for storage and publication of the data Postgree SQL on ArcGis Server is used.

General chart showing formation of special data needed as source data for calculation and analytical module "Risk" is presented on Fig. 4.

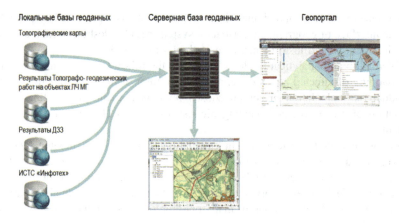

Fig. 4. General chart of special data formation.

5 Conclusion

Worked out calculation and analytical module "Risk" allowed to estimate quantitative evaluation of technogenic risks for the whole linear part of main gas pipeline of Gazprom Transgaz Yekaterinburg, and this became the basis for formation of adequate and sufficient plans for diagnostic operations, technical servicing and repair on linear part of main gas pipeline.

Main directions for further extension of functional capabilities of calculation and analytical module "Risk" are: improvement of its algorithms in order to increase calculating speed, verification and formalization of approaches towards determination of cost characteristics for surrounding objects of linear part of main gas and also realization of technogenic risk calculation possibility for infrastructure sites (gas-compressor station, delivery measuring station).

References

1. Eder, L.V.: The role of transport infrastructure in the energy cooperation between Russia and APRAM. Customs Policy Russia Far East **4**(61), 22–31 (2012)
2. Samsonov, R.O., Bashkin, V.N., Kazak, A.S., et al.: Risk assessment in the impact zones of trunk pipelines. Probl. Risk Anal. **3**(3), 238–249 (2006)
3. Ishkov, A.G., Bashkin, V.N., Akopova, G.S., et al.: Environmental safety during transmission of different types of gas: pipeline, liquefied, compressed and hydrated. In: International Gas Union Research Conference, vol. 3, pp. 1751–1766 (2011)
4. Deneko, Yu.V.: Ecology and environment: present and future. Gas Ind. **7**(693), 68 (2013)
5. Goryunkova, A.A., Galunov, D.V.: Ecological problems of gas industry. Izvestiya Tula State University. Eng. Sci. **11**(2), 292–296 (2014)
6. Oleynik, A.P.: Comparative analysis of accidents on the objects of pipeline transport in Russia and the United States. Bulletin of the Russian University of Friendship of Peoples, vol. 4, pp. 84–91 (2015)
7. Transport and communication in Russia. Rosstat (2016)

8. Unified Gas Supply System of Russia (2018). http://www.gazprom.ru/about/production/transportation. Accessed 22 Oct 2019
9. About Gazprom. Strategy (2018). http://www.gazprom.ru/about/strategy. Accessed 22 Oct 2019
10. Annual report on the activities of the Federal service for ecological, technological and nuclear supervision in (2016). http://www.gosnadzor.ru/public/annual_reports. Accessed 22 Oct 2019
11. The lessons learned from accidents (2016). http://www.gosnadzor.ru/industrial/oil/lessons. Accessed 22 Oct 2019
12. Gostenin, I.A., Viryasov, A.N., Semenov, M.A.: Analysis of accidents on the linear part of main gas pipelines. Eng. J. Don **2**(25), 24–39 (2013)
13. Sageeva, Z.Z., Khayrullin, R.Z.: Analysis of accidents on the linear part of main gas pipelines. Sci. Almanac **4-3**(30), 158–161 (2017)
14. Dmitrievsky, A.N.: The fundamental basis for the innovative development of the oil and gas industry in Russia. Bull. Russian Acad. Sci. **80**(1), 10–20 (2010)
15. Safety rules for hazardous production facilities of main pipelines. Order of Rostekhnadzor 520
16. Methodological basis for conducting hazard analysis and risk assessment of accidents at hazardous production facilities. Order of Rostekhnadzor 144
17. Instructions for the technical diagnosis of underground steel gas pipelines, Order of Rostekhnadzor 47
18. Methodology for the formation of programs for technical diagnostics and repair of objects of the linear part of the main gas pipelines of the UGSS Gazprom. Gazprom (2014)
19. The methodology for calculating reliability indicators for the operation of objects of the linear part of the main gas pipelines of the unified gas supply system of OAO Gazprom. Gazprom (2014)
20. Methodological instructions for conducting a risk analysis for hazardous production facilities of gas transmission enterprises of OAO Gazprom. Gazprom (2009)
21. Recommendations to take into account the impact of technical, technological, natural-climatic and other factors in forecasting accidents at the MG of OAO Gazprom. Gazprom (2007)
22. Methoden voor het bepalen van mogelijke schade. Green Book. VROM (2005). http://www.publicatiereeksgevaarlijkestoffen.nl/publicaties/PGS1.html. Accessed 22 Oct 2019
23. Methods for the calculation of physical effects. Yellow book. VROM (2005). http://www.publicatiereeksgevaarlijkestoffen.nl/publicaties/PGS2.html. Accessed 22 Oct 2019
24. Guidelines for quantitative risk assessment. Purple Book. VROM (2005). http://www.publicatiereeksgevaarlijkestoffen.nl/publicaties/PGS3.html. Accessed 22 Oct 2019
25. Methods for determining and processing probabilities. Red Book. VROM (2005). http://www.publicatiereeksgevaarlijkestoffen.nl/publicaties/PGS4.html. Accessed 22 Oct 2019
26. Alimov, S.V., Nefedov, S.V., Milko-Butovsky, G.A., et al.: Optimization of long-term planning of diagnostics and repair of the linear part of the main gas pipelines in the system of management of the technical condition and integrity of the gas transmission system of OAO Gazprom. News Gas Sci. **1**, 5–12 (2014)
27. Petrova, Yu.Yu., Ovcharov, S.V.: On the collection and preparation of data on environmental objects for the purposes of the management system for the technical condition, integrity of the linear part of the main gas pipelines. News Gas Sci. **1**, 61–62 (2014)

28. Kuimov, S.N., Bausov, S.V., Istomin, A.I., et al.: Informational and analytical support of control processes of technical condition of gas pipelines subject to stress corrosion cracking. News Gas Sci. **3**(27), 131–218 (2016)
29. Bondin, Y.A., Spirin, N.A., Debenko, D.V.: Quantitative assessment of technogenic risks for main gas pipelines in the environment of the geoinformation system ARCGIS DESKTOP. Automation systems in education, science and production AS, pp. 120–127 (2017)

Energy-Saving Oriented Approach Based on Model Predictive Control System

T. A. Barbasova[✉], A. A. Filimonova, and A. V. Zakharov

South Ural State University, 76 Lenina Avenue,
Chelyabinsk 454080, Russian Federation
barbasovata@susu.ru

Abstract. Nowadays the problem of efficient control in energy complexes of industrial enterprises and utilities sector is very important. The article represents the structure of the software and hardware complex for efficient energy consumption control based on the distributed conservation power plant concept. To implement the functions of controlling planning, long-term and short-term forecasting as well as energy consumption control with the use of the data intelligent analysis, a model predictive control module is distinguished within the software and hardware complex. To ensure controllability and observability of the processes, program models of controlled object are applied, which are subject to continuous online updating along with the real controlled object base on the procedure of continuous parameter identification according to the current actual operating data. Furthermore, the problem of optimization of controlling effects according to the technical and economic indicators is solved at each control cycle. The article adduces an example for solving the problem of operating parameters clustering of blast furnace smelting within the effective cluster based on the developed software module. The developed module represents an open software environment for interpretation and execution of the programs source code written in the R programming language.

Keywords: Energy saving · Intelligent data analysis · Energy efficiency · Clustering · Model predictive control

1 Introduction

A priority trend of the state policy in the development of the power infrastructure of the Russian Federation is performance of the works on energy saving and ensuring of energy safety in all the economic activity fields including in housing and public utility sector and energy complexes of industrial enterprises [1].

The energy complexes of industrial companies, including metallurgic ones, represent complex production systems including energy generation, distribution, accumulation and consumption subsystems [2]. The operating modes of these subsystems are conditioned by the operating modes of the main metallurgical production process equipment and involve considerable fluctuations in generation and consumption of energy resources. In the meanwhile, the volumes of energy consumption and recovery of secondary energy resources directly determine the energy intensity of the metallurgical production. In addition, the quality of output products is defined by the technical parameters of energy resources. Therefore, an important practical problem is the provision of the efficient control of energy complexes at metallurgical companies considering the dynamics of the processes of generation, distribution, accumulation and consumption of energy resources along with the impact of parameter deviations on the output product quality [3].

The key challenges preventing from achievement of high technical and economic performance indicators of high-energy complexes are:

- Incomplete observability and controllability of technological and production processes
- Necessity to stabilize processes in extreme boundary conditions being determinative of the maximum technical and economic performance
- Incomplete knowledge about the current conditions of the processes due to their complexity

To improve energy efficiency of the housing and public utility sector and industrial facilities, a software module for the model predictive control system was developed. The module is built based on the conservation power plant concept employing the intelligent data analysis.

2 Related Work

The advanced technologies of energy efficiency control in the housing and public utility sector and industrial facilities are well-covered by the vast literature. However, there exist incompletely solved problems among which the following should be noted:

- The problems of integrated planning and model predictive control of energy saving processes on a scale of a large enterprise or a city [4]
- The problems of complex optimization of technological processes implementing the energy efficiency improvement reserves [5, 6]
- The problems of practical implementation of integrated planning and energy consumption control for particular industries and municipal districts [7, 8]

The development of these ideas applicable to power supply of cities in the USA has resulted in major success in the field of energy saving. Here the works of the "negawatt" idea by Amory Lovins should be noted [9, 10], the practical success was achieved by the power generating companies of Pacific Corporation, Sacramento Municipal District and etc. In the European Union, the positive experience of the city of Hanover as well as that of the developers from the Wüppertal Öko-Institut are worth noting [11, 12]. These works are driven by the concept of a distributed conservation power plant, which is based on the following principles:

1. The principle of duality of efficient energy complexes, in which not only the energy generation, distribution and consumption principles are explicitly distinguished, but also their complementary energy saving processes. Furthermore, energy saving processes are considered as the sources of "negawatt" energy flows, i.e. negative watts with consistently tracing the "passing" ways of negawatt energy flows in the energy complex [13].
2. Building a Conservation Power Plant – the system for integrated planning and management of power resources basically considered as a specific power plan generating the negawatts of power [14].

In Russia, a steady growth of the software market of integrated control systems for composite thermal energy complexes has been observed since 2011 [1].

The key growth factors of the market for intelligent management systems of energy management include the following [14]:

- Adopting a governmental policy in energy efficiency and energy saving resulting in the development of regulatory legal acts obliging to increase saving of energy resources
- The growth of tariffs for energy resources
- The continuing development of housing, road and industrial construction in Russia: launching of new productions, transportation infrastructure facilities
- The obsolete material and technical base of the most of the industrial companies

However, there is also the number of factors producing a negative impact on the development of the market for intelligent energy management systems and restricting its growth, including [1]:

- Limited budgets, insufficient level of project financing
- Unstable economic situation in the country, slowdown in economic growth, freezing or limiting of budget costs
- Imperfection of the legislative base with respect to implementation of the energy service contract mechanisms, frequent changes in tax legislation
- High cost of borrowed funds, frequent changes in the lending terms
- Long payback period of activities in implementation of the intelligent energy management systems (5–10 years), the repayment risk
- Lack of highly-qualified personnel

The investigation of the modern global trends in energy efficiency control showed that the key initiatives of industrial companies in the improvement of energy efficiency and energy saving for the next several years will be as follows [15, 16]: equipment replacement, data analysis, technical supervision, alternative energy sources, intelligent sensors, strategic planning, introduction of software, energy policy changes, preparation of energy consumption reports. The percentage of the above initiatives is shown in Fig. 1.

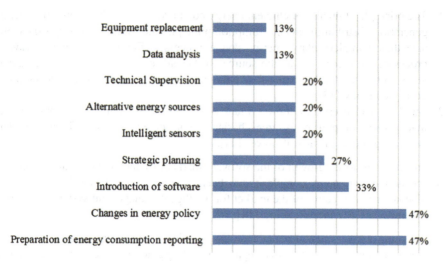

Fig. 1. Key initiatives of industrial companies in improvement of energy efficiency and energy saving.

3 The Software and Hardware Complex of Model Predictive Control

The structure of the software and hardware complex of model predictive control based on the distributed conservation power plant concept is given in Fig. 2.

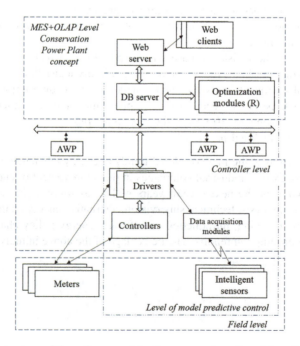

Fig. 2. The structure of the software and hardware complex of model predictive control.

To implement the functions of the manufacturing planning, long-term and short-term forecasting as well as energy consumption control with the use of the data intelligent analysis, a model predictive control module is distinguished in the structure of the software and hardware complex.

To ensure manageability and observability of the processes, program models of controlled object are applied, which are subject to continuous online updating along with the real controlled object base on the procedure of continuous parameter identification according to the current actual operating data. Furthermore, the problem of optimization of controlling effects according to the technical and economic indicators is solved at each control cycle.

Current data in operating the process facilities, which are used for the intelligent data analysis, are located in the database controlled by the Oracle Database DBMS.

The developed module represents an open software environment for interpretation and execution of the source code of the programs written on the R programming language featuring the possibility to restrict the access rights to the file space of the computer, at which the source code is being executed, as well to the database data.

The R programming language is selected for developing the module of model predictive control based on the intelligent data processing: Particularities of the R programming language:

- Availability of built-in functions for statistical data processing
- Convenient implementation of the new developed mathematical algorithms for continuous identification of process parameters of controlled objects both the industrial and housing and utility sector
- A convenient implementation of the new developed mathematical algorithms for detection of the controlling actions

Based on the R programming language, the following functions were implemented in the model anticipatory management module based on the intelligent data processing technology:

- Intelligent data filtering
- Identification of building thermal condition
- Data clustering to detect the operating modes of the process equipment
- Production planning
- Long-term and short-term forecasting of consumption for energy resources: electric energy, natural gas, process steam, water, secondary and other energy resources
- Optimization of consumption of thermal energy resources

The below given is the example of problem solving the effective domains clustering of blast-furnace smelting operating parameters.

4 The Effective Domains Clustering of Values of Blast-Furnace Smelting Operating Parameters

The effective domains of the values of operating parameters are determined based on the set target values of the blast furnace smelting parameters, such as performance, coke consumption, theoretical combustion temperature, indicators of thermal condition of the furnace (*Si* content in cast iron, titanium module, the extent of blast furnace gas usage and etc.).

For example, Fig. 3 represents an effective domain selected through the target function (1):

$$e_g = \alpha_n p_{CI} + \alpha_k b_k^{-1}, \quad \alpha_n, \alpha_k \geq 0, \quad \alpha_n + \alpha_k = 1, \tag{1}$$

where p_{CI} – the relative cast iron production; b_k – the relative coke rate, kg/t; α_n, α_k – the weights of the specific indicators n and k respectively within the composite index reflecting the importance of considering the production and coke saving within the composite target indicator.

The challenge of the problem of clustering the effective domain of operating parameter values is its high dimensionality [17–20]. The number of operating parameters may exceed 70. To simplify solving the problem, the method of breaking down the exact region into two-dimensional sections analytically described by the second-order elliptic regions was employed in the work.

Figure 4 shows an example of the improved efficiency area in the coordinates of "coke rate – *Si* content" at the constraints for the furnace thermal condition.

The basic problem of selecting the effective solutions consists in the selection of the BF controlled parameter values $\{x_i : i \in I_y\}$ at the constraints set for non-controlled parameters $\{x_i : i \in I_N\}$.

The following problem is set up: Finding an admissible value of controlled parameters based on the criterion of the minimum residual error of constraints (2) provided that the non-controlled parameters within the set constraints tend to provide the maximum specified residual error. This problem represents the minimax task of the mathematical programming

$$\min_{\{x_i \in I_y\}} \max_{\{x_i \in I_N\}} E_T^2(\{x_i : i \in I\}), \tag{2}$$

In the Eq. (2) E_T^2 – the total solution residual error considering the process limitations.

The problem will generally be solved by the gradient method. If the recurrent process converges, the generalized solution of the problem (2) under the set constraints is obtained.

Figure 4 shows an example of solving the minimax problem (2) of mathematical programming and finding admissible values of controlled parameters based on the criterion of the minimum residual error of constraints provided that the non-controlled parameters within the set constraints tend to provide the maximum specified residual error. The figure shows the initial calculation point (A), which is located within the

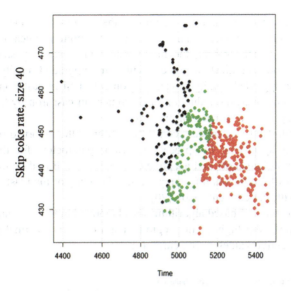

Fig. 3. The improved efficiency area at the constraints for the furnace thermal condition.

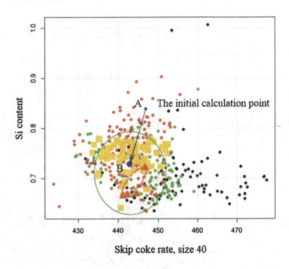

Fig. 4. The example of solving the minimax problem of finding the blast furnace operating mode within the coordinates of "Si content-coke rate".

low-quality region and the found point (B) which is located within the medium quality region, by the thermal condition within the coordinates of "coke rate-*Si* content in cast iron".

Finding a target control vector proceeds from the condition of finding the blast furnace smelting parameters towards approximation to the effective domain of blast furnace process.

The considered methods allow determining the steady-state modes of production processes, efficient by production and resource consumption indicators. However, to ensure the efficient management, the mode transition technology should be known along with the mode stabilization process within the upgraded quality clusters. With switching to the upgraded quality mode, the content of Si in cast iron decreases and gets stabilized, moreover, the reduction of Si content in cast iron by 0.01% results in saving of 3.75 kg of coke per ton of cast iron.

To test the blast furnace process control algorithms, the report forms are prepared on a daily basis specifying the process indicator parameters for the current day, including the target indicators and adaptive control factors. Based on the report forms, the actual values of the factors impacting on the efficiency of the blast furnace process are compared with the efficient mode [20].

The developed model-based algorithms are elaborated in the software and hardware complex to specify the high-quality region in terms of the thermal condition with including the effective values of the current mode.

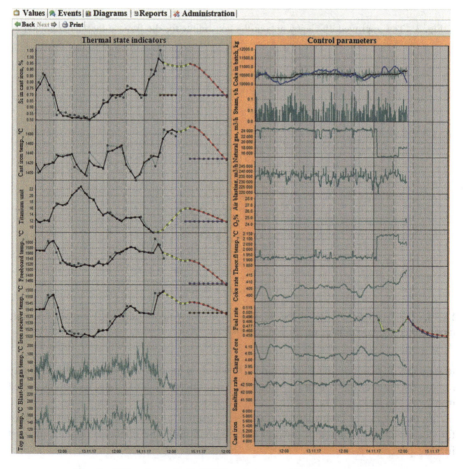

Fig. 5. Video frame in PoliTER software and hardware complex.

A video frame demonstrating the operation the model anticipatory management module within the PoliTER software and hardware complex exemplified by the blast furnace process of a metallurgical company is shown in Fig. 5.

Based on the diagrams for the coke rate shown on the video frame, the feed coke rate is determined considering the current rated feed cast iron.

5 Conclusion

To solve the problem of improving the energy efficiency of the housing and public utility sector and industrial facilities, a software module for the model predictive control was proposed. The module operating principle is based on the conservation power plant concept employing the intelligent data analysis. The developed software module being a part of the software and hardware complex allows for the production planning, long-term and online rating and forecasting as well as management of energy resources: electric energy, natural gas, process steam, water, secondary and other energy resources.

An example is given for clustering problem solving of the effective domains of the values of blast-furnace smelting operating parameters within the effective cluster based on the developed software module.

Acknowledgements. The work was supported by Act 211 Government of the Russian Federation, contract № 02.A03.21.0011.

References

1. Verdantix market research. Green quadrant energy management software. http://www.verdantix.com/index.cfm/papers/Products.Details/product_id/480/green-quadrant-energy-management-software-global-2013. Accessed 10 Dec 2017
2. Eyre, N.: Energy saving in energy market reform—the feed-in tariffs option. Energy Policy **52**, 190–198 (2013)
3. Filimonova, A.A., Barbasova, T.A.: Automated system for simulation of electricity consumption in the iron and steel plant. In: 2017 International Conference on Industrial Engineering. Applications and Manufacturing (2017)
4. Jain, A.: Energy-efficient applications in electrical power systems. In: Energy Conservation Measures, pp. 271–286 (1984)
5. Krause, F., Koomey, J., Olivier, D.: Incorporating global warming externalities through environmental least cost planning: a case study of Western Europe. In: Social Costs of Energy, pp. 287–312 (1994)
6. Klein, J.D., Schmidt, S., Yaisawarng, S.: Productivity changes in the U.S. electric power industry. In: Studies in Industrial Organization Empirical Studies in Industrial Organization, pp. 207–235 (1992)
7. Soares, L.J., Medeiros, M.C.: Modeling and forecasting short-term electricity load: a comparison of methods with an application to Brazilian data. Int. J. Forecasting **24**(4), 630–644 (2008)

8. Andersen, F., Larsen, H., Boomsma, T.: Long-term forecasting of hourly electricity load: identification of consumption profiles and segmentation of customers. Energy Convers. Manag. **68**, 244–252 (2013)
9. Lovins, A.B.: Negawatts. Energy Policy **24**(4), 331–343 (1996)
10. Lovins, A.B.: The Negawatt revolution. Across Board **27**(9), 21–22 (1990)
11. Joskow, P.L., Marron, D.B.: What does a negawatt really cost? Evidence from utility conservation programs. Energy J. **13**(4), 41–74 (1992)
12. Sioshansi, F.P.: Negawatt pricing. In: Electricity Pricing in Transition, pp. 191–206 (2002)
13. Parfomak, P.W., Lave, L.B.: How many kilowatts are in a negawatt? Verifying ex post estimates of utility conservation impacts at the regional level. Energy J. **17**(4), 59–87 (1996)
14. Kazarinov, L.S., Barbasova, T.A.: Case study of a conservation power plant concept in a metallurgical works. Procedia Eng. **129**, 578–586 (2015)
15. Steinberger, J.K., Niel, J.V., Bourg, D.: Profiting from negawatts: reducing absolute consumption and emissions through a performance-based energy economy. Energy Policy **37**(1), 361–370 (2009)
16. Wood, A.J., Wollenberg, B.F., Sheblé Gerald, B.: Power Generation, Operation, and Control. Wiley-Interscience, Hoboken (2014)
17. Omori, Y.: Blast Furnace Phenomena and Modelling. Elsevier Applied Science, London (1987)
18. Kwakernaak, H., Tijssen, P., Strijbos, R.: Optimal operation of blast furnace stoves. Automatica **6**(1), 33–40 (1970)
19. Peacey, J., Davenport, W.: A brief description of the blast-furnace process. In: The Iron Blast Furnace, pp. 1–15 (1979)
20. Kazarinov, L., Shnayder, D., Barbasova, T.: Optimization of the blast furnace operating modes for identification of the areas of unimprovable solutions. Key Eng. Mater. **743**, 363–368 (2017)

Using Method Frequency Scanning Based on Direct Digital Synthesizers for Geotechnical Monitoring of Buildings

D. I. Surzhik[1(✉)], O. R. Kuzichkin[2], and A. V. Grecheneva[2]

[1] Vladimir State University, 87 Gorkogo Street,
Vladimir 600000, Russian Federation
`arzerum@mail.ru`
[2] Belgorod National Research University, 85 Pobedy Street, 308015 Belgorod,
Russian Federation

Abstract. The possibility of using the principle of frequency scanning as an alternative to the spectral-time analysis method for isolating the own frequencies of structures to provide their geotechnical monitoring with the help of an accelerometric phase-metric method is considered in the article. As the devices realizing this method, direct digital synthesizers, are considered, their determining advantages are indicated and the main drawback associated with the presence of a lot of parasitic spectral components in the spectrum of the synthesized signal is indicated. To reduce them, it is suggested to use the method of automatic compensation of phase distortions, its structural realization and frequency synthesizer circuits based on it with various types of regulation are shown. It is shown that the degree of auto-compensation in practice can reach 15 dB and is determined by the amplitude-frequency response of the devices by the phase distortions of the synthesizer, as well as by the conditions for full compensation, which are depending on the type of control used.

Keywords: Geotechnical monitoring · Accelerometric phase-metric method · Own frequencies of constructions · Frequency scanning · Direct digital synthesizers · Automatic compensation of phase distortions

1 Introduction

Geotechnical monitoring is one of the integral components of ensuring the safety of projected buildings and structures of a high level of responsibility and should be carried out both during their construction and during the subsequent exploitation [1]. It includes observations of the subsidence of the buildings, stresses in the foundation and bearing structures of the underground part, deviations, fluctuations of buildings with simultaneous observations of external impacts on the object, including measurements of wind loads, vibration and seismic influences, air temperature, atmospheric pressure, atmospheric precipitations [2].

An informative method for monitoring the state of structures is to control the own frequencies of their designs [3]. One of the variants of its implementation is the use of the accelerometric phase-metric method [4, 5], which makes it possible to excrete the values of the dominant own frequencies of the monitored objects. This method is based on the collection of dynamic data by converting signals from accelerometers placed between control points of structures which having their own technogenic rhythm, in the phase of the sinusoidal oscillation.

For the preliminary processing of accelerometric signals, the method of spectral-temporal analysis is currently widely used [6]. It allows to excrete the main frequency components of the signal on the distributed measuring network of primary accelerometric transducers with minimal phase distortions and consists in synchronous passing of the output signal of the accelerometer through a system of narrowband filters and obtaining the distribution of the amplitude values of the envelopes and their phases at the filters outputs at the corresponding frequencies.

However, in the organization of geotechnical monitoring of structures of high level of responsibility, the method of spectral-temporal analysis becomes unrealizable for a number of reasons [6]. In particular, it has a low speed, does not allow real-time geotechnical monitoring, and also for overlapping even a very narrow frequency range is requires the use of scores of bandpass filters. For example, in order to allotment the own frequencies of a structures in the interval from 1.6 to 4.2 Hz with an accuracy of less than 3%, 30 filters are required.

2 Using the Method of Frequency Scanning Based on Direct Digital Synthesizers

An effective alternative to the method of spectral-temporal analysis can be the use of the principle of scanning by frequency and isolating the own frequencies of structures in a given range of the frequency spectrum. The main requirements for devices, which implementing this method, are to ensure the coherence of scanning signals and their high stability.

One of the attractive options for implementing the proposed method from a practical point of view is the use of direct digital synthesizers (DDS) [7–13]. In the Fig. 1 is shows the block diagram of the device proposed for use, on which the following designations are accepted: CG - clock generator, PA - phase accumulator, ROM - read-only memory, DAC - digital-to-analog converter, LPF - low-pass filter.

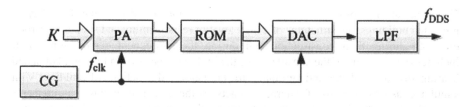

Fig. 1. Block diagram of the DDS.

The DDS is generates the output signal of the required shape and specified frequency f_{DDS}, which is determined by two parameters: the clock frequency f_{clk} and a binary frequency code K, controlled by an automated geotechnical monitoring system. The most important advantages of DDS over other variants of frequency scanning devices are: high accuracy of synthesized signals, extremely high resolution in frequency (up to thousandths of Hz); high speed, easy control, the ability to generate quadrature components of the output signal and low cost of integrated circuits [7, 11]. However, their significant disadvantage at the moment is the insufficient spectral purity of the synthesized signals, which is determined by the presence in the output signal spectrum of a noise component and a scores of discrete parasitic spectral components (PSC). When forming signals with the help of a DDS in the low-frequency region corresponding to ranges of the own frequencies of various structures, in the region of small detuning with respect to the carrier oscillation the discrete components of the output signal spectrum of the DDS due truncation of the phase code and the action of destabilizing factors are most undesirable.

Discrete PSCs caused by the truncation of the phase code are associated with discarding the low-order bits of the phase accumulator when transmitting them in the ROM, which leads to an error in the representation of phase and the appearance of the amplitude inaccuracies when the phase is converted to amplitude [13, 14]. The number of PSCs is determined by two sequences with frequencies:

$$\frac{K}{2^r} \pm n\frac{K}{2^b}, \qquad (1)$$

where r - the bit capacity of the phase accumulator; n - an integer corresponding to the PSC number, $b = r - a$ - number of rounding bits; a - the bit capacity of the ROM.

For $b = 4$, the maximum PSCs level due to truncation of the phase code is independent of the phase accumulator bit capacity and is determined by the number of bits of the ROM [13, 14]:

$$A_{max} = 20\log(2^{-a}) = 6.02a, \text{ dB} \qquad (2)$$

ROMs that are part of the modern DDS, to ensure the required speed are use the value of bit capacity in 14–16 bits. As a result, in the case of a 14-bit ROM the signal-to-noise ratio of the device will be minus 84.28 dB, in the case of a 15-bit ROM - minus 90.3 dB, and in the case of a 16-bit ROM - minus 96.32 dB.

PSCs, caused by the impact of destabilizing factors on the DDS, are related to the instability of its DAC, which is the only analog element of the device [14] and, therefore, is most susceptible to their influence. To the destabilizing factors, manifested by the PSCs in the field of small detunings relative to the carrier oscillation of the DDS, are include: instability of supply voltage, fluctuations of climatic factors (temperature, ambient humidity, atmospheric pressure, etc.), electromagnetic interferences generated by pulsed power supplies and high-voltage lines, as well as mechanical impacts in the form of shock and vibration, under which some radio electronic components can work as converters of mechanical energy into electricity.

PSCs, caused by the impact of destabilizing factors on the DDS, reach their maximum values on the certain time intervals and is characterized by an amplitude value. Their spectrum contains PSCs at certain detunings from the carrier frequency, which can be clearly correlated with certain factors for a given source of signal (for example, frequency of the power line, frequency of vibration).

At the moment, there are two main methods for reducing of PSCs in the spectrum of the output signal of the DDS, are arising from several sources: filtering and randomization. The problems of improving the spectral characteristics of the DDS with the help of filtering are consecrated in the works of F. Kroupa, J. Vankka, K. Halonen, Bar-Giora Goldberg, E. Murphy, C. Slattery, L.I. Ridiko, N.P. Yampurin, L.A. Belov, V.N. Kochemasov, A. Chenakin; with the help of randomization - in the works of Foster Dai Fa, Ni Weining, Yin Shi, C. Jaeger Richard, A.I. Polikarovskih. However, these methods at this moment have limited application and are not effective enough. When filtering is used, the cutoff frequency of the LPF (LPF DDS in Figs. 2, 3, 4 and 5) is adjusted to the maximum output frequency of the DDS, or limited to a value $0.25f_{clk}$, so that the maximum PSC due to the nonlinearity of the DAC was above the sampling frequency and it can be filtered out. As a result, there is always an extremely high probability what discrete PSCs with high amplitudes were falling into the passband of the filter. Randomization of the output signal of DDS is realized by deliberately introducing a random sequence into the low bit of its DAC in the form of jitter of fronts within the clock interval. As a result, the spectrum of the PSCs expands and transforms from a discrete one into a close-to-noise one. However, such a modification of the spectrum, which reduces the level of the PSCs, is equivalent to a significant increase in the phase noises level of the synthesized signals of the device.

3 Application of the Automatic Compensation of Phase Distortions Method to Improve the Spectral Characteristics of Direct Digital Synthesizers

The researches of authors has shown that an effective method of reducing discrete PSCs in the spectrum of the synthesized signal of the DDS is using the method of automatic compensation of phase distortions (ACPD) [15–19]. It has been proved that the unwanted discrete and noise components present in the spectrum of the output signal of the DDS correspond to the parasitic phase modulation of the useful signal. The idea of the autocompensation method for the DDS is that, in the presence of parasitic phase modulation, all components of the spectrum are modulated according to the same law as the synthesized frequency, but with other modulation indices. Since the clock frequency is constant, selecting it in the spectrum of the output signal of the device, it is possible to automatically compensate for phase distortions of the synthesizer output signal at a given frequency.

To isolate phase distortions and form compensating signals, two algorithms are proposed to eliminate differences between the reference and information signal(s) of the ACPD in amplitude and shape while maintaining phase shifts. Both algorithms are implemented structurally by the tracts of formation of control signals of ACPD. The scheme of one of them is shown in Fig. 2. To form the reference signal of the phase detector PD from the output signal of the clock generator CG in the scheme is used the reference tract RT, consisting of the T-trigger Tr1; for the formation of an information

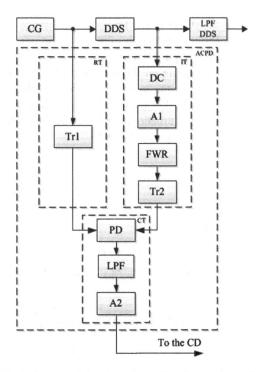

Fig. 2. Block diagram of the formation tract of control signal of ACPD.

signal from the output signal of a digital-to-digital converter DAC - an information tract IT, consisting of a differentiating circuit DC, an amplifier A1, a full-wave rectifier FWR, and a T-trigger Tr2. Further processing of the reference and information signals is carried out in the control tract CT, where phase distortions detection in the PD occurs and their low-frequency filtering in the LPF with subsequent amplification in A2. As a result, a control compensating signal is formed, which is then used to reduce the phase distortions of the DDS in the control device CD.

The transmission coefficient of the phase deviations of the DAC of the DDS to the control output of the CD is determined by the transfer functions by the phase of the links of the tract of the formation of the control signal of the ACPD. Assuming that the transmission coefficient of the information tract by the phase is defined as 0.5, the transmission factor of the amplifier A2 as n_A, denoting the transmission coefficient of the low-pass filter of the control tract in operator form as $M(p)$, and replacing the quasilinear characteristic of detector PD with the corresponding slope K_{PD}, the resulting expression of this transfer function is obtained:

$$H_{\Delta \varepsilon_{DDS} \Delta u} = \frac{\Delta u}{\Delta \varepsilon_{DDS}} = \frac{1}{2} n_A K_{PD} M(p), \qquad (3)$$

where $\Delta \varepsilon_{dds}$ – the phase deviations of the output signal of the DAC of the DDS, Δu – the control compensating signal of the ACPD, p – the Laplace operator.

As the control device of ACPD in the low frequency range it is easiest to use a controlled phase shifter (CPS), the reduction of phase distortions in which is based on the antiphase modulation of the input or output signal of the DDS in accordance with the control signal ACPD. Depending on the location of the CPS relative to the DDS and in which points of the scheme the information about of phase distortions is allocated, developed several types of DDS with ACPD depending on the type of regulation: with the regulation by the perturbation, regulation by the deviation and combined regulation with an adder (Add) - Figs. 3, 4 and 5.

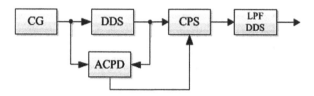

Fig. 3. Block diagrams of the DDS with ACPD and regulation by perturbation.

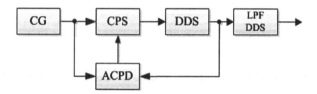

Fig. 4. Block diagrams of the DDS with ACPD and regulation by deviation.

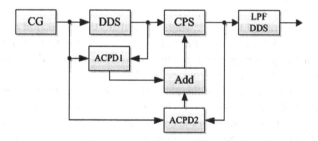

Fig. 5. Block diagrams of the DDS with ACPD and combined regulation.

The DDS with ACPD and the regulation by the perturbation are stable with any characteristics of the components of links, but full compensation of phase distortions in them is impossible. The use of ACPD with a regulation by the deviation allows eliminating this disadvantage: the closed loop of the feedback loop creates conditions for filtering the internal phase deviations of the autocompensator caused by the action of destabilizing factors. However, the use of feedback implies that the device has a static error of compensation. The use of the principle of combined regulation with the use of control circuits by the disturbance and deviation allows us to combine the advantages of both schemes and more flexibly to overcome the main disadvantages of the lasts. So the main suppression of phase distortions is provided in the tract with the regulation by the perturbation (in ACPD1), and the improvement in the quality of the device and the filtering of internal distortions of the all ACPD - in the tract with the regulation by the deviation (ACPD2).

4 Conclusions

It has been theoretically and experimentally established that the degree of automatic compensation of phase distortions, present in the output signal of the DDS, is determined by two factors: the amplitude-frequency response of the devices based by the phase distortions of the DDS and the conditions for full compensation, depending on the type of regulation used. Theoretically, when these conditions are achieved, the phase distortions of the output signal of the DDS are compensated, and the corresponding PSC and noise components are completely eliminated from the output spectrum of the device. However, in practice this is not achievable due to the previously mentioned limitations for each type of regulation. For example, for a DDS with ACPD and the regulation by the perturbation the parameters of the links of autocompensator (LPF and amplifier A2) will have values close to the conditions of full compensation, due to the absence of a feedback loop in the device. The more values of these parameters approach these conditions, the lower the transfer of phase distortions of the DDS to the output of the device. For example, if the gain of A2 deviates from one of the conditions of full compensation (n_{fc}) by 3% the PSCs are reduced by 15 dB, and if deviation is 20%, then by 7 dB - Fig. 6.

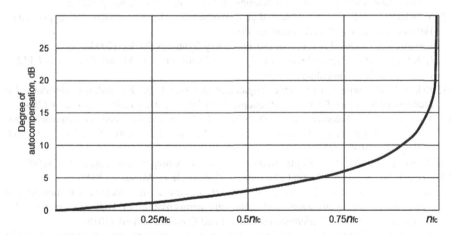

Fig. 6. Dependence of the degree of automatic compensation of the PSCs of spectrum of the output signal of the DDS from the change in the gain factor A2 of the ACPD relative to the conditions for full compensation.

The resulting graphical dependence has two characteristic regions: quasilinear for large detunings from the conditions of complete compensation (0–0.85 nfc) and sharply increasing in the remaining range. This allows us to recommend keeping the values of this coefficient in the band from 0.85 nfc to the maximum attainable in practice values of deviations of this coefficient from the conditions of full compensation in 2–3%.

Acknowledgments. This work was supported by a grant of the Ministry of Education and Science of the Russian Federation № 5.3606.2017/PCH.

References

1. Savin, S.N.: Modern methods of technical diagnostics of building structures of buildings and structures. RDK-print, St. Petersburg (2000)
2. Martins, N., Caetano, E., Diord, S., et al.: Dynamic monitoring of a stadium roof. Eng. Struct. **59**, 80–94 (2014)
3. Salawu, O.S.: Detection of structural damage through changes in frequency: a review. Eng. Struct. **19**(9), 718–723 (1997)
4. Grecheneva, A.V., Kuzichkin, O.R., Dorofeev, N.V., et al.: Application of accelerometers in measuring goniometric systems. Inf. Syst. Technol. **4**(90), 5–10 (2015)
5. Grecheneva, A.V., Kuzichkin, O.R., Dorofeev, N.V., et al.: Application of a phase-measuring method in the inclinometric systems of geotechnical monitoring. In: Proceedings of the IEEE 9th International Conference on Intelligent Data Acquisition and Advanced Computing Systems: Technology and Applications, vol. 1, no. 8095070, pp. 168–171 (2017)
6. Stoica, P., Moses, R.: Cataloging-in-publication data spectral analysis of signals, Library of Congress (2005)
7. Murphy, E., Slattery, K., Vlasenko, Tr.A.: Direct digital synthesis (DDS) in test, measurement and communication equipment. Compon. Technol. **8**, 4p. (2006)
8. Vankka, J., Halonen, K.: Direct digital synthesizers: theory, design and applications. Helsinki University of Technology (2000)
9. Kroupa, V.F.: Phase Lock Loops and Frequency Synthesis. Wiley (2003)
10. Goldberg, B.-G.: Digital Frequency Synthesis Demystified DDS and Fractional-N PLLs. LLH Technology Publishing (1999)
11. Belov, L.A.: Formation of stable frequencies and signal. In: Proceedings Allowance for Students of Supreme Training Institutions. Publishing Center Academy, Moscow (2005)
12. Yampurin, N.P., Safonova, E.V., Zhalnin, E.B.: Formation of precision frequencies and signals. In: Proceedings Allowance. Lobachevsky State University of Nizhny Novgorod, Nizhny Novgorod (2003)
13. Ridiko, L.I.: DDS: direct digital frequency synthesis. Compon. Technol. **7**, 27p. (2001)
14. Koester, W.: Analog-to-Digital Conversion. Technosphere, Moscow (2007)
15. Surzhik, D.I., Kurilov, I.A., Vasilyev, G.S., et al.: Design and mathematical modeling of hybrid frequency synthesizers with automatic compensation of DDS interferences. In: International Siberian Conference on Control and Communications (2015)
16. Surzhik, D.I., Kurilov, I.A., Kuzichkin, O.R., et al.: Modeling the noise properties of hybrid frequency synthesizers with automatic compensation of phase noise of DDS. In: International Siberian Conference on Control and Communications (2015)

17. Vasilyev, G.S., Kuzichkin, O.R., Kurilov, I.A., et al.: Analysis of noise properties of a hybrid frequency synthesizer with autocompensating phase noise of DDS and PLL. In: International Siberian Conference on Control and Communications (2016)
18. Surzhik, D.I., Kurilov, I.A., Vasil'ev, G.S.: Compensation of distortions of the DDS in hybrid frequency synthesizers. Radio Eng. Telecommun. Syst. **4**(20), 13–19 (2015)
19. Kurilov, I.A., Surzhik, D.I., Kharchuk, S.M.: Frequency characteristics of an autocompensator for phase noises of a direct digital synthesizers. Radio Eng. Telecommun. Syst. **1**(17), 12–20 (2015)

Structural and Functional Model of the On-board Expert Control System for a Prospective Unmanned Aerial Vehicle

P. I. Tutubalin and V. V. Mokshin[✉]

Kazan National Research Technical University, 31 Karla Marksa, Kazan 420111, Russian Federation
vvmokshin@kai.ru

Abstract. Currently, such expert systems that are used in the "pilot advisor" mode are widely used in manned aircraft. The main task that these systems allow to solve is that it becomes possible to formulate recommendations when the pilot has an acute lack of time to make appropriate decisions. Manned aircraft are being replaced by unmanned aerial vehicles, which entails changes in the intelligent navigation and control system of such aircraft. The main feature of the corresponding intelligent navigation and control system is the objective need for the formation and implementation of solutions with minimal involvement of personnel of the corresponding unmanned aerial system. The article discusses the structure and model of an on-board expert control system for a promising unmanned aerial vehicle. The structure of the interaction of the proposed promising on-board expert control system for an unmanned aerial vehicle with sources of initial data and consumers of the formed control solutions is presented. The main issues of the organization's work on the formation of the goals of the on-board expert management system are considered. The main issues of operation of an on-board expert control system of a promising unmanned aerial vehicle are considered.

Keywords: Model · Management · System · Unmanned aerial vehicle

1 Introduction

Currently, expert systems are one of the most used in practice components of artificial intelligence systems.

In manned aircraft technology, expert systems are used in the "pilot's adviser" mode, the main purpose of which is to formulate recommendations with a shortage of time for making appropriate decisions.

The main feature of the intelligent navigation and control system of modern unmanned aerial vehicles is the objective need for the formation and implementation of solutions with the minimum participation of the personnel of the corresponding unmanned aerial vehicle complex.

This requires the design and development of special expert systems, different from the existing ones. Such developments should be conducted with the use of well-tested mathematical models and methods, as well as numerical methods [1]. For example, like those used in the following works devoted to the processing of graphic and video information [2–5]. The complexity of the simulated systems and the processes taking place in them [6–8] allows us to conclude that the models and methods used in them fully satisfy modern requirements for the modeling of these or other complex systems.

It is natural to note that in all systems using the models and methods noted in [1–8], it is necessary to apply sufficiently reliable approaches, models and methods to ensure the safety of the information that is circulated and processed in these systems [9]. In this regard, we recommend that attention be paid to a number of the following papers [10–13], in which the basic principles, approaches and methods are sufficiently thoroughly expounded, allowing both to substantially increase the level of information security of a particular system [14–16], and to ensure the necessary level of information security of the said system.

In addition, we note the already common interest in data encryption [17–22], but it is highly specialized, which is detrimental to large systems.

2 Model of the Onboard Expert Control System of an Unmanned Aerial Vehicle

Consider the structure and model of the on-board expert control system for a prospective unmanned aerial vehicle.

We assume that the initial data for the analysis of the current flight and operational situation arrive in the on-board expert control system from a set of sensors of the information and measuring system of an unmanned aerial vehicle $D = \{D_1, D_2, \ldots, D_N\}$ and from the elements of the target equipment that make up the set installed on board $C = \{C_1, C_2, \ldots, C_M\}$.

We will distinguish the following components in the on-board expert control system: E_1 - block of formation of the system's goals; E_2 - block for the formation and analysis of control solutions; E_3 - knowledge base of the system; E_4 - the basis of the facts of the functioning of the system.

Figure 1 shows the structure of the proposed advanced onboard expert control system for an unmanned aerial vehicle and the interaction of the system with the sources of input data and consumers of the generated control decisions. The following designations are used on it: UAV - unmanned aerial vehicle, IIS - information-measuring system, BFG - the block of formation of goals, BFACS - block for the formation and analysis of control solutions, KB - knowledge base, FB - the basis of factors, ACS - automatic control system of UAV, EMCS - executive mechanisms of the control system UAV, OCC - onboard computer complex of UAV.

The structure of the onboard expert control system in the general case will be described using the formalism of binary relations by expressions of the form:

$$Q_1 \subseteq D \times E, Q_2 \subseteq C \times E, Q_3 \subseteq E \times E \qquad (1)$$

where E is the set of elements of the system.

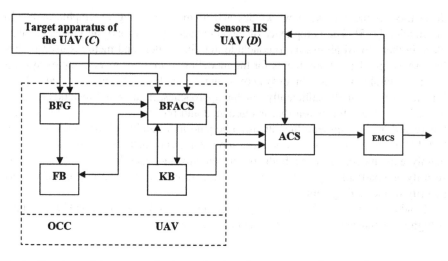

Fig. 1. Structure of the proposed advanced onboard expert control system for an unmanned aerial vehicle.

Let us describe the content and functioning of the elements of the on-board expert control system, beginning with such a traditional element of the expert system as the knowledge base.

In our case, the knowledge base of the on-board expert control system is the main one for choosing control actions on a prospective unmanned aerial vehicle when solving a specific target task.

In contrast to the traditional approach to describing the controlled motion of a prospective unmanned aircraft in a perspective on-board expert control system, it is proposed to use a command-controlled flight of an unmanned aerial vehicle, which in the general case is described by a vector system of differential equations of the form:

$$\dot{x} = f(t, x, K), t \in [t_0, t_k] \qquad (2)$$

with the initial conditions:

$$x(t_0) = x_0, \qquad (3)$$

Here, x is a vector of phase coordinates (flight parameters) of a prospective unmanned aircraft; t_0 and t_k are, respectively, the initial and final moments of the solution time by the perspective unmanned aerial vehicle of the target task; f is a vector function of the marked arguments.

The set of K control commands for a prospective unmanned aerial vehicle entering the right-hand side of Eq. (2) has the form:

$$K = \{k_1(a), k_2(a), \ldots, k_m(a)\}, \qquad (4)$$

where $k_j(a)$ is the name (code number, number) of the j-th control command; a – a vector of parameters describing the maneuvers of a prospective unmanned aircraft during the implementation of control commands, $j = \overline{(1,m)}$.

For example, to implement the maneuver "dive an unmanned aerial vehicle from an altitude h_1 at an angle θ to a horizontal flight at an altitude h_2", the command k_5 – "descent of an unmanned aerial vehicle with a parameter vector" is used $a = (h_1, \theta, h_2)$.

To execute a maneuver "the turn of a perspective unmanned aerial vehicle with radius r at an altitude h", the command k_8 – "right turn with a vector $a = (+1, r, h)$.

Note that at each instant of time $t \in [t_0, t_k]$, one and only one element of the set (4) must be present on the right-hand side of Eq. (2). This means that one particular control command $k_j(a), j \in \overline{(1,m)}$ acts on the unmanned aerial vehicle at each instant of time t.

We introduce the Boolean function:

$$\varpi_j(t) = \{0, 1\}, \tag{5}$$

which takes the value 1, if at the time $t \in [t_0, t_k]$ the on-board expert control system initiates the command $k_j(a)$ and, accordingly, the value 0 otherwise; and in either case, where $j = \overline{(1,m)}$.

Then the condition of using only one control command will look like:

$$\sum_{j=1}^{m} \varpi_j(t) = 1, t \in [t_0, t_k], \tag{6}$$

Taking into account expressions (5) and (6), the model of controlled motion of a prospective unmanned aerial vehicle (2) will be written as

$$\dot{x} = f(t, x, \sum_{j=1}^{m} k_j(a)\varpi_j(t)), t \in [t_0, t_k], \tag{7}$$

In this case, the choice of control actions for an unmanned aerial vehicle can be considered as a choice on the time interval $[t_0, t_k]$ of functions $\varpi_1(t), \varpi_2(2), \ldots, \varpi_m(t)$ satisfying conditions (5), (6) and the requirements of the target task being solved. In this case, the control command of an unmanned aerial vehicle used at time t is determined from the expression:

$$k(t) = \sum_{j=1}^{m} k_j(a) \times \varpi_j(t), \tag{8}$$

The knowledge base of the onboard expert management system is proposed to be formed as a set of rules:

$$\Pi = \{\pi_{ij} | i = \overline{(1,n)}, j = \overline{(1,m)}\}, \tag{9}$$

where each rule π_{ij} has the form:

$$\pi_{ij}: \text{ if } P_i(c(t), d(t)) = 1, \text{ then } \varpi_j(t) = 1, \tag{10}$$

In this expression $P_i(c(t), d(t))$ is a predicate, the arguments of which for each moment of time are the values of the environment characteristics vectors $c(t) = (c_1(t), c_2(t), \ldots, c_M(t))$ received from the target equipment of the prospective unmanned aircraft and the gauge vector of the sensors $d(t) = (d_1(t), d_2(t), \ldots, d_H(t))$ of its information-measuring system.

With the use of this base, the block for the formation and analysis of the control solutions of the system at each instant of time t tests the truth of all the predicates entering into the expressions (9), (10) and extracts the predicate number:

$$i_0 = \arg_i \{P_i(c(t), d(t)) \equiv 1, i = (\overline{1,n})\}$$

which is true at a given time t.

Further, this block with $i = i_0$ performs a rule analysis π_{ij} and allocates the corresponding number $j = j_0$ of the control function $\varpi_j(t) = 1$.

After that, according to the formula (8), the command for control of the prospective unmanned aerial vehicle initiated by the formation and analysis unit of the control solutions is formed.

These actions of the block of the on-board expert control system under consideration are described by a conditional dynamic relation of the form:

$$Q_4(u(t)) \subseteq P \times K, t \in [t_0, t_k], \tag{11}$$

where $P = \{P_i | i = (\overline{1,n})\}$ is the set of predicates of the knowledge base of the system; K is the set of control commands (4); $u(t)$ – applicable condition:

$$u(t) = \{(P_i(c(t), d(t)) = 1) \wedge (\varpi_j(t) = 1)\}, \tag{12}$$

The relation (11) defines a dynamic graph in which at any time $t \in [t_0, t_k]$ there is only one arc (P_i, k_j), $i \in (\overline{1,m})$, $j \in (\overline{1,n})$ for which condition (12) is satisfied.

The composition of the arguments included in the predicates from expressions (10) is represented by the following relations:

$$Q_5 \subseteq D \times P, \ Q_6 \subseteq C \times P, \tag{13}$$

Taking into account the cyclograms for the formation of the values of the vectors $d(t)$ and $c(t)$ with $t \in [t_0, t_k]$ for the relation (13), we can consider both the dynamic relations $Q_5(t)$ and $Q_6(t)$.

For each control command k_{j0} selected at the time t in the block for the generation and analysis of control decisions of the system, an analysis of the effectiveness of its implementation is carried out by predicting the values of the flight parameters of a prospective unmanned aerial vehicle over a time interval $[t, \tau]$. This process is carried out by integration of the system of Eq. (7) with $\varpi_{i0}(t) = 1$ and the corresponding initial conditions of the form (3) in the on-board computer complex of the unmanned aerial vehicle, where $t_0 = t$. In this case, the vector values $d(t)$ are certain components of the vector x_0.

Let be the $x^*(\tau)$ - vector of the required values of the flight parameters of a prospective unmanned aerial vehicle at the moment $\tau \in [t_0, t_k]$ of arrival to the target control unit of the onboard expert control system from the subsystem "Target task" of the intelligent navigation and control system of an unmanned aerial vehicle, $\hat{x}(\tau)$ - the predicted value of the vector of such parameters at this time.

The selected command k_{j0} is considered realizable if the following condition is satisfied:

$$||\hat{x}(\tau) - x^*(\tau)|| \leq \varepsilon,$$

where $||(\cdot)||$ – is the normal of the vector $(\cdot)$; ε – vector of permissible deviations, reflecting the requirements for the accuracy of the solution of the target mission of the UAV.

This control command is transmitted from the block of formation and analysis of control solutions to the ground control point of an unmanned aerial vehicle for its execution, see Fig. 1.

In the ground part of this complex, each command $k_j \in K$ is assigned certain laws of control $z_j \in Z_1$ and stabilization Z_2 that determine the changes in the angles of deviation of the rudders $\delta_{Bj}(t)$, $\delta_{Hj}(t)$, $\delta_{\jmath_j}(t)$, and the "throttle" (traction control) $\delta_{\jmath_j}(t)$ of the engine of a prospective unmanned aerial vehicle, $j = \overline{(1, n)}$. When choosing the automatic control system of the appropriate control and stabilization law, signals for their implementation are generated that are transmitted to the executive mechanisms of the unmanned aerial vehicle for the direct execution of control decisions.

The connection between control commands and the laws that implement them is represented in the form of a relationship:

$$Q_7 \subseteq K \times Z_1 \times Z_2, \quad (14)$$

3 Features of the Development and Use of the Target Shaping Unit of the On-board Expert Control System

Let's consider the issues of creating and using the block for forming the objectives of an on-board expert control system.

We will assume that this block is formed on the basis of the set of S required potential events that should and can occur in the process of solving the target unmanned aerial vehicle of the target task on the time interval $[t_0, t_k]$.

We denote by S_1 and S_0 representations of the onset or not the occurrence of an event $s \in S$.

We introduce into consideration a set $T = \{\tau_1, \tau_2, \ldots, \tau_r\}$ of moments of time in which the occurrence of events of the set S is planned. Note that in the set T conditions can be fulfilled that $\tau_1 = t_0$ and $\tau_r = t_K$. This can correspond to the events "the beginning of the implementation of the target unmanned aerial vehicle by the target task" and "the completion of the fulfillment by the prospective unmanned aerial vehicle of the target task".

The relationship between the sets S and T will be described by the relation:

$$Q_8 \subseteq T \times S, \tag{15}$$

The base of the process of operational goal-setting is the basis of the rules:

$$\Gamma = \{\gamma_s | s \in S\}, \tag{16}$$

in which the following rules are used:

$$\gamma_s : \begin{cases} \text{if } \{\rho_S(c(\tau), d(\tau)) = 1\}, \text{ then } (s = s_1) \\ \text{else } (s = s_0) \end{cases}, \tag{17}$$

These rules, by analyzing the set of predicates $\rho = \{\rho_S | s \in S\}$, whose arguments are the values of the vectors $c(t)$ and $d(t)$ at times $\tau \in T$, allow us to control the process of the onset or non-occurrence of events of the set S.

In this case, it is considered that if the actual set S after the expiration of time $(t_k - t_0)$ consists only of elements of the form S_1, then the target task posed to the perspective unmanned aerial vehicle is completely fulfilled.

If an event of the type S_0 appears at some time $\tau \in T$, then the target formation block must fix it and form for the other elements of the on-board expert control system the goal of transferring this event from the state S_0 to the state S_1 in the minimum time.

To form such goals at each point in time $\tau \in T$, we will form a set:

$$S_0(\tau) = \{s | \rho_S(c(\tau), d(\tau)) = 0, s \in S\},$$

describing the event of the set S that did not arrive at this instant of time.

The fulfillment of the goals for the liquidation of such events in the onboard expert control system is entrusted to the block for the formation and analysis of control decisions and the knowledge base of the system.

When organizing the interaction of the target formation unit with these elements, we will use a conditional dynamic relation of the form:

$$Q_9(\upsilon(\tau) \subseteq \rho \times P, \tau \in T, \tag{18}$$

built on the basis of the condition:

$$\upsilon(\tau) = \{\rho_S(c(\tau), d(\tau)) = 0, s \in S\}$$

The application of this condition means that in the graph describing the relation (18), for each moment of time $\tau \in T$ there are arcs determined by the set $S_0(\tau)$. The subset P of predicates, defined by the relation (18), is used to form control commands that ensure the achievement of goals set by the target formation block.

In the basis of the facts of the system, information reflecting the situations that arose in the previous flights of the prospective unmanned aircraft and other prospective aircraft of this class, as well as the relevant actual and implemented control decisions, should be stored.

We will describe the content of the facts base of the on-board expert system of attitude management:

$$Q_{10} \subseteq S \times K^*, \qquad (19)$$

where $K^* \subseteq K$ is a subset of commands used in the practice of flying an unmanned aerial vehicle with the corresponding actual values of the vectors a^*.

4 The Main Stages of the Functioning of the Onboard Expert Control System for a Pilot Unmanned Aerial Vehicle

It should be noted that the information accumulated in the database based on the results of flights of a prospective unmanned aerial vehicle is used in the block for the formation and analysis of control decisions for the selection of controls by a prospective unmanned aerial vehicle using precedents.

The above expressions (1), (7), (9)–(19) are a structural and functional model of the on-board expert control system of a prospective unmanned aerial vehicle.

The main stages of the functioning of this system are:

- Analysis of the current flight situation in the block of formation of the objectives of the system
- When situations arise such as $s_0 \in S$ the transfer of the objective of operational control of an unmanned aerial vehicle to the block for the formation and analysis of control decisions of the system
- The appeal of the block for the formation and analysis of management decisions to the mere fact for finding control decisions on "precedents". Analysis of the feasibility of "ready-made" solutions. In the absence of such decisions or if they are not realizable, the block for the formation and analysis of management decisions conducts the formation and analysis of new solutions involving the knowledge base and forecasting the consequences of their implementation
- Transfer of the selected or generated solution to the automatic control system of a prospective unmanned aerial vehicle for its implementation

The functioning of the onboard pilot control system for an unmanned aerial vehicle starts at time t_0 and is completed when the current target solution time t_k is reached.

Note that the flight of an unmanned aerial vehicle in the time interval $[t_0, t_k]$ is carried out according to a program located in the "memory" of the automatic control system.

At the same time, the on-board expert control system monitors its implementation, eliminates all possible deviations and manages the process of solving the target task assigned to a prospective unmanned aerial vehicle.

5 The Main Issues of Operation of the On-board Expert Control System

Let's consider the basic questions of operation of the onboard expert control system of a perspective unmanned aerial vehicle.

The operational personnel of the system, which are part of the personnel of the unmanned aerial system, are: (1) a specialist in the management of a prospective unmanned aerial vehicle; (2) expert on the representation of values in the expert system (knowledge engineer); (3) software engineer.

The work of this staff is related to the "training" of the system at the initial stage of its implementation and correction of the content and algorithms of the target formation unit, the facts base, the knowledge base and the block for the formation and analysis of control solutions in the process of operation of the on-board expert control system for a prospective unmanned aerial vehicle.

For these purposes, the current version of the on-board expert control system is stored at the ground control station of the unmanned aerial system, and the current values of the environment parameters and the solutions used by the system with the help of the radio data exchange system from the perspective unmanned aerial vehicle are transmitted to the information display devices and storage devices of the workstation of the perspective management specialist unmanned aerial vehicle. The latter has the ability to correct these decisions in the mode of "manual" control of a prospective unmanned aerial vehicle in real time with their respective memorization. After the flight of a prospective unmanned aerial vehicle, all information received is analyzed by specialists in the board and knowledge representation and the required correction of the corresponding components of the unmanned aerial vehicle is made. The implementation of these changes in the on-board computer complex of an unmanned aerial vehicle is carried out by a software engineer.

The efficiency criteria of the on-board expert control system for a prospective unmanned aerial vehicle is the minimum of adjustments made by its personnel and the maximum degree of complete autonomous solution of the targets assigned to the unmanned aerial vehicle.

With the development of an on-board expert control system for a prospective unmanned aerial vehicle, it is proposed to expand a number of K commands to control the target equipment of an unmanned aerial vehicle and to use algorithms for processing video information received from the corresponding components of this equipment as part of the system goal formation unit.

6 Conclusion

1. The article proposed a model for describing the functioning and analysis of the effective performance of the tasks assigned to a prospective unmanned aerial vehicle operating as part of an unmanned aerial system with the deployment of an automatic control system both on the aircraft and within the ground control complex.

2. The proposed model is constructed from the calculation of the active application for the formation and adoption of managerial decisions regarding the perspective unmanned aerial vehicle of the expert system.
3. The proposed model has a broad focus of its application in the practice of the operation of various information systems, which are not only aviation ones.

References

1. Yakimov, I.M., Kirpichnikov, A., Mokshin, V.V., et al.: The comparison of structured modeling and simulation modeling of queueing systems. Commun. Comput. Inf. Sci. **800** (2017). https://doi.org/10.1007/978-3-319-68069-9_21
2. Yakimov, I.M., Trusfus, M.V., Mokshin, V.V., et al.: AnyLogic, ExtendSim and Simulink overview comparison of structural and simulation modelling systems. In: 3rd Russian-Pacific Conference on Computer Technology and Applications (2018). https://doi.org/10.1109/rpc.2018.8482152
3. Lyasheva, S.A., Medvedev, M.V., Shleymovich, M.P.: The analysis of image characteristics on the base of energy features of the wavelet transform. In: CEUR Workshop Proceedings. Session Image Processing and Earth Remote Sensing, Samara, April 2018, pp. 96–102 (2018)
4. Saifudinov, I.R., Mokshin, V.V., Tutubalin, P.I.: Visible structures highlighting model analysis aimed at object image detection problem. In: CEUR Workshop Proceedings. Session Image Processing and Earth Remote Sensing, Samara, April 2018, pp. 139–148 (2018)
5. Mokshin, A.V.: Adaptive genetic algorithms used to analyze behavior of complex system. Commun. Nonlinear Sci. Numer. Simul. **71**, 174–186 (2018). https://doi.org/10.1016/j.cnsns.2018.11.014
6. Mokshin, V.V., Saifudinov, I.R., Sharnin, L.M., et al.: Parallel genetic algorithm of feature selection for complex system analysis. In: IOP Conference Series: Journal of physics: Conference Series (2018). https://doi.org/10.1088/1742-6596/1096/1/012089
7. Yakimov, I.M., Kirpichnikov, A.P., Mokshin, V.V.: Modeling of complex systems in the simulation environment GPSS W with an extended editor. Bull. Kazan Technol. Univ. **17**(4), 298–303 (2014)
8. Yakimov, I.M., Kirpichnikov, A.P., Matveeva, S.V., et al.: Simulation modeling of complex systems by means of ARIS TOOLSET 6. Bull. Kazan Technol. Univ. **17**(15), 338–343 (2014)
9. Tutubalin, P.I., Mokshin, V.V.: The evaluation of the cryptographic strength of asymmetric encryption algorithms. In: Second Russia and Pacific Conference on Computer Technology and Applications, Vladivostok, September 2017, pp. 180–183. IEEE (2017). https://doi.org/10.1109/rpc.2017.8168094
10. Moiseev, V.S.: A probabilistic dynamic model for the functioning of active protection software for mobile distributed ACS. Inf. Technol. **6**, 37–42 (2013)
11. Tutubalin, P.I.: Optimization of selective control of the integrity of information systems. Inf. Secur. **15**(2), 257–260 (2012)
12. Moiseev, V.S.: General model of a large-scale mobile distributed ACS. Nonlinear World **9**(8), 497–499 (2011)

13. Tutubalin, P.I.: Application of models and methods of stochastic matrix games for ensuring information security in mobile distributed automated control systems. Nonlinear World **9**(8), 535–538 (2011)
14. Tutubalin, P.I.: The main tasks of the applied theory of information security ASU. Sci. Tech. Herald Inf. Technol. Mech. Opt. **39**, 63–72 (2007)
15. Moiseev, V.S.: A two-criteria game-theoretic model with a given ordering of mixed strategies. Bull. Kazan State Tech. Univ. **1**, 40–45 (2005)
16. Gremyachensky, S.S.: Introduction to the game-theoretic analysis of radio-electronic conflict of radio communication systems with radio-suppression means and some estimates of the results of the conflict. VNIIS, Voronezh (1995)
17. ANSI X9.30-1995, Part 1: Public key cryptography using irreversible algorithms for the financial services industry: The Digital signature algorithm (Revised)
18. ANSI X9.30-1993, Part 2: Public key cryptography using irreversible algorithms for the financial services industry: The Secure Hash algorithm 1 (SHA-1) (Revised)
19. ANSI X9.62-1998: Certificate management
20. ANSI X9.63-199x: Elliptic curve key agreement and transport protocols, draft
21. Lenstra, A.K.: Key lengths. Contribution to the handbook of information security. Lucent technologies and technische Universiteit Eindhoven (2004)
22. Lenstra, A.K., Verheul, E.R.: Selecting cryptographic key sizes. J. Cryptol. **14**, 255–293 (2001). https://doi.org/10.1007/s00145-001-0009

Simulation Modelling of the Adaptive System of Structurally and Parametrically Indefinite Object with Control Lag

L. V. Chepak[✉] and Z. D. Pikul'

Amur State University, 21 Ignatevskoye Highway, Blagoveshchensk 675027, Russian Federation
chepak@inbox.ru

Abstract. The problem of simulation modelling of the adaptive system with explicit reference model for structurally and parametrically indefinite object is considered. This system is synthesized using the V.M. Popov's hyperstability criterion. The structurally and parametrically indefinite object has transfer function relative order greater than unity, as well as has the control channel lag. The object under study operates in terms of permanent influence of external uncontrolled disturbance. The negative impact of the control lag is compensated by the connecting the predictor-compensator to the main control loop. The evaluations of the immeasurable state variables of the object obtained at the filter-corrector output are used in the control law. During computational experiments, the selection of control loop and filter-corrector coefficients has been carried out to ensure good quality of the object output tracking the reference signal. The simulation modelling results have shown that the synthesized system satisfies the required properties and achieves the set goal of control during its operation.

Keywords: Structural uncertainty · A priori parametric uncertainty · Control lag · Filter-corrector · Predictor-compensator · Hyperstability criterion

1 Introduction

The structural and parametric uncertainties of the controlled object, change in its parameters over time, lag presence, susceptibility to the influence of external uncontrolled disturbance are far from all difficulties encountered by the developers of the control systems for such objects [1–3]. In robotics, aircraft building, electric power industry there are multimode systems. The normal operation of such automatic control systems implies not only variation of the control plant parameters, but also changes in its structure. The problems of such systems controllability and the optimal control problems of linear composite systems were considered in [1, 4, 5]. The development of control laws for multimode systems, as it shown in [2, 3, 6, 7], can be carried out with the help of adaptive or robust control methods. In modern control theory, there are a number of methods that allow us to synthesize control algorithms for parametrically and structurally uncertain plants [8, 9]. In this paper we use the Popov's hyperstability criterion [7, 10, 11], which contains several consecutive stages of design: firstly, we obtain an equivalent mathematical description of the system under consideration; secondly, we ensure the conditions of strict positive definiteness for linear stationary part of the equivalent system; thirdly, we define explicit form of the control law components which are satisfy the integral inequality. The final stage of the adaptive control systems development is their simulation, with the help of which the dynamic properties of the system and its performance are explored. In addition, during simulation modelling the controller coefficients are selected; the transient response of the system at changing the parameters of the controlled object and external interference is studied; the sampling interval of the control law for discrete-continuous systems is selected.

In this article, the problem of simulation modelling of the adaptive system of control of the structurally and parametrically indefinite object with control channel lag, synthesized basing on the hyperstability criterion, is solved using the results of [10–15].

2 Mathematical Description of the System

The dynamics of the controlled object is described in the operator form as "input-output" model:

$$a(p)y(t) = b_m b(p)(u(t-h) + f(t)),$$
$$p^i(0) = y_{i0}, \ u(\theta) = \phi(\theta), \ \theta \in [-h, 0], \tag{1}$$

where $y(t) \in R$ – object output, $u(t) \in R$ – control; $h = const > 0$ – known permanent lag; $f(t)$ – external interference, $|f(t)| \leq f_0$, $f_0 = const > 0$; $\varphi(\theta)$ – bounded continuous initial function; $p = d/dt$ – operator of differentiation; $b_m = const > 0$ – coefficient; $a(p)$, $b(p)$ – standard polynomials

$$a(p) = p^n + a_{n-1}p^{n-1} + \ldots + a_1 p + a_0,$$
$$b(p) = p^m + b_{m-1}p^{m-1} + \ldots + b_1 p + b_0. \tag{2}$$

The following conditions are supposed to be fulfilled for the controlled object (1):

- Only scalar output of the object (1) is available to direct measurement
- Polynomial $b(p)$ is Hurwitz's
- The controlled object (1) is a priori parametrically indefinite, i.e. the coefficients of polynomials (2) are unknown values depending on the set of unknown parameters belonging to the known limited set
- The controlled object (1) is structurally indefinite, i.e. the degrees of polynomials (2) n, m are unknown numbers
- n_a is known, it is the maximum degree of polynomial $a(p)$
- Relative order of the controlled object (2) $\rho = n - m > 1$ is a known value

We will consider that the structural uncertainty occurs in case of $n_a < n$ and is absent in case of $n_a = n$.

Following [11, 12], to dispose of the structural uncertainty we will use the polynomial:

$$(p+v_0)^{n_a-n}, \quad v_0 = const > 0,$$

and write the equation of the object (1) in the equivalent form

$$c(p)y(t) = b_m d(p)(u(t-h) + f(t)),$$
$$c(p) = a(p)(p+v_0)^{n_a-n} = p^{n_a} + c_{n_a-1}p^{n_a-1} + \ldots + c_1 p + c_0, \quad (3)$$
$$d(p) = b(p)(p+v_0)^{n_a-n} = p^{n_a-\rho} + d_{n_a-\rho-1}p^{n_a-\rho-1} + \ldots + d_1 p + d_0.$$

In the state space, the equation of object (3) will be written as follows:

$$\frac{dx(t)}{dt} = Ax(t) + B(u(t-h) + f(t)), \quad y(t) = L^T x(t),$$
$$x(0) = x_0, \quad u(\theta) = \phi(\theta), \quad \theta \in [-h, 0] \quad (4)$$

where $x(t) \in R^{n_a}$ – state vector, x_0 – initial conditions vector; $u(t) \in R$ – control; h – known permanents lag; $\varphi(\theta)$ – bounded continuous initial function; $f(t)$ – external interference; A – Frobenius matrix of n_a order; L и $B^T = [0,\ldots,0,1]$ – n_a order vectors; $y(t) \in R$ – output.

To obtain the values of immeasurable state variables the following filter-corrector is used in the main loop of the control system:

$$\frac{dx_f(t)}{dt} = A_f x_f(t) + B_f y(t), \quad y_f(t) = G^T x_f(t), \quad (5)$$

where $y_f(t)$ – filter-corrector output; $x_f(t) \in R_{n-1}$ – state vector of the filter-corrector; A_f, G – constant matrix and vectors, their parameters are chosen in a special way [13–16].

The desirable dynamics of the controlled object (1), as well as (4), is assigned in the state space by the explicit reference model:

$$\frac{dx_m(t)}{dt} = A_m x_m(t) + B_m r(t), \quad y_m(t) = G^T x_m(t), \quad (6)$$

where $x_m(t) \in R^{n_a}$ – state vector of the reference model; A_m – Hurwitz matrix; $r(t) \in R$ – reference-input signal; $y_m(t) \in R$ – output of the model, $G^T = [g_0, g_1, \ldots, g_{n_a-1}]$, $B_m^T = [0, 0, g_0^{-1}]$ – n_a order vectors.

The parameters of vector G are selected so that the roots of polynomial $g_{n_a-1}p^{n_a-1} + g_{n_a-2}p^{n_a-2} + \ldots + g_1 s + g_0$ coincide with $(n_a - 1)$ root of the polynomial $\det(pE - A_m)$.

Since the object under study (1) has control lag, we will connect the predictor-compensator to the main control loop [6–8]:

$$\frac{dx_k(t)}{dt} = A_m x_k(t) + B_m(u(t) - u(t-h)), \quad y_k(t) = G^T x_k(t), \tag{7}$$

where $x_k(t) \in R^{n_a}$ – state vector of the predictor-compensator, $y_k(t) \in R$ – output of the model.

The predictor-compensator parameters are specified similarly to the reference model (6).

Let us form the structure of the control law:

$$u(t) = r(t) + u_1(t) + u_2(t) + u_3(t), \tag{8}$$

where $u_1(t)$, $u_2(t)$, $u_3(t)$ are the following control signal components:

$$u_1(t) = \sum_{i=1}^{n} h_i x_{fi}(t) \int_0^t x_{fi}(s)(y_m(s) - y(s) - y_k(s)) ds,$$

$$u_2(t) = \tilde{h}_1 \int_0^t (y_m(s) - y(s) - y_k(s)) ds + \tilde{h}_2(y_m(s) - y(s) - y_k(s)), \tag{9}$$

$$u_3(t) = \tilde{h}_3 u(t-h) \int_0^t u(s-h)(y_m(s) - y(s) - y_k(s)) ds,$$

$$h_i = const > 0, \quad \tilde{h}_1, \tilde{h}_2, \tilde{h}_3 = const > 0$$

Research objective. It is required to select the values of constant coefficients of nonlinear components (9) of the control law (8) and the parameters of the filter-corrector (5) in such a way that in terms of a priori parametric and structural uncertainty of the system (1)–(9) it could provide the satisfaction of the target inequation

$$\lim_{t \to \infty} |y_m(t) - y(t)| \leq \sigma_0 = const > 0, \tag{10}$$

where σ_0 – sufficiently small number.

3 Computational Experiment

The simulation modelling is intended for evaluation of the operation quality of the developed adaptive control system and selecting the values of permanent coefficients of the adaptation loop (9) and filter-corrector (5). In the test example due to the change in the structure and parameters of the controlled object their influence on the dynamic properties of the adaptive control system (1), (2), (5)–(9) has been researched. In a series of computational experiments, we considered the controlled object (1) with relative order, $\rho = 2$, which transfer functions have been as follows

$$y(s) = \frac{b_{11}s + b_{01}}{(s+a_{21})(s+a_{11})(s+a_{01})} \cdot e^{-hs} u(s), \tag{11}$$

$$y(s) = \frac{b_{02}}{(s+a_{12})(s+a_{02})} \cdot e^{-hs} u(s) \tag{12}$$

where coefficients a_{ij} и b_{ij} could take any values from available ranges:

$$\begin{aligned} -0.1 \leq a_{01} < 0; \ 0.1 \leq a_{11} \leq 2; \ 1 \leq a_{21} \leq 3; 0.5 \leq b_{01} \leq 2; \\ 0.45 \leq b_{11} \leq 2; \ -0.2 \leq a_{02} \leq 0; \ 2 \leq a_{12} \leq 5; \ 1 < b_{02} \leq 3, \end{aligned} \tag{13}$$

In one of the experiments, the coefficients have had the values for the object with transfer function (11):

$$b_{11} = 0.8, \ b_{01} = 0.96, \ a_{21} = 2.05, \ a_{11} = 0.6, \ a_{01} = -0.05, \tag{14}$$

For the object with transfer function (12):

$$b_{02} = 2.2, \ a_{12} = 3.56, \ a_{02} = -0.06, \tag{15}$$

Control lag $h = 0.5$ s, initial values $y_{0i} = 0$, $\varphi(\theta) = 0$, $\theta \in [-h, 0]$.

The desired behavior of object (11), (14) or (12), (15) has been set by the explicit reference model (6) with transfer function

$$y_m(s) = \frac{1}{(s+1)(0.5s+1)} r(s), \tag{16}$$

The predictor-compensator (7) has been set by the transfer function of the reference model (16).

The parameters of the filter-corrector (5) are:

$$y_f(s) = \frac{s+1}{(T_0 s + 1)^2} y(s), \tag{17}$$

where time constant T_0, has been selected during simulation modelling out of the conditions:

$$T_0 < \frac{0.93}{(n_a - 2) a_{m, n_a - 1}}, \quad T_0 < \frac{0.465 \cdot a_{m, n_a - 1}}{(n_a - 1) a_{m, n_a - 2}}$$

$a_{m, n_a - 1}$, $a_{m, n_a - 2}$ – are corresponding coefficients of the polynomial $det\ (pE - A_m)$ [17–20].

The best operation quality of systems (1), (2), (5)–(9), (11), (14), (16), (17) and (1), (2), (5)–(9), (12), (15)–(17) has been recorded at $T_0 = 0.001$ s.

The coefficients of the controller (8), (9) at $n_a = 3$ have been selected as

$$h_1 = 200, \ h_2 = 300, \ h_3 = 250, \ \tilde{h}_1 = 150, \ \tilde{h}_2 = 10, \ \tilde{h}_3 = 1.5, \qquad (18)$$

To make the comparison of responses of the researched systems easier the external interferences have been identical, in particular, the reference-input signal has been described by the equation

$$r(t) = 1.11 \sin 0.005t \cdot \sin^2 0.0075t, \qquad (19)$$

and permanent disturbance has been as follows:

$$f(t) = -0.2 \sin 0.005t, \qquad (20)$$

Figures 1, 2 and 3 show the dynamic processes (case $n = 3$), occurring in the adaptive system (1), (2), (5)–(9), (11), (14), (16)–(20); in particular Fig. 1 presents the outputs of the controlled object (11), (14) and the reference model (16); Fig. 2 demonstrates the graph of the error signal of outputs of the controlled object (12), (15) and the reference model (16); Fig. 3 presents the control law (8), (9) with the parameters (18) and external disturbance.

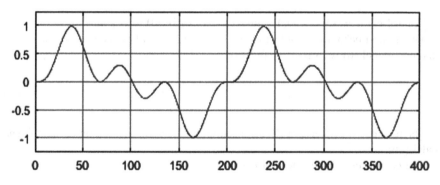

Fig. 1. Behavior of the output of the controlled plant (11), (14) and the main output of the reference (16).

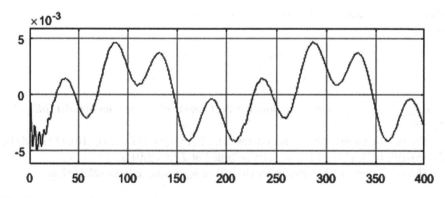

Fig. 2. Error signal of outputs of the controlled object (11), (14) and the reference model (16).

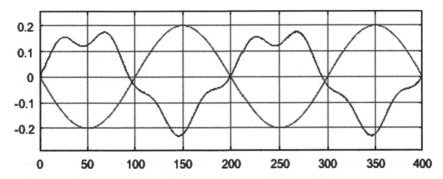

Fig. 3. Dynamics of the control action and external disturbance (20).

Figures 4, 5 and 6 present the transients in system (1), (2), (5)–(9), (12), (15)–(20) in case of $n = 2$. Figure 4 demonstrates the outputs of the controlled object (12), (15) and the reference model (16); Fig. 5 shows the error signal; Fig. 6 presents the control law (8), (9) with the parameters (18) and external disturbance for system (12), (15)–(20).

The graphs of the dynamic processes in the system (1), (2), (5)–(9), (11), (14), (16)–(20) demonstrate good performance of the system, in particular, the mismatch error at the output $(y_m(t) - y(t))$ is sufficiently small quantity, which in the steady state does not exceed 0.47% (Fig. 2). It is obvious that in this case the graphs of the reference and the output of the controlled object $y_m(t)$, $y(t)$ are nearly identical to each other (Fig. 1).

The graphs of the transients, represented at Figs. 4, 5 and 6, make it obvious that the adaptive system (12), (15)–(20) ensures the desired operation mode with the assigned quality in terms of structural-parametric uncertainty and external interference, herewith the mismatch at the outputs of the object (12), (15) and the reference model (16) in steady state does not exceed 0.6%. Thus, according to the results of the simulation modelling the set goal of control (10) is achieved.

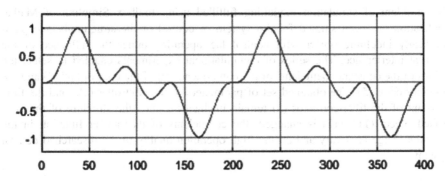

Fig. 4. Behavior of the output of the controlled plant (12), (15) and the main output of the reference (16).

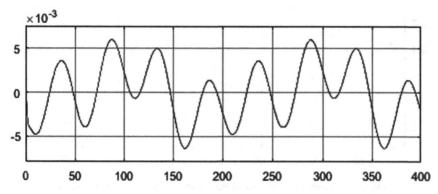

Fig. 5. Error signal of outputs of the controlled object (12), (15) and the reference model (16).

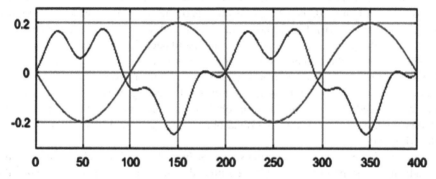

Fig. 6. Dynamics of the control action and external disturbance (20) in the system (12), (15)–(20).

4 Conclusion

The problem of simulation modelling fulfilled using toolbox Simulink of Matlab application has been studied for the system of control of the structurally and parametrically indefinite object with control lag operating under the influence of the external interference. The series of computational experiments enabled to select the values of the constant coefficients of the components (9) of the control law (8) and the filter-corrector (5). The obtained set of parameters of the controller (18) and the time constant of the filter-corrector (5) remain unchanged when the structure of the controlled object (11), (12) is changed, the coefficients of its transfer function in the available range (13) vary and ensure high operation quality of the researched control system.

References

1. Matvejkin, V.G., Muromcev, D.Yu.: Theoretical foundations of energy-saving control of dynamic modes of industrial-technical purpose devices, Mashinostroenie-1, Moscow (2007)
2. Bobcov, A.A., Nikiforov, V.O.: Adaptive output control: subject matter, application tasks and solutions. Sci. Tech. Bull. Inf. Technol. Mech. Opt. **1**(83), 1–14 (2013)
3. Hodgson, S.P., Stoten, D.P.: Robustness of the minimal control synthesis algorithm with minimal phase plant dynamics of unknown order but known relative degree. Int. J. Control **71**, 1–17 (1998)
4. Barsegyan, V.R.: About the problem of optimal control by step-by-lining linear systems with phase limits to intermediate time moments. Learn. Notes EGU **1**, 118–119 (2002)
5. Barsegyan, V.R.: The constructive approach to the research of tasks of control of linear composite systems. Probl. Control. **4**, 11–17 (2012)
6. Rezkov, I.G.: Adaptive controller for a multi-mode object. Autom. Telemechanic **5**, 35–58 (2013)
7. Eremin, E.L.: Adaptive control for dynamic plant on state set of functioning. Inf. Control. Syst. **4**(34), 107–118 (2012)
8. Besekerskij, V.A., Popov, E.P.: Theory of systems of automatic control. Professiya, St. Petersburg (2007)
9. Pupkov, K.A., Egupov, N.D., Barkin, A.I., et al.: Methods of the classical and modern theory of automatic control 5. Moscow State Technological University, Moscow (2004)
10. Eremin, E.L., Chepak, L.V.: Adaptive-robust algorithms of the servomechanism for object with delay in control. Far Eastern Math. J. **1**(4), 141–150 (2003)
11. Eremin, E.L., Chepak, L.V.: The adaptive upreditel-kompensator for the plant with delay of control in systems with the explicit reference models. Syst. Manag. Proc. Fac. Cybern. **5**, 89–92 (2003)
12. Eremin, E.L., Chepak, L.V., Telichenko, D.A.: Synthesis of adaptive systems for scalar plants with delay of control. AmGU, Blagoveshchensk (2006)
13. Eremin, E.L., Chepak, L.V.: Robust control system of affine plant in the scheme with two reference models. Comput. Sci. Control Syst. **3**(41), 121–129 (2014)
14. Eremin, E.L., Chepak, L.V.: The combined control for one class of non-affine plants with time delay. Comput. Sci. Control Syst. **4**(54), 125–134 (2017)
15. Eremin, E.L., Pikul', Z.D., Telichenko, D.A.: Adaptive control system of structural-parametric uncertain objects of one class in a scheme with explicit and implicit reference models. Comput. Sci. Control Syst. **1**(39), 105–115 (2015)
16. Eremin, E.L., Pikul', Z.D., Telichenko, D.A.: Algorithms of adaptive systems for one class of objects with structural and parametric uncertainty with delayed control. Bull. PNU **2**(37), 101–112 (2015)
17. Eremin, E.L.: L-dissipativity of hyperstable control system in structural indignation I. Comput. Sci. Control Syst. **2**(12), 94–101 (2006)
18. Eremin, E.L.: L-dissipativity of hyperstable control system in structural indignation II. Comput. Sci. Control Syst. **1**(13), 130–139 (2007)
19. Eremin, E.L.: L-dissipativity of hyperstable control system in structural indignation III. Comput. Sci. Control Syst. **2**(14), 153–165 (2007)
20. Eremin, E.L.: L-dissipativity of hyperstable control system in structural indignation IV. Comput. Sci. Control Syst. **2**(36), 100–106 (2007)

Experimental Verification of Flux Effect on Process of Aluminium Waveguide Paths Induction Soldering

A. V. Milov, V. S. Tynchenko[✉], and A. V. Murygin

Reshetnev Siberian State University of Science and Technology,
82 Mira Avenue, Krasnoyarsk 660037, Russian Federation
`vadimond@mail.ru`

Abstract. One of the main complicating factors in the automation of control of various technological processes is the presence of measurement errors. This relates also to induction soldering. In relation to this process, this refers to the features of the non-contact temperature sensors use. This brings in a significant error in the results of process parameters measurements. To develop methods for compensation of measurement errors, it is necessary to conduct experiments to determine the flux influence rate on the indications of measuring instruments, which are used in the induction soldering process automation. The article represents the results of an experimental study of flux influence on the technological process of creating permanent connections of antenna-feeder paths based on induction heating. The results of some series of experiments, as well as their statistical processing, are presented. On the basis of these results it is proposed to form methods for compensation of errors introduced by the measuring instrument.

Keywords: Induction soldering · Waveguide path · Statistical processing · Flux · Measurement error estimation · Correction of measurement errors

1 Introduction

Currently, in Russia and all over the world there are a number of enterprises in the rocket and space industry that manufacture spacecraft and their elements. In production at such enterprises, the method of creating permanent joints based on induction heating is widely used. The control of the technological process of creating permanent joints based on induction heating is significantly complicated by the presence in the heating zone of physical phenomena (flux evaporation, change in the emissivity of the material), which make it difficult and sometimes impossible to measure the temperature in the heating zone using non-contact temperature sensors. At the same time, the use of contact sensors is also difficult in the production of spacecraft elements due to the high requirements for the quality of the product surface, as well as the high time required for the installation of contact sensors and human participation in this process.

In this study, the question of an error of measurement instruments is considered. The article presents the results of experiments to assess the influence of the place of information collection on the induction soldering process and the influence of the flux on the correct operation of the measuring devices.

2 Literature Review

Induction heating technology is used to create permanent joints in various branches of engineering [1–4].

One of the obvious examples of the induction heating use is the soldering of aluminum waveguide paths of space vehicles. The production of thin-walled aluminum waveguide paths by induction soldering is a laborious process.

To ensure a quality solder joint, certain conditions must be met:

- The temperature in the maximum heating zones of the soldered elements should be less than their melting point
- The temperature difference in the zones of maximum heating on the various soldered elements should not exceed 10 °C [5–10].

Manual control of the technological process of creating permanent joints based on induction heating is a very complex process. Manual process control leads to the release of a significant number of defective products.

The transition to automated control in [11, 12], allows to improve the control quality of the induction soldering technological process. As part of this work, a hardware-software complex is proposed for controlling the process of induction soldering. The system implements a control algorithm based on a proportional controller. This solution does not allow to achieve optimal process parameters [13, 14].

The authors of [15] developed a mathematical model of heating the assembly elements of the product. The use of mathematical models allows a better study of the technological process. This can allow us to find non-obvious factors affecting the quality of the soldering process control.

One of the ways to improve the control quality of the permanent joints creating technological process based on the induction heating is the use of intelligent control methods, for example, based on fuzzy logic [16]. The fuzzy controller was implemented on the basis of the Matlab software package, using the Simulink module included in its structure. The simulation results showed that under given conditions, such as the heating rate and stabilization temperature, the approach proposed in this work allows us to control the technological process of creating permanent joints based on induction heating with sufficient quality.

The application of flux to the surface of the soldered material has a significant effect on the results of measurements instruments. To reduce the measuring errors, it is necessary to determine experimentally the flux influence on the process of induction soldering. In this study, the research results of the flux influence on the process of induction soldering of aluminum waveguide paths are presented.

3 Materials and Methods

The experiments were carried out on a hardware-software complex for controlling the process of waveguide paths induction soldering. Structurally, the installation consists of the following components:

- A generator
- A matching device
- A set of inductors
- A manipulator-positioner.

The general view of the experimental installation is shown in Fig. 1.

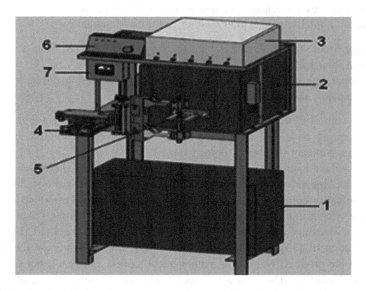

Fig. 1. Soldering installation: 1 – generator; 2 – matching device; 3 – the post block of waveguides soldering management; 4 – manipulator-positioner; 5 – inductor; 6 – remote control; 7 – ammeter.

The temperature reading of the part can be taken in different modes:

- Temperature reading on one side of products without flux
- Temperature reading on one side of products with flux
- Temperature reading on one side of the product, while one pyrometer is aimed at the clean surface of the material, and the other pyrometer is directed to the surface covered with a layer of flux
- Temperature reading at one point from different sides of the waveguide tube without flux
- Temperature reading at one point from different sides of the waveguide tube with flux layer on one side of the tube.

4 Experimental Study

Within the experimental study, 4 series of experiments on induction heating of aluminum waveguide pipes were carried out. The heating of both the pure material of the pipe and the section of the pipe covered with a layer of flux was made.

The characteristics of experiments:

- Number of experiments in one series: 10 pieces
- Size of waveguide tubes and flanges: 58 × 25 mm
- The melting temperature of the flux: from 565 °C to 575 °C
- The stabilization temperature of the soldering process: 580 °C
- Workpiece heating speed: 5 °C/s.
- Power of the magnetic field generator: 7 KW.

The series of experiments differed in the point where the temperature readings were taken.

4.1 Series of Experiments No. 1

In the first series of experiments, temperature readings were taken from one side of the product - the right side, while one of the pyrometers was aimed at a clean area of the surface, and the second – at the area covered with flux.

The location of the temperature reading points is shown schematically in Fig. 2.

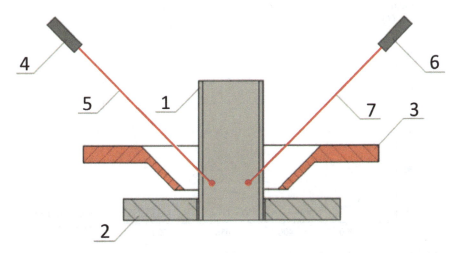

Fig. 2. Layout of pyrometric sensors: 1 – waveguide pipe; 2 – waveguide flange; 3 – inductor; 4 – "non-flux" pyrometer; 5 – "non-flux" pyrometer's beam; 6 – "on-the-flux" pyrometer; 7 – "on-the-flux" pyrometer's beam.

Stabilization of the process occurred at a temperature of 580 °C. Melting of the flux began at a temperature of 540 °C and the end of the flux melting occurred at a temperature of 570 °C.

Figure 3 shows a typical graph of temperatures obtained from pyrometers measuring a clean surface ("non-flux") and a surface covered with flux ("on-the-flux").

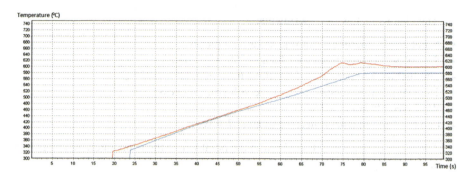

Fig. 3. Average temperature chart for the series of experiments No. 1: red graph – "on-the-flux" temperature of the tube; blue graph – "non-flux" temperature of the tube.

The averaged graph of the temperature divergence in a series of experiments is shown in Fig. 4.

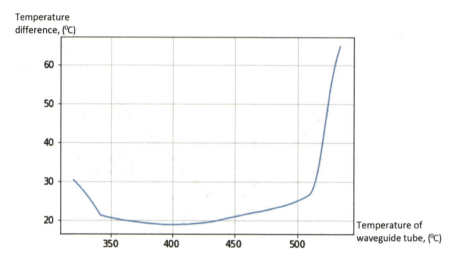

Fig. 4. Average chart of the temperature difference for the series of experiments No. 1.

4.2 Series of Experiments No. 2

The conditions of the experiments in series #2 differ from the series #1 by the side of the temperature reading - the pyrometers are transferred to the left side of the product.

The averaged graph of the temperature divergence is shown in Fig. 5.

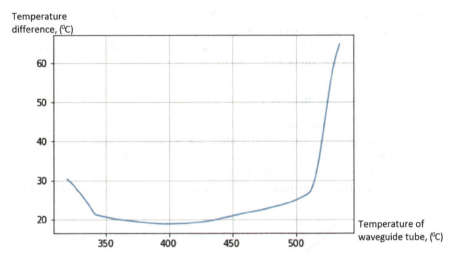

Fig. 5. Average chart of the temperature difference for the series of experiments No. 2.

The results of the experiments of the series No. 2 are the similar to the results in the series No. 1. This is a consequence of the symmetry of the magnetic field formed by the inductor.

4.3 Series of Experiments No. 3

In this series of experiments, the temperature reading points are transferred to the front side of the product.

The averaged graph of the temperature divergence is shown in Fig. 6.

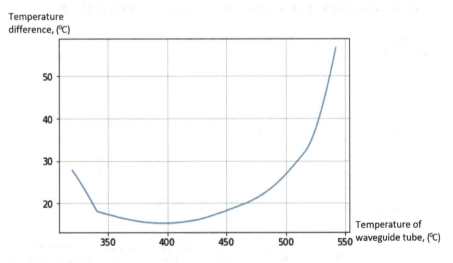

Fig. 6. Average chart of the temperature difference for the series of experiments No. 3.

As a result, the graph in Fig. 6 has slight differences from those in previous experiments. This is due to the fact that the distribution of the eddy currents on the front part of the product has differences in comparison with the lateral parts due to the structural features of the single-turn inductor used for soldering.

4.4 Series of Experiments No. 4

During the series of experiments No. 4, the location of the inductors and, correspondingly, the points of information reading was significantly changed.

Temperature was read at one point from different sides of the waveguide tube. One side was covered with the flux and the other was clean.

The location of the temperature sampling points is shown schematically in Fig. 7.

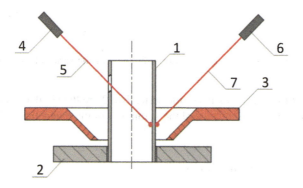

Fig. 7. Layout of pyrometric sensors for experimental series No. 4: 1 – waveguide pipe; 2 – waveguide flange; 3 – inductor; 4 – "non-flux" pyrometer; 5 – "non-flux" pyrometer's beam; 6 – "on-the-flux" pyrometer; 7 – "on-the-flux" pyrometer's beam.

The averaged graph of the temperature divergence is shown in Fig. 8.

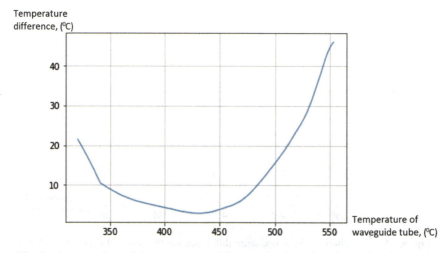

Fig. 8. Average chart of the temperature difference for the series of experiments No. 4.

It can be seen that the graph has a form different from the previous series of experiments. At the same time, the overall trend of changes in the temperature difference remains. Differences can be associated with the inertia of the thermal processes of energy transfer in the section of the tube wall of the waveguide.

5 Discussing

According to the experimental data of the presented experiments, the errors between the readings of the pyrometer "on flux" and "without flux" were calculated.

The correlation coefficients between the results of the experimental series were calculated. Based on the obtained results, a correlation matrix was constructed, which is presented in Table 1.

Table 1. The correlation matrix.

	\multicolumn{4}{c}{Number of experiment series}
	1	2	3	4
1	1	0.980	0.901	0.794
2	0.980	1	0.908	0.784
3	0.901	0.908	1	0.932
4	0.794	0.784	0.932	1

As can be seen from the correlation matrix, the repeatability of the experiments is quite high.

The results of the first experiment are in good agreement with the results of experiments No. 2 and No. 3. With experiment No. 4, there are several greater differences, which are caused by the inertia of thermal processes.

The results of the second experiment completely repeat the first series with a small error related to the presence of the human factor.

The results of the third series of experiments show that even with a significant change in the structure of the magnetic field forming the eddy currents in the soldered product, a characteristic tendency of flux influence on the readings of pyrometric sensors is observed. Occurred differences, in comparison with the first and second series of experiments, are insignificant.

Based on the results of the experiments, the hypothesis was made that the application of flux has a statistically significant effect on the readings of pyrometric sensors in the process of induction heating of the waveguide tube material.

The hypothesis about the statistical significance of the experimental results was verified by the Student's test [17–20]. The probability of confirming the hypothesis of flux influence was:

- For experiment No. 1: 0.9892
- For experiment No. 2: 0.9887
- For experiment No. 3: 0.989
- For experiment No. 4: 0.9907.

As can be seen from the results, the hypotheses for all series of experiments are statistically significant.

6 Conclusion

During the experiments and statistical processing of the results, the repeatability of experiments with flux layer was revealed. From this, we can conclude that there is a statistically significant effect of flux presence on the measurements of pyrometric sensors during induction soldering.

To compensate the error caused by flux meltdown, it is proposed to make an initial correction, calculated on the basis of measurement errors averaging. If necessary, this action can be carried out cascade, with recalculations on the basis of new errors until the least error is achieved.

In the future, it is supposed to use intellectual compensation methods. For example, using the methods of decision trees, as well as their ensembles.

Acknowledgements. The reported study was funded by the President of the Russian Federation grant for state support of young Russian scientists No MK-6356.2018.8.

References

1. Vologdin, V.V., Kushch, E.V., Assam, V.V.: Induction brazing. Mechanical Engineering, Leningrad (1989)
2. Gierth, P., Rebenklau, L., Michaelis, A.: Evaluation of soldering processes for high efficiency solar cells. In: 35th International Spring Seminar in Electronics Technology, pp. 133–137 (2012)
3. Mazon-Valadez, E.E., Hernandez-Samano, A., Estrada-Gutierrez, J.C.: Developing a fast cordless soldering iron via induction heating. Dyna **81**(188), 166–173 (2014)
4. Nishimura, F., Nakamura, H., Takahashi, H.: Development of a new investment for high-frequency induction soldering. Dent. Mater. J. **11**(1), 59–69 (1992)
5. Lanin, V.: High-frequency electromagnetic heating for soldering electronic devices. Technol. Electron. Ind. **5**, 46–49 (2007)
6. Babenko, P.G., Ivanov, I.N.: High-frequency inducers for induction soldering. Weld. Prod. **8**, 47–48 (2013)
7. Slugocki, A.E., Ryskin, S.E.: Inductors for induction heating. Energy, Leningrad (1974)
8. Slugocki, A.E.: Inductors. Energy, Leningrad (1989)
9. Slugocki, A.E.: Induction Heating Plants. Energoizdat, Leningrad (1981)
10. Zlobin, S.K., Mikhnev, M.M., Laptenok, V.D.: Features of the production of waveguide-distributive tracts of antenna-feeder devices of space vehicles. Bull. Sib. State Aerosp. Univ. **6**, 145–157 (2013)
11. Murygin, A.V., Tynchenko, V.S., Laptenok, V.D.: Complex of automated equipment and technologies for waveguides soldering using induction heating. In: IOP Conference Series: Materials Science and Engineering, vol. 173, no. 1, p. 012023 (2017)
12. Tynchenko, V.S., Murygin, A.V., Emilova, O.A., et al.: The automated system for technological process of spacecraft's waveguide paths soldering. In: IOP Conference Series: Materials Science and Engineering, vol. 155, no. 1, p. 012007 (2016)

13. Tynchenko, V.S., Bocharov, A.N., Laptenok, V.D., et al.: Software for the technological process of soldering waveguide paths of spacecrafts. Softw. Prod. Syst. 2(114), 128–134 (2016)
14. Murygin, A.V., Tynchenko, V.S., Laptenok, V.D.: Modeling of thermal processes in waveguide tracts induction soldering. In: IOP Conference Series: Materials science and engineering, vol. 173, no. 1, p. 012026 (2017)
15. Kudryavtsev, I.V., Barykin, E.S., Gotselyuk, O.B.: Mathematical model of heating of the waveguide during transmission of high signal power. Young Sci. **9**, 52–57 (2013)
16. Milov, A.V., Tynchenko, V.S., Murygin, A.V., et al.: Development of a method for controlling induction soldering of waveguide paths based on a fuzzy regulator. Sci. Tech. Bull. Volga Reg. **3**, 118–121 (2013)
17. Grachev, Yu.P., Plaksin, Yu.M.: Mathematical methods of experiment planning. DeLi-Print, Moscow (2005)
18. Gotman, A.: Theory of probability and mathematical statistics. Int. J. Appl. Fundam. Res. **7** (2011)
19. Borovkov, A.A.: Math statistics. Parameter estimation, hypothesis testing. Science, Moscow (1984)
20. Nalimov, V.V., Chernova, N.A.: Statistical methods for planning extreme experiments. Science, Moscow (1965)

Intellectualization of the Induction Soldering Process Control System Based on a Fuzzy Controller

V. E. Petrenko, V. S. Tynchenko[✉], and A. V. Murygin

Siberian State University of Science and Technology, 31 Gazety Krasnoyarskiy Rabochiy Avenue, Krasnoyarsk 660000, Russian Federation
vadimond@mail.ru

Abstract. The article solves the problem of developing an intelligent system for controlling the process of induction soldering of waveguide paths from aluminum alloys. The concept of a control system is proposed, which uses the fuzzy logic methods as a solution of control problem. The following input variables were defined for the fuzzy controller: mismatch of the soldered waveguide elements temperatures; mismatch of the soldered waveguide elements heating rates; estimation of the process control quality by the temperature mismatch signal; estimation of the process control quality by the signal of heating speed mismatch. Output variables of the fuzzy controller were the choice of control algorithm and adjustment of the regulator coefficients. Based on the results of numerical experiments, both the forms of control actions in the system and their parameters were selected. The proposed approach to the formation of control was tested in a series of full-scale experiments on the soldering of waveguide paths. As a result of these experiments the heating curves of the product elements were obtained, according to which it is possible to judge the effectiveness of using intelligent control based on a fuzzy regulator. Application of the proposed approach allows to provide high quality control of the induction heating process and to obtain reliable permanent joints of the elements of waveguide paths.

Keywords: Induction heating · Waveguide path · Fuzzy logic · Technological process automation · Soldering · Intellectual system

1 Introduction

The technological process of thin-walled waveguide paths from aluminum alloys production using an induction heating source is associated with certain difficulties, such as [1, 2]:

- The proximity of the melting points of the base material and solder (less than 100 °C)
- Uneven distribution of energy between assembly elements during induction heating
- Limited active time of flux after melting
- High quality requirements for soldered joints.

The process is complicated by the need for inductors of a complex profile [3, 4]. Detailed description of the requirements for the quality of the soldered joint, as well as the conditions necessary for its preparation, are considered in [5, 6], which describe the early stages of the technology development.

The existing system of two-circuit control of the technological process of waveguide paths induction soldering [7] significantly improved the soldering process, solving some problems [5, 6]. Also, this system has provided the possibility of further technology development, which is possible by direct intellectualization of the existing control system.

The main problem of the control system under consideration lies in the complexity of the adjustment of the technological process, which is caused by the influence of the human factor. There is also the inability to use alternative perspective control algorithms based on the analysis of the results of numerical experiments on mathematical models [8, 9]. In addition, in existing systems there is no possibility of automatically changing coefficients in control units directly during the process, which can avoid any factor not taken into account by the technologist. If the operator mistakes associated with improper adjustment of the initial parameters of the technological process, there is no certainty in the end result of the system. The same uncertainty is introduced by direct intervention in the soldering process by manual control of a number of parameters.

When using fuzzy regulation methods [10], the influence and consequences of the problems discussed above can be partially reduced or completely eliminated. In this regard, the development and implementation of the concept of intellectual control system for the process of induction soldering of thin-walled waveguide paths from aluminum alloys is relevant and in demand within the framework of the technology under consideration.

2 Conceptual Scheme of the Intelligent Induction Soldering System

Intelligent systems find successful practical application in technological processes with strict limitations, which require high accuracy to maintain parameters. Methods of intelligent control allow applying non-standard approaches to regulation, which lets them to dynamically adjust to deviations in process parameters. Also in such intelligent control systems, there is the possibility of self-learning [11–14].

The formation of the concept of an intellectual control system is necessary to better understand its potential. Such a system should provide selection and tuning of control algorithms directly in the process of induction soldering. This can be achieved using various methods of intellectualization.

Within the presented research, authors will use the method of fuzzy logic to form the intellectual core of the automation system for the induction soldering of waveguide paths.

Figure 1 shows a structural scheme of the intelligent control system for the induction soldering process.

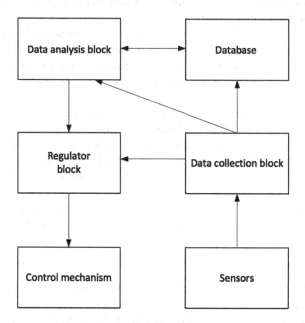

Fig. 1. Structural scheme of the intelligent control system.

The data analysis block is the basis of the intellectual system. Its introduction should determine the potential of intellectualization.

The database stores detailed information about technological processes, communicates with the data analysis unit, if it is necessary for its operation in any interval of the technological process. In addition, this block can contain a knowledge base. The accumulated information on technological processes can be used to select the most appropriate control algorithm, to implement the process of self-learning, to adopt a non-trivial solution, to estimate failures and malfunctions in the equipment.

The data collection block provides the distribution of information between the remaining units of the system, providing the necessary feedback in the control system, as well as collecting, filtering, verifying and converting the data received from the sensors.

The regulator block contains a set of algorithms for controlling the technological process. This unit is a completely independent regulator, which allows to control the process at certain stages and parameter spaces without using the data analysis algorithms. Sensors and actuators are external controls. Further, the authors present a partial implementation of the described concept, on the basis of which it can be judged on the further expediency of the complexity, expansion, processing of existing and integration of new functional blocks of the system.

3 Intelligent Control of Induction Soldering Process on the Basis of Fuzzy Regulator

Fuzzy logic is based on the theory of fuzzy sets. With the help of the theory of fuzzy sets, one can formally define such inaccurate concepts as "high temperature", "young man", "average growth", etc.

Intuitive simplicity and power of the fuzzy logic apparatus allows using it in various control and information analysis systems [15, 16]. In this case, fuzzy logic allows you to connect the power of human intuition and the operator's experience to the management process. Unlike traditional mathematical methods, which require the simulation of precise and unambiguous formulations at each step, fuzzy logic methods offer a different level of abstraction. Numerical data obtained using fuzzy logic are in many ways analogous to statistical distributions, but they are free from some shortcomings, such as:

- Presence of a small number of distribution functions suitable for analysis
- The need to normalize data
- The need to observe the additivity property
- Complexity of substantiating the adequacy of mathematical abstraction for the purpose of describing the behavioral factors observed in actual quantities.

It has been experimentally shown that fuzzy control gives better results, in comparison with those for classical control algorithms [15, 16].

The intelligent system for controlling the process of induction soldering of waveguide paths from aluminum alloys in terms of automatic control systems [17] can be represented by a schema in the following form (Fig. 2):

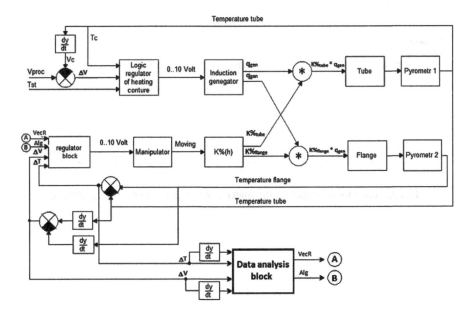

Fig. 2. Generalized scheme of intelligent induction soldering process control system.

The following notations are used in the figure: *Vproc* – heating rate setpoint; *Tst* – stabilization temperature; q_{gen} – generator power transmitted to the product to be soldered; *K%(h)* – generator energy distribution between product elements in percent; *VecR* – vector containing the values of the regulator block coefficients; *Alg* – variable used to switch algorithms in the regulator block.

Control in this system is carried out in two contours:

- Heating rate control contour
- Contour of workpiece position control in the inductor window.

These two loops have a cross-connection. The mutual influence of contours is eliminated by introducing a pulse character of control in the system.

The data analysis block is implemented as a fuzzy controller that manages the internal state of the regulator in the workpiece position control loop, determining the preferred regulation law and gain coefficients for the current process state.

Within the developed fuzzy controller, the following input variables are defined:

- Mismatch of the soldered waveguide elements temperatures
- Mismatch of the soldered waveguide elements heating rates
- Estimation of the process control quality by the temperature mismatch signal
- Estimation of the process control quality by the signal of heating speed mismatch.

As output variables are selected the following:

- Choice of control algorithm
- Adjustment of the regulator coefficients.

Figures 3 and 4 show the terms for, respectively, the mismatch of the temperatures and heating rates of the welded elements.

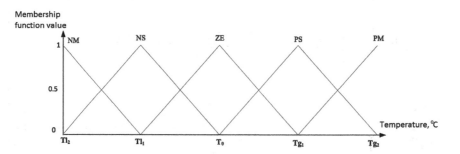

Fig. 3. Term for mismatch of the soldered waveguide elements temperatures.

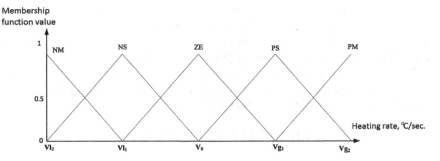

Fig. 4. Term for mismatch of the soldered waveguide elements heating rates.

The values of T_0 and V_0 correspond to the absence of a mismatch.

Explanation of designations on terms: ZE – the value of the error signal does not exceed the permissible deviation or the measurement error level; NS, PS – the value of the error signal beyond the permissible deviations, the moderate deviation of the process parameters; NM, PM – the value of the error signal is large, there is a significant deviation in the parameters of the technological process.

Figures 5 and 6 show the terms for, respectively, the growth rate of the temperatures and heating rates mismatches of the welded elements.

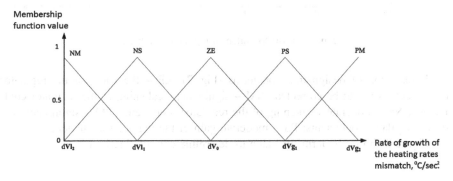

Fig. 5. Term for rate of growth of the soldered waveguide elements heating rates mismatch.

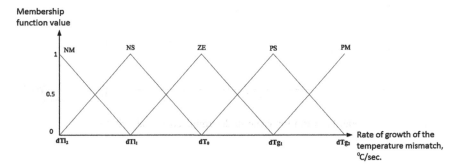

Fig. 6. Term for rate of growth of the soldered waveguide elements temperature mismatch.

Explanation of designations on terms (Figs. 5 and 6): ZE – the quality of process control remains unchanged; NM – the quality of process control worsens sharply; NS – the quality of process control worsens moderately; PS – the quality of process control sharply increases; NS – the quality of process control moderately increases.

Within the developed fuzzy regulator, a comparison of the error signals values at the current step and the previous step of control is used to assess the quality of control. The estimate is described by the following formula:

$$dVoT = |\Delta ToV(t - \tau)| - |\Delta ToV(t)|, \tag{1}$$

where ΔToV – value of the velocity or temperature mismatch signal; t – current process time; τ – frequency of evaluation calculation.

Figure 7 shows the term for value of regulator coefficient.

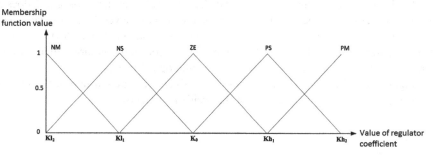

Fig. 7. Term for value of regulator coefficient.

Explanation of designations on terms (Fig. 7): ZE – the value of the regulator coefficients is not to be corrected; NM – significant weakening of the regulator coefficients; NS – moderate weakening of the regulator coefficients; PM – significant gain of the regulator coefficients; PS – moderate gain of the regulator coefficients;

Figure 8 shows the term for value to determine control algorithm.

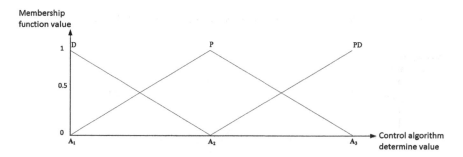

Fig. 8. Term for value to determine control algorithm.

Explanation of designations on term (Fig. 8): P – proportional law of regulation; D – differential law of regulation; PD – proportional-differential law of regulation.

The regulatory laws presented above are described (2), (3), (4), respectively.

As can be seen from the above pictures, the space of input and output variables was divided into 5 ranges, within which linguistic variables are defined [18].

As control algorithms, as a result of experimental study [7], variations of the PID controller [19] are suggested: proportional, differential and proportional-differential regulators.

To control the heating power, it is proposed to apply a temperature control loop. As a result of the search, a dependence for the control effect was found, which is a logical function [7]:

$$u_1(t) = W\,\text{int}(t) + \begin{cases} 0, & \text{if } T_c < T_s, \\ \frac{|\Delta V(t)|}{\Delta V(t)} \cdot k_1, & \text{if } T_c(t) < T_{st} \text{ and } |\Delta V(t)| > V_s \text{ and } T_c(t) > T_s, \\ \frac{|\Delta V(t)|}{\Delta V(t)} \cdot k_2, & \text{if } T_c(t) < T_{st} \text{ and } |\Delta V(t)| > V_s \text{ and } T_c(t) > T_s, \\ -k_3, & \text{if } T_c(t) > T_{st} + T_{\lim}, \end{cases} \qquad (2)$$

where T_s – the sensitivity temperature of the measuring device (pyrometer); $\Delta V(t)$ – deviation of the heating rate from a given; $\Delta V_{proc}(t)$ – heating rate program; $\Delta V_c(t)$ – heating speed of the assembly components; $\Delta V_s(t)$ – the threshold for the allowable deviation of the controlled heating rate from the setpoint; $T_c(t)$ – the temperature of the component of the assembly to be welded, which controls the control; $T_{st}(t)$ – transition temperature; $T_{lim}(t)$ – permissible limit of stabilization temperature rise; k_1, k_2, k_3 – constants of control influence transformation for various stages of technological process; W_{int} – integral component of the regulator.

To control the position of the workpiece in the inductor frame, it is proposed to apply one of three control options.

At the initial stage of heating, as well as in case of too great mismatch of a temperature, it is proposed to use a proportional regulator [18]:

$$u_2(t) = \begin{cases} \frac{|\Delta T(t)| \cdot k}{k_{simple}}, & \text{if } \Delta T(t) < T_s \text{ and } \Delta T(t) > 0, \\ \frac{(\Delta T(t) + T_s) \cdot k}{k_{simple}}, & \text{if } \Delta T(t) \leq 0, \\ \frac{(\Delta T(t) - T_s) \cdot k}{k_{simple}}, & \text{if } \Delta T(t) > 0. \end{cases} \qquad (3)$$

where k_{simple} – coefficient of proportionality; k – overall gain of the control algorithm; $\Delta T(t)$ – the value of the temperature mismatch between the elements of the product; T_s – lower threshold of sensitivity of the temperature sensor.

In the case of a small temperature mismatch, but a significant mismatch in the heating rates of the soldered elements, it is proposed to use a differential regulator [19]:

$$u_2(t) = \frac{\Delta V(t) \cdot k}{k_{simple}}, \qquad (4)$$

In the case of a small temperature mismatch, but a moderate mismatch of the heating rates of the soldered elements, it is proposed to use a proportional-differential controller [20]:

$$u_2(t) = \frac{(\Delta V(t) \cdot k_v - |\Delta T(t)| \cdot k_t) \cdot k}{k_{simple}}, \quad (5)$$

where k_t, k_v – proportionality coefficients.

To form the rule base, training data [21] were obtained, on the basis of which a knowledge base was created, allowing to form the control action with the least error. A fragment of the knowledge base is presented in Table 1.

Table 1. Fragment of the base of fuzzy regulator rules.

Rule number	Term					
	T	V	dT	dV	Alg	K
1	NM	NM	NM	NM	P	PM
2	NS	0	NS	0	PD	PS
3	PM	PS	NS	NS	P	PS
4	PS	PM	ZE	PS	D	ZE
5	NM	ZE	PS	PS	P	PS
6	NM	PM	ZE	ZE	PD	PM
7	NS	PS	ZE	NS	D	PS
8	ZE	ZE	ZE	PS	PD	NM
9	NM	PS	ZE	PM	P	NS
10	ZE	PM	PM	ZE	PD	NS

4 Experimental Study

To approbate and estimate the proposed approach to control the induction soldering process, experimental studies were carried out on various implementations of the control system: single-contour, double-contour and intelligent.

Such formulation of the experiment makes it possible to estimate the effectiveness of proposed intelligent method, in comparison with those already mastered [5, 6].

The experiments were carried out for the process of soldering the tube-flange waveguide assembly with a standard size of 58 × 25 mm.

Figures 9 and 10 show the result of a single- and double-contour control system using.

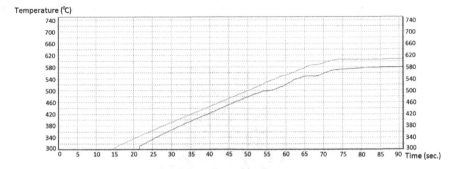

Fig. 9. Soldering chart with single-contour temperature control. Red graph - The temperature of the pipe, Blue graph - The temperature of the flange.

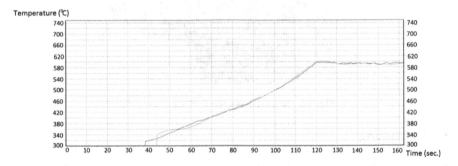

Fig. 10. Soldering process chart with a double-contour temperature and position control. Red graph - The temperature of the pipe, Blue graph - The temperature of the flange.

Figure 11 shows the result of using an intelligent control system.

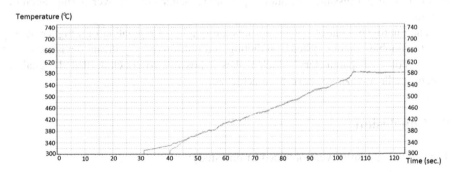

Fig. 11. Soldering process chart with intelligent temperature and position control. Red graph - The temperature of the pipe, Blue graph - The temperature of the flange.

Figure 12 shows the soldered assembly of the waveguide, obtained using an intelligent control system.

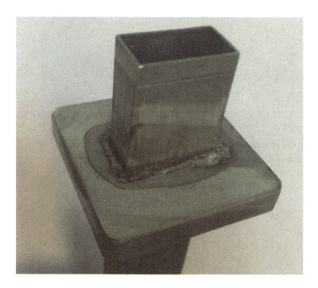

Fig. 12. Soldered assembly of the waveguide.

5 Discussing

From the graphs presented in Figs. 9, 10 and 11, there are differences in the quality of the waveguide path soldering process.

So, when using a single-contour system, the probability of unrecoverable defects occurrence is high, such as: local burns, reflows and deformations of the elements of the waveguide assembly.

With double-contour control, the system can withstand the necessary process parameters in permissible deviations, providing better control over a single-contour system.

The intelligent control system ensures the quality of regulation of a higher quality than the double-contour system, predicting and outrunning significant deviations from the technological parameters, which makes it possible to conclude that the goals of system intellectualization have been achieved.

6 Conclusion

The following results were obtained in this study:

- The concept of an intelligent system for controlling the process of induction soldering has been developed with implementation fuzzy regulation
- The experiments showed the advantages of the proposed intellectual control method.

Acknowledgements. The reported study was funded by the President of the Russian Federation grant for state support of young Russian Scientists No MK-6356.2018.8.

References

1. Silchenko, P.N., Mihnev, M., Ankudinov, A.V., et al.: Ensuring the strength and accuracy of large-sized waveguide-distributive systems of space communication devices. Probl. Mech. Eng. Reliab. Mach. **1**, 112–117 (2012)
2. Vologdin, V.V., Kushch, E.V., Assam, V.V.: Induction brazing. Mechanical Engineering, Leningrad (1989)
3. Slugocki, A.E.: Inductors. Energy, Leningrad (1989)
4. Slugocki, A.E., Ryskin, S.E.: Inductors for induction heating. Energy, Leningrad (1974)
5. Murygin, A.V., Tynchenko, V.S., Laptenok, V.D., et al.: Complex of automated equipment and technologies for waveguides soldering using induction heating. IOP Conf. Ser. Mater. Sci. Eng. **173**(1), 012023 (2017)
6. Tynchenko, V.S., Murygin, A.V., Emilova, O.A., et al.: The automated system for technological process of spacecraft's waveguide paths soldering. IOP Conf. Ser. Mater. Sci. Eng. **155**(1), 012007 (2016)
7. Tynchenko, V.S., Murygin, A.V., Petrenko, V.E., et al.: A control algorithm for waveguide path induction soldering with product positioning. IOP Conf. Ser. Mater. Sci. Eng. **255**(1), 012018 (2017)
8. Murygin, A.V., Tynchenko, V.S., Laptenok, V.D., et al.: Modeling of thermal processes in waveguide tracts induction soldering. IOP Conf. Ser. Mater. Sci. Eng. **173**(1), 012026 (2017)
9. Kudryavtsev, I.V., Barykin, E.S., Gotselyuk, O.B.: Mathematical model of heating of the waveguide during transmission of high signal power. Young Sci. **9**, 52–57 (2013)
10. Burakov, M.V.: Fuzzy controllers. GUAP, St. Petersburg (2010)
11. Demidova, L.A., Kirakovskyi, V.V., Pylkin, A.N.: Algorithms and systems of fuzzy inference when solving problems of diagnostics of urban engineering communications in MATLAB environment. Radio and communication, Moscow (2005)
12. Dudkin, Yu.P., Titov, Yu.K., Filippenkov, R.G.: Fuzzy control of the speed of the free turbine of a gas turbine engine. Bulletin of the Moscow Aviation Institute **17**(6), 55–60 (2010)
13. Gostev, V.I.: Fuzzy controllers in automatic control systems. Radioamator, Kyiv (2008)
14. Minaev, Yu.N., Filimonova, O.Yu., Benameur, L.: Methods and algorithms for solving problems of identification and prediction in conditions of uncertainty in a neural network logical basis. Hotline-Telecom, Moscow (2003)
15. Burakov, M.V., Konovalov, A.S.: Synthesis of fuzzy logic controllers. Inf. Control Syst. **1**, 22–27 (2011)
16. Tynchenko, V.S., Tynchenko, V.V., Bukhtoyarov, V.V., et al.: The multi-objective optimization of complex objects neural network models. Indian J. Sci. Technol. **9**(29) (2016)
17. Aleksandrov, A.G., Palenov, M.V.: The state and prospects of the development of adaptive PID regulators. Autom. Telemech. **2**, 16–30 (2014)
18. Arsenyev, G.N., Shalygin, A.A.: Mathematical modeling of fuzzy regulators based on MATLAB. Inf.-Measur. Control Syst. **9**(5), 26–37 (2011)

19. Chen, H.C.: Optimal fuzzy PID controller design for an active magnetic bearing system based on adaptive genetic algorithms. Math. Struct. Comput. Sci. **24**(5) (2014)
20. Sabir, M.M., Ali, T.: Optimal PID controller design through swarm intelligence algorithms for sun tracking system. Appl. Math. Comput. **274**, 690–699 (2016)
21. Rutkovskaya, D., Pilinskiy, M., Rutkovskiy, L.: Neural networks, genetic algorithms and fuzzy systems. Hotline-Telecom, Moscow (2006)

Internal Combustion Engines Fault Diagnostics

L. A. Galiullin[✉] and R. A. Valiev

Naberezhnye Chelny Institute, 68 Mira Avenue, Naberezhnye Chelny 423812, Russian Federation
`galilenar@yandex.ru`

Abstract. This article describes the methods of diagnosing internal combustion engines (ICE). The conclusion is drawn that the majority of modern methods and ICE diagnostic devices don't solve fully a problem of determination of technical condition of the engine, often are labor-consuming and expensive. The choice of a method and mode of diagnosing of ICE on the basis of external speed characteristics is carried out for what the list of sensors and executive mechanisms of a control system of the engine is defined. The choice of a method of training of fuzzy Sugeno systems on the basis of hybrid neural networks is reasonable. The possibility of identification of difficult dependences by the systems of fuzzy sets on the basis of hybrid networks is proved. Possibilities of systems for fuzzy conclusion on identification of dependences are the basis for algorithms. The assessment of influence of external factors on the accuracy of measurements therefore it is established that the maximum error doesn't exceed 5% is carried out. The experimental studies of metrological characteristics of the diagnostic system have been carried out which showed that the relative errors do not exceed the estimated errors. In this case, a speed characteristic was determined in the entire range of the engine speed.

Keywords: Diesel engine · Neural · Network · Fault diagnostic · Information system

1 Introduction

The estimation of the general condition of the engine is made on the effective parameters of its operation, which include the effective torque and power on the motor shaft, fuel and air consumption, ignition timing, and harmful emissions in the exhaust gases [1]. The work of systems implementing this approach is based on brake and non-brake methods.

Brake methods involve the use of special loading stands with running drums. This method was not widely used due to the high cost of equipment.

Non-brake methods are simpler and do not require the use of special braking devices [2]. In this case, the angular acceleration is measured when the engine is accelerated without an external load from the minimum stable speed to the maximum due to the sharp opening of the injection pump. This method allows carrying out diagnostics in real operating conditions, and equipping modern ICE with electronic control systems - to increase the number of controlled parameters.

The disadvantages of the systems implementing this approach are to a different degree the low accuracy associated with the need for numerical differentiation of the angular velocity variation function, the incompleteness of the parameters to be determined, and a narrow range of rotation frequencies for the characteristics obtained [3].

In view of what has been said, the task of creating a diagnostic system that makes it possible to evaluate the basic performance of ICE over a wide range of engine speed is urgent and requires the development of original methods that go beyond existing approaches.

2 Membership Functions

The main provisions of the theory of fuzzy sets and fuzzy logic are applied. Typical membership functions and operations on fuzzy numbers are used [4]. The concepts of fuzzy and linguistic variables are considered. In this case, the fuzzy variable is determined by the triple <a, X, A>, where a – name of a fuzzy variable; X – the domain of its definition (universe); A – fuzzy set on X, describing the possible values that a fuzzy variable can take.

A generalization of a fuzzy variable is the so-called linguistic variable, defined by a tuple <p, T, X, G, M>, where p – name of the linguistic variable; T – basic term-set of a linguistic variable or the set of its values (terms), each of which is the name of a separate fuzzy variable; X – domain of fuzzy variables that are included in the definition of a linguistic variable; G – a syntactic procedure that describes the process of generating new values for a given linguistic variable; M – a semantic procedure that allows each new value of a given linguistic variable, obtained by procedure G, to be assigned to each meaningful content by forming the corresponding fuzzy set.

The basic configuration of the fuzzy inference system based on the rules of fuzzy products is used, in which conditions and conclusions are formulated in terms of fuzzy linguistic utterances [5].

Thus, the problem of identifying data with fuzzy Sugeno systems with membership functions of a Gaussian type reduces mainly to the selection of the number of terms of the input variable. With an increase in the number of terms of the input variable, the errors do not increase [6]. At the same time, with a decrease in the number of terms of the input variable, the smoothing properties of the fuzzy system approximation will be strengthened [7].

The undoubted advantages of fuzzy systems to identify data can be attributed to the fact that not require special selection output system structure and the form of membership functions, unlike parametric identification methods (exponential, logarithmic, exponential, power and other techniques).

To build the speed characteristic, it is enough to process the information contained in the five signals of the control system [8]. These include the signals of the crankshaft position sensor, the mass air flow, the position of the fuel pump rail, the fuel injection and ignition control systems, etc.

In the developing system, the definition of speed characteristic is based on the nonbrake method. The engine is accelerated by changing the position of the fuel injection pump rail. The signals of the engine control system are continuously measured.

The choice of the diagnostic mode is reduced to providing such engine operating conditions, in which its properties are presented most fully. This mode corresponds to the mode of full fuel supply, when the fuel pump rail is opening as much as possible [9]. This is due, first of all, to the maximum wide frequency range of the engine and the maximum work of inertial forces and frictional forces. In addition, in the real conditions of the diagnosis, to ensure the permanence of the position of the rail (different from 100% of the opening of the rail) is problematic enough.

The hardware of the AIS includes a cable-splitter of the signals of the control system, a coupling device, an input module for analog signals, and a computer. To provide mobility as a computer, a portable personal computer of the "notebook" type was chosen. This makes it possible to carry out diagnostics while the vehicle is moving, when the load is the mass of the car, driven to the crankshaft via the transmission [10]. As an analog input module, an external ADC/DAC E14-440 module from L-card was selected, which was added to the State Register of measuring instruments. The module interacts with the computer via the USB bus, the ADC has a bit capacity of 14 bits, the maximum conversion frequency is 400 kHz [11]. This solution allows for diagnostics in real operating conditions. The software part manages the data collection and processing of the engine control system signals. Processing algorithms are based on the use of fuzzy inference systems as part of hybrid networks, which ensures high accuracy and repeatability of the results of experimental tests in a wide range of engine speed.

The figures for ω, M_H and G_T are shown in the Figs. 1, 2 and 3.

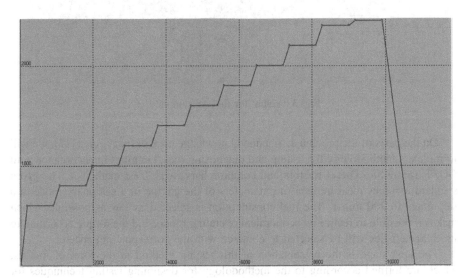

Fig. 1. The figure for the engine speed.

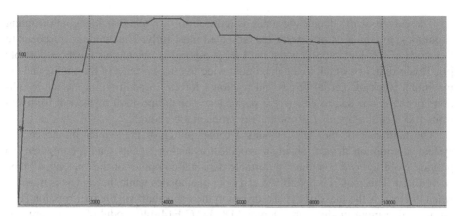

Fig. 2. Figure for moment of loading.

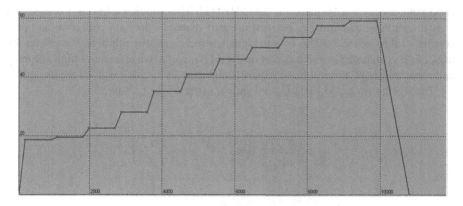

Fig. 3. Figure for fuel consumption.

On the basis of the previously obtained knowledge base of fuzzy rules [12], we will formulate a methodology for testing and diagnosing diesel engines in accordance with GOST 18509-88 «Diesel tractors and combine harvesters. Test methods». The given standard assumes measurement of parameters of the engine in a stationary mode at a step equal to 200 min-1. The fuel consumption is selected as low as possible, which makes it possible to realize an economical operating mode, and the torque according to its characteristics will be selected to coincide with the corresponding speed.

The next stage is the presentation of this testing method in the form of images, which are formed according to the methodology for designing testing techniques for diesel engines, described at [13].

The use of graphical representation of data is convenient for humans, for computer processing it is better to have numeric data. Therefore, the next step is to convert the images into a summary table of parameters.

3 Conclusion

This data is saved in the summary table of parameters (Table 1).

Table 1. Summary table of parameters.

№	Test time, min.	Figure for ω, min^{-1}	Figure for M_H, Nm	Figure for G_T, kg/hour
1	Engine start			
2	15	600	73	18,6
3	30	800	90	19,5
4	45	1000	110	24
5	60	1200	123	30
6	75	1400	125,8	35
7	90	1600	123	41
8	105	1800	115	46
9	120	2000	112,5	50
10	135	2200	110,8	53,5
11	150	2400	110	57,5
12	165	2450	110	59,2
13	180	600	73	18,6
14	Engine stop			

At the next stage, fuzzification is carried out, that is, the conversion of values of the input variables Ai into fuzzy Bi, by linguistic variable [14]. Such transformation is in fact a kind of valuation necessary to translate the given data into subjective estimates. Linguistic variables for translating a value into fuzzy are stored in the knowledge base of fuzzy logic. The result of the work at this stage will be a converted summary table of parameters, in which instead of clear values the membership functions will be located. The result of this stage is shown in the Fig. 4.

The next step is the formation of an approximate fuzzy result in the output block. To do this, fuzzy rules stored in the knowledge base of fuzzy logic were applied [15].

The next stage is defuzzification. Defuzzification means the procedure for converting fuzzy values obtained as a result of fuzzy inference into clear ones, on the basis of which it is possible to conduct engine tests [16].

For defuzzification we use the fuzzy derivation of Sugeno. This is because it is highly accurate and easy to use this algorithm will reduce the processing time of information.

The results of fuzzy inference are control vectors in linguistic variables ω, M_H and G_T.

F_ω = (0; 2,99; 6,48; 11,13; 16,13; 21,1; 26; 30,98; 36; 43,53; 50);
F_{Mh} = (0; 3,04; 7,37; 12,5; 15,17; 21,43; 28; 31,25; 36; 45; 50);
F_{Gt} = (0; 1; 5,7; 10,39; 14,3; 19,32; 24,12; 28,97; 33,97; 41,11; 48).

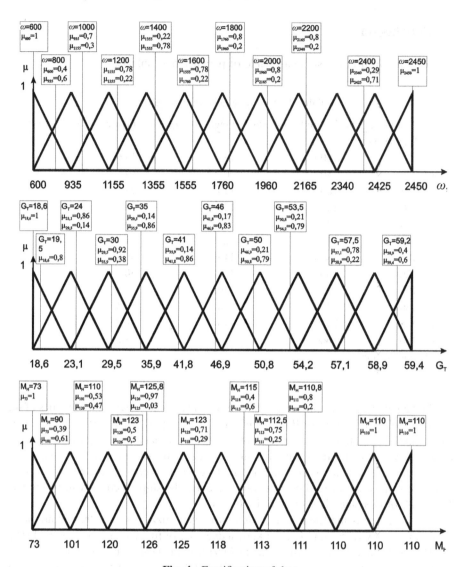

Fig. 4. Fazzification of data.

We calculate the total result vector by three parameters ω, M_H and G_T based on averaging the average value and the method of ranking the characteristics [17]. The result of the control vector over the averaged value will have the form:

F_S = (0; 2,34; 6,52; 11,34; 15,2; 20,62; 26,4; 30,4; 35,3; 43,21; 49,33).

The control vector by the method of paired comparisons will have the following form:

F_R = (0; 2,27; 6,42; 11,2; 15,21; 20,53; 25,81; 30,31; 35,26; 43,01; 49,27).

Using the vector F_ω, which characterizes the engine speed, we will check the degree of adequacy of the control model to the real parameters of the engine [18]. The results of the averaged value of the error of the engine tests are shown in the Fig. 5.

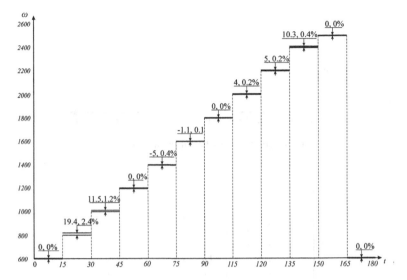

Fig. 5. Calculated and experimental data of the engine speed.

As can be seen from the figure, the maximum absolute error is 19,4 min^{-1}, which corresponds to a relative error of 2,4% at 800 min^{-1}, which is explained by the high non-linearity of this characteristic in this section [19]. To this value it is necessary to add the error of the test stand, equal to 0,5%. From these values, we can conclude that the error in controlling the diesel engine based on the model as a knowledge base of fuzzy rules will not exceed 2.9% [20].

Using the resulting values of ω, MH and GT we construct speed characteristics with these parameters (Figs. 6 and 7). In these figures, Fig. 1 indicates a calculated characteristic, and Fig. 2 indicates an experimental characteristic. The maximum error at the moment of engine load is 3% at 1800 min^{-1}, and the maximum error in fuel consumption is 5% at 1400 min^{-1}.

Calculation of the speed characteristic for M_H and G_T showed that the error of control based on the model using fuzzy logic on these characteristics does not exceed five percent, which satisfies the requirements of GOST 18509-88 «Diesel tractors and combine harvesters. Methods of bench tests».

The accuracy of the methods for obtaining the resulting control vectors with concern to the three parameters ω, M_H and G_T, is determined using the averaging of the mean value and the ranking of the reference parameters based on the testing of diesel engines. The results of the experiment are shown in Fig. 8.

Figure 8 shows the relative error of the tests: dark color - by the average value, and light - by the method of paired comparisons. As can be seen from the figure, the second method is more accurate, in addition to the value of 800 min^{-1}, which is explained by the large nonlinearity in this section of the engine speed.

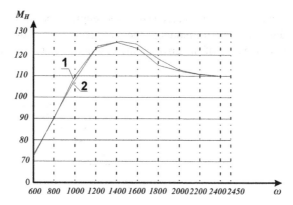

Fig. 6. Speed characteristic for engine load moment.

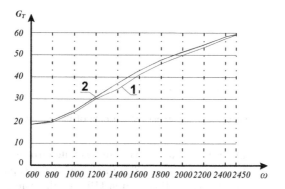

Fig. 7. Speed characteristic for fuel consumption.

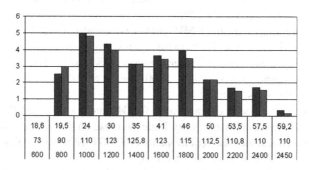

Fig. 8. The error in the methods for determining the resulting vector.

As a result of the work, a hardware-algorithmic complex was developed that makes it possible to carry out experiments to obtain the speed characteristics of an internal combustion engine. This system makes it possible to carry out tests of both a working engine and an engine with various malfunctions in real operating conditions. It can be seen that the average value of the relative deviation for the tests performed does not exceed the estimated relative error.

References

1. Valiev, R.A., Khairullin, A.Kh., Shibakov, V.G.: Automated design systems for manufacturing processes. Russ. Eng. Res. **35**(9), 662–665 (2015)
2. Guihang, L., Jian, W., Qiang, W.: Application for diesel engine in fault diagnose based on fuzzy neural network and information fusion. In: IEEE 3rd International Conference on Communication Software and Networks, vol. 6014398, pp. 102–105 (2011)
3. Galiullin, L.A., Valiev, R.A.: Automated system of engine tests on the basis of Bosch controllers. Int. J. Appl. Eng. Res. **10**(24), 44737–44742 (2015)
4. Galiullin, L.A., Valiev, R.A.: Automation of diesel engine test procedure. In: 2nd International Conference on Industrial Engineering, Applications and Manufacturing (2016)
5. Yang, J., Yu, Y.: The development of fault diagnosis system for diesel engine based on fuzzy logic. In: Proceedings of the 8th International Conference on fuzzy systems and knowledge discovery, vol. 1, no. 6019556, pp. 472–475 (2011)
6. Wei, D.: Design of Web based expert system of electronic control engine fault diagnosis. In: Proceedings of the International Conference on Business Management and Electronic Information, vol. 1, no. 5916978, pp. 482–485 (2011)
7. Zubkov, E.V., Galiullin, L.A.: Hybrid neural network for the adjustment of fuzzy systems when simulating tests of internal combustion engines. Russ. Eng. Res. **31**(5), 439–443 (2011)
8. Galiullin, L.A., Valiev, R.A., Mingaleeva, L.B.: Method of internal combustion engines testing on the basis of the graphic language. J. Fundam. Appl. Sci. **9**(SI), 1524–1533 (2017)
9. Valiev, R.A., Galiullin, L.A., Dmitrieva, I.S., et al.: Method for complex web applications design. Int. J. Appl. Eng. Res. **10**(6), 15123–15130 (2015)
10. Galiullin, L.A.: Automated test system of internal combustion engines. IOP Conf. Ser. Mater. Sci. Eng. **86**(012018), 1–6 (2015)
11. Galiullin, L.A.: Development of automated test system for diesel engines based on fuzzy logic. 2nd International Conference on Industrial Engineering, Applications and Manufacturing (2016)
12. Galiullin, L.A., Valiev, R.A.: Diagnostics technological process modeling for internal combustion engines. International Conference on Industrial Engineering, Applications and Manufacturing (2017)
13. Valiyev, R.A., Galiullin, L.A., Iliukhin, A.N.: Design of the modern domain specific programming languages. Int. J. Soft Comput. **10**(5), 340–343 (2015)
14. Zubkov, E.V., Novikov, A.A.: Regulation of the crankshaft speed of a diesel engine with a common rail fuel system. Russ. Eng. Res. **32**(7–8), 523–525 (2012)
15. Shah, M., Gaikwad, V., Lokhande, S., et al: Fault identification for I.C. engines using artificial neural network. In: Proceedings. of Int. Conf. on process automation, Control and computing (2011). 5978891
16. Valiyev, R.A., Galiullin, L.A., Iliukhin, A.N.: Approaches to organization of the software development. Int. J. Soft Comput. **10**(5), 336–339 (2015)

17. Galiullin, L.A.: Automated test system of internal combustion engines. In: International Scientific and Technical Conference Innovative Mechanical Engineering Technologies, Equipment and Materials, vol. 86, no. 012018 (2015)
18. Li, X., Yu, F., Jin, H. et al.: Simulation platform design for diesel engine fault. In: International Conference on Electrical and Control Engineering, pp. 4963–4967 (2011)
19. Valiyev, R.A., Galiullin, L.A., Iliukhin, A.N.: Methods of integration and execution of the code of modern programming languages. Int. J. Soft Comput. **10**(5), 344–347 (2015)
20. Biktimirov, R.L., Valiev, R.A., Galiullin, L.A., et al.: Automated test system of diesel engines based on fuzzy neural network. Res. J. Appl. Sci. **9**(12), 1059–1063 (2014)

Optimization of Control for the Pneumatic Container Transport System

V. V. Khmara, A. M. Kabyshev[✉], and Yu. G. Lobotskiy

Northern Caucasus Mining and Metallurgical Institute, 44 Nikolaeva Street,
Vladikavkaz 362021, Russian Federation
al.kab.56@yandex.ru

Abstract. The automatic control of any process implies continuous or infrequent monitoring process parameters. At the ore dressing mills and metallurgical plants, such parameters may include composition of feedstock, fluxes, reagents, intermediate products, final tailings and wastewater. Usually, this information is supplied by the automated analytical monitoring system, in which the main component is the automated sub-system for sampling and representative test sample transportation. The article considers the flowchart of a system for container freight delivery via a transport pipeline, designed for delivery of process materials. A control principle for pneumatic conveying subsystems, which builds on pressure control in the transportation pipeline, and a control algorithm for the transportation container delivery process were devised. The proposed engineering solutions allow compressed air pressure to be optimally distributed in long transport lines (up to several kilometers). The noted problem is solved with the help of intermediate stations. Intermediate stations are installed on long sections of the transport pipeline. The scheme of the intermediate station and the corresponding control algorithms are developed.

Keywords: Transport pipeline · Container · Sensor · Control system · Algorithm · Compressed air pressure · Microcontroller

1 Introduction

Robotized mechanisms and equipment make extensive use of positional systems whose functions involve sequential performance of similar process operations. Such equipment includes systems for container freight delivery via transport pipelines.

Container delivery systems are commonly employed in automated systems for analytical process control, in particular, at mining, concentration and iron/steel companies to deliver representative samples of technological products for rapid analysis [1].

The existing pneumatic conveying systems for freight delivery via transport pipelines are characterized by diverse circuit solutions and applications [2–9].

This work is aimed at developing a system for container delivery of process samples for rapid analysis, which is based on modular subsystems controlled by a versatile microprocessor system. Achieving the set target will facilitate the adaptation of the system for a specific process, enhance reliability and simplify in-service maintenance through the use of modular subsystems.

2 Results and Discussion

A pneumatic conveying system for automatic delivery of samples for analysis comprises the following primary devices [10–20]:

- A station for automatic charging of proportionate samples into a transportation container and dispatching the transportation container loaded with samples to a rapid analysis laboratory
- A station for automatic discharging of delivered process samples from a transportation container and returning the empty transportation container to the sampling point
- A system of transport pipelines connecting process sampling points with laboratories that make all the necessary analytical measurements
- Automatic switches for automatic redirection of moving loaded and empty transportation containers from one transport pipeline to another, according to the route assigned.

The flowchart of the developed pneumatic conveying system for container delivery of samples is shown in Fig. 1.

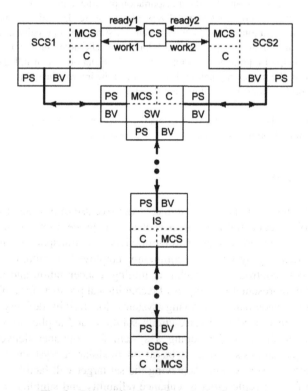

Fig. 1. Pneumatic conveying system flowchart.

The following notation is used in the figure above: SCS1, SCS2 – automatic charging stations to fill transportation containers with technological products; SW – automatic switch; IS – intermediate station; SDS – automatic discharging station to unload delivered technological products; MCS – microcontroller-based automatic station control systems; C – pneumatic compressors; PS – pressure sensors; BV – bypass valves; CS – automatic SCS control system.

Heavy lines in the figure are used to mark transport pipelines.

Pressure sensors (PS) and bypass valves (BV) are mounted on transport pipelines in close proximity to the devices that are part of the system.

Flowcharts, principles and algorithms for operation of the pneumatic conveying system's stations and subsystems are described in [9, 10, 13].

The intermediate station (IS) (one or more) is installed within the long span of a transport pipeline, between the switch and the discharging station. The intermediate station design makes use of modular subsystems and builds on the same principles as the other devices incorporated into the pneumatic conveying system.

The intermediate station flowchart is shown in Fig. 2.

BV is bypass valves installed on a transportation pipeline in the immediate proximity of IS, SCS, SW and SDS, which ensure reliable transportation of a capsule (loaded or empty) via the transportation pipeline by closing BV installed near the station dispatching the capsule, and opening BV installed near the station receiving the capsule. The command is sent to close the bypass valve at the same time with the command to supply compressed air. Once the transporting compressed air is shut off, the bypass valve opens. The bypass valves operate under the control of signals coming from pressure sensors.

IS comprises pneumatic cylinders (PC), electro-pneumatic distributors (EPD) and electro-pneumatic valves (EPV). Pneumatic cylinders move the receiving socket (RS) as appropriate to ensure that empty and loaded transportation containers coming in via transport pipelines are received and dispatched.

Positions of the intermediate station's mechanisms are controlled by magnetic sensors (MS). Signals generated by the sensors are fed to MCS (Fig. 1).

MCS controls the operation of EPD and EPV using the algorithm shown schematically in Fig. 3.

Without the brackets in Fig. 3, are shown the magnetic sensors (MS) and the pneumatic distributors (EPD) involved in moving the container from left to right inside the IS station (Fig. 2). Shown in brackets are devices that operate when transferring a container from the right to the left pipeline.

IS units are controlled by a built-in control system (MCS). The control system flowchart is completely identical to those of SW, SCS и SDS control systems [21].

Operation of the pneumatic conveying system depicted in Fig. 1 starts with the loading of technological products into transportation containers disposed in charging stations (SCS1, SCS2). The loading of technological products into transportation containers generates "ready" signals entering the charging stations control system

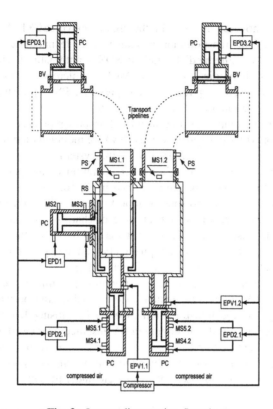

Fig. 2. Intermediate station flowchart.

(CS) that controls the station operation priority by means of "work" signals. CS generates a "work1" signal fed to MCS of SCS1 station to initiate the process of dispatching a container via the transport pipeline to SW. The switch (SW) sends this container on via the pipeline running through IS to the discharging station (SDS). SDS discharges technological products and sends the empty container to the initial point – SCS1. The control system (CS) generates the signal "work2", the process of transporting the container of the SCS2 station is carried out, after that SCS1 starts working again and the whole cycle repeats.

Container transportation in the system under review is possible due to the fact that MCS built into the system blocks monitor compressed air pressure in transport pipelines and control the operation of pneumatic compressors (C), pneumatic valves (EPV) and bypass valves (BV). Figure 4 shows the flowchart of the control algorithm for pneumatic compressors and bypass valves of the transportation system blocks.

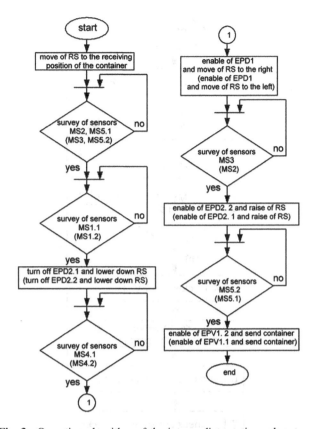

Fig. 3. Operation algorithm of the intermediate station subsystems.

Blocks 1 and 4 are where bypass valves (BV) are enabled and disabled by the selector switches EPD3.1 and EPD3.2 (Fig. 2). In block 5, the intermediate station (IS) functions according to the algorithm shown in Fig. 3. Actions performed in block 5 by devices SW, SCS and SDS are described in [21].

In the algorithm's blocks 7 and 8, a signal received by MCS from pressure sensors is controlled. Pressure sensors generate a signal in response to a pressure decrease in the transport pipeline. This happens when a container is passing the pipeline section fitted with a bypass valve. A signal received from a pressure sensor informs the control system, which is built into the device that has dispatched the container, of the need to disable the compressor. The pneumatic valves (EPV) operate in synchronism with the compressor.

Such organization of the conveying system control process makes it unnecessary to exchange information signals among the system's devices. This is important for long transport lines.

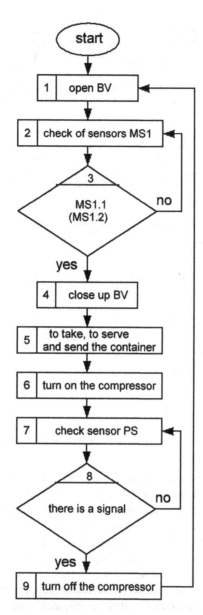

Fig. 4. Flowchart of the control algorithm for pneumatic compressors and bypass valves.

3 Conclusion

The developed flowcharts and algorithms discussed in this article may be employed when building a pneumatic conveying system for process materials at mining, concentration and iron/steel companies.

References

1. Dedegkayev, A.G., Kabyshev, A.M., Lobotskiy, Yu.G., et al.: Conceptual features of the all-purpose system for sequential control of positioners. In: Proceedings of the II International Research and Practice Conference. Present Scientific Research: Procedures, Theory, Practice, pp. 180–193 (2014)
2. Ma, C., Mao, H., Yang, X., et al.: Study on the development mode of urban underground logistics system. Serv. Sci. Manag. Res. **3**(1), 7–12 (2014)
3. Cox, B.: Pneumatic powder transport system. US patent 7464733 B2. 16 December 2008 (2008)
4. Arends, G., Grote, B.J.H.: Underground logistics systems versus trenchless technology. In: Proceedings from the 2nd International Symposium on Underground Freight Transportation by Capsule Pipelines and Other Tube/Tunnel Systems, Delft 28–29 September 1999 (1999). J. Transp. Eng. **12**(4), 300–310
5. Roop, S.S., Roco, C.E., Olson, L.E.: The technical and economic feasibility of a freight pipeline system in Texas – report no. 1519 – 3. Texas Transport Institute, Texas (2002)
6. Hodson, N.F.: Energy saving pipeline capsule goods transport. In: Proceedings from the International Symposium on Underground Freight Transportation by Capsule Pipelines and Other Tube/Tunnel Systems, Arlington, 20–22 March 2008 (2008)
7. Liu, H.: Research, development and use of PCP in the United States of America. Jpn. J. Multiph. Flow **21**(1), 57–69 (2007)
8. Pielage, B.A.: Underground freight transportation. A new development for automated freight transportation systems in the Netherlands. In: Proceedings of the IEEE Intelligent Transportation Systems, pp. 762–767 (2001)
9. Lobotskiy, Yu.G., Khmara, V.V., Kabyshev, A.M.: The principle of the complex systems of capsule pneumatic transport using multi-purpose switches. Modern Appl. Sci. **9**(5), 228–246 (2015)
10. Lobotskiy, Yu.G., Khmara, V.V.: Automated analytical control systems as a basis of technological process management. In: Proceedings of the VII International Research and Practice Conference. A New Word in Science and Practice: Hypotheses and Approbation of Research Results, Novosibirsk, 13 November 2013, pp. 146–151 (2013)
11. Terzini, R.: Container for use in pneumatic transport system. US patent 2011/0142554 A1 (2011)
12. Lobotskiy, Yu.G., Khmara, V.V.: Problems of automatic sampling and delivery of product samples for analysis in processing and metallurgical plants. Int. Sci. Rev. Sustain. Dev. Min. Areas **4**, 44–49 (2014)
13. Khmara, V.V., Lobotskiy, Yu.G.: Optimization of automatic single sampling processes and preparation of averaged samples for transportation for analysis. Int. Sci. Rev. Sustain. Dev. Min. Areas **1**, 42–50 (2014)
14. Lobotskiy, Yu.G., Khmara, V.V.: Methods targeted at enhancement of safety in automatic sample charging into a capsule. Int. Sci. Rev. Sustain. Dev. Min. Areas **2**, 30–36 (2014)
15. Lobotskiy, Yu.G., Khmara, V.V.: Station of automatic sample discharging from a capsule. Int. Sci. Rev. Sustain. Dev. Min. Areas **3**, 35–42 (2014)
16. Kosugi, S., Saitou, K., Matsui, N., et al.: Development of vertical pneumatic capsule pipeline system for deep underground. In: 2nd International Symposium on Underground Freight Transportation by Capsule Pipelines and Other Tube/Tunnel Systems (2000)
17. Lobotskiy, Yu.G., Khmara, V.V.: Automatic switches for changing the route of moving capsules. Int. Sci. Rev. Sustain. Dev. Min. Areas **4**, 27–35 (2014)

18. Dedegkayev, A.G., Kabyshev, A.M., Lobotskiy, Yu.G.: Optimization of the structure and planning the performance of the all-purpose system for sequential control of positioners. In: International Research and Practice Conference Engineering Sciences: Procedures, Theory, Practice, Moscow, 17 June 2014, pp. 45–67 (2014)
19. Liu, H.: Transporting freight capsules by pneumatic capsule pipeline: port security and other issues. In: Invited Presentation at Summer Meeting of U.S. Transportation Research Board, La Jolla, 10 July 2006 (2006)
20. Tatay, S., Magoss, E., Kazmer, S.: Condition monitoring and fault diagnostic of the pneumatic conveying systems. In: International Scientific Conference on Sustainable Development and Ecological Footprint, Sopron, 26–27 March 2012, pp. 1–4 (2012)
21. Khmara, V.V., Lobotskiy, Yu.G., Kabyshev, A.M.: Microprocessor control of devices systems of pneumatic container delivery for the analysis of samples of concentrating and metallurgical production. In: 2nd International Conference on Industrial Engineering, Applications and Manufacturing (2016)

Construction and Movement Control of Integrated Executive Device of Laser-Robot

I. N. Egorov[✉], A. N. Kirilina, and V. P. Umnov

Vladimir State University, 87 Gorkogo Street,
Vladimir 600000, Russian Federation
egorovmtf@mail.ru

Abstract. During the processing and utilization of large-overall metalwork of the complex configuration in the metallurgical, chemical, ship-building and atomic industry application of the universal laser technological complexes (LTC) with optics manipulators - the systems of movement of optical elements on the linear, rotational and complex trajectory is expedient. The actuator of laser-robot is a complexed (integrated) manipulation system consisting of a set of successively arranged transport and technological manipulators. Three-degree transport manipulator operating in permutation mode is carrier for main five-degree technological manipulator, which moves working tool - optical head at fixed transport manipulator. Control of position and orientation of optical head of laser-robot in conditions of structural-technological complexity, non-stability and uncertainty of object is most expedient in class of intelligent control. The offered system of neural network control contains controllers of coordinate and functional conversions and sensors as thickness of the cut material of an object and control of its cutting. The controller of lasing source, changes beam power depending on cutting conditions. Researches of the computer model of the proposed control system have shown its operability and efficiency of its application as a simulator in the development of control programs.

Keywords: Control system · Executive device · Manipulator · Robot · Laser processing · Large-overall object · Model

1 Introduction

One of tendencies of development of the modern technologies of mechanical engineering – is an expansion of a range of application of the laser technological complexes (LTC) and devices.

It allows, in particular: to increase the speed and accuracy of processing of materials that in turn provides minimization of the zone heating and excludes temperature strains; to exclude, owing to a contactless of processing to exclude operations of replacement of the tool; to provide transfer of a laser radiation without loss on the considerable distances; to process any materials including which are not giving in to traditional methods of processing; to provide ecological purity of laser technology.

During the processing and utilization of large-overall metalwork of the complex configuration in the metallurgical, chemical, ship-building and atomic industry application of the universal LTK with optics manipulators - the systems of movement of optical elements on the linear, rotational and complex trajectory is expedient. LTK with manipulators of optics of the last type are named – "laser robots".

2 Urgency

Robotization of operations of processing and utilization of large-size products in nonstationary conditions demands creation of the laser robots in the form of the laser technological centers with the maximal concentration of operations and the integrated manipulation executive system.

The solution to the problem of controlling the position and orientation of the laser-robot optical head (OH) in terms of constructive - technological complexity, unsteadiness and uncertainty virtually precludes the use of classical methods of automatic control theory.

The generalized structure of the mobile autonomous laser-robot Palar-40 for processing of large-overall objects is presented in Fig. 1. It is suggested manipulation systems (MS) of executive devices (ED) of such laser robots to realize as integrated set of the manipulators [1, 2]: transport (TrM) and technological (TM).

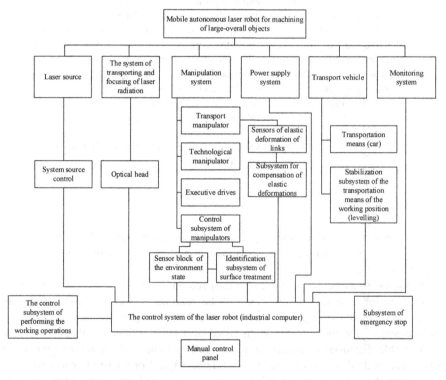

Fig. 1. The generalized structure of the mobile autonomous laser-robot Palar-40 for processing of large-overall objects.

The transport manipulator has three degrees of mobility, it is the carrier for technological manipulator and works in the rearrangement mode. The technological manipulator has five degrees of mobility, it moves the working tool – an optical head while the transport manipulator is stationary and it is the main manipulator. Realization of manipulators according to such scheme will allow to realize the manipulator with a cumulative length of links up to 15 m. Constancy of position of a focus point is provided with the last drive of a translational motion working in the stabilization mode. The angular orientation of an optical head is provided by two angular coordinates.

The process of laser processing from the point of view of control of the movement of OH along the path of the processing, beam focusing, deflection of the beam axis relative to the programmed trajectory and the normal to the treated surface can be called an object coordinate control, and from the point of view of the control of change process parameters of laser processing, it can be defined as the object of parametric control. In general, control of technological process of laser processing can be considered as coordinate-parametric [3] or binary control [4, 5], containing as a control of the movement and orientation of the OH as and control of the parameters of laser radiation.

One of the problems of motion control and orientation of the OH is to coordinate the motions of individual kinematic pairs MS of TrM if it is combined with TM, which provides the required movement of the OH. The another problem of coordination is reduced to the problem of the distribution of forces representing the inverse problem of dynamics when the movement of the OH is completely specified and it is necessary to determine forces and moments as interactive and within the autonomous degrees of freedom of MS.

During movement, positional connection and force interaction are inextricably connected, being essentially positional-force interaction at the second level constantly, and at the first level - depending on the position of the system components and the performed operation.

At the same time, two variants of movement of the considered component of the manipulation system are possible: movement due to a third-party energy source and movement due to energy pumping during position-force interaction with another component. Structural constancy of given connections and interactions at a certain stage of movement can be considered a phase of their state.

Each phase of the state of positional connections and power interactions can be matched to the current state of control in kinematic pairs, which can be considered as the phase of the state of control in the manipulation system.

Then each j-th phase of state of control ($j \in J = \{1, \ldots, k\}$) will be characterized by an "n" - dimensional set $\Psi_j = \{x_1, \ldots, x_n\}$, where k is the number of control phases occurring during the operation, and n is the number of kinematic pairs in the manipulation system.

The finite "k" - dimensional set $\Phi(k) = \{\Psi_1, \ldots, \Psi_\kappa\}$ can be considered a control state space of a manipulation system.

The transfer from one phase of the control state to another will be called as a phase change.

For completeness of representation of the first phase of control in space of its state it is necessary to consider provision of position of static equilibrium before start of movement, and the last - position of equilibrium at end of movement.

Phase transition of control state can be performed by algorithms generated in setting space in accordance with control program, as well as state of position connections and power interactions in working zone and manipulation system.

For each phase and control space there is correspondence:

$$\Psi_j(x); \; \Phi(\kappa) \Rightarrow U(\kappa) \Rightarrow | U(t) | \; T, \qquad (1)$$

where $U_j(t)$ - is "n"-dimensional vector of control actions on actuator drives, generally containing zero components; $|U(t)|$ - is "k × n" dimensional block vector of control actions formed during the operation.

On the basis of (1) it is possible to write operator expressions for state variables of kinematic couples as:

$$Q_j(p) = W_J(p) \cdot U_j(p); \; |Q(p)| = |W(p)| \cdot |U(p)|, \qquad (2)$$

here $Q_j(p)$ - "n" - dimensional vector of the control and not controlled kinematic pairs for j-th phases of control; $|Q(p)|$ "k × n" - dimensional block vector of kinematic pairs in the process of operation execution; $W_J(p)$ and $|W(p)|$ - the transfer matrixes determined by kinematic structure of a handling system, structure of the drive and a type of state variable of a kinematic pairs (movement, speed or force-torque interaction).

For a simple open kinematic chain, the matrices $W_J(p)$ and $|W(p)|$ have a diagonal view.

If the positional connections and force interactions occurring in the work zone in performing some operation are represented in the inertial coordinate system by some "s" - dimensional vector R(t), the task of constructing control of the manipulation system of the robot performing this operation in general form can be formalized as:

$$\forall R_j(t) \in R(t) \, \exists \, \Psi_j(x) \in \Phi(\kappa) \Rightarrow Q_j \, \exists \, U_j(Q_j, t) \in U(t) \forall D_j \to D_{\max(min)}. \qquad (3)$$

here D_j is a criterion for optimizing of the solution of the problem for the j-th phase of the movement.

At kinematic control, the vector of position or high-speed control impacts on a handling actuating system by an analytical or iterative solution of the return problems of kinematics forms. In case of dynamic control forming of the control moments of drives can be executed on the basis of kinematic control with compensation not of stationary dynamic effects of an actuating system in space of generalized coordinates and their derivatives with inclusion in structure of a system, for example, of a reference model of an actuating system, or use of moment regulators.

Assuming existence in each drive unit of the robot of position and moment feedback couplings, and possibilities of the combined position-force control structural synthesis of the state space of control, can be considered as a problem of creation of a system with variable structure depending on quantitative values of dynamic indicators of an actuating system and requirements to its quality at different stages of the movement.

The most difficult is the problem of creation of the state space of management during the work with the imposed communications.

In this case it is necessary to find displays of a non-stationary six-measured vector of position communications and a six-measured vector of power interactions of the working tool with object of works for different stages of the movement (operation execution) on space of generalized coordinates for the purpose of determination of its structure from a condition of execution of working operation at minimization, for example, of the managing moments.

The report is considered as an impedance motion control based on the calculated hinge moments, and the hybrid control involving a combination of positional movement control with position-force control [6, 7], compensating the forces occurring in interaction.

The control algorithm is non-coincident, and TrM and TM can be built based on the linearization and decomposition of the mathematical description of MS laser robot with the introduction of nonlinear feedback and nonlinear transformations of state variables [8, 9].

Systems with dynamic coordinator (proofreader) [10] and position-force controllers are another form of a system of coordinated control two-manipulated ED of laser-robot.

The coordinator generates the nominal control of the global power and/or process feedback. Compensation of deviations of OH from the nominal trajectory is provided by additional supervisor.

At the solution of the task of control of integrated MS of the laser robot in the conditions of technological indeterminacy the option of use of intellectual system is considered. According to the Fig. 2 the block diagram of an intellectual neural network control system of an optical head movement of the considered laser–robot is represented during performing of operation of utilization cutting in the conditions of indeterminacy thickness of an object material.

The technological controller of formation of parameters of processing defines a task for the controller of planning of a trajectory, for the controller of formation of standard speed in space of a task and for the controller of laser source. The controller of planning of a trajectory develops parameters of points of a trajectory, which transform to the generalized coordinates of angles of rotation in the neurocontroller 1 transformation of coordinates. Signals from the last enter on the positional controller, which is connected with the block of neurocontrollers of speed regulators. The controller of formation of standard speed forms a task of the required processing speed. Transformation of standard speeds to the generalized coordinates happens to the help of the neurocontroller 2 transformations of speed from which signals enter on the block of neurocontrollers of speed regulators. Control signal from the block of neurocontrollers of speed enters to electric drives of MS, which realized the movement of the system and the executive kinematic chains. Information on the position of an output point of manipulating system, and speed of movement of this point is feeded by means of sensors and move for correction of control signals on the block of neurocontrollers of speed and the block of positional controllers.

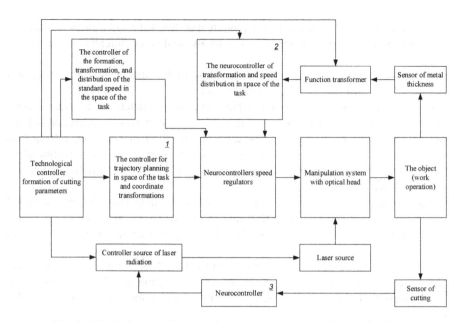

Fig. 2. Block diagram of a control system of movement of an optical head.

The manipulating system has mechanical communication with an optical head on which the beam moves from lasing source. Parameters of a laser radiation are formed by the controller of lasing source. As cutting is carried out in the conditions of indeterminacy of thickness of metal, the sensor of its measurement moves ahead of an optical head and in the functional converter there is a transformation of a signal from this sensor in cutting speed parameters.

At the same time there is a subtraction of the speed given initially in the technological controller and speed of thickness demanded depending on indications of the sensor during process of laser processing.

The main technological feed-back is the feed-back on a condition of the guaranteed cutting through material realized by means of the sensor, the neurocontroller 3 and the controller of the lasing source, which changing of beam power depending on cutting conditions. The neurocontroller 3 can be trained for various characters of a cutting through material and for various materials on a model of cutting process.

Neurocontrollers in a contour of speed of drives are intended for rapid fine tuning of coefficients of regulators in the function to task its standard value and a signal of a technological feed-back.

At the beginning of movement, the necessary data are formed including:

- The flow chart of surfacing processing (on the display of the monitor the type of a surface, a route of processing and a point of installation of the TrM are set
- The processing conditions of an object (speed and accuracy of processing, the processed material etc.)
- Modes of driving and control of manipulators (character of movement, sequence of driving of manipulators, type of control etc.)

In the mode of distance or automatic control, the TrM is moved to an initial point. At the same time, the TM is in the transport situation defined from a condition of a minimum of static loads on an output link of the TrM and convenience of a supply to object of works.

Experimental studies of motion control of the optical head (OH) were carried out by computer simulation using the 3D model of the robot laser executive system (Fig. 2), constructed using the SimMechanics second generation library and computer models of controllers, sensors and other system components. Checking of technical solutions of exhaust gas, sensors and separate units was carried out on the automated laser technological complex LK-5V equipped with multi-channel CO_2 laser, manipulator of the product and five-step manipulator of moving the OH. After that, a conclusion of the work.

3 Conclusion

1. The structural-algorithmic system management software provides a solution to the problem of controlling the position and orientation of the optical head laser-robot in terms of constructive - technological complexity, unsteadiness and uncertainty.
2. If the surface has the irregular curvilinear shape and its situation in relation to the laser - to the robot is not defined, the analytical task of a trajectory of driving even in an approximate look is almost impossible, and use of systems of a location is accompanied by the considerable difficulties. In this case parameters of a trajectory of driving can be measured the laser robot, working in the self-studying mode which can be combined with operation realization.
3. For the purpose of reduction of time of calculations when tutoring it is necessary to use multiplex systems, and the simulator of driving of manipulating system allowing to analyses, besides, emergence of singularities and to exclude them in the mode of replication of the control program.
4. The computer model of the executive system of the laser robot, which is executed taking into account its dynamic properties, can be such simulator.
5. Researches of computer model of the offered control system, showed operability and effectiveness of application it as the simulator when developing of the control programs.

References

1. Egorov, I.N., Umnov, V.P.: Principles of construction and control of manipulating systems of laser-robots. Mechatron. Autom. Control **11**, 29–34 (2004)
2. Egorov, I.N., Umnov, V.P.: Multipurpose manipulating executive systems of the robotic technological centers. Probl. Mech. Eng. Autom. **2**, 111–115 (2012)
3. Zemlyakov, S.D., Rutkovsky, V.Yu.: Coordinate-parametric control. Determination of the possibility of problems. Autom. Remote Control **2**, 107–115 (1976)
4. Emel'yanov, S.V.: Binary Automatic Control Systems. MNEPU, Moscow (1984)

5. Emelyanov, S.V., Korovin, S.K.: New Types of Feedback: Management Under Uncertainty. Fizmatlit, Moscow (1997)
6. Egorov, I.N.: System position-force control technology. Mechatron. Autom. Control **10**, 15–20 (2003)
7. Egorov, I.N.: Position-force control system of the interconnected manipulators laser robots. Actual problems of protection and security: Proceedings of the VII all-Russian STC, vol. 4, pp. 88–92 (2004)
8. Bejczy, A.K., Tarn, T.Y.: Dynamic control of robot arms in task space using nonlinear feedback. Automatisierungstechnik **10**, 374–388 (1988)
9. Tarn, T.Y., Bejczy, A.K., Yun, X.: New nonlinear control algorithms for multiple robot arms. IEEE Trans Aerosp. Electron. Syst. **24**(5), 571–583 (1988)
10. Ahmad, S., Guo, H.: Dynamic coordination of dual-arm robotic systems with joint flexibility. In: IEEE International Conference on Robotics and Automation, vol. 1, pp. 332–337 (1988)

Design for Adaptive Wheel with Sliding Rim and Control System

A. Ivanyuk[✉], D. Marchuk, and Yu. Serdobintsev

Volgograd State Technical University, 28 Lenina Avenue,
Volgograd 400005, Russian Federation
ivanyuk_aleksei@mail.ru

Abstract. In this paper, an automatic wheel diameter control process for mobile robotic complexes is proposed. The proposed design features an adaptive wheel with a sliding rim, strain gauges and a microprocessor. Specifically this is achieved by equipping the adaptive wheel with a microprocessor, which inputs data from the strain gauges located on the inside of the elastic tyre, and the outputs of the microprocessor are connected to the actuator with the ability to control is movement. The obtained reading from the strain gauge is translated using approximating functions. The software and algorithm for determining the need to change the driving characteristics (in particular, the diameter) of the adaptive wheels has been developed. The program checks whether the wheel can negotiate the obstacle, based on the defined height of the wheel. The cut-off is set at half the height of the wheel, and if this is exceeded, the program will report a need to increase the wheel diameter and a new diameter is displayed. This program has application in robotics, namely for use of adaptive wheels on devices such as recon robots, lunar rovers, rescue robots, and robots for exploration of geological reserves. This creates the potential for combined modes of motion and self-diagnosis of the wheels.

Keywords: Adaptive wheels and locomotors · Mobile robotic complexes · Automated control system · Strain gauge

1 Introduction

Mobile robots are increasingly being used for rescue, military and reconnaissance purposes. The development and exploration of lunar terrain, exploration of geological reserves and other tasks solved by these technical systems require different speeds, clearance, etc. from mobile systems. Accordingly, it is necessary to select the most suitable type of locomotor [1]. Wheeled vehicles have the best mobility and speed of movement. At the same time, there is the problem of the ratio of the robots' speed and ability to travel across uneven terrain is still a challenge [2]. Therefore, there is a need for wheels that change their characteristics and modes depending on the conditions of use, and algorithms for controlling mobile robotic complexes in case of technical vision failure [3].

Based on statements from the National Association of Market Participants of Robotics, the focus for technological research up to 2020 is as follows:

- Movement in a dynamic environment (including extreme manoeuvring)
- Navigation under conditions of electronic countermeasures
- Highly efficient transportation vehicles for personnel and cargo
- Autonomous operations of robots (underwater, on land, and air)
- Robotic transport in air and water, on rugged terrain and public roads

Based on the above, it is necessary to design wheels capable of changing their characteristics and modes, depending on the conditions of the terrain being overcome. Additionally there is a prerequisite for algorithms for controlling mobile robotic complexes in case of technical vision failure, for autonomous movement from one given place to another, without human participation in the control system [4].

2 Adaptive Wheel Design Development

The prototype design focuses on the design of wheels with a metal rim which is capable of changing its axial position relative to the hub [5].

Existing analogues and prototypes are examined for an estimation of challenges and tendencies of their solutions.

A wheel design with a sliding rim is known [USSR patent 821228, IPC: B60B 3/02, 1995], where the rim contains movable and fixed parts, the wheel design also includes a hub, pin and device for changing the width of the rim. The latter consists of a power mechanism, guides and the on-off control. This on-off switching mechanism is made in the form of an electromagnetic pusher of the power transfer unit, which contains a crenelated ring on the pin of the wheel, and a gearwheel, which is kinetically connected to the power mechanism, crenelated ring and the base of the electromagnetic pusher.

A wheel with increased cross-country ability [USSR patent 473625, IPC: B60B 9/00, 1975] contains a locking hub mounted on the axle, with an elastic spheroid-shaped tyre. The locking hub is made of two separate spring-loaded parts, secured at the ends of the axle, which can be displaced longitudinally. The elastic tyre is made up of multiple longitudinal flat strips separated by elastic spacers, each of which with its ends is rigidly attached to the rims of the half-hub.

The disadvantage in the above listed devices is the lack of adaptation of the wheel dimensions to the obstacles to be overcome in real time, along the vehicle's travel.

The closest technical solution is a wheel with a sliding rim [USSR patent 439414, IPC: B60B 3/02, 1975]. The wheel contains a hub mounted on the pin with a rim flange bearing an axially movable spigot which also has a projecting rim. Between the two projecting rims are an elastic tyre and a device which changes the distance between them. The spigot is mounted on the hub on the splines. The above-mentioned device which changes the distance is made in the form of a sealed cover attached to the lip of the spigot and connected by means of a bearing to the pull rod of the actuator located inside the trunnion.

However, the disadvantage of the device described above is the inability to automatically adjust the wheel to the dimensions of obstacles in its path.

The objective of the present study is to improve the manoeuvrability and ability to travel over rough terrain of mobile systems which use wheels with sliding rims.

This is resolved through automatic control of wheel diameter while the vehicle is in motion, based on the dimensions of the obstacles in its path [6].

Specifically this is achieved by equipping the adaptive wheel with a microprocessor, which inputs data from the strain gauges located on the inside of the elastic tyre, and the outputs of the microprocessor are connected to the actuator with the ability to control is movement.

The distinctive advantage of this design is that it becomes possible to automatically control the wheel diameter based on the size of the obstacles in its path, all while the vehicle is in motion.

A diagram of the adaptive wheel with sliding rim is presented in Fig. 1.

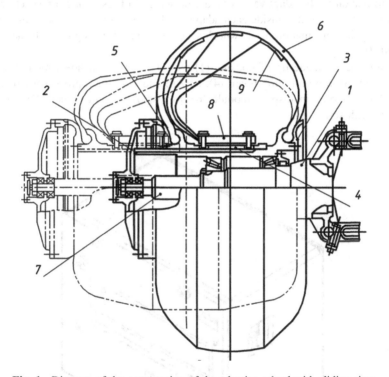

Fig. 1. Diagram of the construction of the adaptive wheel with sliding rims.

The adaptive wheel with sliding rim consists of a hub (2) mounted on a trunnion (1) with a rim flange (3) bearing an axially movable spigot (4) with a different rim (5), an elastic tyre (6), placed between named rim flanges (3) and (5) and an actuator (7). The inputs of the microprocessor (8) are connected to the outputs of strain gauges (9) located on the inside of the elastic tyre (6), and the outputs of the microprocessor are connected to the power mechanism (7) which changes the wheel diameter [7].

The adaptive wheel with sliding rim is designed and functions as follows:

The strain gauges (9) (bending resistors) are located on the inside of the elastic tyre, calibrated in such a way that when there is no pressure on that portion of the tyre, the strain gauge gives a level of resistance which is taken as the baseline value. When pressure is applied to the tyre, the gauges will output a lower resistance, as they become less deformed. The gauge sensor readings are transmitted to the microprocessor (8), where the readings are compared and location of contact with tyre is determined to calculate the geometric parameters of the obstacle. If, after contact with the ground, the resistance reading from the strain gauge is decreasing, that means there is an obstacle in the way of the wheel, and it is necessary to calculate if it can be overcome. To do this, the wheel diameter is increased. The signal from the microprocessor is fed to the actuator (7), located inside the trunnion (1). The actuator (7), in turn, reduces the distance between the rims (3) and (5), by moving the spigot (4) relative to the hub (2).

The three-dimensional model and the drawing of the designed device for measuring the deformation of the adaptive wheels are presented in Figs. 2 and 3 respectively [8].

This device is located inside each adaptive wheel, is responsible for measuring and controlling the deformation of the wheel, sending a controlling signal to the actuator as needed [9].

The device is made up of housing (1) with a cover (2), inside which is a power supply unit (3), a prototyping board (4), an Arduino Uno microprocessor (5), and for ease of use, a plug-in display module (6), to display current values. The device is fastened with screws (7).

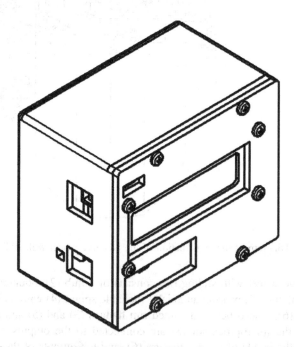

Fig. 2. Three-dimensional model of the designed device for measuring the deformation of the adaptive wheels.

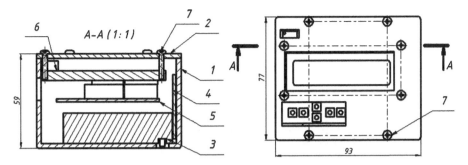

Fig. 3. Drawing of the designed device for measuring the deformation of the adaptive wheels.

3 Management System Development

The algorithm for defining requirement to change diameter of the adaptive wheels is described in more detail. Various machine vision systems are used to determine presence of obstacles in the path of transport systems (TS) or robotic complexes (RC). When these systems are working correctly, the obstacle is identified, distance to it and its size are measured, and this information is sent to the controlling microprocessor (e.g. Arduino Uno). After this a command is sent to the actuating unit of the adaptive wheel to be adjusted in which it is necessary to change a parameter, for example, to increase the diameter of wheels on the right side of the vehicle to allow it to pass the obstacle without having to change direction of travel [10].

In the event of machine vision failure or loss of signal transmission, it is imperative that the vehicle automatically determine the presence of obstacles and the need to vary the parameters of the adaptive wheels. Since the speed of most such vehicles (lunar rovers, scout robots, etc.) is quite low, the proposed solution is to automatically determine the height of the obstacle based on values obtained from the strain gauges located inside the adaptive wheel tyres.

The stress-strain state of the tyre is calculated from values received from the strain gauges. A program was developed (Fig. 4) specifically for this design, to determine presence of obstacles and whether the vehicle can continue on its route, or needs to adjust the wheel diameter to traverse them.

The program is designed to measure the stress-strain state of the exterior of adaptive wheels on mobile robots (MR) using strain gauges. Based on readings from the gauges, the microprocessor determines the presence and height of the obstacles to be overcome. Thus the program provides an algorithm for changing the running characteristics of the MR. This program has application in robotics, namely for use of adaptive wheels on devices such as recon robots, lunar rovers, rescue robots, and robots for exploration of geological reserves [11].

The program calculates the change in deformation of the adaptive wheel exterior from the force acting on it. The obtained reading from the strain gauge is translated using approximating functions and is displayed on the monitor connected to the Adruino Uno microprocessor.

Fig. 4. Output window which displays calculated values.

The program then checks whether the wheel can negotiate the obstacle, based on the defined height of the wheel. The cut-off is set at half the height of the wheel, and if this is exceeded, the program will report a need to increase the wheel diameter and a new diameter is displayed.

The program was tested using data which simulated readings from strain gauges located inside the tyre of an adaptive wheel. The values were random numbers in a given range. Readings from the last (sixth) sensor simulate contact of that section of the wheel with an obstacle. The range of values was taken from experimental data from calibrated sensors and simulates realistic conditions as thoroughly as possible. When an obstacle is detected, its height is compared with the height of half of the wheel, and when this is exceeded, the change in diameter required to traverse the obstacle is calculated [12].

If the analogue input with the number "0" is not connected to anything, then an arbitrary analogue noise on it will provide different initial numbers passed to the randomSeed() function each time the program is run. Subsequently, this will allow the random function to generate different values.

When there are no obstacles present in front of a specific section of the adaptive tyre, the message "no barriers" is displayed.

The height of the obstacle (k) at the point of contact with the tyre of the adaptive wheel is calculated based on the dimensions of the wheel and the elastic components. Specifically, it is determined based on the number of sections which make up the carcass of the wheel and the number of gauges (in the case under study – 6 gauges), the angle between them (60°), and the diameter of the wheel (comparison is based on radius r = 63 mm). When this threshold value is exceeded, the new diameter of the adaptive wheel is calculated. When the threshold is not exceeded, and it is possible to overcome the obstacle with the current diameter of the adaptive wheel, the message "it is possible to move" is displayed. The work is performed within research - "Development of model of adaptive wheels and an algorithm of automated management of their parameters, for improvement of passability and maneuverability of robotic complexes of land and space basing" (SP-1100.2018.3).

4 Conclusion

This paper describes a proposal for an adaptive wheel with a microprocessor which receives readings from strain gauges located inside the elastic tyre, and controls the activator. This allows for automatic control of the diameter of the adaptive wheel while the vehicle is travelling, based on the size of the obstacles in its path. Using automatic adaptive wheels makes it possible to combine manoeuvrability and speed with load distribution on loose ground for wheeled and tracked vehicles, as well as overcoming obstacles exceeding the size of the wheels of autonomous robots, both in walking and insectomorphic robots. A combination of modes and self-diagnosis of the wheel is also achieved.

An algorithm for automatic control of the main parameters of the adaptive wheels of a robotic complex has been developed, which makes it possible to perform diagnostics on the wheel and determine the height of an obstacle, and independently change its parameters.

References

1. Dubovska, R.: The quality control of machining process with CAD/CAM system support. In: 8th International Conference of DAAAM Baltic Industrial Engineering, pp. 27–32 (2012)
2. Sinha, R., Paredis, C.J., Khosla, P.K.: Integration of mechanical CAD and behavioral modeling. In: Proceedings of IEEE/ACM, pp. 31–36 (2000)
3. Karnopp, D.C., Margolis, D.L., Rosenberg, R.C.: System Dynamics: Modeling, Simulation, and Control of Mechatronic Systems, Hoboken, New Jersey (2012)
4. Serdobintsev, Yu.P., Makarov, A.M., Ivanuk, A.K.: Deformation of instrument housings under external pressure. Russ. Eng. Res. **37**(2), 675–678 (2017)
5. Bat, M.I., Dzhanelidze, G.Yu., Kelzon, A.S.: Theoretical mechanics at examples and problems. Lan, St. Petersburg (2013)
6. Migilinskas, D., Ustinovicius, L.: Computer-aided modelling, evaluation and management of construction projects according to PLM concept. Lecture Notes in Computer Science, pp. 241–250 (2006)
7. Jambor, J.: Quality of production process with CAD/CAM system support. DAAAM Int. Sci. Book **11**, 277–286 (2012)
8. Santos, A.D., Ferreira, J.: Duarte and Ana Reis. J. Mater. Process. Technol. **119**, 152–157 (2007)

9. Serdobintsev, Yu.P.: The use of strain gauges for changing the diameter of adaptive wheels of robotic systems. In: Advanced Systems and Control Tasks, Management and Information Processing in Technical Systems, pp. 558–565 (2017)
10. Ivanyuk, A.K., Panchenko, M.V., Ryabchuk, V.A., et al.: Device for measuring deformation during testing of structures for strength. RU Patent 172816, 25 July 2017 (2017)
11. Serdobintsev, Yu.P.: Analyzing the stress-strain state of object bodies subjected to hydrostatic pressure using modern CAD systems. In: 2nd International Conference on Industrial Engineering, Applications and Manufacturing (2016). https://www.doi.org/10.1109/ICIEAM.2016.7911678
12. Spiridonova, I.A., Grinchenkov, D.V.: On a mathematical modeling of a complex mechanical system by formal and by classical methods. J. Russ. Electromechanics **5**, 75–79 (2013)

Improvement of Acoustic-Resonance Method and Development of Information and Measuring Complex of Location of Deposited Pipelines

S. O. Gaponenko[✉], A. E. Kondratiev, and N. K. Andreev

Kazan State Power Engineering University, 51 Krasnoselskaya Street, Kazan 420066, Russian Federation
sogaponenko@yandex.ru

Abstract. The modern control methods location of buried pipelines include: magnetic method, infrared thermography, radiation method, radio wave method, electromagnetic method, acoustic location, resonance-acoustic profiling. These methods have several disadvantages related to the limited in scope, i.e., they are not applicable to ducts of non-metallic materials and have a weak selectivity control location of buried pipelines when in proximity with the controlled object undergoes a lot of metal engineering communications and a variety of runs – communication cables, power cables, pipe heating, water and gas supply. An acoustic location method is applicable to control the location of pipelines from non-metallic materials but has a number of disadvantages associated with the difficulty of carrying out search works. The above mentioned determines the relevance of the location of buried pipelines made of different metallic and non-metallic materials. The development of an improved method based on the analysis of acoustic resonant characteristics as well as the development, implementation and testing of a measuring complex with the best performance in comparison with the prototype, allows to solve this problem. The aim of research is the development of a measuring complex on the basis of advanced acoustic-resonance control method location of buried pipelines made of different metallic and non-metallic materials.

Keywords: Diagnostics · Acoustic-resonance method · Vibroacoustic signal · Information and measuring complex · Spectrum · Device for monitoring the location of buried pipelines · Harmonic oscillations · Pipeline

1 Introduction

Modern technologies allow us to show a pipeline or cable route by using 3D modeling with GPS binding. However, the problem of location of hidden communications on the area of constructions or other works is highly relevant. In case of non-availability of the information about laid hidden communications and possible accidental damage to power cables and pipelines the laying of new communications with the digging of trenches and canals or laying hidden communication by the method of horizontally directed drilling (HDD) are often can't be realized or realizing with considerable financial expenses. Moreover, the scope of contractors tasks is expanding - it's not only construction or repairing works, it's also includes providing of detailed information (maps, plans and diagrams) about the location of engineering networks. To reduce the occurrence of risks its necessary to realize the monitoring of location of engineering communication.

Nowadays task of controlling the location of hidden pipeline laid on the ground real time is being resolved by devices, based on the electromagnetic location method. Well-known and widely used locators "the success of the AG-308.10 M", "Alternative AG-401", "SR-20", teacher assistant correlation "Iskor-405", cameltransferencoding "Athlete TEK-500A", etc. However, the electromagnetic method does not allow to control reliably and selectively locations of the buried pipelines, when in close proximity to the controlled object passes a lot of metal utilities and various routes, such as communication cables, power cables, heat, water and gas supply pipes. In addition, this method cannot be used to search for non-metallic objects.

The above limitations of the field of application and disadvantages of existing devices make the task of developing an improved method, as well as the development, implementation and testing of information-measuring complex, which has better characteristics compared to prototypes, urgent.

Known is a comprehensive method for detecting non-metallic pipelines (polymer-reinforced pipes) and damages on them according to the invention RU No. 2328020, IPCG01V3/08, 20.04.2007, which consists in generating sound vibrations in the pipeline that cause mechanical vibrations of the metal pipe armature in a magnetic field of the Earth. The electric E and magnetic H components of the emerging electromagnetic radiation, the ground temperature and the noise level emitted by the medium transported through the pipe are measured [1, 2].

The disadvantage of this method is the difficulty in determining the location of pipelines, due to the presence of a number of monitored parameters, as well as the inability to search for metal pipelines.

The object of the present invention is to provide a simple method that provides high reliability and selectivity in determining the location of both non-metallic and metal pipelines.

The improved acoustic resonance method is described in patent for invention No. 2482515 [3] and consists in the generation of resonance sound waves in the cavity of the desired object, while the delineation of this object is accomplished by moving the sensing element (microphone or piezoelectric sensor) above the search zone.

The proposed method makes it possible to simplify the control of the location of buried pipelines, because the resonant frequency of the desired object is excited, the selectivity of the control increases. The method also provides high reliability of monitoring the location of buried both nonmetallic and metallic pipelines [3–8].

A device for determining the location of buried pipelines of different materials.

The method is implemented in a device for monitoring the location of buried pipelines made of various materials [3–8] for utility model patents No. 120784, No. 120785 [4–8], the principle of which consists in recording sensitive elements of sound signals from a leak or impulse waves, which are created by an additional generator connected to the pipeline.

Thus, this solution makes it possible to simplify the process of monitoring the location of buried pipelines and leaks, reducing the number of scanning operations, and also ensuring reliable detection of pipelines of various diameters and materials [4–16]. To test the proposed method of monitoring the location, an information and measurement complex was created in the work, which consists of an experimental installation and software.

Development and creation of an information and measuring complex.

Development and manufacture of an experimental installation in laboratory conditions.

The structural diagram of a laboratory installation for monitoring the location of buried pipelines is shown in Fig. 1 [4–8] and includes excitation device (acoustic emitter), piezoelectric sensors, ADC-DAC and personal computer. The acoustic radiator is supplied with a harmonic signal of the oscillation frequency within the range from 100 to 1100 Hz.

Using the digital-to-analog converter (DAC) 2, the PC output signal is converted to an analog form. For reception of a vibroacoustic signal in system piezoelectric sensors of mark KD-35 are applied. The received signal by a piezoelectric sensor is converted into a digital code from an analog signal to an ADC and analyzed in a personal computer [4–8]. The appearance of the laboratory setup is shown in Fig. 2.

Development and creation of software for the operation of the information-measuring complex.

In the graphical application development environment "LabVIEW" software was created "Software complex for monitoring the location of hidden hollow objects at their resonance frequency" (see the state program No. 2013610546, No. 2012661393) [9–21]. The program "Software complex for monitoring the location of hidden hollow objects at their resonant frequency" is designed to control the location of buried pipelines of various diameters and materials at their resonance frequency.

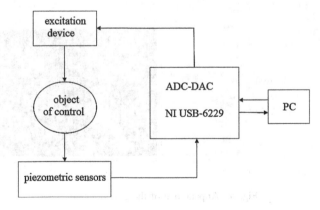

Fig. 1. Block diagram of the laboratory installation.

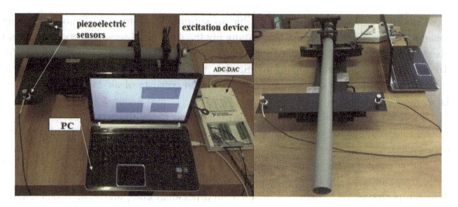

Fig. 2. Photo of the laboratory installation.

The program provides the following functions:

- Selection and generation of a resonant frequency in the cavity of a hidden hollow object (the pipeline under investigation)
- Search and reception of the generated signal by sensitive elements (piezoelectric sensors) over the search area
- Converting the received signal into a spectrum in real time and saving it in.txt format to a personal computer

In Fig. 3 shows the appearance of the program panel "Software complex for monitoring the location of hidden hollow objects at their resonance frequency" ("Generator" tab).

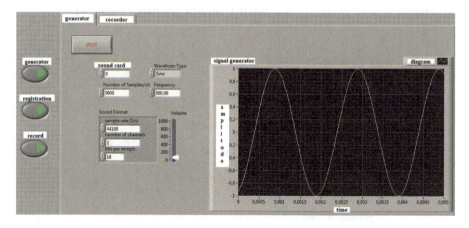

Fig. 3. Appearance of the generator tab.

In Fig. 4 shows the appearance of the program panel "Software complex for monitoring the location of hidden hollow objects at their resonant frequency" (tab "Registrar").

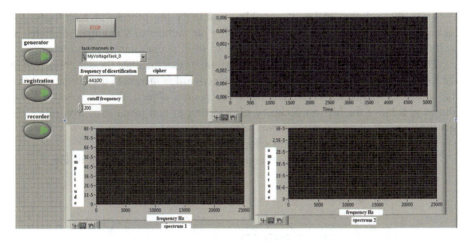

Fig. 4. Appearance of the tab "Registrar".

2 Research Methods

Method for determining pipelines location.

In this note, we recommend a way to determine the location of the covered pipelines, see Fig. 5. The technical result is achieved by the fact that in the pipeline the sound wobbles are generated with a resonant frequency in the spectrum from 100 to 1100 Hz with the support of a dynamic emitter 4, which is installed on the valve space. With the support of the sensitive substance 7, tuned to the resonant frequency of the pipeline, the amplitude of the desired signal is determined, and after that the exploration of the pipeline 6 will be realized by motion of substance method 7 over the territory in the direction of maintaining the greatest amplitude of the ground swings at this resonant frequency [3–8].

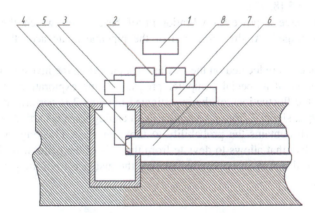

Fig. 5. A device for performing the method of analyzing the condition of the pipeline (1-PC (personal computer); 2-digital-to-analog converter; 3-amplifier; 4-speaker; 5-pipeline input; 6-pipeline; 7-signal receiving device; 8-analog-to-digital converter).

Thus, this solution makes it possible to simplify the process of monitoring the location of buried pipelines and leaks, reducing the number of scanning operations, as well as to ensure reliable detection of pipelines of different diameters and materials.

The proposed method highlights the probabilities that facilitate the determination of the location of pipelines. The selectivity of the control is increased due to the fact that the resonance frequency is excited from the desired object. The proposed method allows to achieve the highest accuracy of detection of non-metallic or iron pipes. To test the above method, an experimental device was created.

Description of the experimental setup.

Exterior view of the information and measuring complex is shown in Fig. 6.

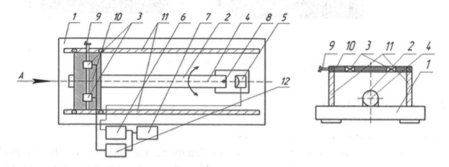

Fig. 6. Information and measuring complex (1 – base; 2 – test pipeline; 3 – signal receiver (microphone); 4 – longitudinal axis of the test pipeline; 5 – acoustic emitter; 6 – digital-to-analog converter; 7 – personal computer; 8 – swivel mounting of the test pipeline to the base; 9 – device for moving the sensing element; 10 – carriage; 11 – guide carriage; 12 – analog-to-digital converter).

The principle effect of the installation, see Fig. 6, consists in recording acoustic signals by microphones 3, which are excited in the investigated pipeline 2 by an acoustic emitter 5 [8, 12].

The units have a powerful vibration-proof base 1, on which the investigated pipeline 2 is hinged. At the same end of the pipeline 2, an acoustic radiator 5 is installed.

Microphones 3 are located on the carriage 10, placed on the investigated pipeline 2. With the support of a special computer program 7, the exploration of the resonant frequency of the fluctuations of the investigated pipeline 2 is performed. Reorganization of an output signal 7 from a computer in an analog form is carried out by DAC 6. The carriage 10 contains the probability to move in ways 11 along the investigated pipeline 2, in fact that allows to deviate from the axis of the paths with the support of the hinge 8. In this way, it is possible to vary the distance from the pipeline 2 to the microphone 3.

3 Results of a Research

The estimation of the error of the information-measuring complex was carried out by a probabilistic-statistical method in accordance with GOST R 8.736-2011, which provides for the determination of the error in the characteristics of the laws of distribution of errors in measuring instruments included in the complex. For the developed complex, the absolute error was $\Delta c = \pm\, 0{,}0139$ V at a confidence probability $P = 0{,}95$ for the average values of the vibration amplitude $\tilde{A} = 0{,}298$ V. The results obtained with the help of the proposed laboratory installation and software showed the operability and high reliability of the results obtained. Laboratory studies were conducted without taking into account the soil.

4 Discussion and Conclusions

1. Developed and implemented in the form of an information and measuring complex consisting of devices for exciting and receiving forced oscillations, an improved acoustic resonance method for monitoring the location of buried pipelines.
2. Algorithmic and software-technical support of control processes of the information-measuring complex and processing of vibroacoustic signals in the LabVIEW environment was developed and created.
3. To test the proposed method and software, a laboratory installation was set up to monitor the location of buried pipelines made of various materials and laboratory tests were performed.
4. The developed technique for monitoring the location of buried pipelines using the advanced acoustic resonance method and the finite element modeling method and the information and measuring complex have been tested in field trials at the Scientific and Technical Center "GEOKOM", which is confirmed by the relevant act of implementing scientific research.

References

1. Shumaylov, A.S., Gumerov, A.G., Moldavanov, O.I.: Diagnostics of Main Pipelines. Nedra, Moscow (1992)
2. Kanevskiy, I.N., Salnikova, E.N. Non-destructive testing methods. DVGTU, Vladivostok (2007)
3. Kondratiev, A.E., Gaponenko, S.O.: Pipeline location determination method. RU Patent 2482515, 20 May 2013 (2013)
4. Gaponenko, S.O., Kondratiev, A.E.: Model unit for the development of the method determining the location of hidden pipelines. University Bulletin. Problems of energy 7-8:123–129 (2014)
5. Gaponenko, S.O., Kondratiev, A.E.: Perspective methods and techniques for hidden channels, cavities and pipelines searching with a vibration-acoustic method. Bull. North-Caucasian Fed. Univ. **2**(47), 9–13 (2015)

6. Gaponenko, S.O., Kondratiev, A.E., Zagretdinov, A.R.: Low-frequency vibration-acoustic method of determination of the location of the hidden canals and pipelines. Procedia Eng. **150**, 2321–2326 (2016)
7. Nazarychev, S.A., Gaponenko, S.O., Kondratiev, A.E.: Determination of informative frequency ranges for buried pipeline location control. Helix **8**(1), 2481–2487 (2017)
8. Kondratiev, A.E., Gaponenko, S.O., Zagretdinov, A.R., et al.: Measuring - diagnostic complex to locate hidden pipelines. RU Patent 127203, 20 April 2013 (2013)
9. Kondratiev, A.E., Zagretdinov, A.R., Gaponenko, S.O.: Software complex for determine the location of the hollow objects at their resonant frequency. RU Certificate of state registration of computer software 2012661393 (2012)
10. Kondratiev, A.E., Zagretdinov, A.R., Gaponenko, S.O.: Software complex for detect objects in their hollow resonant. RU Certificate of state registration of computer software 2013610546 (2012)
11. Gaponenko, S.O., Kondratiev, A.E., Kamardin, A.S.: Resonance 2015. RU Certificate of state registration of computer software 2015612259 (2015)
12. Gaponenko, S.O., Kondratiev, A.E.: Device for developing a method for determining the location of the hidden pipes. News High. Schools Issues Energ. **7–8**, 123–129 (2014)
13. Gaponenko, S.O.: Variants of registration and analysis of useful vibro-acoustic signal in LabVIEW software product. Herald North Caucasus Fed. Univ. **5**(44), 8–15 (2014)
14. Gaponenko, S.O., Kondratiev, A.E.: Advanced methods and techniques of covert channel search, cavities and conduits vibroacoustic method. Herald North Caucasus Fed. Univ. **2** (47), 9–13 (2015)
15. Gaponenko, S.O., Kondratiev, A.E.: Measuring and diagnostic equipment for locate hidden pipelines. News High. Schools Issues Energ. **3–4**, 138–141 (2013)
16. Gaponenko, S.O., Kondratiev, A.E.: Device for calibration of piezoelectric sensors. Procedia Eng. **206**, 146–150 (2017)
17. Vinogradova, N.A., Listratov, Y.I., Sviridov, E.V.: Development of Application Software with LabVIEW. Publishing House MPEI, Moscow (2005)
18. Zagretdinov, A.R., Gaponenko, S.O., Serov, V.V.: The concept of evaluation technical condition with HHT-transform vibro-acoustic signals, **3** (2015). The North Caucasian Centre of Science of the Higher School
19. Zagretdinov, A.R., Kondratiev, A.E., Gaponenko, S.O.: The method of calculation informative harmonics of vibroacoustic signals applied to control the multilayer composite constructions. Eng. Bull. Don **4**, 28–37 (2014). The North Caucasian Centre of Science of the Higher School
20. Saifullin, E.R., Ziganshin, Sh.G., Vankov, Y.V., et al.: Neural network analysis of vibration signals in the diagnostics of pipelines. J. Fundam. Appl. Sci. **9**(2S), 1139–1151 (2017)
21. Ziganshin, Sh.G., Vankov, Yu.V., Gorbunova, T.G.: Reliability of thermal networks for city development. Procedia Eng. **150**, 2327–2333 (2016)
22. Gaponenko, S.O., Kondratiev, A.E., Ibadov, A.A.: Mathematical modelling of vibrations of an elastic shell under the influence of ground. In: International Science and Technology Conference (2019)

Fuzzy Control of Underwater Walking Robot During Obstacle Collision Without Pre-defined Parameters

V. V. Chernyshev[✉], V. V. Arykantsev, and I. P. Vershinina

Volgograd State Technical University, 100 Universitetskiy Avenue,
Volgograd 400062, Russian Federation
vad.chernyshev@mail.ru

Abstract. Industrial exploration of sea bottom requires special underwater technical devices, which moved at sea bottom. Walking machines in seabed conditions, which characterized by sophisticated shape and low bearing ability of the ground, have higher traction properties and possibility, compared with traditional vehicles. In the paper discussed some results of investigation of controlled movement dynamics of walking robot with cyclic mover during obstacle collision without pre-defined parameters. At the base of integration of information from video sensors, was offered situational method to organization behavior of the robot in conditions of incomplete and ambiguous understanding of the current situation and workspace. At the base of fuzzy movement control algorithms were developed typical leg motions, autonomously performed by the robot with the purpose of exclusion of emergency situations. Interactive analysis of the device control system as software model of control object showed, that majority of typical situations on the control can be solved without operator intervention. Experimental verification of gained results of mathematical modeling at the base of underwater walking robot MAK-1 was held. During underwater experiments dynamics of walking device, traction properties and possibility were investigated. In research of possibility were passed local obstacles both straight movement mode and in special maneuvering mode. Some attention during the tests was given to tuning of the robot fuzzy control system in training mode. Results of the work can be demand in underwater walking robotic systems development for underwater technical works and for new industrial technologies of seabed resources reclamation.

Keywords: Mobile robots · Seabed robots · Walking robots mechanics · Dynamics of controlled movement · Fuzzy control · Mathematical modelling · Industrial seabed resources reclamation

1 Introduction

Investigation and industrial exploration of seabed resources require special underwater technical equipment. Existing robotic systems, which moves at sea bottom (underwater bulldozers, self-propelled bottom reclamation devices, cable-packer etc.) have, as a rule, tracked mover [1–3]. However, conditions of sea or river bottom, which characterized by sophisticated shape and low bearing ability of the ground, often makes traditional mover types unusable [4, 5]. Walking machines seems more suitable for bottom works [6–8]. Tests of walking robot "Vosminog" (Fig. 1a) in real swamp and underwater tests of walking device MAK-1 (Fig. 1b), showed, that they have higher abilities on shape and ground passability, in compare with tracked and wheeled machines [9–12]. Also, they have no force for movement resistance [13]. Robots were developed in VSTU for testing of method of movement control for underwater robotic systems and optimization of parameters their walking mechanisms at development stage.

Fig. 1. Tests of walking robot "Vosminog" in a swamp (a) and underwater walking device MAK-1 tests (b).

However, underwater tests of walking robot MAK-1 revealed some problems, connected with device control. During movement control was used the next method. An operator was not at the workspace and controlled autonomously the working machine on the visual information, transmitted from on-board video sensors. At the same time, there was no task of useful information determining by algorithms of picture processing in on-board video cameras signals to determine workspace characteristics. In straight movement mode operator intervened in movement control only if it was necessary (he could change the velocity, direction of movement, and some abilities on correction of the step parameters). In special maneuvering mode, for example during obstacle passing, control was performed by operator in manual mode. However, as underwater tests showed, visibility under water significantly limited – less than 1–3 m. That's why in straight movement mode (with velocities up to 1.5 m/s) operator, as a rule, had no time to react in case of collision with obstacle. In result were cases when feet were broken because of accelerated motion of support points in transfer phase. Also, it is possible to lose stability of the device.

In the paper results of theoretical and experimental investigations on development of fuzzy control algorithms of underwater walking robot MAK-1 movement in in conditions of incomplete and ambiguous understanding of the current situation.

2 Walking Mechanism

In mover of MAK-1 (Fig. 1b) was applied cyclic walking mechanism (Fig. 2a) with convertible support point trajectory (Fig. 3) and passive foot control system [14, 15]. Last one provides rising of toe in transfer phase. At interval of points 10–22 of the trajectory (Fig. 2b) foot is in contact phase (filled points). In the end of interval—new one with points 22–24, foot 4 turned on the angle α_1 to support link 2. In transfer phase at points 24–7 of the trajectory, angle α_1 is not changed. If foot is ski-shaped, it is possible that heel E will be in contact phase. In this case angle α_1 will decrease to α_2. After transfer—at points 7–10 of trajectory, foot falls on the ground. Angle α_2 decreases to α_3, and angle φ_4 become equal to π. Toe rising can be observed in reverse movement too (Fig. 2c). Passive adaptation of the foot allows to pass obstacles with height more than step length by 2 times.

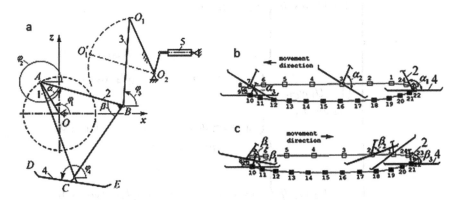

Fig. 2. Walking mechanism scheme (a) and its support point trajectory with featured positions of foot in straight (b) and reverse (c) movement of machine: 1 – winch; 2 – crank rod; 3 – rocking arm; 4 – foot; 5 – linear drive of rocking arm supporting point shifting; 6 – damper.

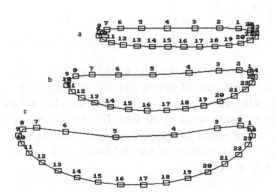

Fig. 3. Transformation of support point trajectory of underwater device MAK-1 after sifting of suspension point: (a) straight movement mode; (b), (c) special maneuvering mode.

For decreasing of impact load in walking mechanism introduced an additional damping unit, which provide dissipative connection of the foot with support link of walking mechanism (Fig. 2a). The damper, in addition to decreasing of impact load influence, also decrease influence of random factors to software movement. Moment of viscous friction foot in hinge M_R determined by viscous friction coefficient in damper and angular speed of foot relatively to support link.

Movement of foot at the cycle determined by value and direction of absolute speed of support points C and angular speed of support links. "Software" angular speed of feet with 2 counter-phase 4-linked walking mechanisms calculated as

$$\omega_{41} = \begin{cases} (1-U_1)\omega_{21} \text{ if } \begin{cases} z_{D1} > z_{C2} \text{ and } z_{E1} > z_{C2} \\ \text{or} \\ z_{D1} = z_{C2} \text{ and } v_{C1z} + \omega_{21} l_4 \cos\varphi_{41} > 0 \\ \text{or} \\ z_{E1} = z_{C2} \text{ and } v_{C1z} - \omega_{21} l_4 \cos\varphi_{41} > 0, \end{cases} \\ \dfrac{(1-U_1)v_{C1z}}{l_4 \cos\varphi_{41}} \text{ if } \begin{cases} z_{D1} = z_{C2} \text{ and } v_{C1z} + \omega_{21} l_4 \cos\varphi_{41} \leq 0 \\ \text{or} \\ z_1 = z_{C2} \text{ and } v_{C1z} - \omega_{21} l_4 \cos\varphi_{41} \leq 0, \end{cases} \end{cases}$$

$$\omega_{42} = \begin{cases} (1-U_2)\omega_{22} \text{ if } \begin{cases} z_{D2} > z_{C1} \text{ and } z_{E2} > z_{C1} \\ \text{or} \\ z_{D2} = z_{C1} \text{ and } v_{C2z} + \omega_{22} l_4 \cos\varphi_{42} > 0 \\ \text{or} \\ z_{E2} = z_{C1} \text{ and } v_{C2z} - \omega_{22} l_4 \cos\varphi_{42} > 0, \end{cases} \\ \dfrac{(1-U_2)v_{C2z}}{l_4 \cos\varphi_{42}} \text{ if } \begin{cases} z_{D2} = z_{C1} \text{ and } v_{C2z} + \omega_{22} l_4 \cos\varphi_{42} \leq 0 \\ \text{or} \\ z_{E2} = z_{C1} \text{ and } v_{C2z} - \omega_{22} l_4 \cos\varphi_{42} \leq 0, \end{cases} \end{cases} \quad (1)$$

where ω_{4k} – angular speed of k-th walking mechanism of mover ($k = 1, 2$), U_k – unit function which described state of mover and can be equal only to two values: 1 in contact phase and 0 in transfer phase; ω_{2k} – angular speed of supporting link; φ_{4k} and ω_{4k} – turning angle angular speed of the foot; l_4 – half of foot length; v_{Ckz} – vertical component of absolute speed of support points; $z_{Dk} = z_{Ck} + l_4 \sin\varphi_{4k}$ and $z_{Ek} = z_{Ck} + l_4 \sin(\varphi_{4k} + \pi)$ – coordinates of toe and heel of k-th foot in coordinate system, connected with body of device.

3 Movement of the Robot

Shape passability depends on initial position of feet at ground. That's why succeed obstacle passing depends on behavior algorithm during the passing. For this purpose, is necessary to build a model of the process. However, in our case this is difficult because of majority of possible situations. In this case more appropriate method of situational control [16–19], where each class of possible situations, corresponds to own control solution. After combining of different conditions, contained in Eq. (1) and introduction some linguistic variables (obstacle high, not high, deep, not deep, wide, not wide, etc.), were formed rules of robot behavior if a form of fuzzy decision as "If A ..., then B...". For example, situations, showed in Table 1, arrows show movement direction of the robot. For situation at (Table 1a), "if $U_1= 0$ and $U_2= 1$" (1-st foot in transfer, 2-nd in contact) and $z_{D1} > z_{C2}$, $z_{E1} > z_{C2}$ (toe and heel of 1-st foot are higher than support point of 2-nd foot) and $\Delta\omega_{41} > 0$ (positive relative angular speed of 1-st foot, relatively to support link of same leg), then obstacle is "high" and cannot be passed in straight movement mode. In this case necessary to stop robot and give control to operator. If with the same conditions $\Delta\omega_{41} < 0$, (Table 1b), then obstacle is "not high" and movement can be continued.

For situation from (Table 1c) $U_1 = 0$ and $U_2 = 1$ (1-st foot in transfer, 2-nd one in contact) and $z_{D1} > z_{C2}$, $z_{E1} = z_{C2}$ (toe of 1-st foot higher than support point of 2-nd foot, heel of 1-st foot at the same level as support point of 2-nd foot) and $\Delta\omega_{41} > 0$. In this case during feet changing there is "software" foot motion, there is no obstacle and it is possible to continue movement in autonomous mode.

For situation from (Table 1d) $U_1 = 0$ and $U_2 = 1$ (1-st foot in transfer, 2-nd one in contact) and $z_{D1} > z_{C2}$, $z_{E1} < z_{C2}$ (toe of 1-st foot higher than support point of 2-nd foot, heel of 1-st foot lower than support point of 2-nd foot) and $\Delta\omega_{41} > 0$. In this case obstacle is "not deep" and it is also possible to continue movement in autonomous mode.

If with the same conditions $\Delta\omega_{41} = 0$ (there is no relative angular speed of 1-st foot, relatively to support link of the same leg), (Table 1e), then obstacle is "deep" and it is necessary to stop the robot.

Interactive analysis of control system behavior of walking device in typical situations was conducted at the base of developed fuzzy rules of robot behavior and dynamic models of controlled movement of underwater walking robot. For typical situations was determined sequence of motion of robot limbs. Geometric abilities of walking mover on obstacle sometimes cannot be realized because of limited power of electric drive. That's why during modelling of typical situations were determined value and structure of energy costs at each stage of movement. It is necessary to take under consideration, that shape passability depends on initial position of support points of walking mechanisms at the ground. That's why possible different situations during passing of the same obstacle. In case (Table 2a) obstacle was not detected, but was passed. In case (Table 2b) obstacle was not detected during 1-st step, but was passed and during 2-nd step was detected. In case (Table 2c) during 1-st step heel of 1-st foot contacted with opposite wall of the pit. In case (Table 2d) the foot during 1-st step foot was not contacted with anything.

Table 1. Possible situations of robot movement.

Possible situations	Possible control solutions
a) Characteristic of the situation: obstacle is "high"	Emergency stop, step behind, increasing of step height, next attempt to pass the obstacle in autonomous mode with low speed or transfer of control to the operator
b) Characteristic of the situation: obstacle is "not high"	Continuation of autonomous movement in straight mode
c) Characteristic of the situation: there is no obstacle	Continuation of autonomous movement in straight mode
d) Characteristic of the situation: obstacle is "not deep"	Continuation of autonomous movement in straight mode
e) Characteristic of the situation: obstacle is "deep"	Emergency stop, step behind, increasing of step length, next attempt to pass the obstacle in autonomous mode with low speed or transfer of control to the operator

Interactive analysis of control system behavior of robot showed, then majority of typical situations allow to control the movement without an operator. With the purpose to verify results of the analysis were conducted experiments in conditions of real water objects at the base of walking robot MAK-1.

Table 2. Different situations during passing the same obstacle.

Possible situations	Characteristic of the situation
a) Characteristic of the situation: There is no obstacle	Continuation of autonomous movement in straight mode
b) Characteristic of the situation: Obstacle is "not wide"	Locking of the damper of 1-st foot and continuation of autonomous movement in straight mode
c) Characteristic of the situation: Obstacle is "deep" and "not wide"	Emergency stop, increasing of step length, next attempt to pass the obstacle in autonomous mode
d) Characteristic of the situation: Obstacle is "deep" and "wide"	Emergency stop, step behind, increasing of step length, next attempt to pass the obstacle in autonomous mode with low speed or transfer of control to the operator

4 Experimental Verification of the Analysis

During the tests was checked system working capacity of walking robot and was investigated his design influence on controllability, dynamics, maneuverability, traction properties and passability. In underwater tests for investigation of dynamics and traction properties of walking device there was used the method, adapted for underwater conditions [20], based on video recording of the walking device movement process with frame by frame processing of video on computer. For determining of limits of ground passability were recorded conditions, when passability loss was occurred, because of low bearing ability ground. For that purpose, was chosen hardest, from the passability point of view, squares of bottom. During shape passability investigation were passed different types of slopes and obstacles. Obstacles were passed both in straight movement mode and in special maneuvering mode. In last case parameters of the step were corrected by an operator.

In development of coordinated control of the legs of underwater walking device were used methods of control, based on fuzzy information perception and fuzzy logic. Similar methods allow to control robotic systems in live mode, using experience of expert, introduced in the system in form of affiliation functions and rules of fuzzy output. However, during exploitation of the robot in unknown environment that knowledge can be unusable. Thus, during underwater experiments some attention was paid to tuning of fuzzy control system of the robot in training mode [21]. Setting of adaptive fuzzy control was performed with a help of comparing the data on the situation and the appropriate control. In training mode (Fig. 4) at the same time with the information about situation from video cameras of the robot, have been saves the data about control signals, generated by the operator. Telemetry data, gained from sensor system of the robot and information about control signals from operator were recorded in a file.

Fig. 4. Training process of the MAK-1 walking robot.

After the tests was performed an analysis of recorded data together with video of movement process. It allows to connect relative motions of the robot legs with his absolute movement. In the file were selected parts, related to key motions of the robot, for example, obstacle passing. The training consists in iterative selecting of parameters of adaptive fuzzy output system, which provide smallest discrepancy between autonomous movement system and system, controlled by operator.

5 Conclusion

The underwater tests approved reliability of main results of mathematical modeling.

Thus, offered situational method of organization of underwater tests for walking robots with cyclic type of mover, based on integration of information from video sensors of the walking robot and fuzzy algorithms of his leg motion control, allows to make easier the process of leg control in conditions of incomplete and ambiguous understanding of the current situation and workspace, conditioned by underwater environment.

Results of the work can be demand in underwater walking robotic systems development for underwater technical works and for new industrial technologies of seabed resources reclamation.

Acknowledgments. Work partially supported by RFBR, research project No. 19-08-01180, 16-08-01109, 15-41-02451 and the scholarship of president RF SP-5102.2018.1.

References

1. Verichev, S., Jonge, L., Boomsma, W.: Deep mining: from exploration to exploitation. In: Minerals of the Ocean 7 and Deep-Sea Minerals and Mining, pp. 21–24 (2014)
2. Nautilus Minerals. http://www.nautilusminerals.com
3. Amphibious Bulldozer (2018). http://www.komatsu.com/CompanyInfo/views/pdf/201312/Views_No20_amphibious_bulldozer.pdf. Accessed 20 Sept 2018
4. Hong, S., Kim, H.W., Choi, J.S.: Transient dynamic analysis of tracked vehicles on extremely soft cohesive soil. In: The 5th ISOPE Pacific/Asia Offshore Mechanics Symposium, pp. 100–107 (2002)
5. Kim, H.W., Hong, S., Choi, J.S.: Comparative study on tracked vehicle dynamics on soft soil: single-body dynamics vs. multi-body dynamics, pp. 132–138 (2003)
6. Pavlovsky, V.E., Platonov, A.K.: Cross-country capabilities of a walking robot: geometrical, kinematical and dynamic investigation. In: Theory and Practice of Robots and Manipulators (2000). Proceedings of the 13th CISM-IFToMM Symposium, pp. 131–138
7. Silva, M.F., Machado, J.A.T.: A literature review on the optimization of legged robots. J. Vib. Control **18**(12), 1753–1767 (2012)
8. Yoo, S.Y., Jun, B.H., Shim, H.: Design of static gait algorithm for hexapod subsea walking robot: Crabster. Trans. Korean Soc. Mech. Eng. A **38**(9), 989–997 (2014)
9. Briskin, E.S., Chernyshev, V.V., Maloletov, A.V., et al.: On ground and profile practicability of multi-legged walking machines. In: Climbing and Walking Robots, pp. 1005–1012 (2001)
10. Briskin, E.S.: The investigation of walking machines with movers on the basis of cycle mechanisms of walking. In: The IEEE International Conference on Mechatronics and Automation, pp. 3631–3636 (2009)
11. Chernyshev, V.V., Arykantsev, V.V., Gavrilov, A.E., et al.: Design and underwater tests of subsea walking hexapod MAK-1. In: Proceedings of the ASME 2016 35th International Conference on Ocean, Offshore and Arctic Engineering, p. 9 (2016)
12. Briskin, E.S.: Problems of increasing efficiency and experience of walking machines elaborating. In: Ceccarelli, M., Glazunov, V. (eds.) Advances on Theory and Practice of Robots and Manipulators, pp. 383–390. Springer, Cham (2014)
13. Chernyshev, V.V., Gavrilov, A.E.: Traction properties of walking machines on underwater soils with a low bearing ability. In: Minerals of the Ocean 7 and Deep-Sea Minerals and Mining 4, pp. 21–24 (2014)
14. Chernyshev, V.V., Arykantsev, V.V., Kalinin, Ya.V.: Underwater tests of the walking robot MAK-1. In: Human-Centric Robotics, pp. 571–578 (2017)
15. Chernyshev, V.V., Arykantsev, V.V., Kalinin, Ya.V.: Passive foot control in cyclic walking mechanism. In: Industrial Engineering, Applications and Manufacturing, p. 5 (2017). https://doi.org/10.1109/icieam.2017.8076189
16. Volodin, S.Y., Mikhaylov, B.B., Yuschenko, A.S.: Autonomous robot control in partially undetermined world via fuzzy logic. In: Ceccarelli, M., Glazunov, V. (eds.) Advances on Theory and Practice of Robots and Manipulators, pp. 197–203. Springer, Cham (2014)

17. Pryanichnikov, V.E., Andreev, V.P.: Intellectualization of special mobile robots, including return algorithm to a zone of stable RC. In: Proceedings of the XXI International Conference on Extreme Robotics, pp. 46–49 (2012)
18. Andreev, V.P., Pryanichnikov, V.E., Prysev, E.A.: Multi-access control of distributed mobile robotic systems based on networking technologies. In: Annals of DAAAM and Proceedings of the 21st International DAAAM Symposium, pp. 15–16 (2010)
19. Pyanichnikov, V.E.: Algorithmic base for remote sensors of mobile robots. Mechatron. Autom. Control **10**(91), 10–21 (2008)
20. Chernyshev, V.V.: Modeling of the dynamics of the walking machine with the cyclic propulsors as system solids with elastic and damping relations. In: The 3rd Joint International Conference on Multibody System Dynamics, p. 9 (2014)
21. Briskin, E.S.: Research of the energy-efficient methods of fuzzy movement control multilegged walking robot "Ortonog" in a real environment. In: Network Cooperation in Science, Industry and Education Proceedings, pp. 149–150 (2016)

Complex Application of the Methods of Analytical Mechanics and Nonlinear Stability Theory in Stabilization Problems of Motions of Mechatronic Systems

A. Ya. Krasinskiy[✉] and E. M. Krasinskaya

Moscow State University of Food Production,
12 Vrubelya Street, Moscow 125080, Russian Federation
krasinsk@mail.ru

Abstract. The control system of the mechatronic device must ensure the formation and implementation of the control action for the realization of the specified behavior (operating mode) of the working body. Of particular importance is the stability of this mode of operation. The vast majority of existing control formation algorithms are traditionally oriented towards the use of such methods when the conclusion about asymptotic stability is obtained on the basis of the negativity of the real parts of all the roots of the characteristic equation of a closed system. But, firstly, for entire classes of technical devices (for example, for manipulators with geometric constraints), with any control method, the part of the roots of the characteristic equation always remain zero. Secondly, the operating modes (for example, steady-state motions) have the property of non-asymptotic stability with respect to certain degrees of freedom, according to which control actions can be omitted. But for the justified use of the properties of non-asymptotic stability, it is necessary to develop a complete non-linear model in the form that allows analysis of the structure of non-linear terms from the point of view of the theory of critical cases. This article develops the application of analytical mechanics methods to simplify the construction of complete nonlinear dynamics models of complex mechatronic systems taking into account transient processes in executive drives. The effectiveness of the proposed approach is proved by solving the stabilization problem for Ball and Beam as a mechatronic system with a geometric constraint.

Keywords: Critical case · Control · Stability · Stabilization

1 Introduction

When studying the behavior of any dynamical system, one should keep in mind that in reality no single phenomenon can be represented in its pure form. However accurately the forces acting on the system and its parameters have been determined, some minor (in comparison with the taken into account) factors, some perturbations, always remain unaccounted for. These perturbations, however small they are, affect the motion of the system, especially if this motion is unstable. Thus, only the stable movements retain their general character. Therefore, only stable motions more or less correctly describe the actual movements (operating mode) of the system. Consequently, it is obvious that the study of the stability of the desired behavior of the system is a necessary stage in modeling the dynamics of any object, and the urgency of studying stability, not only from general theoretical positions, but also from the point of view of applications, is undoubtedly true. Equally urgent are the problems of stabilization - the task of determining such control actions under which the necessary motion becomes stable, i.e. the real object will move similarly to this (model) motion. The task of stabilization is the task of constructing regulatory impacts that ensure the sustainable implementation of the desired process. It closely connects with the problem of sustainability and is the development of the problem of sustainability in the application to controlled systems [1].

The question of the stability of motions of various systems is most often solved by considering the so-called first-order equations, that is, those equations that are obtained from the initial system of equations of the disturbed motion, when the terms in the right-hand sides of the system are discarded by terms higher than the first order. However, in many cases that are very important from the point of view of applications, consideration of only the first approximation of the stability problem does not solve. Such cases are called Lyapunov critical cases.

For steady motions (i.e., motions described by autonomous systems of differential equations), they are characterized by the fact that the characteristic equation of the first approximation system, with the exception of roots with negative real parts, necessarily has zero and purely imaginary roots.

The investigation of critical cases is one of the most complex sections of stability theory. A hundred years ago Lyapunov wrote about critical cases ([2], p. 13): "… cases of this kind are very diverse, and in each of them the task gets its own special character, so that there can be no talk of any common methods its decisions that would apply to all such cases …". Lyapunov attached great importance to the problem of stability in critical cases and considered some of these cases for steady and periodic motions. A special role in the theory of critical cases is played by the so-called reduction principle, which goes back to Lyapunov [1]. It consists in the transition from the study of the stability of a complete system to the study of only a truncated system containing only critical variables [2–4]. In one form or another, this principle is used in almost all works relating to the study of critical cases. In the process of using the principle of information in the works of various authors was subjected to further development and improvement.

The application of the principle of reduction in the general case is associated with overcoming significant difficulties, since it requires linear and non-linear changes of variables. These transformations lead the equations of the perturbed motion first to the special form [2–4] of the theory of critical cases (for the selection of critical variables), and then to the form for which it is possible to construct Lyapunov-Chetaev functions that solve the problem.

2 Description of the Proposed Approach. Formulation of the Problem

In the general case, it is very difficult to determine the constraints for the initial equations of perturbed motion from the conditions of theorems of the theory of critical cases formulated in new variables. At the same time, for problems of stability (and stabilization) of steady motions of mechanical systems, it becomes possible to apply the theory of critical cases without actually performing the above changes in variables. The simplest and easily applicable in such problems are the so-called special [2, 3] (essentially special [4]) cases. This is a special category of critical cases in which, in addition to the aforementioned arrangement of the roots of the characteristic equation, it is required that the right-hand sides of the equations of the perturbed motion vanish when the noncritical variables vanish (in the presence of purely imaginary or multiple for elementary zero-root divisors, in a truncated system). Exactly such exceptional cases from the mathematical point of view are naturally obtained in many practical problems of stability (and stabilization) of the motions (operating modes) of mechatronic systems, when the real parts of all the remaining roots of the characteristic equation (except roots with zero real parts) are negative.

The use of methods of analytical mechanics to simplify the analysis of the structure of equations from the standpoint of the theory of critical cases can greatly facilitate the consideration of problems of stabilization of mechatronic systems in which asymptotic stability cannot be established by studying only the first-approximation equations. Proceeding from this, the formulation of the problem of stabilization of steady motions to non-asymptotic stability and the development of rigorous methods for their solution with the fullest possible use of the properties of the stability of proper motions to reduce the dimension of the vector of stabilizing control and the volume of measuring information sufficient for its formation are relevant and important.

One of the important classes of mechatronic systems of this kind are manipulators with geometric constraints. In this paper, the efficiency of the proposed approach is shown using the example of a rigorous solution of a fairly simple mechatronic system. Taking into account the total nonlinear geometric coupling, one more previously unknown equilibrium position with a non-zero rotation angle of the drive wheel was detected. A complete study of the problem of stabilization of equilibria with known zero (Fig. 1) and new non-zero angles (Fig. 2) of rotation of the driving wheel is given.

The device "Ball and Beam" consists of a mechanical part and a controlled electric drive [5–10]. The position of the ball $r(t)$ is regulated by the angle $\alpha(t)$ of changing the tilt of the trough and stabilizes at a certain predetermined position r_0. The angle of rotation of the drive wheel is denoted by $\theta(t)$. The control will be carried out by additional voltage at the motor armature.

The nonlinear constraint equation [10–13] has the form:

$$(L(\cos \alpha - 1) + d(1 - \cos \theta))^2 + (L \sin \alpha + l - d \sin \theta)^2 = l^2, \qquad (1)$$

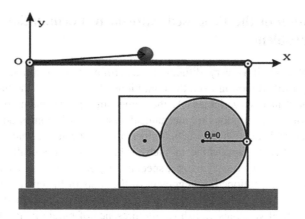

Fig. 1. Zero equilibrium.

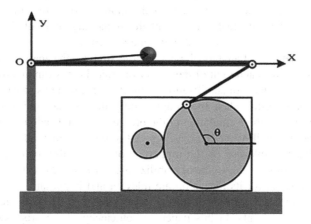

Fig. 2. Non-zero equilibrium position.

where $L = OA$ is the length of the trough, $l = AB$ is the length of the lever, d is the radius of the wheel. The excess coordinate can be taken as the angle of inclination of the trough $\alpha(t)$, and the angle of rotation of the wheel $\theta(t)$.

Obviously, the equilibrium position is possible only at $\alpha = 0$, when the trough is horizontal. Setting $\alpha = \alpha_0 = 0$ and assuming that we obtain from (1) the values of the angle θ corresponding to all equilibrium positions of the system:

$$\theta_0 = 0;\ \theta_1 = 2\arcsin\left(\frac{l}{\sqrt{l^2+d^2}}\right)$$

Note that, without the application of a constant torque generated by the drive motor, there is no equilibrium in the "Ball and Beam" design discussed here. The constant moment for a given equilibrium position is a non-zero non-conservative positional force. Consequently, in the equations in the form of Shulgin [14] should distinguish components that depend on the linear terms of the expansion of the kinematic coupling coefficient and compare them for two different equilibrium positions. These components vanish for a known equilibrium position when the angle of rotation of the drive wheel is zero. Then it is possible to use [14–16] the Lagrange equations of the second kind, excluding the dependent and coordinate, and the velocity with the aid of a linearized coupling. It is now necessary to establish the values of these components for the previously unknown, equilibrium position of the "Ball and Beam" system and to show the incorrectness of the linearization of the connection for the study of this equilibrium position.

For different equilibrium positions the structure of the equations of the disturbed motion will be different. In addition, for systems with geometric constraints, as in the "Ball and Beam" system, dependent coordinates can be selected in various ways.

The system models will differ. In the "Ball and Beam" system, the redundant coordinate can be chosen in two ways and the Schulgin equations will have different forms. To obtain the same initial perturbed state with different choice of the dependent coordinate and comparison of the control results (for the same equilibrium), it is necessary to match the initial perturbations of the angular variables due to the presence of the constraint (1).

The purpose of this paper is to perform a comparative analysis of the structure of the first-approximation equations for two equilibrium positions and two options for selecting an excessive coordinate, and to compare control actions, as functions of time, for both equilibrium positions.

To build a complete model of "Ball and Beam" as a mechatronic system, it is necessary to add the equation of a controlled electric drive to an executive DC direct-current motor with independent excitation. The control action is the additional voltage on the armature winding of the drive motor. With the chosen method of implementing control actions, the mathematical model of the object is a system of indirect control, which, as is known [16], significantly expands the capabilities of the control system. In the proposed stabilization method, the characteristic equation for the perturbed motion necessarily has zero roots in a number equal to the number of geometric constraints. If the real parts of the remaining roots are negative, then for any control method there is a critical case. Then according to the theory of critical cases [2–4], in the equations of disturbed motion it is necessary to perform a linear replacement, which allocates variables-excess coordinates to which these zero roots correspond. This transformation in the general case changes the structure of linear positional forces in the first approximation of the perturbed motion equations.

The linear controlled subsystem [17] does not include the critical variable and essentially depends on the choice of the redundant coordinate. The control action depends on the variables of the selected subsystem. Its coefficients will be determined by the method of Krasovskii [1] of the corresponding linear-quadratic problems as a linear function of phase variables. A sufficient condition for the solvability of the stabilization problem for the distinguished linear controlled subsystem is the controllability condition (by the Krasovskii theorem on stabilization by the first approximation [1] (Theorem 1, p. 509)). In the complete nonlinear system, closed linear control found for the selected controlled subsystem, asymptotic stabilization will be provided for all variables, including the redundant coordinate, to which the zero root corresponds (according to Theorem 1 [12, 18]).

The numerical solution is carried out using the Repin-Tretyakov procedure [1]. For this, it is necessary to compile a program in which an additional module is introduced that takes into account the matching of the initial perturbations of the dependent and independent coordinates because of the presence of the geometric constraint equation in the form (1), and also assumes a different choice of the redundant coordinate. In the software product systems with redundant coordinates were not considered. When compiling the program in this work, it is planned to construct transient diagrams for two cases of selecting an excessive coordinate and comparing control actions, as functions of time, for both equilibrium positions.

3 Construction of the Mathematical Model of the Ball and Beam System

The kinetic and potential energy of the mechanical part of the system, including the ball and rotor of the motor with the reducer, is taken in the form [12]:

$$T = \frac{1}{2}\left(m(r\dot{\alpha})^2 + m\frac{(r\dot{r})^2}{r^2 - R^2} - 2mR\frac{r\dot{r}\dot{\alpha}}{\sqrt{r^2 - R^2}} + J\left(\frac{r\dot{r}}{R\sqrt{r^2 - R^2}} - \dot{\alpha}\right)^2 + J_0\dot{\theta}^2\right)$$

$$\Pi = mg\left(\sqrt{r^2 - R^2}\sin\alpha + R\cos\alpha\right)$$

where R is the radius of the ball, m is its mass, J is its moment of inertia, J_0 is the moment of inertia of the entire system, reduced to the engine. Expressions of the kinetic and potential energies are made taking into account the size of the ball, which is necessary if we consider equilibrium and motion near the origin. Dynamics of a direct current DC motor with independent excitation describes the Kirchhoff equation [19]:

$$L_a\frac{di_a}{dt} + R_a i_a + k_3\frac{d\theta}{dt} = k_1 e_v, \quad e_b = k_3\frac{d\theta}{dt}, \qquad (2)$$

where e_v is the voltage at the output of the amplifier supplying power to the motor's armature winding, e_b is the voltage of the anti-emf, k_1 is the power converter coefficient, k_3 is the motor constant, L_a is the inductance of the armature winding, R_a is its resistance, and the current in the armature circuit.

With respect to the coordinates α and r, only potential forces act on the system:

$$Q_r = -\frac{\partial \Pi}{\partial r} = -mg\frac{r}{\sqrt{r^2 - R^2}}\sin\alpha;$$

$$Q_\alpha = -\frac{\partial \Pi}{\partial \alpha} = -mg\left(\sqrt{r^2 - R^2}\cos\alpha - R\sin\alpha\right).$$

And only non-potential force is applied to the coordinate θ [12] $Q_\theta = k_2 i_a - b_0\dot{\theta}$, where b_0 is coefficient of rotation resistance, k_2 is the electromechanical constant of the motor. Differentiating Eq. (1) with respect to time, we obtain kinematic equation.

From this equation, we can express the velocity $\dot\alpha$ or $\dot\theta$ depending on which version of the dependent coordinate we are considering. The equation of the constraint will be taken in the form (3) or (4).

$$\dot\alpha = B^I(\alpha,\theta)\cdot\dot\theta,\ B^I(\alpha,\theta) = \frac{d\,(L\sin(\alpha-\theta)+(L-d)\sin\theta+l\cos\theta)}{L\,(d\sin(\alpha-\theta)+(L-d)\sin\alpha+l\cos\alpha)} \quad (3)$$

$$\dot\theta = B^{II}(\alpha,\theta)\cdot\dot\alpha,\ B^{II}(\alpha,\theta) = 1/B^I(\alpha,\theta) \quad (4)$$

We write Shulgin's equations for the "Ball and Beam" system in two cases. In the first, we assume that the dependent angle is the slope angle α of the chute. In the second case, the dependent coordinate is the steering angle θ of the wheel. The equations of the perturbed motion will differ in different cases.

Remark 1. In what follows we shall call the equilibrium position zero to which the values of the coordinates $r = r_0 = const$, $\alpha = \alpha_0 = 0$, $\theta = \theta_0 = 0$ correspond. The second position $r = r_0 = const$, $\alpha = \alpha_0 = 0$, $\theta = \theta_1$ is called non-zero. From the formulation of the problem $0 < r_0 < L$.

Remark 2. If we use a simplified approach and use the linearized equation $\alpha = \frac{d}{l}\theta$ instead of the nonlinear Eq. (1), the second (nonzero) equilibrium position is lost.

Remark 3. The control coefficients for the same equilibrium position for a different choice of the redundant coordinate will be different. Since, respectively, the redundant coordinate input, the sets of phase vector components entering the controlled subsystems are different. Thus, in total we obtain 4 different stabilization problems: for two equilibria, there are two variants of introducing an excessive coordinate.

4 Case I. Dependent Coordinate α

We use (3) to exclude the dependent velocity from the kinetic energy. In this case, the Shulgin equations will be written in the form

$$\begin{cases} \frac{d}{dt}\frac{\partial L^{I*}}{\partial \dot r} - \frac{\partial L^{I*}}{\partial r} = 0; \\ \frac{d}{dt}\frac{\partial L^{I*}}{\partial \dot\theta} - \frac{\partial L^{I*}}{\partial \theta} - B^I(\alpha,\theta)\frac{\partial L^{I*}}{\partial \alpha} = Q_\theta; \end{cases}$$

To these equations, it is also necessary to add the equation of the motor (2), and also to take into account the Eq. (3). We obtain the system:

$$\left(m+\frac{J}{R^2}\right)\left(\frac{r^2}{r^2-R^2}\ddot r - R\frac{rB^I}{\sqrt{r^2-R^2}}\ddot\theta - \frac{R^2 r}{(r^2-R^2)^2}\dot r^2\right)$$
$$-\left(R\left(m+\frac{J}{R^2}\right)\frac{r\left(\frac{\partial B^I}{\partial\alpha}B^I+\frac{\partial B^I}{\partial\theta}\right)}{\sqrt{r^2-R^2}} + mrB^{I2}\right)\dot\theta^2 = -\frac{mgr\sin\alpha}{\sqrt{r^2-R^2}};$$

$$((mr^2 + J)B'^2 + J_0)\ddot{\theta} - R\left(m + \tfrac{J}{R^2}\right)\tfrac{rB'}{\sqrt{r^2-R^2}}\ddot{r} + 2mrB'^2\dot{r}\dot{\theta} + (mr^2 + J)B'\left(\tfrac{\partial B'}{\partial \alpha}B' + \tfrac{\partial B'}{\partial \theta}\right)\dot{\theta}^2$$
$$+ R^3\left(m + \tfrac{J}{R^2}\right)\tfrac{B'}{(r^2-R^2)^{3/2}}\dot{r}^2 = k_2 i_a - b_0\dot{\theta} - mgB'\left(\sqrt{r^2 - R^2}\cos\alpha - R\sin\alpha\right);$$
$$L_a \tfrac{di_a}{dt} + R_a i_a + k_3 \dot{\theta} = k_1 e_v; \quad \dot{\alpha} = B'(\alpha, \theta)\dot{\theta};$$
(5)

The value of the system parameters (5) in the equilibrium position (r_0, $\alpha = 0$, $\theta = \theta_i$):

$$\begin{cases} 0 = k_2 i_a^0 - mgB'(0, \theta_i)\sqrt{r_0^2 - R^2}; \\ R_a i_a^0 = k_1 e_v^0; \end{cases} \Rightarrow \begin{cases} i_a^0 = \tfrac{mgB'(0,\theta_i)\sqrt{r_0^2-R^2}}{k_2}; \\ e_v^0 = \tfrac{R_a i_a^0}{k_1} = \tfrac{mgR_aB'(0,\theta_i)\sqrt{r_0^2-R^2}}{k_1 k_2}; \end{cases} \quad (6)$$

We introduce perturbations:

$$\begin{cases} r = r_0 + x_1; \ \dot{r} = x_2; \ \theta = \theta_i + x_3^I; \ \dot{\theta} = x_4^I; \\ i_a = i_a^0 + x_5; \ \alpha = x_6^I; \ e_v = e_v^0 + u_i^I; \end{cases} \quad (7)$$

Here u_i^I is the control. The subscript denotes the equilibrium position, and the upper index indicates the choice of the dependent coordinate.

Remark 4. With different choice of the redundant coordinate, not only the equations of motion, but also certain components of the phase vectors will be different. For the dependent coordinate α we will denote the increments of these coordinates by the superscript I, and in the case of the dependent coordinate θ - the index II.

Having isolated the first approximation in the system of Eq. (5), we write it in the normal form

$$\dot{x}^I = H_i^I x^I + S u_i^I + X^{I(2)}, \quad (8)$$

where $x^I = (x_1, x_2, x_3^I, x_4^I, x_5, x_6^I)'$ is a phase vector, H_i^I is a matrix whose nonzero elements are calculated by formulas

$$h_{21} = -mg\tfrac{RB'^2}{(m(r_0^2-R^2)B'^2+J_0)}; \quad h_{23} = -\tfrac{mg}{r_0}\tfrac{RB'(r_0^2-R^2)}{(m(r_0^2-R^2)B'^2+J_0)}\tfrac{\partial B'}{\partial \theta};$$
$$h_{41} = \tfrac{-mg}{\sqrt{r_0^2-R^2}}\tfrac{B'r_0}{(m(r_0^2-R^2)B'^2+J_0)}; \quad h_{43} = \tfrac{-mg\sqrt{r_0^2-R^2}}{(m(r_0^2-R^2)B'^2+J_0)}\tfrac{\partial B'}{\partial \theta};$$
$$h_{44} = \tfrac{-b_0}{(m(r_0^2-R^2)B'^2+J_0)}; \quad h_{45} = \tfrac{k_2}{(m(r_0^2-R^2)B'^2+J_0)};$$

$$B' = B'(0, \theta_i), \ \tfrac{\partial B'}{\partial \alpha} = \tfrac{\partial B'}{\partial \alpha}\bigg|_{(0,\theta_i)}, \ \tfrac{\partial B'}{\partial \theta} = \tfrac{\partial B'}{\partial \theta}\bigg|_{(0,\theta_i)}, \ S = (0, 0, 0, 0, s, 0)', \ s = \tfrac{k_1}{L_a}.$$

$X^{I(2)}$ is the vector of nonlinear terms. To isolate the variables to which the zero roots of the characteristic equation correspond, we can apply (according to the theory of critical cases) a linear substitution [2–4]:

$$x_6^I = B^I x_3^I + z^I, \quad (9)$$

the coefficients of the variable will change as follows:

$$h'_{23} = h_{23} + h_{26}B^I = -\frac{mg}{r_0} \frac{B^I \sqrt{r_0^2 - R^2}\left(m(r_0^2 - R^2)B^{I2} + J_0 + R(m + \frac{J}{R^2})\sqrt{r_0^2 - R^2}\left(B^I \frac{\partial B^I}{\partial \alpha} + \frac{\partial B^I}{\partial \theta}\right)\right)}{\left(m + \frac{J}{R^2}\right)\left(m(r_0^2 - R^2)B^{I2} + J_0\right)};$$

$$h'_{43} = h_{43} + h_{26}B^I = -mg \frac{\sqrt{r_0^2 - R^2}}{m(r_0^2 - R^2)B^{I2} + J_0}\left(B^I \frac{\partial B^I}{\partial \alpha} + \frac{\partial B^I}{\partial \theta}\right);$$

Now, for the controlled subsystem, according to [17], we must choose a subsystem:

$$\dot{w}^I = M_i^I w^I + N u_i^I; \; w^I = (x_1, x_2, x_3^I, x_4^I, x_5), \quad (10)$$

The controllability condition [1] for system (10)

$$\text{rank}\left(N \; M_i^I \cdot N \; M_i^{I2} \cdot N \; M_i^{I3} \cdot N \; M_i^{I4} \cdot N\right) = 5, \quad (11)$$

$$M_i^I = \begin{pmatrix} 0 & 1 & 0 & 0 & 0 \\ h_{21} & 0 & h'_{23} & h_{24} & h_{25} \\ 0 & 0 & 0 & 1 & 0 \\ h_{41} & 0 & h'_{43} & h_{44} & h_{45} \\ 0 & 0 & 0 & h_{54} & h_{55} \end{pmatrix}, N = \begin{pmatrix} 0 \\ 0 \\ 0 \\ 0 \\ s \end{pmatrix}$$

is satisfied. By the method of Krasovskii [1], the coefficients of the stabilizing control $u_i^I = K_i^I w^I$ are found by solving the corresponding linear-quadratic stabilization problem. The analysis is performed taking into account the dynamics of the executive motor for a different choice of the excess coordinate. The complete coincidence of transient processes for both equilibria is shown, in spite of the fact that the stabilizing controls are determined from different control subsystems and depend on different phase variables because of the different choice of the dependent coordinate.

The first approximation of Eq. (8) depends on the quadratic terms of the expansion of the geometric constraint equation (or the linear terms of the expansion of the kinematic coupling), which can affect the stability of the equilibrium position. The condition for the admissibility of the transition to linearized bonds is the vanishing of the indicated terms in the equilibrium position in question. For the Ball and Beam system, the partial derivatives have the following meanings:

Dependent Coordinate α:

1. Zero Equilibrium

$$B^I(0,0) = \frac{d}{L}, \; \left.\frac{\partial B^I}{\partial \alpha}\right|_0 = \left.\frac{\partial B^I}{\partial \theta}\right|_0 = 0.$$

2. non-zero equilibrium

$$B^I(0,\theta_1) = \frac{d}{L}\cdot\frac{l^2+d^2}{d^2-l^2}, \left.\frac{\partial B^I}{\partial \alpha}\right|_{(0,\theta_1)} = \frac{2d^2l}{L}\cdot\frac{(2dl-d^2-l^2)}{(l^2-d^2)^2}, \left.\frac{\partial B^I}{\partial \theta}\right|_{(0,\theta_1)} = 0$$

Dependent Coordinate θ:

1. Zero Equilibrium

$$B^{II}(0,0) = \frac{L}{d}, \left.\frac{\partial B^{II}}{\partial \alpha}\right|_0 = \left.\frac{\partial B^{II}}{\partial \theta}\right|_0 = 0$$

2. Non-zero equilibrium

$$B^{II}(\alpha_1,\theta_1) = \frac{L}{d}\cdot\frac{d^2-l^2}{l^2+d^2}, \left.\frac{\partial B^{II}}{\partial \alpha}\right|_{(0,\theta_1)} = -2Ll\cdot\frac{(2dl-d^2-l^2)}{(l^2+d^2)^2}, \left.\frac{\partial B^{II}}{\partial \theta}\right|_{(0,\theta_1)} = 0$$

For numerical calculations, take the following parameter values: mass of the ball m = 0.064 kg; the radius of the ball R = 0.0254 m; the length of the gutter L = 0.425 m; the length of the shoulder l = 0.34 m; the radius of the drive wheel.

$$u_0^I = -1435\cdot x_1 - 720\cdot x_2 + 708\cdot x_3^I + 22\cdot x_4^I + 84.5\cdot x_5.$$

The case of a nonzero equilibrium:

$$u_1^I = 1993\cdot x_1 + 817\cdot x_2 + 898\cdot x_3^I + 32\cdot x_4^I + 99\cdot x_5.$$

5 Case II. Dependent Coordinate θ

In this case the problem is solved similarly, but now the phase variables and, accordingly, the perturbations

$$\begin{cases} r = r_0 + x_1; \dot{r} = x_2; \alpha = x_3^{II}; \dot{\alpha} = x_4^{II}; \\ i_a = i_a^0 + x_5; \theta = \theta_i + x_6^{II}; e_v = e_v^0 + u_i^{II}; \end{cases}$$

$$u_0^{II} = -1434.4\cdot x_1 - 720\cdot x_2 + 2508.3\cdot x_3^{II} + 79.2\cdot x_4^{II} + 84.5\cdot x_5.$$

The case of a nonzero equilibrium

$$u_1^{II} = 1986 \cdot x_1 + 857 \cdot x_2 - 2451.5 \cdot x_3^{II} - 88.4 \cdot x_4^{II} + 98 \cdot x_5$$

The graphs of some transient processes are shown in Fig. 3, 4, 5 and 6.

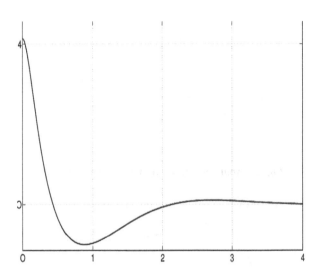

Fig. 3. Trough angle $x_3^{II}(t)$ (rad).

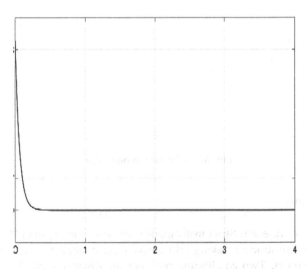

Fig. 4. Stabilizing control $u_0^{II}(t)$.

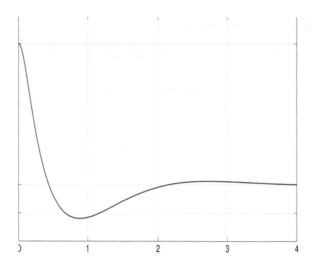

Fig. 5. Angle of rotation of drive wheel $x_3^l(t)$ (rad).

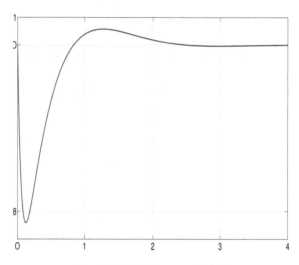

Fig. 6. Stabilizing control $u_0^l(t)$.

6 Conclusion

On an example of research Strict mathematical models of the system "GBB 1005 Ball and Beam" are constructed, taking into account the complete nonlinear equation of geometric constraint. Two equilibrium positions are distinguished. The principal difference between the equations of the first approximation of perturbed motion in the neighborhood of the known and new, previously unexplored, equilibrium positions is shown. In two cases of choosing an excessive coordinate by the method of Krasovskii [1] obtained linear control, which solves the stabilization problem. The conclusion about asymptotic stability with respect to all variables for a complete nonlinear system follows from the theorem proved in [12, 18]. At the same time, for different dependent variables, control is a function of different phase variables. The obtained results demonstrate [20] the importance of taking quadratic terms into account in the expansion of the equations of geometric constraints in the construction of a mathematical model of systems with redundant coordinates.

References

1. Krasovskiy, N.N.: Problems of Stabilization of Controlled Motions. Science, Moscow (1967)
2. Lyapunov, A.M.: Collection op. Academy of Sciences of the USSR, Moscow - Leningrad (1956)
3. Malkin, I.G.: Theory of Stability of Motion. Science, Moscow (1952)
4. Kamenkov, G.V.: Fav. Proceedings. Moscow (1972)
5. Aguilar-Ibanez, C., Suarez-Castanon, M.S., de Jesús Rubio, J.: Stabilization of the Ball on the Beam system by means of the inverse Lyapunov approach. Math. Probl. Eng. **2012** (2012). https://doi.org/10.1155/2012/810597
6. Koo, M.S., Choi, H.L., Lim, J.T.: Adaptive nonlinear control of a Ball and Beam system using centrifugal force term. Int. J. Innov. Comput. Inf. Control. **8**(9), 5999–6009 (2012)
7. Keshmiri, M., Jahromi, A.F., Mohebbi, A., et al.: Modeling and control of Ball and Beam system using model based and non-model-based control approaches. Int. J. Smart Sens. Intell. Syst. **5**(1), 14–35 (2012)
8. Nonlinear, Yu.W.: PD regulation for Ball and Beam system. Int. J. Electr. Eng. Educ. **46**(1), 59–73 (2009)
9. Rahmat, M.F., Wahid, H., Wahab, N.A.: Application of an intelligent controller in a Ball and Beam control system. Int. J. Smart Sens. Intell. Syst. **3**(1), 45–60 (2010)
10. Andreev, F., Auckly, D., Gosavi, S., et al.: Matching, linear systems, and the Ball and Beam. Automation **38**, 2147–2152 (2002)
11. Krasinskaya, E.M., Krasinsky, A.Ya., Obnosov, K.B.L.: On the development of scientific methods of MF Shulgin's school in application to problems of stability and stabilization of equilibria of mechatronic systems with redundant coordinates. Collect. Sci. Methodol. Artic. Theor. Mech. **28**, 169–184 (2012)
12. Krasinskaya, E.M., Krasinsky, A.Ya.: On the stability and stabilization of the equilibrium of mechanical systems with redundant coordinates. Sci. Educ. **03** (2013). https://doi.org/10.7463/0313.0541146
13. Krasinsky, A.Ya., Krasinskaya, E.M.: Modeling of GBB 1005 Ball and Beam stand dynamics as a controlled mechanical system with redundant coordinate. Sci. Educ. **01** (2014). https://doi.org/10.7463/0114.0646446
14. Shulgin, M.F.: On some differential equations of analytic dynamics and their integration. Scientific works of SAGU 144, p. 183 (1958)
15. Raus, E.J.: Dynamics of the System of Solids. Nauka, Moscow (1983)
16. Kuntsevich, V.M., Lychak, M.M.: Synthesis of Automatic Control Systems with the Help of Lyapunov Functions. Nauka, Moscow
17. Krasinsky, A.Ya.: On a method for studying the stability and stabilization of nonisolated steady motions of mechanical systems. In: Selected Works of the VIII International Conference "Stability and Oscillations of Nonlinear Control Systems", Institute for Control Sciences, Moscow, pp. 97–103 (2004)

18. Krasinsky, E.M., Krasinsky, A.Ya.: On a method for studying the stability and stabilization of steady-state motions of mechanical systems with redundant coordinates. In: Proceedings XII All-Russian Meeting on the Control of WSP-2014, Moscow, 1–19 June 2014, pp. 1766–1778 (2014)
19. Zenkevich, S.L., Yushchenko, A.S.: Fundamentals of Manipulation Robots. MSTU, Moscow (2004)
20. Krasinsky, A,Ya., Krasinskaya, E.M.: On the admissibility of the linearization of the equations of geometric constraints in problems of stability and stabilization of equilibria. Collect. Sci. Methodical Artic. **29**, 54–65 (2015)

Determining Magnetorheological Coupling Clutch Damping Characteristics by the Rotary Shafts Shear Deformation

S. N. Okhulkov[1], A. I. Ermolaev[1], A. B. Daryenkov[2], and D. Yu. Titov[2(✉)]

[1] Institute of Mechanical Engineering of RAS, 85 Belinsky Street,
Nizhny Novgorod 603024, Russian Federation
[2] Nizhny Novgorod State Technical University, 24 Minin Street,
Nizhny Novgorod 603155, Russian Federation
dm_titov@list.ru

Abstract. The article shows an approach to determining the damping properties of a magnetorheological coupling clutch by shear deformation of rotating shafts. The irregularity of the shaft rotation speed is caused by the irregularity of the torque of the diesel engine and the synchronous machine and is manifested due to the cyclical operation of the diesel engine. The article describes an approach for estimating the elastic and damping properties of a magnetorheological coupling clutch that connects the shafts of a diesel engine and a generator at a power plant of an autonomous object. This estimate is based on the determination of the ratio of torque from torsional vibrations on the driven and driving rotating shafts. According to the calculations, the magnetorheological coupling clutch should dampen the torsional vibrations of the diesel shaft three times at an irregularity of its rotation frequency of 2.657%. This causes a decrease in momentary irregularity of the rotation frequency of the generator shaft to 0.886%. This eliminates emergency operation of the diesel engine and synchronous generator, which can significantly extend the life of the power plant of an autonomous facility.

Keywords: Diesel generator plant · Torsional oscillations · Magnetorheological fluids

1 Introduction

Single electrical power plant (SEPP) systems of a standalone object [1–4] have been widely used in water, road and railway transport. Herewith traction (propulsion) electric motors together with other users are connected to SEPP. The use of SEPP allows to increase the safety and simplify the electrical power system maintenance of a standalone object as a result of decreasing its components number. Diesel engines and modular construction internal combustion engines (ICE) [5–7] are widely used as primary motors. As a rule SEPP is built on the basis of constant speed diesel-generator plants (DGP) [1–4].

Research and development of SEPP on the basis of variable speed DGP is a new technical trend in Small Energy. Analysis shows that there are few domestic and foreign scientific publications dealing with this topic. It is particularly true concerning the research of dynamic rating with a single electrical power supply system based on variable speed DGP and synthesis of systems regulating electric propulsion channels [1–4].

2 Defining the Task of Diagnosing the Irregularity in Shafts Speed Rotation of DGP SEPP of a Standalone Object

Research of SEPP dynamic ratings of standalone objects is one of the basic tasks at SEPP powerful diesels of standalone objects technical maintenance as the number of cylinders increases, the construction becomes more sophisticated and there is antivibrator-torsional oscillations damper of rotary crankshaft [8–10]. Embedded systems, running during operation allow to identify faults at early stages of SEPP and DGP operation [11].

When diagnosing DGP the crankshaft angular speed is measured by induction sensors. The sensors are a part of electronic control system and are installed in nearly all modern DGP [5–7]. When creating DGP diagnosis systems calculation methods and determining couple unbalance of DGP synchronous electric generator shaft rotation are of special interest [12, 13].

When the diesel is running at engine pulses a big amount of energy is released and the pressure sharply increases. As a result of this the crankshaft rotational motion by connecting rods takes place. As soon as the number of cylinders is limited the torque imposes the shaft at certain moments, which affects the rotational rate [8, 9]. The change of crankshaft rotational rate of DGP 16-cylinder diesel in time t is shown in Fig. 1 (τ - cylinders activation period).

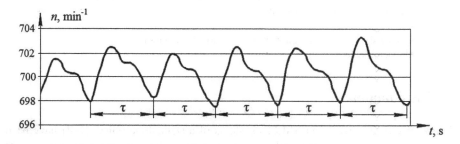

Fig. 1. Hydromount with MRT functional chart with a coaxial choking channel, controlled by rotating magnetic field.

Degree of irregularity δ of DGP crankshaft rotational rate reaches 0.7% and is determined by measuring timespans τ, during which the crankshaft rotational motion to the preset angle takes place.

Degree of irregularity δ of DGP crankshaft rotational rate is also determined by the difference ratio of rotational rate maximum n_{max} and minimum n_{min} to the crankshaft mean speed of rotation n_{cp} [8, 9, 14]:

$$\delta = (n_{max} - n_{min})/n_{cp}, \qquad (1)$$

where n_{max}, n_{min} and n_{cp} are dimension of crankshaft rotation rates [rpm].

The irregularity of rotational rate is caused by the diversity of energy source and detector torques and particularly emerges due to the cycling of engine operation expressed by driving forces affecting its cylinders periodically through a certain output angle. The diesel crankshaft output angle during one period of non-steady motion depends not only on instantaneous angular speed of revolution but also on the system torsional oscillations "crankshaft – flywheel – generator rotor".

DGP SEPP diesel load is a synchronous electric generator its rotor as a rule being connected with the diesel crankshaft by the coupling clutch which damps axial, radial and angular displacements with the minimal reaction forces due to its elastic and dissipative characteristics [5–9]. Rotary shafts coupling clutch in DGP can be a magnetorheological one using magnetorheological fluids (MRF) in inertial magnetorheological transformers (MRT) of vibration and shocks damping systems [15, 16].

The task of evaluating elastic and damping characteristics of magnetorheological coupling clutch to decrease rotary shafts variable torques in stationary and transient operating conditions of the electric generator is acute and meets contemporary machine building demands [13, 17].

3 Faults in SEPP Diesel Electric Power Plants

In standalone electric power supply systems as an electric energy source as a rule DGP is used in which two automatic control systems are employed: the automatic control system (ACS) of diesel rpm and the automatic control system of electric generator voltage [1–4]. The purpose of the first automatic system is to stabilize the diesel rpm. The purpose of the second one is to stabilize the generator voltage [1–4].

The diesel engine of the generator power plant (Fig. 2) differs from the classical scheme of the diesel-electric installation by the presence in the control system of the diesel engine speed of the torque and torsional oscillations meters 8.

Diesel 1 rotates synchronous generator rotor 2. The voltage frequency at the output of the synchronous generator 2 is proportional to the rotor rpm, thus the voltage value being proportional to the electric generator exciting current.

The crankshaft rpm control 6 determines the synchronous generator voltage output frequency 2 and affecting the control gears 3 of diesel 1, sustains the diesel constant crankshaft rpm 1, which stabilizes the electrical generator output frequency voltage 2 in all operating modes.

Synchronous generator voltage regulator 7 affecting its exciting current sustains the value of the output voltage being nearly the same at all admissible loads. The torque and torsional oscillations meter 8 measures the torque and couple unbalance on the

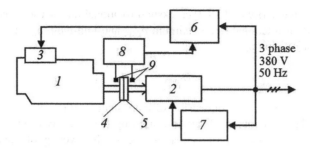

Fig. 2. Scheme of a diesel-electric installation of a power station with a torque meter and torsional oscillations.

synchronous generator rotor shaft 2, as well as protects diesel 1 and electric generator 2 from emergency operating modes which allows to significantly decrease the number of emergency situations and to extend the DGP SEPP life-cycle.

At the diagnosis and automatic control of DGP the diesel crankshaft momentary angular speed 1 is measured by induction sensors 9 [5–7], their values of crankshaft rpm being taken from the driving plate 4 and the driven one 5 of the plate coupling, connecting the diesel crankshaft 1 and the traction synchronous generator rotor shaft 2.

Incorporating into DGP SEPP torque and torsional oscillations meter 8 allows to protect DGP along the torques of the diesel rotary shafts 1 as well as the synchronous generator 2 by comparing the current momentary imbalances on the diesel shafts 1 and the electrical generator 2 with their threshold values [5–7].

When creating the systems of automatic control, diagnosis and the protection features of rotary shafts torque 1 and the electric generator 2 DGP it is necessary to calculate and estimate coupling clutch elastic and damping characteristics, which damps variable imbalances moments on the diesel shafts 1 and the synchronous generator 2 in both stationary and transient operating conditions [13, 18].

The classical DGP schemes have basic faults [1–4]:

- The necessity to function at the constant speed, required to stabilize the output voltage, forces us to reject the modes providing fuel-burn improvements and increasing the diesel life-cycle. At the load of lower than 20% from the nominal one and at sustaining the diesel constant speed it is impossible to run the DGP at length, as there occurs oil coking in engine
- At abrupt changes in electrical load at the starting moment takes place a significant change in the generator voltage (up to 20% from the nominal), which cannot be counteracted by the ACS electric generator voltage. It is connected with the electric generator characteristics and practically is not eliminated by the ACS
- In DGP ACS there is no torque and torsional oscillations meter, having the information about variable rotary shafts torques in both stationary and transient operating conditions of the electric generator
- To eliminate the above faults which are characteristic when creating the traditional scheme of the DGP SEPP of a standalone object it is necessary to solve the following tasks

- To create the energy source, generating the constant frequency and quantity voltage (with an admissible error) with the best dynamic characteristics at the condition of admissibility of diesel crankshaft variable speed
- To create ACS optimization diesel shaft rpm at the minimum fuel consumption
- To create ACS optimization diesel shaft rpm with the torque and torsional oscillations meter, having the information about variable rotary shafts torques in both stationary and transient operating conditions of the electric generator which allows to protect DGP along the torques

To solve the first and second task is possible using the electric generator of any type (synchronous generator, synchronous generator with permanent magnets, asynchronous generator, direct-current generator) connected to the electronic unit transducing the generator voltage into three-phase voltage of the required value and frequency. To solve the second task it is necessary to use the computer technology having the proper software [1–4].

This article considers the third task – the estimate of elastic and damping characteristics of the magnetorheological clutch coupling to decrease the variables of rotary shafts in electric generator stationary operating conditions [17].

4 The Estimate of Elastic Damping Characteristics of Magnetorheological Coupling Clutch

When determining elastic and damping characteristics of DGP magnetorheological coupling clutch of electric power plants standalone objects the following method of determining the ratio of torques and torsional oscillations on the driving and driven DGP SEPP rotary shafts is used.

At this approach we consider that the driving disc of magnetorheological coupling clutch, connected with the diesel crankshaft is an input element and the magnetorheological coupling clutch driven disc connected with the synchronous electric generator rotor shaft is an output element of the mechanical DGP SEPP chain. At this we consider that the diesel crankshaft is adduced to the synchronous generator rotor shaft [9, 14, 18, 19].

Further on we consider that the shafts rotate with the angular speed $\omega_{вр}$ and when rotating affected by torques M_X^+ and M_X^- there appear torsional oscillations of frequency $\Omega_{кр}$, at this the values of the shafts elements swirling do not remain constant and equal to the swirl angle φ_x but are measured with the linear speed $v_л = \Omega_{кр} \cdot r$, where r – is the DGP adduced shaft element radius [17].

Basing on [17, 19, 20] the peak strains ε_{max} and voltage σ end-areas (circular cross-sections) of the adduced diesel crankshaft and the electric generator shaft are expressed as follows

$$\varepsilon_{max} = \frac{v_\pi \tau}{h} = v_\pi \sqrt{\frac{2\rho(1+\mu)}{E}} = \frac{v_\pi}{V}, \qquad (2)$$

and the voltage

$$\sigma = G\varepsilon_{max} = \frac{Gv_\pi}{V} = Gv_\pi\sqrt{\frac{2\rho(1+\mu)}{E}}, \qquad (3)$$

where h and τ — sampling intervals with respect to the coordinates and time; E — elasticity modulus; G — modulus of shearing; μ — Poisson's ratio; ρ — shaft material density; V — mechanical oscillations propagation speed within the shaft body.

According to $A = (G\tau^2)/(\rho h^2) = 1$, the expression from [17], the time step to the adduced shafts of circular cross-sections is expressed as

$$\tau = h\sqrt{\rho/G}\, a.$$

The adduced diesel crankshaft end-areas swirling and the electric generator one is:

$$T_{\text{сж}} = L\sqrt{\rho/G},$$

where L — is the length of adduced rotary shafts DGP base areas.

As known from [11, 17], the mechanical oscillations propagation speed in solid bodies is expressed by

$$V = \sqrt{E/\rho}.$$

Mechanical oscillations of the adduced shafts elements are torsional waves, being clearly transversal but not dispersion ones and their speed does not depend on the wave length [17]. Thus, if the solid bodies swirl (in this case — the shafts elements) the speed of mechanical oscillations propagation in them is expressed as

$$V = \sqrt{\frac{G}{\rho}} = \sqrt{\frac{E}{2\rho(1+\mu)}}.$$

The scheme of torsional wave propagation in the elements of DGP cross-sections of adduced shafts is shown in Fig. 3.

The swirling time of the adduced shafts is [17]:

$$T_s = T_{\text{сж}} = L/V, \qquad (4)$$

As we can see the peak strains and voltage of the adduced rotary shafts elastic elements of the DGP mechanical chain are expressed as (2) and (3) being proportional to the linear speed, i.e. when the torques variables are applied they appear to be the following functions $\varepsilon = f(v_\pi)$ and $\sigma = f(v_\pi)$, and as $v_\pi = \Omega_{кр} \cdot r$, the required values are the functions of the adduced shafts torsional oscillations elements, i.e. $\varepsilon = f(\Omega_{кр})$ and $\sigma = f(\Omega_{кр})$.

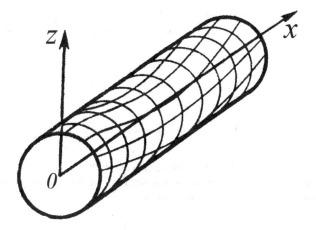

Fig. 3. The scheme of torsional wave propagation along the cross-section shaft length.

For example, for the DGP adduced electric generator rotor shaft at the torsional oscillations $\Omega_{\kappa p}$ = 1.884 rad/s and the shaft radius r = 0.06 m the linear swirling speed of the shaft elements will be:

$$v_\pi = \Omega_{kp}\, r = 1,884 \cdot 0,06 = 0,11304 = 113,04 \cdot 10^{-3}\, \text{m/s}.$$

In case the electric generator rotor shaft material has the following characteristics: shear modulus G = 8·10^{10} n/m²; ρ = 7.8·10^3 kg/m³, and the grid pitch is chosen as equal to h = 0.06 m, then according to [17] the time step will be

$$\tau = 0,06\sqrt{\frac{7,8 \cdot 10^3}{8 \cdot 10^{10}}} \approx 18,735 \cdot 10^{-6}\,\text{s}.$$

If the electric generator rotor shaft elements swirl at the speed of v_π = 113.04·10^{-3} m/s, from the expression (2) get the following peak strains of all the rotor shaft elements:

$$\varepsilon_{\max} = \frac{v_\pi\, \tau}{h} = \frac{113,04 \cdot 10^{-3} \cdot 18,735 \cdot 10^{-6}}{0,06} = 35,296 \cdot 10^{-6}.$$

Thus peak strains of all the rotor shaft elements are not significant. So, every unit of rotor shaft elements is equal to 35.296 μm. The peak strains of all the synchronous electric generator rotor shaft elements are [17]:

$$\sigma = G\varepsilon_{\max} = 282,374 \cdot 10^3\, \text{N/m}^2.$$

The peak swirl of all cross-sections of the elements on the electric generator rotor shaft being relative to each other we find from the expression, replacing $tg\alpha$ with the peak strains ε_{max}:

$$\varepsilon_{max} = tg\,\alpha = \frac{(\phi_{x,i+1} - \phi_{x,i})r}{h};$$

$$\Delta\phi_{max} = (\phi_{x,i=1} - \phi_{x,i}) = \frac{\varepsilon_{max}h}{r}.$$

Using the data of the above example: $\varepsilon_{max} = 35.296 \cdot 10^{-6}$; $r = 0.06$ m; $h = 0.06$ m, the peak reciprocal swirl of all the cross-sections of the elements on the electric generator rotor shaft relative to each other is

$$\Delta\phi_{max} = (\phi_{x,i+1} - \phi_{x,i}) = \frac{\varepsilon_{max}h}{r} = 35,296 \cdot 10^{-6} \text{rad}.$$

In angular units the peak strains in all the shaft elements at the cross-sections angle of rotation are $\varphi_{max} \approx 7.28''$.

The peak electric generator shaft swirl angle length is $L = 23,666 \cdot 60 = 1420$ mm will be:

$$\Delta\phi_{в.max} = 23,666 \frac{\varepsilon_{max}h}{r} = 835,315 \cdot 10^{-6} \text{rad}.$$

In angular units the peak synchronous electric generator strain of $L = 1420$ mm length will be:

$$\Delta\phi_{в.max} \approx 23,666 \cdot 7,28'' = 172,288''.$$

The maximum time of electric generator shaft swirl of 1420 mm length at $\Omega_{кр} = 1.884$ rad/s and the period of torsional oscillations $T_{кр} = 3.333$ s will be

$$T_{з.max} = T_{сж} = \frac{\Delta\phi_{в.max}}{2\pi} \cdot T_{кр} = 443,3 \cdot 10^{-6} \text{ s}.$$

The nominal admissible torque on the electric generator rotor shaft is expressed by:

$$M_{кр.н} = \frac{\Delta\phi}{L} G \int_S R^2 dS. \tag{5}$$

Integral in expression (5) is a polar inertia moment of the shaft cross-section being radius R of the electric generator rotor. The polar moment of the shaft cross-section inertia of the electric generator rotor, having a uniform cross-section is constant along the shaft length and does not depend on the length of its base section (base $L = 1420$ mm):

$$J_p = \int_S R^2 dS.$$

For the electric generator shaft cross-section of radius $R = 0.06$ m the polar inertia moment is expressed by: $J_p = \pi \cdot R^4/2$ [20]; electric generator shaft shear modulus being $G = 8 \cdot 10^{10}$ N/m^2; electric generator shaft material density $\rho = 7.8 \cdot 10^3$ kg/m^3; rigidity $GJ_p = G \pi R^4/2 = 1{,}6286 \cdot 10^6$ N·m^2.

The torque of electric generator rotor shaft torsional oscillations at frequency of the torsional oscillations $\Omega_{\kappa p} = 1.884$ rad/s is determined by the expression:

$$M_{\kappa p.\mathfrak{z}.2.} = \frac{\Delta \phi_{\mathfrak{s}.\max}}{L} GJ_p = 958{,}024 \text{ N} \cdot \text{m}.$$

For DGP with 16-cylinder diesel engine the crankshaft rpm irregularity reaches 2.657% (Fig. 1).

The degree of DGP shaft rpm irregularity δ is determined by the difference quotient of the peak one $n_{\max} = 709.8$ rpm and the minimal $n_{\min} = 691.2$ rpm to the crankshaft mean speed $n_{cp} = 700$ rpm [8, 9, 14]:

$$\delta = (n_{\max} - n_{\min})/n_{cp} \approx 0{,}02657,$$

which is 2.657% from the diesel crankshaft mean speed.

At the diesel shaft cyclic mean speed 11,666 Hz and its irregularity 2.657% the cyclic and angular torsional oscillations frequencies take a value $f_{\kappa p2.65} = 0.31$ Hz and $\Omega_{\kappa p2.65} = 1.9468$ rad/s.

At the DGP crankshaft torsional oscillations cyclic frequency $f_{\kappa p2.65} = 0.31$ Hz the torque affected by them on the end-areas (elements) takes a value $M_{\text{кр.кол.2.65}} \approx 960$ N·m.

The evaluation of elastic and damping characteristics of magnetorheological coupling clutch to decrease torsional torques variables of DGP rotary shafts in stationary operating conditions is based on determining the ratio of torques from the torsional oscillations on the DGP SEPP driving and driven rotary shafts (Fig. 2).

At the torques calculated values from the torsional oscillations on the adduced driven and driving shafts end-areas we can calculate the torques ratio $M_{\kappa p.\mathfrak{z}\mathfrak{z}09}$ and $M_{\kappa p.\kappa o \pi.2.65}$:

$$\frac{M_{\text{кр.эг90}}}{M_{\text{кр.кол.2,65}}} = \frac{319{,}36 \text{Н} \cdot \text{м}}{960 \text{Н} \cdot \text{м}} \approx 0{,}333. \tag{6}$$

The coupling magnetorheological clutch of the DGP must be 3 times damped by the torsional oscillations of the crankshaft of the diesel engine with a non-uniformity of its rotational speed of 2.657% (which is 9.543 dB) in accordance with the ratio (6) for twisting moments from the torsional oscillations of the shafts of the DGP. It means that the momentary irregularity of DGP SEPP electric generator shaft rpm should decrease and reach 0.886%.

According to the relation (6), for torque from the torsional oscillations of the rotating shafts of the DGP, different torques from the torsional oscillations act on the driving and driven shafts of the connecting magnetorheological clutch - on the leading cranked shaft of the DGP $M_{кр.кол.2.65} = M_{кр.к.в}$ and the driven shaft of the electric generator $M_{кр.кол.0.9} = M_{кр.э.г}$. The connecting magnetorheological coupling completely transfers the power of the rotating crankshaft to the shaft of the synchronous generator of the DGP.

The ratio of the moments $M_{кр.кол.2.65}$ и $M_{кр.эг.0.9}$ at the input and output of the connecting magnetorheological clutch is calculated by the torque and torsional vibration meter included in the automatic control system for optimizing the diesel engine speed of the DGP SEPP of autonomous objects (Fig. 2).

In case DGP crankshaft is directly connected with the electric generator rotor shaft, the torque from the rotary DGP crankshaft on the rotary electric generator shaft endangers the rotor shaft durability, which can lead to electric generator breakdown as a result of rotor shaft eccentricity increase and the emerging vibration [12, 13].

For the electric generator rotor shaft to function within the work range of rpm without losing the durability, the electric generator shaft is connected with the DGP diesel crankshaft through the magnetorheological coupling clutch (Fig. 2). Magnetorheological coupling clutch provides the diesel crankshaft connection with the electric generator shaft and additionally dampers the appearing peak torsional oscillations due to its elastic and dissipative characteristics [9, 10]. By this electric generator shaft rotation regularity is reached, providing its safe operation in the SEPP standalone object.

5 Conclusion

The ratios obtained and the calculation sequence presented allow to consider the effect of DGP crankshaft torsional oscillations at the preset irregularity of its rotation rate. With the help of the obtained ratios it is possible to estimate the elastic and damping characteristics of the magnetorheological coupling clutch to decrease the rotary shafts torques variables in electric generator stationary operation conditions. The obtained ratios can also be recommended to carry out strength calculations, to determine the diesel-generator capacities, and to estimate their generated electricity quality. However, the results obtained need to be corrected due to the necessity of considering the effect of system forced oscillations.

Acknowledgements. The presented research results were obtained with the support of grants from the President of the Russian Federation for state support of young Russian scientists (MK-590.2018.8).

References

1. Daryenkov, A.B., Khvatov, O.S.: High-performance standalone electric power plant. Trans. NNSTU **77**, 68–72 (2009)

2. Khvatov, O.S., Daryenkov, A.B., Tarasov, I.M.: Diesel power generation plant with the shaft variable speed. Bullet. Ivanovo State Power Inst. **2**, 53–56 (2010)
3. Khvatov, O.S., Daryenkov, A.B.: Knowledge-based tools to control the high-performance variable speed diesel power plant. Izvestiya Tula State University. Eng. Ind. Technol. Sci. **4**, 126–131 (2010)
4. Tarpanov, I.A., Khvatov, O.S., Daryenkov, A.B.: High-performance diesel power generation plant with the shaft variable speed based on doubly-fed electric machine. Drive Technol. **5**, 14–19 (2010)
5. Teplovoz 2TE10L: Transport, Moscow (1974)
6. Rudaya, K.I., Loginova, E.U.: Locomotives. Electrical equipment and schemes. Transport, Moscow (1991)
7. Akimov, P.P.: Ship power plants. Transport, Moscow (1980)
8. Veinsheidt, V.A.: Ship internal combustion engines. Sudostroenie, Leningrad (1977)
9. Istomin, P.A.: Torsional oscillations in ship ICE. Sudostroenie, Leningrad (1968)
10. Alekseev, V.V., Bolotin, F.F., Kortyn, G.D.: Oscillations damping in ship's drive line components. Sudostroenie, Leningrad (1973)
11. Genkin, M.D., Sokolova, A.G.: Vibroacoustic Diagnostics of the Machines and Mechanism. Mashinostroenie, Moscow (1987)
12. The reference book on electrical machines. Energoatomizdat, Moscow (1989)
13. Shubov, I.G.: Noise and Vibration of Electrical Machines. Energoatomizdat, Leningrad (1986)
14. Tverskyh, V.P.: Torsional oscillations of power plant shaft line. Research and calculation methods, system elements and excitation elements, vol. 1. Sudostroenie, Leningrad (1969)
15. Gordeev, B.A., Erofeev, V.I., Sinev, A.V., et al.: Vibration protection systems using the time lag and dissipation of rheologic environment. Fizmatlit, Moscow (2004)
16. Gordeev, B.A., Okhulkov, S.N., Plekhov, A.S., et al.: Magnetorheological fluids application in machine building. Privolzhsky Sci. J. **4**, 29–42 (2014)
17. Makvecov, E.N., Nartakovsky, A.M.: Mechanical effects and protection of communications-electronic equipment. Radio and connection, Moscow (1993)
18. Chelomey, V.N.: Vibrations in Engineering: Reference Book. Mashinostroenie, Moscow (1980)
19. Tverskyh, V.P.: Power plant torsional oscillations calculation. Mashgiz, Moscow (1953)
20. Birger, I.A., Shor, B.F., Shneiderovich, R.M.: Strength calculation of the machine elements. Mashinostroenie, Moscow (1966)

Automation in Foundry Industry: Modern Information and Cyber-Physical Systems

M. V. Arkhipov(✉), V. V. Matrosova, and I. N. Volnov

Moscow Polytechnic University, 38 Bolshaya Semyonovskaya Street,
Moscow 107023, Russia
maksim_av@mail.ru

Abstract. The automation of the foundry industry in the framework of the forthcoming sixth technological system evolves through the introduction of: cyber-physical systems, industry internet, IoT industry, smart production, 4.0 industry, cloud calculations and neural networks. Currently, in the area of the foundry industry the manual labour still prevails at the stages of materials finishing processing. Cyber-physical systems based on manipulation robots are very efficient in solving tasks of grinding parts after casting. The positioning problem solution for precise approach of a part to the surface of a processing tool with addition of force control makes the system more complicated. The equipment of the manipulation robot with a control system on the basis of a neural network controller is considered assuring the solution to the inverse kinematics problem taking into account the force of processed part interaction with a grinding disc of the abrasive tool. The comparison of analytical and experimental solutions has shown that the precision of the abrasive machining is approximately uniform in the limits of normative values. In this case the complexity of the development of the control algorithm is significantly lower if the neural network control method is used.

Keywords: Manipulation robot · Abrasive machining · Positioning-force control · Neural networks

1 Introduction

Recently, there is a growth of the potential and a trend of foundry industry development in the modern conditions of beginning of development of works in the framework of the fifth technological system, and it is expected that in the next decade there will be a transition into the sixth technological system [1].

At the modern stage of discussions on the technological development, new terms are being added, that informatively yet had no direct relationship to the foundry industry, and in the public opinion are connected with innovative development, newest technologies and new markets yet under development. These terms are: cyber-physical systems, industry internet, IoT industry, smart production, 4.0 industry et al. The task of this article is not only to introduce, but also to connect this set of terms with the development discourse of the foundry industry in particular, but with traditional machine building technologies in general. The need and importance of this step is

© Springer Nature Switzerland AG 2020
A. A. Radionov and A. S. Karandaev (Eds.): RusAutoCon 2019, LNEE 641, pp. 382–392, 2020.
https://doi.org/10.1007/978-3-030-39225-3_41

determined by a known phenomenon of relation of technologies progress with activity of these technologies discussion in the professional community.

All the mentioned terms are beginning to be used equally in the area of knowledge of the foundry industry automation, however, the concept of the "cyber-physical system" should be underlined. The definition of cyber-physical systems in the foundry industry may be formulated in the following way - it is a production in which robot-technical complexes, technological equipment with integrated information systems are used providing an opportunity of making a decision without a human involvement in real time [2]. Reference sources define the content of this term through a set of other terms or similar new or more familiar concepts. So, in cyber-physical systems of foundry industry it is assumed to use simultaneously a few of the following new technological trends: modelling and simulation, big data systems, adaptive robots manipulators [3], cloud calculations, Internet of things, 3D printing and augmented reality.

Each of these trends, as applied to the foundry industry, as expected, will become the whole direction of development. From positions of cyber-physical systems approach, the most technological direction of development of the foundry industry automation is the task of using adaptive robot-technical complexes both for main and auxiliary operations. In conjunction with high complexity and significant amount of manipulations, the maximum effect may come from the application of manipulation robot-technics for automation of post-casting technological operations, such as stripping sprues by abrasive wheels.

Let us note that in the conditions of the fifth technological system with mass production and high level of production automation, there is no obvious need to use cyber-physical systems. The actuality of these systems grow rapidly in changing now production and economic conditions and in expectation of transition to the sixth system with its low serial and even individual production not concentrated inside large foundry shop or plant.

2 Foundry Industry Automation in the Terms of the Fifth System and 4.0 Industry

The sixth technological system introduces into the modern production processes solutions related to new principles and approaches to automation. One of hi-tech areas where a share of manual labour is still high is machining of parts after casting. Let us consider in more detail terms defining modern automation trends due to implementation of information and cyber-physical systems.

Industry 4.0 in the area of automation uses the following concepts: - autonomous robots (equipped with information system for making decisions on choosing algorithms for solving the following tasks: the quality of the moulding materials, of the charge; the rejection of castings; fettling sprues; trimming deflashing; synchronization of the conveyor, casting and smelting complexes; co-interaction with humans in the working zone); - modelling and simulation (technology of working with a digital prototype of a product (CAD/CAM/CAE); making real time decisions and without human involvement; application of new computational strategies); - cloud calculations (transfer of information and calculation infrastructure to external high capacity cloud network systems; joint operation and data exchange between territorial distributed partners); - large data bases (data storage for automatic decision making); - Internet of things (automatic identification of objects or processes and decision making on using typical operations on them at any production stage); - augmented reality (virtual visualization of any process on any stage as a reference sample of the result that has to be obtained); - neural networks and artificial intelligence technologies (assurance of automatic decision making on performing both separate operations and processes in general based on accumulated data and knowledge).

The examples of complex tasks for cyber-physical systems in the area of the foundry industry due to use of feedback principles are parameters control of molding mixture preparation, melting and quenching. The information systems are also using principles of the cyber-physical systems, in particular, multiple virtual testing of technological process with a choise of an optimal one. The examples are systems of computer modelling of foundry processes that could now be functioning without a human participation.

The further development of the cyber-physical systems will be determined by a degree of implementation of systems in which decision making is done without a human participation.

3 The Cyber-Physical System of Grinding Machining

After completion of foundry technological processes of making materials (castings, blanks, parts), there is a transition to technological processes of the materials processing up to a finished product. Out of materials processing, the most labour consuming is finishing treatment that has to be a subject to automation using manipulation robots [4]. After the procedure of cleaning obtained after casting materials, their final treatment is done by grinding - removal from their surfaces sprues of metal, burrs and other irregularities. The main complexity of grinding process automation using manipulation robots is the diverse nature of the products, that's why traditionally blanks were grouped by dimensions, form and other design parameters.

Grinding of different types of blanks may be automated using adaptive robotic complexes equipped with industrial rigs with abrasive tools and information system of position and force process control of finishing treatment [5]. The common distinguishing feature of adaptive robots is the presence of manipulator with a stable grip that supply and keep blanks in the process of grinding by abrasive wheel. The additional mean allowing to broaden capabilities of this cyber-physical system is force measurement system.

The technological equipment of the cyber-physical system on Fig. 1 includes: vacuum grip system I, the feeding conveyor with the storage II, grinding machining system with force control III, robot manipulator IV, conveyor V for finished products transportation in boxes to the warehouse.

Functional diagram of automation of production line for edging castings according to the known envelope is shown on Fig. 1.

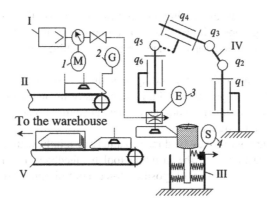

Fig. 1. The cyber-physical system of grinding machining.

From the conveyor II from the device of one-piece delivery (G) 2 blanks are gripped by the robot manipulator with ejector installed on its flange (E) 3 with an elastic suction cup as part of the vacuum grip I. During this process, the blank is fixed on the grip, positioned and smoothly moved till the contact with the grinding disk of the grinding machining system III. During the rotation of the end profile of the workpiece along the whole contour, the constant pressing force to the surface of the grinding disk is provided, due to it grind off excess metal is grinded off. The force control is provided by one component force probe (S) 4. Machined workpieces are placed into the storage on the conveyor V, the products are put into boxes in the storage; after filling the box, it is transported to the finished products warehouse.

To prevent the conveyor II overloading, there is a provision for keeping determined tact of their discharge by the automatic control system of a workpiece reaching its removal place; the system consists of a delivery device with the end sensor. The conveyor II is equipped with controlled electric motor acting as regulator to exclude an overflow of parts at the delivery point [6]. In case of a delay at the machining stage, a signal is sent to the conveyor for waiting of the workpieces delivery. After finishing the grinding technological operation, the signal of turning on the conveyor II is sent.

For control of the pressure level in the pneumatic line, there is a manometer (M) 1 that turns on air supply into the storage when the pressure level is below 4 atm. The maintenance of the set pressure level in the system is needed for holding the workpiece by the vacuum grip and elimination of slipping. The displacement of the point of retention of the workpiece is provided by excluding a situation of the workpiece drop out from the grip by controlling joint coordinates $q_1..q_6$ of the manipulator [7].

4 Position/Force Control of Cyber-Physical System of the Grinding Machining

In the modern foundry industry, automated systems with position/force adaptation in the process of profiled multifaceted castings are used with the grinding machining equipment [8].

The known methods of continuous control of the grinding machining may be grouped into three groups: without feedback, e.g. positioning; with feedback on one or a few parameters (velocity, force); adaptive using machined learning (neural network method).

The position control during grinding machining operations is done with rigid positioning of processed products and on conditions of manual control of the grinding tool wear. Rigid positioning and continuous manual control do not exclude a possibility of making defective products.

The positioning control may be done separately or jointly with measuring the force of interaction of casting with the grinding tool by feedback method. Action of the feedback is based on the use of position/force control intended for assurance of relations:

$$F \to F_o, X \to X_o, \qquad (1)$$

where F, F_0 – the force of interaction of casting with the grinding tool, current and set up values, correspondingly [9]; X, X_0 – the trajectory of the processed casting movement in the point of the contact with the grinding tool, current and real, correspondingly [10].

The system of the position/force control consists of position sensors on the robot IV (encoders) and one-component force sensor 4 (Fig. 1). The signal received from the force sensor is fed into the robot control system comparing set up force with the real one and developing controlling action for the manipulator.

The casting fixed in the vacuum grip is moved along two-dimensional trajectory on Fig. 2. Movement along the whole casting contour happens in the continuous contact with the compliant abrasive tool fixed via elastic elements with a stationary foundation. The force of the workpiece pressure onto the grinding disk is transmitted through an elastic element (spring) to the sensor attached to the casing of the stationary foundation. Removal of the excess metal on the casting in the process of the grinding machining is done due to the force of elastic element between the grinding tool and the stationary foundation.

The output signal u of the sensor is related to movements in the elastic element transmitted from the grinding tool by the expression:

$$u = s_1 \xi + u_0, \qquad (2)$$

where s_1 – proportionality factor, u_0 – sensor zero drift (its output signal without the load), ξ – movement in the elastic element. The force acting on the sensor is calculated by its signal u taking into account dynamic effects:

$$f = s^{-1}(u - u_0), \qquad (3)$$

where $s = s_1/k$ – the sensor sensitivity coefficient, k – the stiffness of the sensor.

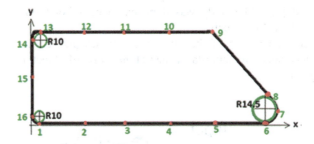

Fig. 2. The profile of the casting subject to the grinding machining.

The feedback on the forces is matched with the check of discrete conditions, for example, in case of exceeding the force threshold value, the manipulator is slowed down of stopped [11]. If forces become inside set up range, it allows to meet conditions of the formula 1.

The third type of control contains control unit that continuously reacts on the force sensor signal change that may be considered as a controller with analog transducer the input of which receives the information from the sensor (sensors), and output of which is manipulator control signals. The relation for such controller parameters is provided as the result of learning by neural network.

5 Information Control and Management of Grinding Machining Process

According to the technological system concept 4.0, it is possible to assume that in the future the classical systems will be probably replaced completely by neural systems. In managing robots, it is related first of all to such levels as task formalization, planning of movements and adjustments of actuators [12].

The main indicator of a robot technical system intellectuality may be considered their ability to solve independently set up tasks based on multiple sensors signals.

One of the promising directions of computer technics development is creation and improvement of neural network devices including neural computers, specialized neural network cards and other devices implementing neural network paradigm [13].

The development of information technologies allows currently to model neural network devices on network and personal computers. Such modelling (emulation) allows to build different neural network structures and to implement neural network algorithms. The development of neural emulators for personal computers makes available implementation of neural network paradigm for the grinding machining.

The following neural network emulators are known [14]: Neural Network and Fuzzy Logic (MatLab, Mathworks); NeuroView+, NeuroEmulator, NeuroControl (Alfa System); SNNS (Stuttgart University), etc.

The purpose of this report is the solution of the inverse kinematics problem for ABB IRB-140 manipulator taking into account signal from force sensor in the system of grinding machining of material due to use of neural network package Neural

Network Wizard [15]. The study was done based on robot technics laboratory of Moscow polytechnical university.

The experimental testing of the inverse kinematics problem for ABB IRB-140 manipulator was done by multi-layer neural network of direct proliferation with feedback (Fig. 3).

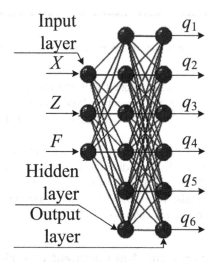

Fig. 3. Diagram of the neural network for the task of workpieces grinding machining by the manupulation robort.

On Fig. 2, the circles represent neurons, X, Y, F, – input values; $q_1 \ldots q_6$ – output values.

In the process of the problem solution, the position of the manipulator grip was determined in the space with the processed part taking into account its force interaction F with grinding wheel on conditions that manipulator's generalized coordinates $q_i \ldots q_6$ ($i = 1..N$) are known [16]. In the experiment, 30 trials were done, in which the increment of the generalized coordinates q_1 and q_6 was 1°. In the process of movement, the values of coordinates X, Y were fixed, values of the coordinate Z and the grip orientation were kept constant.

The solution of the direct kinematics problem was found by analytical method based on using transition matrix with the use of projections and application of the engineering package for mathematical calculations Mathcad and by experimental method - by direct use of IRB-140 robot and taking required values of absolute coordinates using the RobotStudio software environment.

The study of tables with experimental and analytical values allows to make the following conclusions: in the experiment there is a positioning error of 0.005 mm - the flange moves the workpiece close to inputted coordinates. In the analytical solution, the positioning error of 0.005 mm is, on the contrary, absent, the flange moves the workpiece precisely along the inputted coordinates.

For solving the inverse problem of the robot control with the help of neural networks, obtained experimental data are used allowing to take into account the uncertainty factors in the grinding machining.

The experimental setup, on which studies of the effectiveness of the control algorithms were conducted, is shown on Fig. 4.

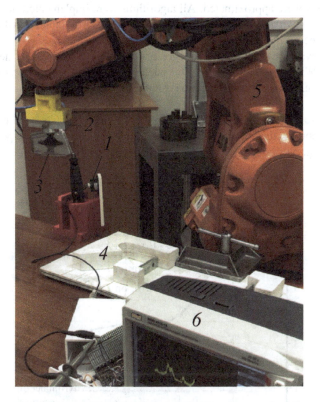

Fig. 4. The experimental setup of the grinding machining based on manipulation robot IRB-140, where 1 – force sensor, 2 – grip, 3 – part, 4 – storage, 5 – manipulation robot, 6 – oscilloscope.

In this work, the experimental testing of PFC problem was done by ABB IRB-140 manipulator with the use of neural network of direct proliferation. The influence of the number of learning examples onto the solution precision was studied. The software implementation of the neural network of multilayer type with feedback was used in the experiments. Networks with 5, 10, 15, 20 neurons on one hidden layer were used.

The learning was done using Neuro Pro 0.25 and Neural Network Wizard 1.7 packages.

Experimental values of cartesian coordinates and forces were inputted, the outputs were rotation angles for the manipulator links. After this, actual positions of the grip were obtained from the solution of the direct kinematics problem corresponding to obtained values of the generalized coordinates, and average errors for rotation angles, positions and forces were calculated.

During the experiment, one dimensional data for 30 pares were used in learning, and the results were approximated. All algorithms were implemented in the Neuro Pro 0.25 software environment developed by the Institute of computational modelling of SD RAS (Siberian Division of Russian Academy of Sciences).

This software product is the manager of learning artificial neural networks operating in MS Windows environment. This package provides the following functionality: loading from external storages, creation, tuning, testing, storage of results into external storages.

As the result, a number of dependencies was obtained, reflecting the relations of the number of neurons, the number of learning examples with error on the position (angle) and force (Fig. 5).

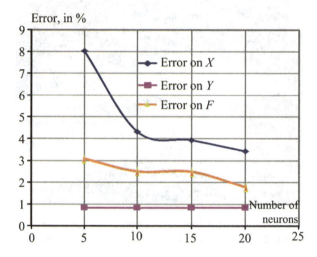

Fig. 5. Dependence of position error for coordinates X, Y and force F from the number of neurons.

The obtained dependences show that with the increasing number of neurons in the network the errors are reduced significantly. Reduction of errors in the solution depends on the neural network realization, that's why there is a possibility of reaching very high solution accuracy in case of using specialised neural controllers. However, the software implementation of the neural network allows also to solve many problems of real time manipulator control, because the number of operations required for obtaining one solution is significantly less than in any iteration method.

6 Conclusion

The concepts of the industry technological system 4.0 are beginning to be actively implemented into production processes. Growing informatization rates require new computational capacities which are suggested to implement based on automated devices equipped with control systems with functions of cloud calculations, technologies of neural networks, etc.

Artificial neural networks are important extension of the calculation concept. Based on it, controllers are created fulfilling functions that previously were an exclusive prerogative of the human being. Reliability, high speed performance, possibility of the solution of poorly formalized problems make it attractive to use neural networks for the control of manipulation robots, in particular, for solving the problem of grinding machining.

References

1. Volnov, I.N., Evseev, S.Y.: The main directions of the strategic development of the foundry in the conditions of the fifth technological wave. Foundry Prod. **6**, 27–33 (2017)
2. Golovin, V., Arhipov, M., Zhuravlev, V.: Robotics in Restorative Medicine. Robots for Mechanotherapy. Lap Lambert Academic Publishing, Saarbrucken (2012)
3. Vukobratovich, M., Stokic, D., Kirchansky, N.: Non-adaptive and Adaptive Control of Manipulative Robots. Mir, Moscow (1989)
4. Arkhipov, M.V., Rachkov, M.Yu., Golovin, V.F., et al: Control device for technological tools of a manipulator. RU Patent 173686, 05 September 2017
5. Solomentsev, Yu.M.: Industrial Robots in Machine Building: A Collection of Diagrams and Drawings. Mashinostroeniye, Moscow (1986)
6. Matrosova, V.V., Bebenin, V.G.: Automated Electric Drive. Moscow Polytech, Moscow (2017)
7. Arkhipov, M.V., Vartanov, M.V., Mishchenko, R.S.: Control of Manipulation Robots. Mospolitech, Moscow (2017)
8. Egorov, I.N.: System of position-force control of technology robots, mechatronics. Autom. Control **10**, 15–20 (2003)
9. Arkhipov, M.V., Golovin, V.F., Leskov, A.G., et al.: Training of the robot by demonstration of motion taking into account the deformation of the environment. News of SibFU Tech. Sci. **10**(171), 213–227 (2015)
10. Leskov, A.G., Golovin, V.F., Arkhipov, M.V., et al.: Training of robot to assigned geometric and force trajectories. In: 4th Workshop on Medical and Service Robotics, vol. 39, pp. 75–84 (2016)
11. Gorinevsky, D.M., Formalsky, A.M., Shneider, AYu.: Control of Manipulation Systems Based on Information on Forces. Fizmatlit, Moscow (1994)
12. Kukuy, D.M., Odinochko, V.F.: Automation Of Foundry Industry: Textbook. New Learnings, Minsk (2008)
13. Egupov, N.D.: Methods of Robust, Neuro-Fuzzy and Adaptive Control. MSTU, Moscow (2001)

14. Yarushkina, N.G.: Basics of Fuzzy and Hybrid Systems Theory. Finance Stat., Moscow (2004)
15. Vartanov, M.V., Arkhipov, M.V., Petrov, V.K.: Experimental research of conditions of assemblability in active robotic assembly. Mach. Tools Instrum. **4**, 14–16 (2017)
16. Vzhesnevskiy, E.A., Arhipov, M.V., Orlov, I.A.: Problems and tasks of tactile sensing in manipulation robotics when working with compliant objects. Intellectual systems, management and mechatronics, pp. 184–189 (2017)

Research of Non-sinusoidal Voltage in Power Supply System of Metallurgical Enterprises

R. V. Klyuev[✉], I. I. Bosikov, and A. D. Alborov

North Caucasian Institute of Mining and Metallurgy, 44 Nikolayeva Street,
Vladikavkaz 362021, Russian Federation
kluev-roman@rambler.ru

Abstract. Non-ferrous metallurgy enterprises are large consumers of electrical energy. The specifics of non-ferrous metals production involve the use of significant non-linear loads (6–12 phase valve inverters, induction furnaces, welding machines, etc.). This is up to 60% of the total load on the enterprise. Therefore, an important and urgent problem is the study, analysis and calculation of the quality indicators of electrical energy consumed. Much attention that is being paid to the problem of quality of electrical energy (QE) in recent years is explained by a significant economic damage due to QE indicators deterioration. Improving the basic indicators is a part of the energy- and resource-saving policy, the role and importance of which is now increasing all the time as a part of the energy program in the whole community. Based on the comprehensive energy audit in a non-ferrous metallurgy enterprise, the authors investigated coefficients characterizing the voltage non-sinusoidality: ratio of n-th harmonic component of voltage ($k_{U(n)}$, %) and ratio of non-sinusoidal voltage (k_U, %). According to the obtained samples coefficients, the authors performed a rank analysis of the spectrum of the most powerful electrical energy consumers (noevaya caste) higher harmonic (HH) voltage and determined the correlation coefficients between $k_{U(n)}$, k_U and fraction of the noevaya caste electrical energy consumers (W, %). In addition, the contribution made by consumers in non-sinusoidality voltage at the points of common coupling (PCC) with the system is defined. The obtained results are introduced into the process at the enterprises of non-ferrous metallurgy for the production of hard alloys and zinc.

Keywords: Power quality · Non-sinusoidality · Harmonic · Voltage · Rank distribution

1 Introduction

In accordance with the Presidential Decree "On approval of the priority directions of development of science, technology and engineering in the Russian Federation and the list of critical technologies of the Russian Federation" dated July 9, 2011, in which one of the priorities is "Energy, energy efficiency, nuclear energy" around industrial complex of the Russian Federation should occur major changes related to the modernization of outdated equipment and the transition to more efficient production technology, and primarily related to an increase in energy efficiency. Therefore, the study and solution of problems in this area is of particular relevance and importance.

The issues related with electrical energy quality analysis in power supply systems (PSS) of industrial enterprises acquire special importance during the implementation of energy efficiency policy in the Russian Federation. QE is regulated in accordance with the requirements of government standard and the right of consumers to electric power quality is reflected in the article 542–543 of the Civil Code of the Russian Federation.

QE deterioration is primarily connected with the emergence of non-sinusoidal modes in PSS of enterprises during operation of nonlinear consumers. Distortion of sinusoidal currents mode and voltages is caused by operation of rectifiers, induction furnaces, rolling mills, etc. Emerging HH cause additional losses of active power in all PSS components, accelerated aging of insulation of electric motors, transformers, cables, power factor deteriorates, disrupted control devices, computer hardware, electrical energy meters.

For sources of QE deterioration in the case of non-compliance with laws in force are provided economic sanctions. Therefore, the development of methods of a complex estimation of QE, which allows to monitor indicators of QE in on-line mode with the technological process of the enterprise is relevant.

Analysis of the works in the field of research qualitative indexes of electrical energy consumption has shown the lack of addressing the issues related to the analysis of non-sinusoidal voltage in the non-ferrous metallurgy enterprises with a large proportion of non-linear loads [1–14].

In this regard, development of appropriate scientific-methodological foundations based on a comprehensive study of indicators characterizing non-sinusoidality of voltage on the non-ferrous metallurgy enterprises is needed.

Research objectives:

- Development of methods for determining the actual customer contribution (ACC) and systems (ASC) in the non-sinusoidality voltage through active experiment - short-term switching of power transformers main step down substation (MSDS) in parallel operation mode
- Development of methods of experimental study of amplitude-frequency characteristics of the HH voltage in PSS of non-ferrous metallurgy enterprises
- Conducting rank analysis of the noevaya caste spectrum of HH on voltage

2 The Results of the Pilot Higher Harmonic Components of the Voltage in the Power Supply System of Non-ferrous Metallurgy Enterprises

As a result of the energy audit was formed cluster of problems whose solution is aimed at energy savings and improve of the efficiency and reliability of equipment. QE study is based on the developed method of experimental investigation and analytical calculation of HH in PSS. Experimental study of HH and QE was carried out using the power consumption analyzer of type AR-5, integrated control unit type PKK-57 and the PKE Energotester PSS enterprises for the production of hard alloys and zinc [15–17].

Based on the HH analysis of all castes consumers of hard alloy production was constructed rank HH distribution, as shown in Fig. 1.

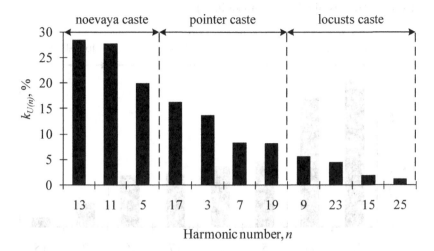

Fig. 1. Rank distributions of HH for all castes consumers of hard alloy production.

Figure 1 shows that the HH are ranked as follows: HH with $n = 13, 11, 5$ – noevaya caste; HH with $n = 17, 3, 7, 19$ – pointer caste and HH with $n = 9, 23, 15, 25$ – locusts caste.

Figure 2 shows the dependence of k_U and W on the rank of consumers ($k_U = f(\text{rank})$, $W = f(\text{rank})$) for consumers noevaya (rank 1–6) and pointer (rank 7–11) distribution of castes.

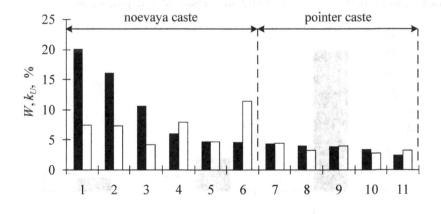

Fig. 2. Dependence of $W = f(\text{rank})$, $k_U = f(\text{rank})$.

The correlation coefficients between $k_{U(n)}$, %, k_U, % and W, % are 0.56 and 0.37 respectively.

Based on the analysis of HH for all castes consumers of zinc production was planned rank HH distribution, as shown in Fig. 3.

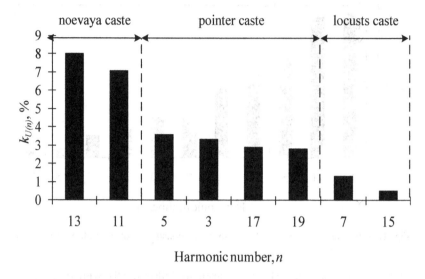

Fig. 3. Rank distributions of HH for all castes consumers of zinc production.

Figure 3 shows that the HH are ranked as follows: HH with $n = 13, 11$– noevaya caste; HH with $n = 5, 3, 17, 19$ – pointer caste and HH with $n = 7, 15$ – locusts caste.

Figure 4 shows the dependence of k_U and W from consumers rank ($k_U = f$ (Rank), $W = f$ (Rank)) for consumers noevaya (valve converters, rank 1, induction furnaces, rank 2; slip furnace, rank 3) the distribution castes of zinc production.

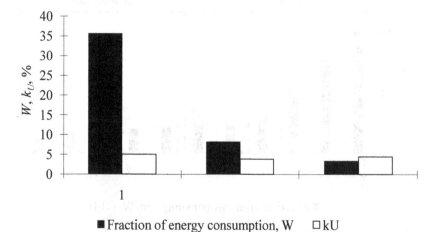

Fig. 4. Dependence of $W = f$ (Rank), $k_U = f$ (Rank).

The correlation coefficients between $k_{U(n)}$, %, k_U, % and W, % equal to 0.77.

3 Determination of the Contribution Made by Consumers in Non-sinusoidality Voltage at the Points of Common Coupling (PCC) with the System

An important and urgent task in the power quality analysis is to identify the level of emission of higher harmonic components of voltage and current in PSS of enterprise. The contribution made by consumers in non-sinusoidality voltage at the points of common coupling with the system is defined on the basis of the developed method [18–28].

The developed method is based on the active experiment transformer short-term inclusion of main step down substation for parallel operation, and allows to determine the actual contribution of the consumer (ACC) and systems (ASC) at the point of common coupling.

Measurement and calculation algorithm:

- The currents $I_{T(n)1}$ and the voltage $U_{T(n)1}$ in the normal operation of the transformer are measured (index of 1 indicates the original, the initial value of current and voltage, n - harmonic number)
- The currents $I_{T(n)2}$ and the voltage $U_{T(n)2}$ after a brief turn on the second transformer for parallel operation are measured (index 2 indicates the value of current and voltage after the switching of the second transformer)
- The resultant resistance $Z_{rr(n)}$ of the PCC when the second transformer, ACC, ASC on the calculated values of voltage changes ($\Delta U_{T(n)} = U_{T(n)2} - U_{T(n)1}$) and current ($\Delta I_{T(n)} = I_{T(n)2} - I_{T(n)1}$) are determined

The method allows to calculate the resulting resistance of n-th harmonic component of voltage $Z_{rr(n)}$ based on the reading differences of $\Delta U_{T(n)}$ and $\Delta I_{T(n)}$ of the PCC before and after the switching two transformers for parallel operation: $Z_{rr(n)} = \Delta U_{T(n)}/\Delta I_{T(n)}$. When $Z_{rr(n)} < 0$, $Z_{rr(n)} = Z_{c(n)}$ (consumer resistance). When $Z_{rr(n)} > 0$, $Z_{rr(n)} = Z_{s(n)}$ (system resistance). ACC and ASC in non-sinusoidality voltage in PCC are determined by the expressions:

$$U_{ACC(n)} = \frac{\left(U_{T(n)} + Z_{c(n)} \cdot I_{T(n)}\right)}{Z_{s(n)} + Z_{c(n)}} \cdot Z_{s(n)}; \quad U_{ASC(n)} = \frac{\left(U_{T(n)} - Z_{s(n)} \cdot I_{T(n)}\right)}{Z_{s(n)} + Z_{c(n)}} \cdot Z_{c(n)}.$$

The calculation results obtained from zinc production enterprise are shown in Table 1.

The error of calculations does not exceed 7%, which confirms the efficiency and sufficient determination accuracy for engineering calculations of ACC and ASC based on the method developed of active experiment.

Table 1. Value $Z_{s(n)}$, $Z_{calc.(n)}$, ACC and ASC on n-th harmonic component.

n	$Z_{s(n)}$, Om	$Z_{calc.(n)}$, Om	U_{ACC}, V	U_{ASC}, V
5	1.064	1.216	37.421	16.579
11	2.339	2.674	83.877	29.423
13	2.764	3.160	81.659	34.341
17	3.614	4.132	49.175	3.525
19	4.039	4.618	22.291	1.809

4 Conclusion

1. The presence of HH PSS significantly reduces the reliability and work period of electrical equipment. Thus, the reduction in the average time between failures is: for power transformers up to 10–15%, motors of up to 20%, cables up to 30–50%, capacitor banks up to 100%.
2. Reducing HH level in PSS can be provided by optimizing the circuit design of PSS and implementation of technical means, in particular narrow-band resonant filters, filter-compensated (FC) and filter-symmetrize (FS) units.
3. The method of determining the ACC and ASC in non-sinusoidality of voltage through active experiment - short-term switching power transformers MSDS in parallel mode. Relative calculation error was 6.5%.
4. The method of experimental study of amplitude-frequency characteristics of HH voltage in PSS of non-ferrous metallurgy enterprises was developed.
5. The rank analysis of the spectrum of noevaya caste voltage HH (n = 11, 13, 5) was carried out, and it was found that the correlation coefficients between $k_{U(n)}$, k_U and shares of noevaya and pointer caste electrical energy consumers (W, %) for the production of hard alloy are equal to 0.56 and 0.37 respectively; for zinc production the correlation coefficient between $k_{U(n)}$, k_U and shares of noevaya caste of electrical energy consumers is 0.77, which limits the range of the task of calculating the required filter-compensated devices.
6. Russia's accession to the World Trade Organization necessitates further research of HH in PSS of industrial enterprises and the development of methods of calculation and optimum distribution of FS in PSS non-ferrous metallurgy.

Acknowledgements. The results were reflected in the Grant of the President of the Russian Federation for support of young scientists: MK-1324.2007.8 on "Research and development of mathematical models of power quality in the non-ferrous metallurgy enterprises".

References

1. Mahela, O.P., Shaik, A.G.: Power quality improvement in distribution network using DSTATCOM with battery energy storage system. Int. J. Electr. Power Energy Syst. **83**, 229–240 (2016)

2. Hernandez, J.C., Ortega, M.J., De la Cruz, J., et al.: Guidelines for the technical assessment of harmonic, flicker and unbalance emission limits for PV-distributed generation. Electric Power Syst. Res. **81**(7), 1247–1257 (2011)
3. Norouzi, H., Abedi, S., Jamalzadeh, R., et al.: Modeling and investigation of harmonic losses in optimal power flow and power system locational marginal pricing. Energy **68**, 140–147 (2014)
4. Bhonsle, D.C., Kelkar, R.B.: Analyzing power quality issues in electric arc furnace by modeling. Energy **115**(1), 830–839 (2016)
5. Geng, Y.H.: Assessing the harmonic emission level from one particular customer. University of Liege (1992)
6. Mahmoud, A.A., Shults, R.D.: A method for analyzing harmonic distribution in A.C. power system. IEEE Trans. Power Appar. Syst. **101**, 1815–1824 (1982)
7. Forrester, W.: Networking in harmony. Electrical contractor, pp. 38–39 (1996)
8. Yang, H.G.: Assessment for Harmonics Emission Level from a Distorting Load. FSA, Belgium (1997)
9. Huddart, K.W., Brewer, G.L.: Factors influencing the harmonic impedance of power system. In: Conference on HVDC transmission, Manchester (1966)
10. Arrilaga, J., Bradley, D.: Harmonics in Electrical Systems. Energoatomizdat, Moscow (1990)
11. Zhezhelenko, I.V.: Higher Harmonics In Systems Industrial Enterprises. Energoatomizdat, Moscow (1984)
12. Zhezhelenko, I.V.: Indicators of Power Quality and Control of Industrial Plants. Energoatomizdat, Moscow (1986)
13. Lipsky, A.M.: The Quality of Electrical Energy Supply Industry. High School, Kiev (1985)
14. Kartashov, I.I., Tulsky, V.N., Chamonov, R.G., et al.: Quality Management of Electrical Energy. MEI Publishing House, Moscow (2006)
15. Klyuev, R.V., Bosikov, I.I., Youn, R.B.: Analysis of the functioning of the natural-industrial system of mining and metallurgical complex with the complexity of the geological structure. Sustain. Dev. Mountain Territ. **8**, 222–230 (2016)
16. Vasiliev, I.E., Klyuev, R.V., Vasiliev, E.I., et al.: Methods of calculating the stability of the power supply system 6 kV with a nonlinear load non-ferrous metallurgy. Audit Financ. Anal. **4**, 448–456 (2012)
17. Bosikov, I.I., Klyuev, R.V.: System analysis methods for natural and industrial system of mining and metallurgical complex. IPC IP Copanova A.Ju, Vladikavkaz (2015)
18. Du, X., Liu, Y., Wang, G., et al.: Three-phase grid voltage synchronization using sinusoidal amplitude integrator in synchronous reference frame. Int. J. Electr. Power Energy Syst. **64**, 861–872 (2015)
19. Boudebbouz, O., Boukadoum, A., Leulmi, S.: Effective apparent power definition based on sequence components for non-sinusoidal electric power quantities. Electr. Power Syst. Res. **117**, 210–218 (2014)
20. Morsi, W.G., El-Hawary, M.E.: Selection of suitable fuzzy operators for representative power factor evaluation in non-sinusoidal situations. Electr. Power Syst. Res. **81**(7), 1381–1387 (2011)
21. Kaczmarek, M.: A practical approach to evaluation of accuracy of inductive current transformer for transformation of distorted current higher harmonics. Electr. Power Syst. Res. **119**, 258–265 (2015)
22. Agrawal, S., Mohanty, S.R., Agarwal, V.: Harmonics and inter harmonics estimation of DFIG based standalone wind power system by parametric techniques. Int. J. Electr. Power Energy Syst. **67**, 52–65 (2015)

23. Bayliss, C.R., Hardy, B.J.: Power quality - harmonics in power systems. In: Transmission and Distribution Electrical Engineering, pp. 987–1012 (2012)
24. Fresner, J., Morea, F., Krenn, C., et al.: Energy efficiency in small and medium enterprises: lessons learned from 280 energy audits across Europe. J. Cleaner Prod. **142**(4), 1650–1660 (2017)
25. Tso, C.K.F., Liu, F., Liu, K.: The Influence factor analysis of comprehensive energy consumption in manufacturing enterprises. Procedia Comput. Sci. **17**, 752–758 (2013)
26. Cagno, E., Trianni, A.: Evaluating the barriers to specific industrial energy efficiency measures: an exploratory study in small and medium-sized enterprises. J. Cleaner Prod. **82**, 70–83 (2014)
27. Anisimova, T.: Analysis of the reasons of the low interest of Russian enterprises in applying the energy management system. Procedia Econ. Finan. **23**, 111–117 (2015)
28. Kluczek, A., Olszewski, P.: Energy audits in industrial processes. J. Cleaner Prod. **142**(4), 3437–3453 (2017)

Choice of Wind Turbine for Operation in Conditions of Middle Ural

A. Valtseva[✉] and K. Karamazova

Ural Federal University, 19 Mira Street, Yekaterinburg
620002, Russian Federation
a.i.valtseva@urfu.ru

Abstract. Wind energy is one of the fastest growing most developing renewable sources, however, not all areas have sufficient wind speed for the installation of powerful wind turbines. The operational experience of foreign wind power stations confirms the possibility of accurate predicting the generation of electricity for different periods, which can become the basis for resolving the issues of the need for redundancy, regulation of active and reactive power, and the need for various resources in the operation of the wind power plant. There is always a power reserve in the power system of Russia, although there are scarce power systems; coverage of the deficit occurs due to energy flows from other energy systems. In some areas there is no sustainable electricity supply, so most of the electricity is generated by diesel generators, and the cost of fuel increases because of the difficulties of delivery. The variant of placement of a small wind electric installation will allow possible to produce clean and cheap electric power, in comparison with imported diesel fuel, at relatively low operating costs. This article discusses horizontally-axial wind turbine «Wind Energy», capable of producing power at low wind speeds.

Keywords: Renewable energy · Wind energy · Small wind turbines · Wind speed

1 Introduction

According to Rosstat more than 60% of electricity is generated at thermal power plants. Wherein the efficiency of domestic stations is 32–33%. Large coal-fired power plants burn millions of tons of coal per year, it has negative effect on the environmental situation not only in the station area. At present, the main share of energy used by mankind is the chemical energy of the burning reaction of natural fuels [1]. Then the chemical energy of this reaction is converted either into mechanical (internal combustion engines) or into electrical energy (thermal power plants). The disadvantage of existing methods of energy conversion is small efficiency. Especially large energy losses occur at the stage of conversion of heat into mechanical work [2].

Technological innovations are largely aimed at more efficient use of energy. Examples are the best insulation of houses leading to a reduction in energy costs for heating [3] increasing the efficiency of engines with a corresponding increase in mileage with the same total consumption of gasoline. Wherein the increasing use of alternative energy sources reduces the need for fossil fuels for energy and industry [4]. The development and creation of more efficient energy conversion devices does not give instant results in energy savings, it takes quite a long time for their widespread adoption [5]. The reaction of the economy to innovative technologies has a considerable time inertia.

All the potential wind energy for realization over the year on the surface of the Earth is approximately equal to 1.2 · 1013 kWh (for comparison, the total consumption of all types of energy resources on Earth is ≈ 7 · 1013 kWh per year). Calculations of specialists show [6] that in most countries wind power plants could provide up to 10% of all necessary electrical power when using for this purpose up to 1% of the country's territory.

2 Wind Power: Theoretical Justification for the Application

The wind is a pure source of energy, infinitely renewable and very reliable. The development of wind energy around the world in recent years is very rapid: in 2015 wind energy has become the leading source of new renewable energy in Europe and the US and the second largest in China. In 2015 a record 63 GW of wind generation has been introduced, and at present the total installed capacity of wind power generation is 433 GW [7].

Basically new capacities were installed in countries that are not members of the economic cooperation organization led by China [8]. Wind energy remains the cheapest and most reliable source of energy from all renewable energy sources. Wind energy plays a very important role in satisfaction the electricity demand in a growing number of countries, including Denmark (42% of the demand in 2015), Germany (more than 40% in four lands), Uruguay (15.5%). The challenges for the industry include insufficient level of development of the network infrastructure and even the winding up of wind generation [9].

The energy density of the wind flow is distributed unevenly: in the lower layers of the atmosphere, it increases with increasing altitude is one of the problems of effective use of wind [10]. At an altitude of 10 m under optimal conditions for the arrangement of the wind wheel, its average value is about 300 W/m^2, while at an altitude of 50 m it is about 700 W/m^2 [11].

Power plants usually use wind in the surface layer at an altitude of 50–70 m, less often - up to 100 m from the Earth's surface, therefore the characteristics of the motion of air masses in this layer are of the greatest interest. In the future, as the corresponding technical means are created, it may be possible to use jet streams typical of the troposphere [6]. The wind always changes the speed by its nature [12] it is recommended [13] to have energy storage devices, or wind turbines are connected to the electrical network [14], which is considered by the authors as the main option for using wind farms to meet the demand for electricity at peak loads residents of developing cottage settlements [15].

Wind power plants can be divided into several classes, differentiating them according to the type of work with the network. There are individual, designed to meet small needs in the power source and not having a rigid binding to regulatory requirements related to power supplies; their power is less than 50 kW. [6] There are also hybrid windmills, working in conjunction with other comparable in power sources of energy, they are intended for the uninterrupted supply of electricity to consumers of nominal power. Wind turbines from 50 to 500 kW power refers to stand-alone installations that generate industrial power and are used to power standard sources that are not connected to a common network. Wind turbines operating in parallel with a powerful electric grid, from 200 kW to 5 MW, belong to the system (network class).

Wind turbines differ in their design and in terms of output. The larger the wind power plant, the more energy it can generate, however, not all regions have the same wind speeds and therefore the using of powerful wind turbines is territorially limited. The optimal size of wind turbines has long been the subject of discussion [16]. Machines of high power have large-scale advantages in their design and some characteristics, but they lose to small wind turbines according to technical and economic indicators [17]. Advantages of small wind turbines are their relatively small mass, the cost, flexibility in creating wind farms from a large number of wind turbines, higher reliability of the wind farm - failure of several aggregates does not have a significant effect on the power of the wind farm [18].

It is noted [6] that multi-aggregate wind power stations of smaller unit capacity are less complex machines. They allow better smoothing out the gusts of the air currents due to the spatial dispersal of individual aggregates while working for the general consumer. However, wind installations can't be closer to each other than 30 diameter of the wind wheel [6] at lower distances, the aerodynamics of the flows are significantly distorted, which can lead to a significant drop in useful power. This restriction also applies to high-power wind turbines, for example, the necessary distance of at least 35 diameters from each other is equal for units with a rotor diameter of 60 m.

3 Features of the Wind Flow in Yekaterinburg and the Proposed Solutions

According to the map of Russia's wind potential [2], the wind potential of the Sverdlovsk region is very modest - the average annual wind speed varies from 2.5 m/s at an altitude of 10 m and up to 4.5 m/s at an altitude of 50 m, so powerful wind power plants at such wind speeds can't be used [3]. The schemes of small and medium-power wind turbines operating in parallel with thermal units of approximately equal power have the greatest practical value in the near future. In this case, two options are possible: the wind farm acts as a base station or duplicates the power of the main installation.

The key task in choosing a particular scheme of wind installations on the market is determining the location of the installation, its optimal design, configuration and composition of the power supply system [19], taking into account the climatic conditions of the area of expected operation, as well as the load characteristics. The power supply system should provide a high indicator of the guaranteed power supply of the consumer in question, have acceptable dimensions and cost, high reliability, long service life and minimum maintenance costs. Compliance with the above requirements should ensure the competitiveness of such systems in comparison with traditional technical solutions - laying long line cables or in the absence of electrical networks with autonomous power from gasoline, gas and diesel electric generators.

Taking into account the experience, the introduction of small-capacity wind turbines is an advantageous measure for an agro-industrial complex that does not require large expenditures. The introduction of stand-alone small installations up to 30 kW can provide full or partial power supply to small or full-scale residential, office, small production facilities to create infrastructure in remote areas [19].

In order to know the average wind speed in the city Yekaterinburg in 2016, it was used data which include information about the wind speed for each season. The methodology of determining the energy indices of the wind energy potential is based on the regularities of geographical, altitudinal and temporal (multi-year, seasonal, daily) distributions of wind characteristics in the territory of Russia [5]. Data is presented in Table 1.

Table 1. Average annual wind speed in Yekaterinburg.

Wind speed by months, m/s	I	2,7
	II	2,8
	III	2,8
	IV	2,6
	V	2,8
	VI	3,1
	VII	2,5
	VIII	2,5
	IX	2,4
	X	2,6
	XI	2,6
	XII	2,8
Average annual wind speed, m/s		2,68

Based on the data in Table 1, the average annual wind speed in 2016 was 2.68 m/s, this wind is insignificant for wind power, and especially for horizontal-axis installations capable of generating serious power [2]. However, there is undoubted interest in the velocities of the wind flow at other heights, with increasing altitude the wind speed also increases [20].

For the accuracy of calculations is necessary to calculate the vertical wind profile, that is the law of wind shear - the estimated change in wind speed over the height above the ground [19], this dates are in Table 2 [12].

Table 2. Average wind speed at different heights in Yekaterinburg.

h, m		10	20	30	40	50
Wind speed by months, m/s	I	2,7	3,4	3,9	4,1	4,4
	II	2,8	3,5	4,0	4,3	4,4
	III	2,8	3,5	4,0	4,4	4,7
	IV	2,6	3,3	3,8	4,2	4,5
	V	2,8	3,5	4,2	4,7	5,0
	VI	3,1	3,8	4,4	4,9	5,25
	VII	2,5	3,2	3,7	4,3	4,5
	VIII	2,5	3,2	3,7	4,3	4,5
	IX	2,4	3,1	3,5	3,7	3,9
	X	2,6	3,3	3,7	3,9	4,2
	XI	2,6	3,3	3,7	3,9	4,2
	XII	2,8	3,5	3,95	4,2	4,4
Year		2,68	3,4	3,85	4,2	4,5

Based on Table 2 and on the vertical wind profile, it is possible to calculate [21] the frequency of wind speeds, which is presented in Table 3.

Table 3. Repeatability of wind speeds in Yekaterinburg at different altitudes.

h, m	10	20	30	40	50
Average annual wind speed, m/s	2,68	3,4	3,85	4,2	4,5
0,5	40	10	6	4	3
1	45	25	13	9	8
2	85	36	29	25	17
3	70	74	65	62	54
4	30	54	69	71	76
5	27	43	51	50	65
6	15	37	38	39	43
7	10	23	24	26	28
8	15	19	20	22	23
9	5	15	17	18	19
10	9	11	14	17	9
11	4	6	1	12	6
12	3	4	3	4	5
13	3	2	2	2	3
14	2	2	1	1	2
15	1	2	1	1	2
16	1	1	1	1	1
17	0	1	1	1	1

Prevailing wind speed in the highest number of days in the year is in the range from 2 m/s to 4 m/s, speed of 5 m/s is at an altitude of 20 m and 30 m for a sufficient number of days (43 and 51, respectively). The operation of the potential wind is a serious technical problem [5].

Based on the calculations, it can be concluded that the optimal height for the using of the wind power plant is 20–30 m, however the wind power plant should have a low starting speed: at these altitudes, the average annual wind speed is 3.4 to 3.85 m/s. The discussed below installation can be placed at this height [14].

To generate electricity for household using, authors recommend wind power station Energy Wind 5 kW as this wind turbine set is suitable for home and it is able to provide comfortable family living in a countryside heated house. It is important to note that this wind farm can be modified at its discretion: the height of the mast is regulated; it is possible to select an inverter designed for a large peak load. With an average wind speed of 2–4 m/s, the output will be between 300 and 400 kW per month [22]. Maximum simultaneous load is possible to connect to such an electrical system - 6 kW. With the lights on and working computers, a washing machine, a refrigerator, a TV and an electric kettle, it will allow to include, for example, a microwave oven, that is, an electrical grid can be used in fact without any restrictions.

Helium accumulators included in this set can serve 10–12 years, as in contrast to acid batteries, their resource is at least 2 times greater [23]. And also they are not afraid of complete discharge and negative temperatures.

The diagram of the Energy Wind generator is shown in Fig. 1.

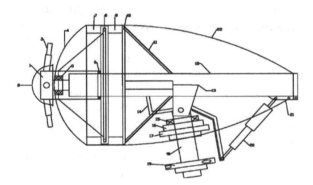

Fig. 1. Internal circuit of the generator Energy Wind: 1 – generator's shaft; 2 – fairing; 3 – blade swing; 4 – front fairing; 5 – front bearings; 6 – back bearing; 7 – front stator; 8 – rotor disk with magnets; 9 – back stator; 10 – power ring; 11 – power ring frame; 12 – generator power axis; 13 – mounting bracket of the mast pipe; 14 – emphasis; 15 – thrust bearing; 16 – upper polyurethane sleeve; 17 – shim; 18 – mast pipe; 19 – lower centering polyurethane sleeve; 20 – generator deflection damper; 21 – collar of tail assembly; 22 – back firing.

4 Conclusion

At the moment, the growth rate of the wind energy market in Russia is significantly behind countries with developed wind energy. Recently, work began on the construction of the Kola Wind Farm, a wind farm of Enel Russia, located in the Murmansk Region. Putting the Kola wind farm into operation is planned for the end of 2021. This will be the largest renewable energy project in the Arctic Circle. The wind farm will be able to produce about 750 GWh per year, while avoiding the release of about 600,000 tons of carbon dioxide into the atmosphere. The wind farm will be equipped with 57 turbines and located on a total area of 257 ha. The total installed capacity of this wind farm is 201 MV, which will undoubtedly make this station the most powerful in Russia. However, not all regions of Russia can build such powerful stations because of the low average wind speed. Considering the peculiarities of the wind flow in Russia, the authors propose the installation of low-power wind power units capable of generating electricity even at low wind speeds. Such wind turbines are needed in decentralized settlements, where there are power outages. Also, such a wind turbine can reduce the harm caused by the use of diesel engines.

The authors of the article relying on the above calculations of the wind speeds for different heights in the city of Yekaterinburg, consider that the using of this wind turbine is expedient considering its low starting speed, small dimensions and ease of installation and operation.

Given that this wind farm has its starting wind speed of 2 m/s and small dimensions, it can be confidently said that the using of the Wind Energy installation is possible within the city of Yekaterinburg and its environs [4]. This becomes actual, considering the pace of construction of new microdistricts, cottage settlements and other places where for some reason there are problems with centralized power supply. According to calculations with competent operation, the payback period of the Wind Energy wind farm will be 5 years.

References

1. Rosa, A.: Fundamentals of Renewable Energy Processes. Elsevier Inc, Stanford (2005)
2. Bezrukih, P.P., Karabanov, S.M.: The state of renewable energy in 2015. Energy Abroad **1**, 2–31 (2016)
3. Bezrukih, P.P., Karabanov, S.M.: The state of foreign renewable energy and prospects of development. Energy Abroad **6**, 2–23 (2016)
4. Nikolaev, V.G.: Resource and Feasibility Study of Large-Scale Wind Power Development. Atmograph, Moscow (2011)
5. Borisenko, M.M., Stadnik, V.V.: Atlases of Wind and Solar Climate of Russia. GGO Cojekova, St. Petersburg (1997)
6. Heping, L., Jing, S., Xuili, Q.: Empirical investigations on using wind speed volatility to estimate the operating probability and power output of wind turbines. Energy Convers. Manag. **67**, 8–17 (2013)
7. Akkoziev, I.A., Deriugina, G.V., Tjagunov, M.G., et al.: Wind turbines at the site hybrid energysystems. Part 2. Definition power of wind turbines modeling of the vertical profile and wind speed. Bull. KRSU **16**, 120–123 (2016)

8. Alehina, E.V.: The main aspects of wind energy. Izvestiya TulGU **12**, 8–12 (2013)
9. Gurov, V.I., Karinbaev, T.D., Shabarov, A.B.: New features of wind energy systems. Energy: Econ. Technol. Ecol. **5**, 32–35 (2010)
10. Alekseev, B.V.: Wind energy and its issues. Energy Abroad **5**, 31–47 (2010)
11. Ignatyev, S.G.: Development of methods for assessing wind power potential and calculation of annual performance wind turbines. Altern. Energy and Ecol. **10**(90), 49–72 (2010)
12. Nilolaev, V.G.: Effectiveness of methods to forecast wind power potential energy and economic performance of wind power stations in Russian Federation. Small Energ. **1**, 12–17 (2010)
13. Vasiljev, Y.S., Bezrukih, P.P., Elistratov, V.V., et al.: Assessment of Resources of Renewable Energy Sources in Russia. SPbGU, St. Petersburg (2008)
14. Nikolaev, V.G.: Development of technology determining potential wind energy in Russia. Sci. Tech. Statements **2**, 68–76 (2011)
15. Debiev, M.B., Popov, G.A.: System classification the factors determining the choice of locations for wind energy. Vestnik AGTU: Control Comput. Eng. Comput. Sci. **2**, 15–21 (2011)
16. Hou, P., Enevoldsen, P., Weihao, H.: Offshore wind farm repowering optimization. Appl. Energy **208**, 834–844 (2017)
17. Allen, D.J., Tomlin, A.S., Bale, C.S.: A boundary layer scaling technique for estimating near-surface wind energy using numerical weather prediction and wind map data. Appl. Energy **208**, 1246–1257 (2017)
18. Scott, S., Capuzzi, M., Laston, D.: Effects of aeroelastic tailoring on performance characteristics of wind turbine systems. Renew. Energy **114**, 887–903 (2017)
19. Turyan, K.J.: Power of wind turbines with a vertical axis rotation. Aerosp. Eng. **8**, 105–121 (1988)
20. Sabolic, D., Zupan, A., Malaric, R.: Minimization of generation variability of a group wind plants. J. Sustain. Dev. Energy Water Environ. Syst. **4**, 466–479 (2017)
21. Salim, M.H., Shuenzen, K.H., Grave, D.: Including trees in the numerical simulations of the wind flow in urban areas: should we care? J. Wind Eng. Ind. Aerodyn. **144**, 84–95 (2015)
22. Pollicino, F.: Minimizations of risks of wind power plants. Stahlbau **84**, 703–709 (2015)
23. Baranov, N.N.: Renewable Sources and Methods of Energy Conversion. MPEI, Moscow (2012)

Development Trend of Electrification and Small-Scale Power Generation Sector in Russia

K. V. Selivanov[✉]

Bauman Moscow State Technical University, 5 Vtoraya Baumanskaya Street, Moscow 105005, Russian Federation
selivanov_kv@mail.ru

Abstract. Purpose: The purpose of the research is consideration of trends and prospects of supplying electric power to remote, technically and technologically isolated regions of Russia. Classification of Russian energy consumption areas. Analysis of the current methods of supplying electric power to remote regions. Determination of the trend of changing the existing power generating facilities of remote regions. Determination of autonomous electrification regions of Russia. Juxtaposition of the effectiveness and use range of various autonomous power generation sources, in particular, alternative (renewable) power generation sources. Methods: Static data analysis. Numerical and analytical methods of electrical engineering, multi-criteria solution optimization methods. Results: A map of Russian territory electrification from the centralized power supply system and autonomous power generation sources has been made. Energy consumption areas have been classified. A model for calculation of power generation sources efficiency has been created. Dependence of the generated energy on the population with year-by-year breakdown has been shown. It has been proved that at present energy to the major part of newly electrified regions of Russia is supplied by autonomous power generating facilities. Practical significance: An allowable distance for efficient electrification of small consumers from the centralized power supply lines has been shown. A calculation model for comparing the efficiency of electrification means has been given.

Keywords: Electrification · Power supply · Power supply areas · Power lines · Alternative energy sources

1 Introduction

Electric power is a universal form of energy. The universal nature of electric power manifests itself in its ability to be transferred over great distances and easy conversion to any other types of energy, such as mechanical, thermal, light energy and any other energy form. At present energy in its unchanged state is also increasingly used in the metal industry, computing and microwave technologies.

2 Global Energy Sector Growth

Electric power is a universal form of energy. The universal nature of electric power manifests itself in its ability to be transferred over great distances and easy conversion to any other types of energy, such as mechanical, thermal, light energy and any other energy form. At present energy in its unchanged state is also increasingly used in the metal industry, computing and microwave technologies [1, 2].

Electric power is an essential resource for the economy [1]. Development and growth of the industry and global population wealth is impossible without the increase of the generated power amount. If we bring the total global amount of energy [3, 4] calculated as electric power amount into correlation with yearly population growth, we obtain the diagram shown in Fig. 1.

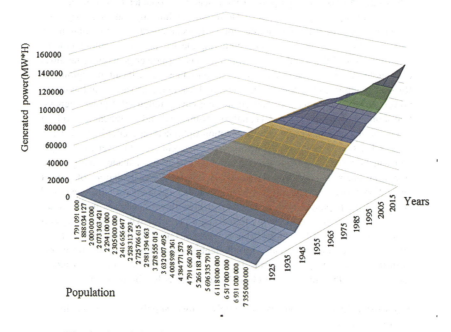

Fig. 1. Correlation between generated energy and global population.

The diagram shows that the amount of consumed energy calculated as amount of electric power keeps growing and is in direct correlation with the population growth [5, 6].

3 Electrification of Russia

Russian population growth results in exploration of new remote territories [7, 8]. Cities, production facilities and infrastructure under construction need electric power [8]. Power can be supplied to new territories either by expanding the centralized power supply system of Russia [9], or by creation of new or expanding the existing autonomous power supply areas. In course of our country electrification areas with various electrification principles emerged [2].

These power consumption areas can be conventionally divided into three major classes:

- Developed centralized (unified) power supply area
- Areas that do not belong to the common power supply system and have their own centralized energy hub
- Isolated areas with small energy hubs and (or) own small-scale power generation sources [10]
- The trends of the used power supply methods are shown in Fig. 2(a, b). It is apparent that in terms of newly electrified areas in Russia creation of autonomous power generation facilities prevails

Autonomous power consumption areas are classified by the degree of fuel and energy resources availability:

- Areas with their own major fuel and energy resources and mature fields, able to satisfy their need in electric power on their own (Khanty-Mansi, Nenets and Yamal-Nenets Autonomous Districts, Komi Republic, Murmansk and Tyumen Regions, etc.)
- Areas without their own major fuel and energy resources, but where resources are available due to well-developed infrastructure and major power lines from neighboring "donor" regions
- Areas without their own fuel and energy resources and with hindered power supply

At the first stage electrification in our country was carried out by construction of high-capacity generating power plants and their unification into the centralized power supply system of Russia, and construction of power lines. Construction of power line for transmission of energy from the centralized power supply system is currently used to a less extent, as Fig. 2 shows.

The main reason is a poor effectiveness of electrification by installation of power lines for remote low-capacity consumers. The low capacity of electrified areas is conditioned by the absence of major construction sites, power-consuming facilities and other major power consumers in the newly electrified territories. Two basic criteria at transmission line construction are remoteness and capacity of an electrification object. GOST R 54149-2010 "Power quality limits in the public power supply systems" determines the requirements to the acceptable voltage variation in power lines, which should be within ±5% [11]. The voltage deviation from the preferable value can be expressed by formulas 1 and 2:

$$\Delta U\% = \frac{P \cdot l \cdot 10^5}{c \cdot S \cdot U^2 \cdot \gamma}, \qquad (1)$$

where: ΔU is voltage deviation from the nominal value, %; P is transmission power, kW; l is transmission line length, m; c is a factor indicating power loss in power lines; S is wire section, mm^2; U is nominal transmission line voltage, V; γ is specific wire material conductivity, siemens (m/(Ohm*mm2)).

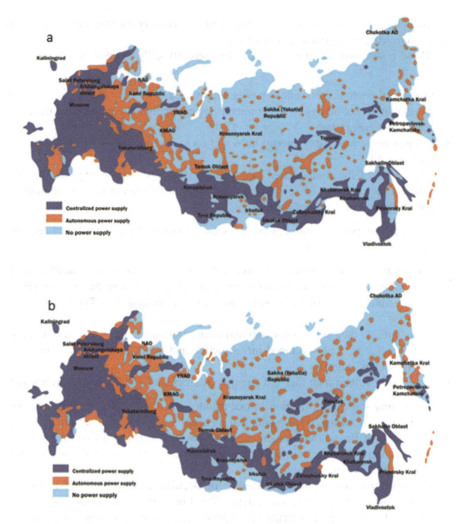

Fig. 2. Variations in the structure and capacity of electric power consumption in Russian regions: (a) – 2000 year; (b) – 2015 year.

$$l = \frac{c \cdot S \cdot \gamma \cdot U^2 \cdot \Delta U}{P \cdot 10^5}, \qquad (2)$$

The dependence of the possible length of low power lines on transmitted power is shown in the diagram (Fig. 3).

It is apparent that the use of a higher distribution voltage enables to extend transmission line length while the same load and wire section are preserved. Unfortunately, unlimited voltage increase in impossible, because in this case power lines will be operated with a low load factor, being almost idle. It is one of the reasons for increasing the share of areas electrified using the autonomous power supply method.

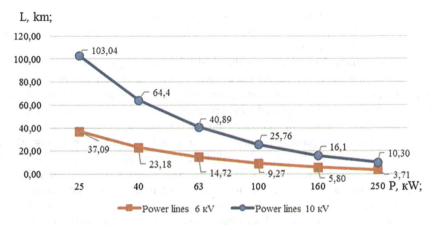

Fig. 3. Length of power lines depending on the transmitted power.

The autonomous power consumption areas are of a high industrial significance, they locate such important economy branches as agricultural sector (agricultural lands, active farming, cattle breeding, reindeer breeding, fur farming, fishery and hunting lands) and materials sector (extraction of ore, precious metals, hydrocarbons and timber harvesting). The autonomous power consumption areas are supplied with energy from small-scale power generation sources.

The small-scale power generation sector widely uses diesel power plants due to a wide range of generated power (from 2 kW to 2.5 MW); a respective capacity diesel power plant is selected depending on the required energy consumption [12]. A low specific fuel consumption per generated power unit and abundance of diesel fuel increase the cost effectiveness of the generation method. Diesel power plants advantages as opposed to other alternative power sources include a long operational life (up to 40 000 machine hours of continuous operation at the average), high mobility, easy mounting and operation, high reliability and repairability.

Despite their long successful operation life, diesel power plants have a number of operational drawbacks and difficulties. A low performance of diesel power plants, the need in professional maintenance at operation and the rising diesel fuel prices, as well as logistics costs required for their operation considerably increase the cost of the generated energy [13].

Attempts to solve the arising diesel power plant operation problems found expression in the development of alternative power generation methods and their application as small-scale power generation sources. Among most widespread alternative power sources there are solar batteries, small-scale hydraulic power plants on mountain rivers, wind electric power generators, tide power plants, biofuel driven generators, etc. None of the above has enjoyed wide distribution and occupied a substantial share in the total amount of generated energy [14].

The main drawback of the majority of alternative power generation means is the need in binding power plants to power generation sources. Whereas diesel power plants are easy to transport to a power consumption place together with the fuel, wind, tide, hydraulic and other power plants can be located only in certain places, which are normally far away from the consumers. The generated power is not constant and depends on external factors.

Foreign countries' experience in the field of solar energy conversion to electric power has shown that at present the use of solar batteries as electric power generators is inefficient and unprofitable; in some cases, solar batteries, having exhausted their lifespan and dismantled or replaced with new ones, appear not to have paid off in terms of the equivalent amount of generated power. Despite the unprofitability of solar energy generation, at present a number of prominent scientists including Zhores Alferov – a Nobel Prize Winner, believe that the future of the power generation sector is solar energy conversion. The solar energy generation sector will develop and in due course will be the main power generation sector; even today its peak performance may reach 40%, whereas at the beginning of solar energy generation research in 1954 the performance was 6%. Today in Russia the share of energy generated by conversion of solar energy is next to nothing.

Analysis of power supply means of remote and isolated consumers and small-scale power generation sources has shown that at present the main share of power generation in the sector is ensured by diesel power plants. The increase of diesel power plant performance, assurance of the quality of generated power and reliability of power plants as main generators of the small-scale power generation sector are the aspects to consider for industrial production and scientific thought.

4 Electrification Efficiency Benchmarks

To determine power generation efficiency, it is necessary to distinguish the factors, efficiency depends on, and connect them with the formula.

Power supply efficiency can be expressed with the following formula (3):

$$E_{el.} = \frac{P - P_{not\,cons.}}{(V_{non-ren.\,res.} \times k) + N_{depr.} + C_{cost\,of\,the\,service}} \quad (3)$$

In the formula E_d is power supply efficiency (a non-dimensional value applied for comparison of various power supply means); P is generated power (kW, MW); $P_{not\,cons}$ is not consumed power (kW, MW); $V_{non-ren.\,res.}$ is the amount of non-renewable resources spent on power generation (TFOE); k is a factor to be selected depending on the non-renewable resource type; N_{depr} is amortization (rub); $C_{cost\,of\,the\,service}$ is maintenance cost at operation (rub.).

It is apparent that the methods of power generation of the cheapest fuel with the minimum investments to generating facilities are most efficient. Another necessary factor of efficient power generation is full or next to full consumption of the generated power, since in case of untimely generation a high efficiency is not achieved, and the funds spent on the generation are wasted. If alternative power sources are used, the amount of non-renewable resources spent on the generation of unused energy is zero, but amortization and maintenance cost remains the same, and these costs normally exceed similar deductions for typical power plants (diesel, thermal power plants, diesel generators, etc.). To supply energy to remote low and medium capacity consumers, small-scale power generation facilities are normally used, because they do not require a costly construction of power lines and due to a small number of required transformers and the possibility to adapt to locally available energy resources, including renewable energy resources. The use of renewable energy sources with next to zero cost of energy resources in combination with considerable capital intensity reduction due to a smaller distance between power generation means and the consumers makes small-scale power generation means highly efficient. Another essential factor is capital intensity required to arrange power supply to a certain electrification object. The capital intensity includes construction of generation facilities, construction of power lines, installation of transformers (step up, step down, distribution transformers) and, certainly, operation cost. To supply energy to remote low and medium capacity consumers, it is usually more profitable to use small-scale power generation sources due to a lower capital intensity required for power lines construction, a small number of transformers and the option of adaptation to the use of locally available energy resources, including renewable energy resources. The use of renewable energy sources with next to zero cost of energy resources in combination with considerable capital intensity reduction due to a smaller distance between power generation means and the consumers makes small-scale power generation means highly efficient.

5 Conclusion

At present power to remote regions of Russia is supplied from autonomous power sources, with their share being over 8% of the total amount of power generated in Russia. The use of alternative power generation means is possible in case of their hybridization and coordination with standard power generation methods in isolated areas. These are primarily diesel and gas power generators. The first rule of electrical engineering says that electric power must be generated at the moment it is consumed, which is impossible through the use of alternative power sources only, since their output depends on external factors.

In the future the share of autonomous power generation methods will grow, and the major part of such power sources will be operated on typical fuels (oil, gas, coal); however, the proper hybridization of the standard and alternative power generation methods can considerably improve power generation volume and efficiency.

References

1. Elistratov, V.V.: Renewable Power Generation. Politechnical University Publishers, St. Petersburg (2016)
2. Vainzikher, B.F.: Power Engineering in Russia 2030, target edn. Alpina Business Books, Moscow (2008)
3. Federal Law 35-FZ dated 26.03.2003 (as amended on 29.07.2017). "On the electric power industry"

4. Makarov, A.A., Fortov, V.E.: Tendencies of the development of world energy energy strategy of Russia. Bull. RAN **3**, 195–208 (2004)
5. Surinov, A.E.: Statistical Yearbook of Russia. Rosstat, Moscow (2002)
6. Surinov, A.E.: Statistical Yearbook of Russia. Rosstat, Moscow (2015)
7. Bystritsky, G.F.: General Energy (Thermal and Electric Power Generation). KNORUS, Moscow (2014)
8. Selivanov, K.V.: Small-scale power generation as a means to ensure energy security of Russia forest complex today. View of young researchers. Moscow State Forest University **01**, 217–220 (2017)
9. Handbook on renewable energy resources of Russia and local fuels. Indicators for territories: reference book. Energy Publishing and Analytical Center, Moscow
10. Selivanov, K.V.: Analysis of methods of small distributed power supply. Int. Res. J. **01**(55), 107–110 (2017)
11. GOST R 54149: Power quality limits in the public power supply systems (2010)
12. Tarlakov, Ya.V.: Performance of forest complex diesel power plants operated on biofuel. Dissertation, Moscow state forest university, Moscow (2013)
13. BP Statistical review of world energy (2016) https://www.bp.com. Accessed 20 May 2018
14. Elistratov, V.V.: The present state and trends of development of research in the world. Int. J. Altern. Energy Ecol. **1–3**, 84–100 (2017)

Principles of Energy Conversion in Thermal Transformer Based on Renewable Energy Sources

Y. A. Perekopnaya[✉], K. V. Osintsev, and E. V. Toropov

South Ural State University, 76 Lenina Avenue,
Chelyabinsk 454080, Russian Federation
k.perekopnaya@gmail.com

Abstract. The method of cooling and conditioning of industrial rooms is proposed, combined with the electricity generation for the company's own needs. The developments are based on the basic and fundamental theories of heat and mass transfer, as well as on the principles of the action of centrifugal electric drive and power generating aggregates. The installation which implements the scheme in conditions of high solar radiation intensity can be used to reduce the enterprise's energy dependence from electrical supply from the central distribution networks. Two options for the installation are considered: in the daytime and at night. In addition, in peak modes of electrical generation of the plant, depending on its capacity, the excess of generated capacities can be given to external networks. Which definitely shows its economic benefits. Technological solutions used by the authors are applied for the first time and have a commercial perspective in the countries of Central and East Asia, and also in the Middle East and North Africa.

Keywords: Solar panel · Refrigerant · Low-boiling liquid · Turboexpander · Compressor · Industrial air conditioning

1 Introduction

In climatic conditions with average annual temperatures of ambient air above zero, a number of difficulties arise in enterprises, for example, in the construction, metallurgical, oil and gas industry and agro-industry. Energy systems of human life support in an environment with extremely high heat engineering and low humidity parameters should be reliable in operation. Developments in the field of improving reliability are known [1]. However, there are factors affecting the efficiency of an industrial enterprise as a whole. For example, the issues of energy and resource saving can be considered from different points of view. In particular, it is possible to create a reliable, efficient energy-technological complex based on renewable energy sources, which simultaneously generates cold and electricity for the enterprise's own needs.

2 Domestic and World Analogues

It is known a method of generating coldness for industrial enterprises on the basis of drafts and compressor aggregates, which includes the heating line and evaporation of the refrigerant when it passes through the heat exchanger. In this case, the chilled air enters the rooms and provides the working conditions for the personnel of the enterprise [2]. The disadvantage of the method is an increased electrical consumption of the fan drive and compressors, which significantly reduces the efficiency of the installation implementing this method.

It is known a method of generating electric power for industrial [3] enterprises in an installation including an energy steam generator, where a low boiling point liquid can be a heat carrier. The vapor of the liquid is fed to the turbo-electric generator, where the liquid condenses and returns to the thermal circuit [4]. The disadvantage of the method is the low reliability of the steam generator and the increased costs of carrying out major repairs. The closest analogue is the method of combined generation [5] of electric power, heat and cold in the trigeneration cycle [6] of a hybrid thermal power plant for industrial enterprises [7], which including a gas piston or micro-gas turbine plant [8], a peak water boiler and an absorption refrigeration machine. The advantage of this combining, surely, is to reduce the cost of generating heat and cold, and resource saving of organic fuels. The disadvantage of the method is still the same - increased power consumption from the external network, thereby reducing the efficiency of the entire hybrid thermal power plant [9].

3 Study Objectives

The technical task of the research is the creation of an energy-technological complex based on renewable energy sources by the combined production of cold and electric power for own needs, the principles of its operation are based on the fundamental theories of heat and mass transfer, and also on the principles of the action of centrifugal electric drive and electric generating aggregates.

4 Principles of the Method Implementation

For solve of the task considered two options: day and night time. At the daytime with a high intensity of solar [10] radiation the low-boiling heat-carrier, according to the laws of heat exchange during free convection, heats up and, with the using of vertical pipes, rises upwards due to the difference of densities between liquid and vapor [11]. Vertical pipes are made with fins and assembled in panels. The shape of the cross-section and its size depend on the type of heat carrier and its mass flow, the intensity of solar radiation. The height of the pipes depends on the design of the prefabricated panels. In ideal case in the horizontal plane of the panels are installed along the perimeter of the production room, and the height of the panels coincides with the height of the room. At the bottom and at the top of the panels are installed manifold collectors of liquid and saturated wet steam, respectively. From the upper collectors, a saturated wet steam rises through the bypass pipes to the drum set at the upper mark of the room. The drum is equipped with a louvered separator in which the moisture separates from the steam. Moisture excess returns to the lower collectors, and dry saturated steam leaves the top of the drum and is sent to the turboexpander [1]. The potential energy of the steam goes into the mechanical energy of the shaft rotation, and then into the electrical energy in the electric generator, which is connected to the turboexpander shaft by a transition clutch. After the turbo-expander, a condensate forms, which is sent to the lower collectors of the panels. The shaft of the turboexpander is connected through a coupling with the shaft of a centrifugal compressor in the process scheme of cold supply. In this scheme the heated atmospheric air from the chilled room or production hall is taken up by the fan and sent to the heat exchanger-evaporator [12] of the refrigerant (for example, freon, carbon dioxide, ammonia). In the heat exchanger, the liquid refrigerant compressed in the compressor evaporates and absorbs a portion of the air warmth. At the same time, the air is cooled down to the necessary temperature for the working conditions or the operating conditions of the production halls. The economic benefit of operating the equipment according to the proposed method in the daytime is obvious, since the electric power to the centrifugal compressor drive is not expended, and the compressor works due to mechanical connection with the turbo-expander [13]. It should be noted that depending on the mass flow in the external power circuit of the turboexpander [14] in the daytime may be more necessary in the internal circuit. In this case, electricity excess is released to the external network.

In the transitional period, when the intensity of solar radiation decreases, the electric drive of the centrifugal compressor is partially activated. And even in this case there is an economic benefit remains from the application of two circuits: external and internal.

At night time, in the absence of solar radiation in the thermal spectrum, the electric [15] drive operates at full power. It should also be noted that the necessity of the compressor operation at the nominal mode is mandatory only for the case of constant cooling of the production rooms. In the case of ensuring a temperature acceptable in the conditions of labor protection in the rooms for the work of the personnel of the enterprise and under the condition of a one- or two-shift technological cycle without the constant presence of personnel in workplaces at night, the compressor may be switched off by the automation system or not worked at full capacity.

5 Scientific Novelty of the Method

For the first time the authors proposed a method for operating an energy technology complex based on renewable energy sources for a combined production of cold and electric power for own needs. The principal technological scheme of the energy technological complex has been developed.

6 The Research Methods

The research methods are based on fundamental theories of heat and mass transfer, and also on the gas dynamic theory of the centrifugal electric drive operation and electricity generating machines of aggregates. The calculations were carried out according to the normative documents approved by the Ministry of Energy.

7 The Scheme for Implementing the Method of Energy Conversion in a Thermal Transformer Based on Renewable Energy Sources

For Consider the principle of the energy-technological complex based on renewable energy sources, Fig. 1.

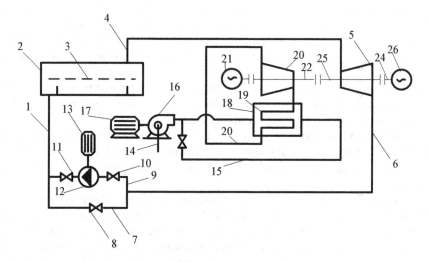

Fig. 1. Scheme of the proposed method for combined generation of cold and electricity based on renewable energy sources (designations hereinafter).

The low-boiling heat-transfer medium circulates in the panel system of the tubes 1 operating on the principle of a solar collector. Panels of the pipe system 1 are installed along the vertical walls of the production room. At the daytime, the heat carrier rises due the difference in density due the arising forces of natural convection by heating the electromagnetic (thermal) radiation of the sun. Drum 2 is installed on the building roof, it separates the droplets of the heat carrier from the steam in the separator 3. A dry saturated steam 4 is formed, it is fed to the turboexpander 5. The turboexpander shaft 25 is connected to the electric generator shaft 26 that generates electricity for the company's own needs. After the turbo-expander 5 steam is condensed and returns again to the cycle via conduit 6 as a condensate. During the daytime, the condensate is sent through the open valve 8 through the pipeline 7. At the night time the pump 12 turns on with the electric drive 13 with the open valves 10 and 11. During the transient time, the pump 13 operates not in the nominal mode due to the use of a frequency-controlled electric drive 13. The industrial air conditioning scheme consists of two circuits: a cooling air circuit 15 and a heated refrigerant [16] circuit 20. The air from the atmosphere 14 is pumped by a fan 16 with an electric drive 17 and supplied to a heat exchanger 18. The heat exchanger 18 consists of tubes 19 in which freon is heated. The freon vapor is compressed in the compressor 20 with the electric drive 21. The compressor shaft 22 is connected to the turboexpander 5 shaft 25.

8 The Basic Calculation Dependencies

The operation principle of vertical pipe panels is based on thermogravitational convection, which is the main type of free convection.

The heat exchange problem can be viewed from two sides: from the outside of the panels and from the inside of the vertical pipes. First of all, we determine the parameters that determine the heat exchange inside [17] the pipes:

$$Ar = \frac{g l_0^3 (\rho_0 - \rho)}{\rho_0 v^2} \quad (1)$$

where g – the gravitational acceleration, m/s^2; ρ_0 – the density of the liquid, kg/m^3; ρ – the vapor density of the liquid, kg/m^3; l_0 – the determining size (in our case, height), m; v – the kinematic viscosity coefficient, m^2/c.

Thus, with constant properties of the coolant [18], the height of the pipe panels becomes the determining size.

At the next, consider free convection and radiation from the outside. In the presence of sufficient wind forces and temperature differences, it is necessary to take into account the average heat transfer from the side of the vertical wall:

$$\overline{Nu_i} = \left(\frac{0,4 \Pr}{1 + 2\sqrt{\Pr} + 2\Pr} \right)^{0,25} Ra^{0,25} \quad (2)$$

where $Ra = Gr \cdot Pr$, i.e, the dependence is determined by the physical properties of the air and the height of the vertical wall.

The nature of the spectrum is taken into account when calculating the spectral radiation density according to Planck's law:

$$E_{0\lambda} = \frac{c_1}{\lambda^5} \frac{1}{\exp\left(\frac{c_2}{\lambda T}\right) - 1} \quad (3)$$

λ – the wavelength, m; c_1 – constant, $c_1 = 2\pi h c_0 = 3{,}74 \cdot 10^{-16}$ W·m^2; c_2 – constant, $c_2 = h c_0/k = 1{,}44 \cdot 10^{-2}$ m·K; $k = 1{,}38 \cdot 10^{-23}$ J/K; T – temperature, K.

Compression and expansion [19] in the compressor and turboexpander are respectively described by equations:

$$N_1 = \left| \frac{n_1}{n_1 - 1} (p_1 V_1 - p_2 V_2) \right| \tag{4}$$

$$N_2 = \left| \frac{n_2}{n_2 - 1} (p_3 V_3 - p_4 V_4) \right| \tag{5}$$

where p_1, V_1, p_2, V_2 – the refrigerant parameters at the inlet and outlet of the compressor; Pa, m^3; p_3, V_3, p_4, V_4 – parameters of low-boiling heat-transfer agent at the inlet and outlet of the turboexpander; Pa, m^3; n_1, n_2 – indicators of polytropic processes of compression and expansion in the internal and external circuits.

Savings from the application of an external heating heat carrier circuit in vertical solar panels are estimated from the equation:

$$\Delta N = |N_1 - N_2| \tag{6}$$

9 The Scheme for Implementing the Method of Energy Conversion in a Thermal Transformer Based on Renewable Energy Sources

The practical importance of the operation of the equipment according to the proposed method during the day is essential. Electricity is not expended on the drive of the centrifugal compressor, and the compressor works by mechanical connection with the turbo-expander [14].

It should be noted that depending on the mass flow in the external power circuit of the turboexpander in the daytime may be more necessary in the internal circuit. In this case, electricity excess is released to the external electrical network.

10 Conclusion

The technical task of the research was solved. The scheme of the energy-technological complex based on renewable energy sources for combined production of cold and electric power for own needs is developed, the principles of its operation are based on the fundamental theories of heat and mass transfer, and also on the principles of the action of centrifugal electric drive and electric generating aggregates.

In addition, in peak modes of power generation of the plant, depending on its capacity, the excess of generated capacities can be delivered to external networks. Technological solutions used by the authors are applied for the first time, have a commercial perspective.

References

1. Sa, C., Chen, L., Hou, L.: Effect of impeller blade profile on the cryogenic two-phase turbo-expander performance. Appl. Thermal Eng. **126**, 884–891 (2017)
2. Roman, R., Hernandez, J.I.: Performance of ejector cooling systems using low ecological impact refrigerant. Int. J. Refrig. **34**(7), 1707–1716 (2011)
3. Borunda, M., Jaramillo, O.A., Dorantes, R., et al.: Organic Rankine cycle coupling with a parabolic trough solar power plant for cogeneration and industrial processes. Renew. Energy **86**, 651–663 (2016)
4. Verde, G.: Thermal operating machine. Int. J. Mech. Prod. Eng. Res. Dev. **7**(5), 311–322 (2017)
5. Kalogirou, S.A.: Solar thermoelectric power generation in cyprus: selection of the best sys. Renew. Energy **49**, 278–281 (2013)
6. Askari, B., Sadegh, O., Ameri, M.: Energy management and economics of a trigeneration system considering the effect of solar PV, solar collector and fuel price. Energy. Sustain. Dev. **26**, 43–55 (2015)
7. Besagni, G., Mereu, R., Inzoli, F.: Ejector refrigeration: a comprehensive review. Renew. Sustain. Energy Rev. **53**, 373–407 (2016)
8. Wang, Y., Lior, N.: Performance analysis of combined humidified gas turbine power generation and multi-effect thermal vapor compression desalination systems. Desalination **196**(1–3), 84–104 (2006)
9. Cocco, D., Cau, G.: Energy and economic analysis of concentrating solar power plants based on parabolic trough and linear Fresnel collectors. Proc. Inst. Mech. Eng. Part A: J. Power Energy **229**(6), 677–688 (2015)
10. Pugsley, A., Zacharopoulos, A., Mondol, J.D., et al.: Global applicability of solar desalination. Renew. Energy **88**, 200–219 (2016)
11. Verde, G.: A novel configuration layout for a vapor compression reverse cycle. J. Fundam. Appl. Sci. **9**, 1211–1224 (2017)
12. Chavez, E.A., Vorobiev, Y., Bulat, L.P.: Solar hybrid systems with thermoelectric generators. Sol. Energy **86**(1), 369–378 (2012)
13. Galoppi, G., Secchi, R., Ferrari, L., et al.: Radial piston expander as a throttling valve in a heat pump: focus on the 2-phase expansion. Int. J. Refrig **82**, 273–282 (2017)
14. Ferrara, G., Ferrari, L., Fiaschi, D., et al.: Energy recovery by means of a radial piston expander in a CO2 refrigeration system. Int. J. Refrig **72**, 147–155 (2016)
15. Chen, J.: Thermodynamic analysis of a solar-driven thermoelectric generator. J. Appl. Phys. **79**(5), 2717–2721 (1996)
16. Saidur, R., Masjuki, H.H., Hasanuzzaman, M.: Performance investigation of a solar powered thermoelectric refrigerator. Int. J. Mech. Mater. Eng. **3**(1), 7–16 (2008)
17. Dai, Y.J., Hu, H.M., Ge, T.S., et al.: Investigation on a mini-CPC hybrid solar thermoelectric generator unit. Renew. Energy **92**, 83–94 (2016)
18. Sixsmith, H., Valenzuela et al.: Small turbo-brayton cryocoolers. In: Advances in Cryogenic Engineering. vol. 33, pp. 827–836 (1988)
19. Ino, N., Machida, A., Tsugawa, K., et al.: Development of externally pressurized thrust bearing for high expansion ratio expander. In: Advances in Cryogenic Engineering. vol. 37, pp. 817–825 (1991)

Poultry Wastes as Source of Renewable Energy

M. V. Zapevalov[1], N. S. Sergeev[1(✉)], and Yu. B. Chetyrkin[2]

[1] South Ural State Agrarian University, 75 Lenina Avenue,
Chelyabinsk 454080, Russian Federation
s.n.st@mail.ru
[2] South Ural State University, 76 Lenina Avenue,
Chelyabinsk 454080, Russian Federation

Abstract. The study was conducted with the objective of efficiency raising of poultry manure utilization. In order to achieve this objective, the following actions have to be done: developing technological regulations for manure processing with the production of fuel bricks and organomineral fertilizer; doing research and development works on creation of fuel bricks and organomineral fertilizer production lines, and on manure processing workflow for the large-scale implementation. Manure processing complex is designed and backed up. It includes manure collection tank where manure gets from poultry houses and then gradually proceeds to the further processing; liquid collection tank where liquid goes during manure dehydration; fuel bricks and organomineral fertilizer production lines. With manure moisture content of 65–70%, the productiveness will be around 5 tons per hour. If the complex works twenty-four – seven, it is proved that around 43000 tons will be processed a year. Thus, the production of organomineral fertilizer with up to 32% nutrients concentration will be around 1500 tons, while the production of fuel bricks with the calorific value up to 18 MJ/kg will be around 8000 tons. An estimated annual economic benefit from the manure processing complex is around 20 million rubles; the repayment period is not more than 1.5 years. If the cost of fertilizer is 5000 rubles per ton, in which 320 kg is an active ingredient, it will be cost-efficient to use it for cropping. This manure-utilizing technology gives a company energy-saving, agrochemical, economical, ecological and social benefits.

Keywords: Technology · Poultry manure · Processing complex · Organomineral fertilizer · Fuel bricks · Fertilizer characteristics · Processing efficiency

1 Introduction

By the beginning of the 21st century, the world has intensified the research on looking for new, alternative sources of energy. In particular, the possibility of using solar radiation, wind energy, water energy of small rivers and reservoirs, geothermal energy, biomass energy, etc. Now the annual economic potential of alternative energy sources exceeds production volume of all organic fuel types and constitutes within 200 bln tons of conventional fuel. According to the analysts' projections, the share of alternative energy sources in total volume consumption will be up to 21% and equal with the share of natural gas use by 2030. At the same time, much attention is given to the use of biomass, first of all, it refers to all wastes of woodworking, pulp and paper industry, utilities and agricultural sector.

All manufactures produce wastes. However, the higher production standards are, the more attention is paid to utilization. Utilization is waste processing in order to get new in-demand product. Agriculture is the main source of organic wastes, including agricultural plant residue, animal manure, poultry manure, etc. Chelyabinsk Oblast is in top three Russian regions in terms of eggs and poultry production. As a result of the existing poultry farms modernization and setting up new farms, every year the poultry stock is increasing. However, it also leads to the amount of poultry manure increase, the annual output of which is close to 1.5 million tons. If poultry is kept in cages, manure moisture content is about 65–70%. In this case, manure is classified as hazardous substance of the 3rd class, so it demands a particular approach to its utilization [1, 2]. Nowadays, the standards of eggs and meat production at poultry farms are very high. At the same time, in most cases poultry manure utilization technology doesn't meet modern ecological and economic requirements. Manure storage facilities are often filled with hundreds thousands tons of hazardous substance, causing harm to the environment. If such poultry manure is used as fertilizer, expenditures, connected with its application and transportation to long distances, are not paid back with yield increase, as nutrient content of such manure is very low [3–7].

Crop farming demands effective fertilizer. It is also necessary that this fertilizer has low price, high nutrient content, long period of activity and could be applied with the existing equipment [8].

Burning issue of agriculture is providing energy. It is well-known that the cost of agricultural goods production mostly consists of energy costs due to expensive energy materials. Electricity, oil-products, natural gas and solid fuel are getting more and more expensive, which leads to the production costs increase, and, as a result, all agricultural goods production becomes unprofitable. With a large amount of organic wastes it would be possible to produce solid and gas fuel for agriculture's own needs.

The research objective is to increase poultry manure utilization efficiency. In order to accomplish the objective, the following tasks were set and solved.

- Developing technological regulations for poultry manure advanced non-waste processing with production of fuel bricks and complex organomineral fertilizer
- Doing research and development works and setting up a production line for fuel bricks and organomineral fertilizer production
- Working over manure processing workflow for the large-scale implementation

2 Materials and Methods

Modern level of agriculture development, both in Russia and abroad, requires new directions on setting-up non-waste production with high sanitary-veterinary level. For further crop residues application, animal and poultry manure, in order to preserve the main nutrients in them and to improve physical and chemical features, available for the plant processing techniques are applied. All known methods of processing can be divided into four separate methods: biological, physical, chemical and mixed (Fig. 1). The mentioned methods contain 15 ways. Each of the ways can be implemented in 2–3 different technologies. Thus, at present time, more than thirty different technologies of livestock and poultry waste processing are known [9–11].

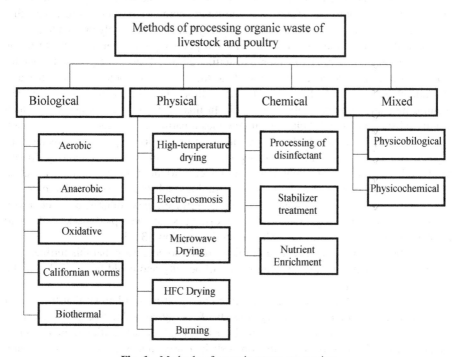

Fig. 1. Methods of organic waste processing.

The largest number of organic waste processing technologies are based on the biological method [12]. A deeper understanding of the factors influencing processes in industrial methods of biological processing, allows taking into account technological peculiarities of choice and observance of optimum modes and parameters of equipment operation.

The biological method of processing is based on aerobic [13], anaerobic, oxidative and biothermal methods, as well as the method of using Californian worms.

The same method can be performed by different technologies, indoors or directly on open platforms. If we consider that biological processes are based on the laws of microbiology, their application in the technology of livestock and poultry manure utilization allows better and more effective way to use the biological processes patterns of different components biofermentation in which the main ingredient is organic matter.

Without knowing the fundamentals and details of composting technology, in a biothermal way of processing organic compounds, the fermentation process can lead to unknown and even negative results. Instead of valuable organic fertilizers, at best, a ballast material can be obtained, and at worst - significant amounts of additional ecologically hazardous waste with a large number of different weeds seeds and a pathogenic microflora seedbed.

One of the rational methods of processing organic waste is physical, which is based on changes in physicomechanical features, in particular, on the removal of moisture content [14–16]. For this purpose, various methods of drying are applied. High-temperature drying is the most noteworthy. Drying is made in the dryers of a drum type at a temperature of 800–900 °C. During thermal processing of livestock or poultry manure, it is dehydrated to the moisture content of 12–14%, decontamination from pathogenic bacteria, viruses, helminth eggs, release from viable seeds of weeds, fluff and feathers. As a result, there is a dried substance containing essential nutrients (NPK), which can be used as organic fertilizer. However, due to the fact that this fertilizer contains a small number of nutrients (3–4%), it is required to make larger doses, which leads to increase in transport costs in order to provide the plants with the necessary quantities. The cost of using such fertilizer does not pay off the increase in yield of cultivated culture.

In order to solve the problem of poultry manual utilization and provide crop-growing farms with effective fertilizers, the technology of poultry manure processing with the production of complex organomineral fertilizer and fuel bricks has been developed be authors.

The complex consists of manure collection tank where manure gets from poultry houses and then gradually proceeds to the further processing; liquid collection tank where liquid goes during manure dehydration, fuel bricks and organomineral fertilizer production lines. All production equipment is united into one production process.

After poultry manure is removed from poultry houses, it is loaded into the dump carrier vehicle, then proceeds to the processing complex, and transferred to the manure collection tank. From the collection tank 3 tons of manure per hour proceed to the dehydration tank and 2 tons per hour proceed to the gasification tank. Manure dehydration results in 1200 kg of dried manure with moisture content up to 15% per hour. Dried manure is cooled down, chopped and delivered to the collection tank. A part of

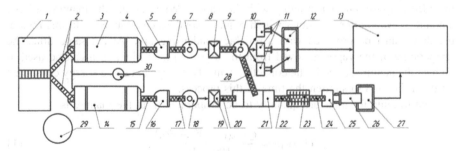

Fig. 2. The structure of production line of the poultry manure processing complex: 1 – manure collection tank; 2 – scraping conveyor; 3 – manure dehydration tank; 4, 6, 9, 15, 17, 20, 22, 24, 28 – conveyor; 5, 16 – cooling station; 7, 10, 18 – collection tank; 8, 19 – centrifugal rotor chopper; 11 – briquetting machine; 12 – bricks packing table; 13 – storage; 14 – manure gasification tank; 21 – batcher; 23 – mixing tool; 25 – granulating machine; 26 – conveyor belt; 27 – granules packing table; 29 – water tank; 30 – gas collector.

dried chopped manure is used in organomineral fertilizer production, another part in fuel bricks production. Finally, fuel bricks are packed and delivered to the storage. As a result of manure pyrolytic action, burning gas appears. A part of the gas is used for pyrolytic action maintenance, another part is delivered to manure dehydration tank. Pyrolytic action results in 160 kg of ash per hour. Ash is cooled down, chopped and mixed with dried manure. End organomineral product consists of 25% of dried manure and 75% of ash. This mixture is granulated, packed and delivered to the storage.

Figure 2 below shows the structure of production line of the liquid manure processing complex with the productiveness of 5 tons per hour.

3 The Results and Debating the Results

The proposed technology of manure processing in the first stage involves removal of moisture content from it by drying method.

This occurs as the result of thermal and mass transfer processes combinations, flowing both at the surface and inside the wet manure, contributing to its dehydration from 70–75% to 14–15%. As a result, structural-mechanical, chemical, biochemical and rheological changes occur in the manure. The speed of these changes and the degree of their completion depend on many factors. The main factors are the way of heat supply to the dried material and drying mode. Since manure contains a high percentage of organic matter, its drying prevents the flow of unwanted phenomena, such as firefanging, decomposition with the formation and emissions of ammonia, oilseeds compounds, etc.

Poultry manure is paste disperse material, which means convection or contact drying must be applied to it. The most common way is convection drying, when dryable material is heated by gaseous drying agent (heated air, fuel gases or their mixture), which directly contacts the manure. The same agent is used to remove water vapor out of the dryer. The speed of manure drying by gaseous agent depends on outer and inner heat and mass exchange intensity, as moisture content on the evaporation surface depends on them. With this way, some of the heat energy is not used for its intended purpose, but is lost along with the vapors. When drying the manure, a large amount of evaporating moisture, which requires condensation, appears. In this regard, the dryer must be provided with a reliable condensing unit. At the same time, the temperature of the vapor removed from the dryer is rather high, so it can be used to preheat the manure stored in the collection tank. This helps to reduce energy consumed on its drying and to shorten the time of this process.

The formula for moisture quantity to be evaporated from the manure is:

$$Q_{WEVAP} = \frac{Q_{MDAY} \cdot (W_I - W_D)}{100 - W_D}, \qquad (1)$$

in which Q_{WEVAP} – water evaporated per day, tons; Q_{MDAY} – manure processed by the dryer per day, tons; W_I – initial manure moisture content, %; W_D – dried manure moisture content, %.

For reasons of expediency, the dryer should work around the clock. With its hourly capacity of 3.0 t/h, 72 tons of wet manure will be processed per day, while about 51.0 tons of water will evaporate. This will result in 21 tons of dry manure per day.

The process of liquid manure drying is very power-intensive. According to the research of Russian National Scientific Research and Technological Institute of Poultry Farming, the heat capacity of dry poultry manure is about 0.42 kcal/kg.deg. The heat consumption for evaporation of 1 kg of moisture depends on the heat content and moisture content of the air.

At an air temperature before heating in the dryer $T_{A\ INP} == 20$ °C, the moisture content is $P_A = 60\%$, the heat content of the air is $I_{A\ INP} = 10.25$ kcal/kg, the moisture content of the air input in the dryer is $d_{A\ INP} = 0.01$ kg/kg.

The heat content of dry air heated to a temperature of $T_H == 850$ °C is $I_{A\ H-D} = 230$ kcal/kg, the moisture content will not change and will be $d_{A\ H-D} = 0.01$ kg/kg.

On exit from the drying drum, the dry air heat content is $I_{A\ EX} = 230$ kcal/kg, and its moisture content is $d_{A\ H-D} = 0.308$ kg/kg.

As the drying process includes heating the material and evaporation of moisture, the heat consumption for evaporation of 1 kg of moisture will be the following:

$$q_{EVAP-D} = \frac{I_{A\ EX} - I_{A\ INP}}{d - d_{A\ INP}}, \qquad (2)$$

in which q_{EVAP-D} is the amount of heat, kcal/kg.

Taking into account the heat and moisture content of the air before and after heating in the dryer, the amount of heat necessary for evaporation of 1 kg of moisture is about 716 kcal/kg.

The required amount of heat for heating 1 kg of evaporated moisture is the following:

$$q_{H-D} = G_{MD} \cdot \frac{100 - W_I}{W_I - W_D} \cdot (T_{MD} - T_{MW}) - G_W \cdot T_W, \qquad (3)$$

in which G_{MD} is the heat capacity of the dry manure, kcal/kg degrees ($G_{MD} = 0{,}42$ kcal/kg. degrees); T_{MD} is the temperature of the dry manure, degrees; T_{MW} is the temperature of the wet manure, degrees; G_W is water heating capacity, kcal/kg °C (1 kcal/kg °C); T_W is the temperature of water, °C.

As manure is pre-heated before getting to the manure collection tank for drying, the temperature of the wet manure entering the drying drum will be about 30 °C, and the temperature of dry manure after processing by the dryer will be $T_{MD} = 120$ °C. Taking into account preheating of manure, the amount of heat for heating 1 kg of evaporated moisture is about 2.5 kcal/kg.

When drying the manure, there is also external heat loss, the approximate amount of which can be calculated by multiplying the amount of evaporated moisture by 10 kcal. In our case, the loss of heat to the environment will be about 7.0 kcal/kg. Thus, the total heat consumption will be about 726 kcal/kg. With the 95% efficiency of the dryer, the total heat consumption for evaporation of 1 kg of moisture from manure will be about 764.0 kcal/kg. The total hourly heat consumption will be 1619680 kcal/h.

The amount of fuel (natural gas) burned to obtain the amount of heat necessary for manure drying is calculated with this formula:

$$G_{GAS} = \frac{q_H}{g_{GAS}}, \qquad (4)$$

in which G_{GAS} is the amount of gas burnt during drying of the manure, m³/h; q_H is the heat consumption per hour, kcal/h; g_{GAS} is gas heating value, kcal/m³.

If the heating value is about 8040 kcal/m³, its consumption will be 201.4 m³/h.

The required volume of the drying drum is:

$$V_{DR} = \frac{Q_{MEVH}}{A}, \qquad (5)$$

in which Q_{MEVH} is the amount of moisture evaporated per hour, kg/h; A is the drum intensity on evaporated moisture, kg (m³/h).

With the drum intensity of 80 kg (m³/h), dryer volume should be at least 26.5 m³. Assuming the drum diameter of 1.5 m, its length should be about 12 m.

Taking into account the results of the researches, the prototype model of a dryer has been constructed and has passed some industrial tests (Fig. 3).

Fig. 3. The prototype model of a dryer.

The suggested manure processing technology makes it possible to process up to 5 tons of wet manure per hour and to produce up to 0.7 MW of thermal power, 1150 kg of fuel bricks and 210 kg of organomineral fertilizer per hour. Mixing dried poultry manure, which nutrient content is 8.4%, with ash, which contains 41.3% of nutrients, results in complex organomineral fertilizer, which contains about 320 kg of active material. This organomineral fertilizer can be compared to solid mineral fertilizer in terms of nutrient content.

The analysis of mineral fertilizer costs in Chelyabinsk Oblast and its active material content shows that an average cost of one kilogram of active material is 41.5 rubles. On the basis of the cost of 1 kg of active material of mineral fertilizer, one ton of organomineral fertilizer, which contains 320 kg of active material should cost about 13 thousand rubles. However, estimates suggest that 1 kg of active material of fertilizer results in 5–7 kg of wheat yield increase. So, in order to cover the costs of fertilizer, wheat purchase price should be 8–6 rubles per kg. Taking into account storage costs of fertilizer, transportation, ground application, post-harvest handling of extra yield due to the usage of fertilizer and minimal profit, purchase price of wheat should be 12–10 rubles per kg. Wheat purchase price is usually much lower than the figures above, so mineral fertilizer application becomes economically unprofitable. It means that the price of mineral fertilizer in domestic market is too high, which makes it impractical to use. Our estimates suggest that in order to make poultry manure processing economically efficient, using the suggested technology, the cost of 1 kg of active material of fertilizer should be up to 16–17 rubles. From this perspective, if organomineral fertilizer contains 32% of nutritive, its price will be about 5 thousand rubles per ton.

The heating effect of fuel bricks is the same as the one of mineral coal. Bricks take minimum storage space and can be automatically transferred for burning. They can be used for heating in the existing boiler-plants without any reconstructions, while their burning is relatively harmless to the environment. Burning of fuel bricks results in ash, which can be used as fertilizer. Considerable proportion of fuel bricks is consumed in European countries. In recent years interest in fuel bricks in Russia has been growing. Depending on materials that are used in bricks production, their market price is 4–6 thousand rubles per ton. In contrast, a ton of mineral coal costs 3.1–3.9 thousand rubles. We suggested that 3 thousand rubles per ton would be an economically efficient cost.

120 tons of manure with moisture content 65–70% result in about 1500 tons of organomineral fertilizer and about 8000 tons of fuel bricks. Sales revenue from these products will be around 31.5 million rubles. With 11.5 million rubles of direct costs, the calculated economic impact from poultry manure utilization using the suggested technology is about 20 million rubles. If the complex costs 25.5 million rubles, the payback period is not more than 1.5 years. Poultry manure processing produces the following effects.

- Energy effect. Production of fuel bricks which are superior to traditional energy resources (wood, coal, etc.)
- Agrochemical effect. Getting efficient packaged organomineral fertilizer and restoration of natural fertility
- Economical effect. Making profit from fertilizer and fuel bricks sales. Diversification of agricultural production. Lowering costs of mineral fertilizer for own needs. Lowering costs of utility rooms' heating due to self-production of solid fuel
- Ecological effect. Poultry manure is utilized in flux, so manure storage facilities are not necessary. As a result, less pollution of the environment and croplands

- Social effect. New workplaces. Living and working in the countryside becomes more attractive. A new source of income appears. Better infrastructure of the countryside, increased awareness and better communication lines of local people with territorial form of government

4 Conclusion

1. The suggested technology of poultry manure utilization makes it possible to process all poultry manure from poultry houses in flux and get pyrolitic gas, fuel bricks, heating energy and organomineral fertilizer.
2. Utilization of poultry manure, which is hazardous substance, helps to fix an ecological problem and production of in-demand competitive products. Utilization of 120 tons of manure with 65–60% moisture content by one production line per day will result in 6000 MW of thermal power, 8000 tons of fuel bricks, and 1500 tons of organomineral fertilizer per year. Annual economic benefit for business will be 25 million rubles per year.
3. Organomineral fertilizer production makes it possible to solve the problem of natural fertility. If the cost of fertilizer is 5000 rubles per ton, in which 320 kg is an active ingredient, it will be cost-efficient to use it for cropping.

References

1. GOST 31461-2012. Poultry manure. Raw material for organic fertilizer production. Technical conditions. Standartinform, Moscow
2. Fisinina, V.I., Lysenko, V.P.: Technologies and technical facilities for manure processing at poultry farms. Voskhod-A, Moscow (2011)
3. Krasnitsky, V.M.: The use of poultry manure in agriculture of Western Siberia. Omsk State Agrarian University, Omsk (2016)
4. Sinyavsky, I.V., Chinaeva, Y.Z., Kalganov, A.A.: Agro-ecological and microbiological evaluation of the effectiveness of organo-mineral fertilizer derived from bird droppings. Agro-Ind. Complex Russia **24**(5), 1134–1140 (2017)
5. Teuchezh, A.A.: The application of poultry manure as organic fertilizer. Polythematic Netw. Electron. Sci. J. Kuban State Agrarian Univ. **128**, 914–931 (2017)
6. Gusakova, N.N.: Prospects for the use biotransformiroetsa poultry manure in crop production. Agric. Res. J. **3**, 16–19 (2016)
7. Tabakova, E.V., Titova, V.I.: Assessment of the prospects of using dried poultry manure as fertilizer. Environmental problems of using organic fertilizers in agriculture. VFASC, Vladimir (2015)
8. Zapevalov, M.V.: Automated processes of preventive plant care. Technological and technical support for using organomineral fertilizer and for chemical processing of seeds. Lap Lambert Akademik Publishing, Saarbryukken (2011)
9. Shustov, A.A., Rusinov, A.V.: Perspective technology of bird droppings utilization. Technogenic and natural safety materials of the IV all-Russian scientific-practical conference, pp. 155–158 (2017)

10. Ivanov, A.N., Belov, V.V.: Methods and installations for the disposal of bird droppings. Education and science: modern trends, pp. 151–167 (2017)
11. Alekseenko, V.A., Halldin, V.A., Ivanov, D.V.: Rational technology for processing poultry litter. Sci. Rev. **10**, 75–78 (2016)
12. Ohjtnikov, S.I.: Biologic reception of bird droppings utilization. In: Livestock Breeding in the European North: Fundamental Problems and Prospects of Development Abstracts of the International Conference Barents Euro-Arctic region (1996)
13. Kovalev, D.A., Kovalev, A.A.: Advanced technology of anaerobic processing of poultry manure. Bull.-Russian Sci. Res. Inst. Mech. Animal Husbandry **3**(27), 115–118 (2017)
14. Tomasev, V.B.: Technological and technical feasibility of energy saving drying poultry manure with the use of the heat of fermentation, Moscow (1993)
15. Sivitsky, D.V., Deep, I.S.: Improving the efficiency of drying systems with a combined supply of energy for drying poultry manure. In: Actual Problems of Energy, Agriculture Proceedings of the II International Scientific-Practical Conference, pp. 74–75 (2011)
16. Alekseenko, V.A., Halldin, V.A.: Device for drying poultry manure. Sci. View **8**, 89–92 (2016)

Application Features of Mathematical Model of Power System for Analysis of Technical and Economic Indicators of Reactive Power Compensation Device

D. V. Ishutinov[✉], V. I. Laletin, and E. N. Malyshev

Vyatka State University, 41 Preobrazhenskaya Street,
Kirov 610020, Russian Federation
ishutinov@vyatsu.ru

Abstract. The application of the mathematical model which takes into account the parameters of the electric power supply system elements and the shape of real reactive load graphs is considered. The model includes a static condenser of reactive power compensation. The assumptions made in the development of the model are specified. The implementation of the main elements included in the model is disclosed. The compensation device implemented in the model makes it possible to take into account the following switching algorithms of the capacitor banks stages: 1: 1: 1, 1: 2: 2 and 1: 2: 4. The model has an additional data measurement system which includes a block for calculating the saved power and energy losses as well as an accounting block of the number of condenser switching. The article presents the results of the power supply system study in accordance with the actual daily baseload. The analysis of the results led to the conclusion about the preferred use of the algorithm 1: 2: 2. The advantages of this algorithm are significantly reduced energy losses and a smaller nomenclature of the equipment typical size.

Keywords: Electric power supply system · Mathematical model · Reactive power compensation device · Capacitor banks · Algorithm · Data measurement system · Power losses

1 Introduction

Solving the problem of energy saving and increasing energy efficiency is among the priority directions of the development of modern science, technical and technology. The problems of reactive power compensation are acute in connection with the growth of electricity tariffs [1]. As a rule the need for compensating arises for those energy users who have extended electric networks [2–5].

Energy saving and energy efficiency when compensating RP are achieved by reducing the losses of active capacity, which are caused by the circulation of reactive power. It is known that the main losses caused by the RP circulation are excreted in power-reducing transformers (10/0.4 kV) and high-voltage supply lines (6, 10 kV) [1, 6].

At present there are the means and methods of static RP compensation [6–8]. The tasks of determining the optimal capcitor banks locations are mainly solved [9–12]. However, engineering methods for calculating the optimum degree of RP compensation which take into account the discreteness of the nominal values of power factor capacitors are not widely spread. Therefore, the number and nominal power of capacitor banks stages (CB) when designing the reactive power compensation devices (RPCD) are often chosen on the basis of the accumulated experience.

2 Formulation of the Problem

The work [13] shows that the most effective is the RPCD with three stages of regulation. However, the calculation carried out by this method requires refinement, since it is based on the use of average values of active and reactive loads. In addition, we need to assess the economic effect of reducing the loss of active energy. The authors developed [14] a mathematical model of automatic RPCD. It allows to investigate the algorithm of its work [15], but it does not take into account the parameters of the real power supply system.

The model proposed in [16] reflects the operation of a real power supply system with a compensation device, but it does not include the model of an automatic control system for this device. The model [17] does not allow analyzing the losses of power and energy in the electric network of the consumer. There is a program for calculating the values of reducing losses in the electric grid of the consumer [18], however, it does not allow to take into account real electrical load graphs, as it works with average values.

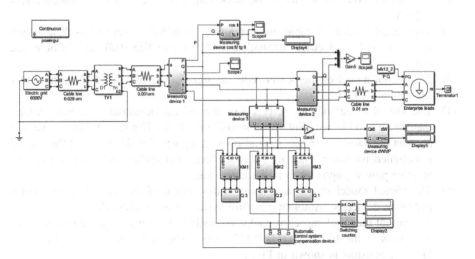

Fig. 1. The structural diagram of power supply system model with RPCD.

The purpose of this article is to develop an imitation mathematical model of a section of the power supply system of an enterprise with an automatic capacitor-based RPCD.

The model should take into account the parameters of the elements of the power supply system and the form of real graphs of the reactive load, which will allow us to analyze the wear resistance of the switching equipment and conduct a complex economic analysis.

3 Development of the Power Supply System Model

The proposed model of the power supply system includes the previously created model of RPCD [14]. When developing the power supply system model, the following assumptions are made:

- The electric network of 6 kV has infinite power
- The transformer substation conventionally has one step-down transformer
- Cable power lines are represented by active elements with resistance equal to the active resistance of the line
- Fast processes, in particular processes when switching capacitor banks, are reflected in real time. Slow processes (modes of consumption and energy generation) use 1: 360-time scale.

The structural diagram of the power supply system model with RPCD is shown in Fig. 1.

The model includes the following elements, which names are given in Fig. 1:

(1) The virtual model of an electric network with a voltage of 6 kV – "Electric grid 6000 V";
(2) The virtual models of cable lines of a 6 kV distribution network – "Cable line 0,029 Ω" and an electrical network 0,4 kV: "Cable line 0,001 Ω", "Cable line 0,01 Ω";
(3) The virtual model of the power reducing transformer of a 6/0.4 kV transformer substation – "TV1";
(4) The virtual model of electricity consumers of the industrial enterprise, implemented on two blocks: "Enterprise loads" and "P Q". The first allows to take into account the nature of energy consumption at a given time interval and the second is designed for commissioning into the block "Enterprise loads" real active and reactive power graphs taken at the enterprise;
(5) The virtual model of the automatic control block of the reactive power compensation device is "Automatic control system compensation device" [14];
(6) The virtual model of capacitor banks, consisting of several stages of power factor capacitors, - the blocks "Q1", "Q2", "Q3". The power factor capacitor with bleeding resistors is shown in Fig. 2;

Application Features of Mathematical Model of Power System 437

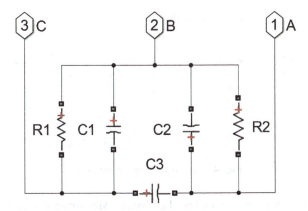

Fig. 2. The model of the power factor capacitor.

(7) The model of capacitor contactors – "KM1", "KM2", "KM3" (Fig. 3). The contactor model includes four elements. The element "KM11" simulates the main contacts, and "KM12" together with "R" - auxiliary contacts with increased resistance. The "Delay" element forms an algorithm for the operation of the capacitor contactor - at first, auxiliary contacts with increased resistance are closed, as a result, the current of the capacitor charge is limited, and then they are shunted with the main contacts with some time delay.

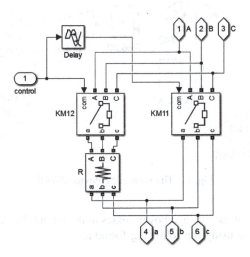

Fig. 3. The model of the capacitor contactor.

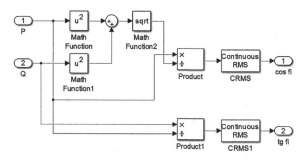

Fig. 4. The measuring device cos/tg.

(8) Information-measuring complex. The block "Measuring device cos/tg" (Fig. 4) calculates the phasor power factors cosφ and tgφ. The blocks "Measuring device 1", "Measuring device 2" and "Measuring device 3" provide the measurement of active, reactive power and current at the output of the power transformer, at the point of energy distribution between consumers, as well as in the circuit of the capacitor banks of the RPCD.

The block "Measuring device dW/dP" (Fig. 5) is intended for calculating the value of the day saved average active power loss ΔP_{mid} and the day active energy loss ΔW_Q, caused by the circulation of reactive power;

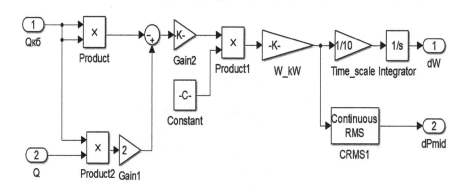

Fig. 5. The measuring device dW/dP.

The calculation of the average power losses at the output 2 – "dPmid" of the meter (Fig. 5) is performed using the following formula:

$$\Delta P_{mid} = \frac{1}{T}\int_0^T \left[\frac{R_E}{U^2}\left(2 \cdot Q(t) \cdot Q_{CB}(t) - Q_{CB}^2(t)\right)\right] dt, \qquad (1)$$

where $Q(t)$ – reactive power circulating in the consumer network; $Q_{CB}(t)$ – reactive power of the RPCD.

The calculation of energy losses at the output 1 – "dW" of the meter (Fig. 5) is performed by the formula:

$$\Delta W_Q = \frac{1}{T}\int_0^T \Delta P(t)dt, \qquad (2)$$

where $\Delta P(t)$ – saved losses of active power due to the reactive power circulating in the customer's network.

(9) "Switching counter" (Fig. 6) allows to calculate the number of switchings of CB sections for the given algorithm of the operation of RPCD, which allows to analyze the switching wear resistance. The "Switching counter" model is composed of three event counters. The increase in the state of the event counter occurs along the front and the drop in the input-output signal of the "Automatic control system compensation device" when the capacitor banks is switched on or off.

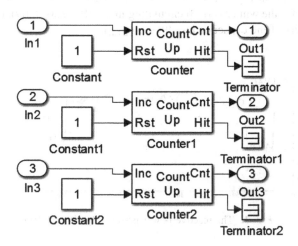

Fig. 6. The model of the switching counter.

Thus, the proposed model of the power supply system (Fig. 1), in addition to the RPCD model, takes into account the characteristics of the power step down transformer, cable lines, specialized condenser contactors and the capacitor banks.

The model allows to investigate the following technical characteristics of the power supply system of the industrial enterprise and the projected RPCD: medium-unweighted power factor (cosφ и tgφ) for the selected measurement interval; current values of active, reactive power (P and Q), the current in static and switching modes of operation.

With the help of this model it is also possible to determine the saved losses of active energy in the supply cable line (6/10 kV network) and the windings of the power low-voltage transformer, caused by the circulation of reactive power, and to carry out a complex economic analysis.

The presence of the switching counter in the model makes it possible (for the given algorithm of the RPCD operation) to estimate the service life of the capacitor contactors, based on the known value of their switching wear resistance.

4 Experimental Studies

In the practical part the study with the help of the developed model of the power supply system for the woodworking enterprise is presented [19]. The results for one transformer substation are given below as an example. The calculation of the parameters of the capacitor banks of RPCD was carried out in accordance with the method proposed by the authors [20]. In this case, for the algorithm of the automatic control system of RPCD 1: 1: 1, the RP value of the minimum stage is defined as 150 kVAr, for the algorithm 1: 2: 2 - 80 kVAr, and for the algorithm 1: 2: 4–60 kVAr.

The solid lines in Figs. 7, 8 and 9 show the daily graphs recorded during monitoring (typical for the winter period) circulating in the RP network (Q), the broken lines show the graphs of the capacitor installation (Q_{CB}).

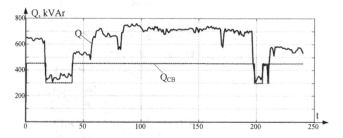

Fig. 7. The reactive power graphs (algorithm 1:1:1).

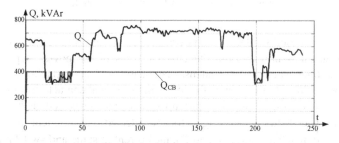

Fig. 8. The reactive power graphs (algorithm 1:2:2).

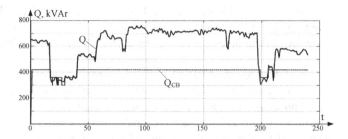

Fig. 9. The reactive power graphs (algorithm 1:2:4).

For the algorithm 1: 1: 1, the weighted average power factor value for the RPCD operation was equal according to the simulation results to cosφ = 0,96.

For the algorithm 1: 2: 2, the weighted average power factor was – cosφ = 0.95, and the number of switching of capacitors that was fixed visually increased. At the same time, the number of local areas of undercompensation and overcompensation decreased.

When the algorithm 1: 2: 4 was applied, the weighted average power factor was - cosφ = 0.96.

The results of calculating the number of switching of the capacitor contactors and losses of saved active energy ΔW_Q (per day) due to the use od RPCD at substations are presented in Table 1.

Table 1. Results of calculating quantity of switching and economy of electricity per day.

Algorithm of RPCD operation	Q_{st}, kVAr	Q_{CB}, kVAr	Contactor number 1	2	3	ΔW_Q, kW/h
1:1:1	150	450	7	1	1	60,8
1:2:2	80	400	17	1	1	61,2
1:2:4	60	420	19	9	1	58,8

5 Conclusion

Based on the results of the simulation, it can be concluded that even with a complex configuration of the load curve to maintain the $cos\varphi$ at a given level (according to the requirement of legislation documents tg φ = 0.34 for 0.4 kV networks), it is sufficient to use capacitor banks with three control stages. The smallest stage of the CB works most of the time. This confirms the conclusions on the optimal, from the point of view of capital expenditures, number of stages made on the basis of the analytical calculations [18].

The results of the calculation of the number of switching operations obtained during the simulation (Table 1) allow estimating the service life of the capacitor contactors. It is known that the switching wear resistance of the capacitor contactor is

100 .. 200 thousand cycles with the turn-on frequency 180 .. 200 per hour. The application of any of the algorithms for RPCD with three stages of regulation guarantees the service life several times exceeding the payback period of the installation – 1,8 year.

The authors found that the use of RPCD with the algorithm of operation 1:2:2 is the most preferable as, on the one hand, there are less (in comparison with the algorithm 1: 1: 1) local areas of undercompensation and overcompensation, and, on the other hand, (in comparison with algorithm 1: 2: 4) there is a smaller nomenclature of standard sizes of the equipment. At the same time, the amount of saved energy losses is higher.

The values of the average active energy daily losses (taking into account the real schedule of the reactive load) obtained during the modeling allow us to clarify the additional costs of these losses, which for the example considered in the article is 1253 EUR per year.

The formulated conclusions are confirmed by the results of the studies carried out by the authors at other industrial sites.

References

1. Elizondo, M.A., Samaan, N., Makarov, Yu.V., et al.: Literature survey on operational voltage control and reactive power management on transmission and sub-transmission networks. In: Proceedings of IEEE/PES General Meeting, pp. 1–5 (2017)
2. Skornyakov, V.A., Fedyaev, A.A., Sergeevichev, A.V.: Rational compensation of reactive power of consumers of wood processing industries. Syst. Methods Technol. **1**, 73–77 (2017)
3. Malyuk, E.G.: Reactive power and features of reactive power compensation in the networks of the housing and utilities sector. Energy Installations Technol. **1**, 57–62 (2017)
4. Petrov, A.A., Shchurov, N.I.: Organization of reactive power compensation at subway substations. Sci. Technol. Innovation, 239–240 (2016)
5. Korolev, N.I., Chernykh, N.A.: Peculiarities of reactive power compensation in the city electric grid. Energy-XXI Century **1**, 47–58 (2017)
6. Paulose, A.P.S., George, A.M., Jose, L., et al.: Reactive power compensation using SVC. Int. Res. J. Eng. Technol. **3**(6) (2016)
7. Semenova, L.A., Inzhevatova, A.O., Salimov, R.M.: On the automation of the choice of optimal power and the location of reactive power compensation devices. Energy: state, problems, prospects, pp. 15–18 (2016)
8. Semenenko, D.V., Filonov, S.A.: Reactive power compensation system based on static capacitor installations. Energy-XXI Century **4**, 43–46 (2014)
9. El-Fergany, A.A., Abdelaziz, A.Y.: Capacitor placement for net saving maximization and system stability enhancement in distribution networks using artificial bee colony-based approach. Int. J. Electr. Power Energy Syst. **54**, 235–243 (2014). https://doi.org/10.9790/1676-10418290
10. Duque, F.G., Oliveira, L.W., Oliveira, E.J., et al.: Allocation of capacitor banks in distribution systems through a modified monkey search optimization technique. Int. J. Electr. Power Energy Syst. **73**, 420–432 (2015)
11. Devabalaji, K.R., Yuvaraj, T., Ravi, K.: An efficient method for solving the optimal sitting and sizing problem of capacitor banks based on cuckoo search algorithm. Ain Shams Eng. J. (2016). https://doi.org/10.1016/j.asej.2016.04.005

12. Popov, V.V., Komarichina, D.I.: Development of the method for selecting reactive power compensation devices according to the condition of minimum total costs. Electr. Eng. Electr. Power Eng. **1**, 77–82 (2013)
13. Kondratiev, Yu.V., Tarasenko, A.V.: Determination of parameters of reactive power compensation devices. Int. J. Appl. Fundam. Res. **4**, 344–347 (2016)
14. Ishutinov, D.V., Malyshev, E.N., Slastikhin, N.S., et al.: The simulation model of the reactive power compensation device. Bull. TulGU. Ser.: Tech. Sci. **1**, 518–524 (2017)
15. Ishutinov, D.V., Grudinin, V.S.: Controlling the reactive power compensation device. RU Patent 2015613730 (2015)
16. Katyara, S., Shah, M.A., Izykowski, J., et al.: Power loss reduction with optimal size and location of capacitor banks installed at 132 kV grid station qasimabad Hyderabad. Present Problems Power Syst. Control **7**, 53–64 (2016)
17. Pionkevich, V.A.: Mathematical modeling of static compensator of reactive power for solving voltage regulation problems in power systems. Bull. Irkutsk State Tech. Univ. **12**, 192–197 (2015)
18. Kuznetsov, A.V., Rebrovskaya, D.A., Argentova, I.V.: A software model for estimating the reduction in power losses in a network organization when reactive power in the customer's network is compensated. Ind. Energy **6**, 48–54 (2016)
19. Ishutinov, D.V.: Analysis of energy consumption modes of a woodworking enterprise. Society, science, innovation, p. 21 (2013)
20. Ishutinov, D.V., Malyshev, E.N., Slastikhin, N.S., et al.: The choice of parameters of the capacitor banks of the reactive power compensation device based on economic criteria. Bull. TulGU. Ser.: Tech. Sci. **1**, 536–542 (2017)

On Technogenic Impact of Electromagnetic Components of Rectified Current and Voltage on Environment

K. Kuznetsov and A. Zakirova[✉]

Ural State University of Railway Transport, 66 Kolmogorova Street,
Ekaterinburg 620034, Russian Federation
azakirova@usurt.ru

Abstract. The harmful effect of the electromagnetic field on the electrical personnel from the sources which are rectifier converters used in the converter traction substations is reviewed. The harmonic composition of the rectified current and voltage of rectifier converters is analyzed. It is shown that the rectifier converter is a source of harmonic components of alternating current and voltage, has a spectrum of magnetic and electrical stresses, the harmful effect of which on personnel has not been studied to date. The main analytical relations allowing to estimate the spectrum of higher harmonic components from the rectified current and voltage by means of a calculation are given, the harmonic analysis of currents and voltages at the input and output of the rectifier of traction substations is performed. When comparing theoretical calculations and experimental measurements of magnetic field strengths, it was concluded that the effect of the magnetic and electric fields of the individual harmonics of the rectified current and voltage are harmful.

Keywords: Magnetic field · Electric installation · Rectified current · Rectifier converter · Traction substations

1 Introduction

Rectifier transformers AC to DC are widely used in various sectors of national economy, including traction electric transport network [1–4]. Rectifier converters of traction substations of direct current belong to the class of loads that have a non-linear volt-ampere characteristic, distorting the voltage curve of supplying electrical systems.

Non-linear loads consume from the network a non-sinusoidal current, which can be represented as the sum of sinusoidal leaving, called harmonics.

It is known [5, 6], that rectified current and voltage contain, in addition to a constant component, a series of harmonics of alternating current. In previous years, the presence in the traction networks of the harmonic composition of the variable components was a big problem due to the existence of wire-based electrical communication lines that were subjected to the intensive interfering components of the harmonic composition of the currents and voltages of the alternating current. In recent years, the development of communication lines using fiber optics has led to almost complete displacement from the practice of using wire communication lines. This circumstance solved the problem of electromagnetic compatibility in the part of the interfering effect of AC harmonics from the rectified current and voltage on the communication line made on the basis of fiber optics, which reduced the attention of scientific and technical specialists to this problem.

The problem of technogenic impact on the environment and human harmonic composition of AC from rectifier converters was first raised in recent years [7–9], which was the reason for carrying out a series of studies in this field.

The problem of electromagnetic compatibility with the harmonic composition of alternating current from rectifier converters to a wide range of computer technologies that are increasingly used in electrification and communication facilities has not been removed from the agenda.

Still the theoretical ideas about rectifier converters as sources of the harmonic spectrum of the variable components [10] do not always coincide with the experimental studies of the spectral composition and magnitudes of these components. This is due to a number of circumstances, the main ones are:

- The difference between the rectifier converter supply voltage and the ideal sinusoidal voltage, i.e. the presence of AC harmonics in the composition of supply voltage
- Presence in the traction network of direct current, consuming the energy of the rectifier converter, alternating current components and voltage induced from parallel AC lines of 50 Hz
- Occurrence of frequency components of the current consumed by direct current locomotives with inverter rectifiers and asynchronous motors.

2 Theoretical Research

The difference of the supply voltage from the sinusoidal shape creates conditions for the appearance of AC harmonics with frequencies of a multiple of 50 Hz in the 50 Hz supply voltage. It is obvious that these harmonics of alternating current will also be converted into a direct current in the rectifier converter, which will cause the appearance in the rectified voltage and current of the variable components proportional to the frequency of the primary voltage harmonic. In general, the frequency k of the harmonic f_k with a symmetrical supply voltage is determined as follows:

$$f_K = q \cdot f_c, \tag{1}$$

$$q = n \cdot m, \tag{2}$$

where m – a number of pulses of the converter unit; n – natural number of integers from 1 to ∞; f_c – frequency of alternating current in the supply network, Hz.

The frequency of the emerging harmonics of the EMF for a 6-phase bridge (6-pulse) rectifier converter for the frequency of the 50 Hz mains supply network, which is widely used in traction networks of railway transport in this case, can be represented as the following series: 300, 600, 900, etc. every 300 Hz and theoretically to infinity.

In the case of the presence of harmonics of the primary supply voltage, whose frequency is proportional to 50 Hz (100, 150, 200... etc.), additional harmonics will arise in accordance with the relation (1), in which instead of the network frequency it is necessary to substitute frequency, for example 2 harmonics (100 Hz) $f_c = f_{g2}$. In this case, the considered rectifier will further generate harmonics with frequencies 600, 1200, 1800... Hz, and if there are harmonics 25 Hz in the supply voltage – 150, 300, 450... Hz.

The voltage curve at the output of the rectifier is pulsating. The instantaneous value of the rectified voltage is the sum of the constant component U_d and the variable component consisting of an infinite series of harmonics:

$$U_m = U_d + \sum_{n=1}^{\infty} U_{do} \sin(n\Theta + \phi_n), \qquad (3)$$

where U_d – constant component, B; U_{do} – amplitude of the n-th harmonic, B; n – ordinal number of the rectified voltage harmonic; Θ – phase angle of the first harmonic; φ_n – initial phase of the n-harmonic.

Table 1 shows the frequencies of the harmonics of the voltages generated by rectifier converters of different designs with a sinusoidal supply voltage. In this table (+) frequency present, (−) frequency is absent.

Table 1. The presence of a spectrum of higher harmonic components in the rectified voltage curve in the idle mode of rectifiers.

Constructive straightening circuit	Frequency, Hz							
	300	600	900	1200	1500	1800	2100	2400
	q = 6	q = 12	q = 18	q = 24	q = 30	q = 36	q = 42	q = 48
6-pulse	+[a]	+	+	+	+	+	+	+
12-pulse	−[a]	+	−	+	−	+	−	+
24-pulse	−	−	−	+	−	−	−	+

In a number of works [11–20] it was shown that each higher harmonic of the rectified voltage of the order of $n_U = m \cdot n$ in the alternating current of the converter corresponds to two higher current harmonics with ordinal numbers:

$$n_I = m \cdot n \pm 1 = q \pm 1, \qquad (4)$$

where m – number of ripples of the rectified voltage curve; $n = 0, 1, 2, 3...$ natural number.

Table 2 shows the components of the spectrum of higher harmonic rectified current for various design schemes of rectifier converters. In this table (+) frequency present, (−) frequency is absent.

Table 2. The spectrum of higher harmonic components in the curve rectified current.

Straightening diagrams	Frequency, Hz							
	250	350	550	650	850	950	1150	1250
	$m = 5$	$m = 7$	$m = 11$	$m = 13$	$m = 17$	$m = 19$	$m = 23$	$m = 25$
6-pulse	+b	+	+	+	+	+	+	+
12-pulse	−b	−	+	+	−	−	+	+
24-pulse	−	−	−	−	−	−	+	+

3 Experimental Research

Picture 1 shows the measured spectrum of the electric field strength in the vicinity of the 6-pulse rectifier converter of the traction substation, and in Figs. 1 and 2 the spectrum of the magnetic field strength. The measurements were carried out by the device "Ecophysics MI PKF-09-001", which provides for automatic measurement of the modules of the EP and MP voltage at any orientation of the sensor.

It is obvious that the frequency components correspond in the first case to the harmonics of the rectified voltage, and in the second case to the harmonics of the rectified current.

It is interesting to compare the spectral composition of the harmonics of the rectified voltage and the current of a 6-pulse rectifier converter (RC), obtained theoretically and experimentally (Tables 3 and 4). In this tables (+) frequency present, (−) frequency is absent.

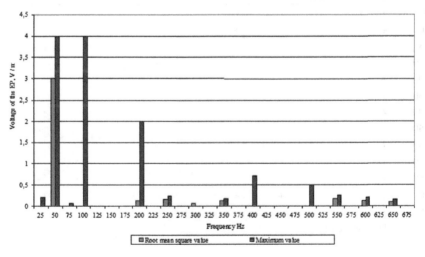

Fig. 1. The frequency spectrum of the higher harmonic components of the electric field strength (rectified voltage).

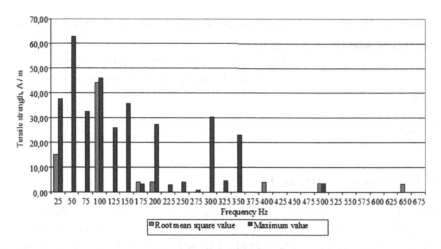

Fig. 2. The frequency spectrum of the higher harmonic components of magnetic field induction (rectified current).

Table 3. Comparison of experimental measurements and theoretical spectra of harmonics of the rectified voltage up to 1000 Hz of the 6-pulse RC.

f harm (Hz)	Under the strains of the rectified voltage, experimental research	Perfect 6-pulse RC	Perfect 6-phase power supply system RC 50 Hz	The presence of a harmonic of 100 Hz, 6-phase power supply system RC	6-pulse RC, other harmonics
	Theoretical research				
25	+c				+
50	+		+		
75	+				+
100	+			+	
200	+			+	
250	+				+
300	+	+		+	
350	+				+
400	+			+	
500	+			+	
550	+				+
600	+	+	+	+	
650	+				+
>700	Do not measure				

Table 4. Comparison of the theoretical spectrum of harmonics of rectified current with the results of experimental measurements.

f harm, Hz	Under the RC buses, the experiment	6-pulse RC	6-phase power supply system RC 50 Hz	The presence of a harmonic of 100 Hz, 6-phase power supply system RC	6-pulse RC, other harmonics
	Theoretical research				
25	+^{d}				+
50	+		+		
75	+				+
100	+		+	+	
125	+				+
150	+		+		
175	+				+
200	+		+	+	
225	+				+
250	+	+	+		
275	+				+
300	+		+		
325	+				+
350	+	+	+		
400	+		+		
500	+		+	+	
650	+	+	+		
>700	Do not measure				

Analysis of the data shown in the tables shows that the list of frequencies in the spectrum of the measured electric and magnetic field strengths contains components that are not part of the most probable sources:

- 6 pulse rectifier converter with a perfectly sinusoidal AC voltage of an industrial frequency of 50 Hz
- 6-phase scheme of the secondary voltage of the traction transformer, which supplies power to the RC by bridge circuit
- 6-pulse rectifier converter with a 6-phase AC voltage of 50 Hz, containing a spectrum of AC harmonics

On the basis of the analysis carried out, it is possible to present the spectrum of the harmonics of the electromagnetic field strengths near the rectifier converter of an alternating current of an industrial frequency into a rectified constant in the form of several terms of the spectra:

$$S_E = S_W^{BI\!I} + S_E^{\sim} + S_E^{BI\!IS\sim}, \tag{5}$$

$$S_H = S_H^{BII} + S_H^{\sim} + S_H^{BIIS\sim}, \tag{6}$$

where S_E, S_H – spectra of the harmonic components of the intensities, respectively, of the electric and magnetic fields of the RC; S_E^{BII}, S_H^{BII} – spectra of strengths of electric and magnetic fields of RC at sinusoidal supply voltage; $S_E^{\sim}$, $S_H^{\sim}$ – spectra of the strengths from the non-sinusoidal supply voltage and the supply voltage of the RC; $S_E^{BIIS\sim}$, $S_H^{BIIS\sim}$ – spectra of the strengths of the electric and magnetic fields of the RC during the transformation of the variables of the harmonics of the supply voltage to the rectified direct current.

4 Conclusion

It follows from the considered analysis that in the harmonic spectrum of the supply rectifier converter the alternating current voltage may contain some additional harmonics, in particular, with a frequency of 25 Hz, the presence of which, together with the harmonics known from theoretical calculations, creates a sufficiently large frequency spectrum of the harmonic composition of the electromagnetic field components near the RC.

This circumstance, on the basis of relation (4), shows that for each RC in different electrical installations the harmonic spectrum of alternating current of different composition can be contained, which makes it difficult not only to detect the spectral composition of the magnetic field (current) and electric field (voltage), but also hinders the application of technical protection due to the variety of their frequency characteristics.

For papers published in translation journals, please give the English citation first, followed by the original foreign-language citation [6].

References

1. Bey, U.M.: Traction substations. Transport, Moscow (1986)
2. Teroganov, E.V., Pyshkin, A.A.: Electricity supply of railways. USURT, Yekaterinburg (2014)
3. Kuznetsov, K.B.: Basics of electrical safety in electrical installations. FPSFI of APE Educational and methodological center for education in railway transport (2017)
4. Burkov, A.T.: Electronic engineering and converters: study guide for higher educational institutions of railway transport. Transport (1999)
5. Kuznetsov, K.B., Zakirova, A.R.: Higher harmonic components of rectifiers magnetic fields and their adverse health effects. Procedia Eng. **129**, 415–419 (2015)
6. Kuznetsov, K.B., Zakirova, A.R.: Assessment of harmful health effects of AC rectifier converters harmonic components. Procedia Eng. **129**, 420–426 (2015)
7. Kuznetsov, K.B., Zakirova, A.R.: Probability of occurrence of professionally conditioned disease of workers. Electr. Saf. **2**, 26–33 (2015)
8. Kuznetsov, K.B., Zakirova, A.R.: Estimation of EMF at workplaces of electro technical personnel of traction power supply. Transp. Urals 3(38), 112–118 (2013)

9. Zakirova, A.R., Kuznetsov, K.B.: Estimation of electromagnetic environment and probability of occurrence of professionally conditioned disease. Herald Ural State Univ. Railway Transp. **4**(24), 82–89 (2014)
10. Bader, M.P.: Electromagnetic compatibility: textbook for higher educational institutions of railway transport. Educational methodical complex of the Ministry of Railways (2002)
11. Khohlov, Y.I., Gizzatullin, D.V.: Modeling of electromagnetic processes in a compensated rectifier with voltage feedback based on autonomous voltage inverter with PWM. Bull. South Ural State Univ. Ser.: Energetics **11**(111), 32–38 (2008)
12. Sokolov, S.D., Bey, U.M., Guralnik, Y.D., et al.: Semiconductor converter units of traction substations. Transport (1979)
13. Chetvergov, A.V., Maslov, G.P., Pozdnyakov, O.I., et al.: Report № 960 on research work: experimental design development, research and operational development of twelve-pulse rectifiers of traction substations (1980)
14. Shalimov, M.G.: Twelve-pulse semiconductor rectifiers of traction substations. Transport (1990)
15. Schwab, A.J.: Electromagnetic compatibility. Energoatomizdat (1995)
16. Arrillaga, J., Bradley, J., Bodger, P.: Harmonics in electrical systems. Energoatomizdat (1990)
17. Treivas, M.D.: Higher harmonic components of the rectified voltage and their reduction in traction substations of direct current. Transport (1964)
18. Tamazova, A.I.: Unbalance of currents and voltages caused by single-phase traction loads. Transport (1965)
19. Bader, M.P.: Electromagnetic compatibility. Part 3. Harmonic analysis of influencing currents and voltages. Tr. MIIT (1999)
20. Kosarev, A.B.: Fundamentals of electromagnetic safety of railroad power supply systems. Intex (2008)

Engineering Solutions for the Use of Air Enriched with Oxygen in Power Engineering Complexes of Power Plants Operating on Secondary Gases

E. B. Agapitov, M. S. Sokolova[✉], and A. E. Agapitov

Nosov Magnitogorsk State Technical University, 38 Lenina Avenue,
Magnitogorsk 455000, Russian Federation
margo88k2017@mail.ru

Abstract. The article analyzes the efficiency of using oxygen-enriched air to control combustion in a steam boiler of a steam-blowing power plant of PJSC "Magnitogorsk Iron and Steel Works". In the boilers of a steam-blowing power plant, a mixture of three gases is burned: natural, coke and blast furnace. In the case of a change in the ratio of gases, the combustion conditions and the geometric characteristics of the flame change. In some cases, this leads to an undesirable increase in the heat load in the boiler elements, especially in the superheater. Since the task of boilers of a steam blower power plant is to provide blast furnace production with air blast enriched with oxygen, the work of turbo expanders is directly related to the rhythm of the blast furnace. In the case of reducing the load of the blast furnace, there is an excess of air blast, which can be used to solve other problems, in particular, to control the combustion process in the boiler furnace. The article presents the results of numerical simulation of thermal fields in the furnace of an energy boiler of a steam-blowing power plant using oxygen enriched air.

Keywords: Steel plant · Energy · Fuel and energy resources · Electricity · General and specific consumption of electricity

1 Introduction

The power systems of an industrial enterprise are developing, being motivated by the dynamics of the development of the whole enterprise, improving the equipment and logistics of the energy system itself, and also by changing the economic situation both inside the enterprise and outside it. One of the tasks to be solved at the metallurgical enterprise is to increase the efficiency of using the purchased fuel, which is implemented on the following principles:

- Increase in the energy efficiency of technological units due to the use of appropriate equipment and technologies, leading to a reduction in the absolute amount of fuel consumption in the technological unit itself
- Use of energy resources, which arise as a result of the technological process in metallurgical units. These are the so-called secondary energy resources (SER),

which can be used outside the above-mentioned unit [1]. The implementation of this path leads or should lead to a reduction in the consumption of the purchased fuel, which is replaced in power plants for SER, with practically unchanged fuel consumption in the source unit

One of the ways to reduce the volume of purchased fuel is to enrich the air blast with oxygen, but until recently the effectiveness of its application was considered only for solving the problems of metallurgical technologies, due to its relative high cost. However, the development of membrane technologies, which allows obtaining air enriched to 31–35% with low cost, has set the task of expanding the areas of its use [2].

In the blast furnace industry, air blast is widely used at metallurgical plants, enriched with oxygen to a level of 27–29%. For various technological reasons, the volume of blast furnace blast is periodically changing, which leads to the short-term occurrence of excess oxygen in the oxygen supply networks. In view of the specifics of the operation of oxygen production stations associated with the impossibility of operational control of production, there is a need to use oxygen for other technological processes [3–5].

2 Mathematical Description

In the immediate vicinity of blast-furnace production there are steam-blowing power stations (SBPS) that provide blast-furnace blast and produce electricity. At the station, energy boilers are usually installed, using secondary gases, such as blast furnace and coke, as fuel. Air enriched in oxygen is compressed using turbo-blowers with a steam drive. The dynamics of changes in the amount of oxygen used by typical blowers is shown in Fig. 1.

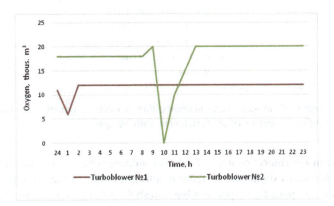

Fig. 1. Dynamics of oxygen supply to blast furnaces in PJSC "Magnitogorsk Iron and Steel Works".

These data show that it is possible the formation of "peak" of excess oxygen in an amount of 6000–20000 m^3/h, which can be used for 3–5 h.

The simplest way is to use oxygen in power boilers, but no experience of such use has been found in the literature. The analysis of theoretical and practical work on the use of oxygen-enriched air for the combustion of low-calorie industrial gases in other technologies has shown that the geometric characteristics of the flare, its emissivity, temperature and composition of the gaseous medium significantly change. This causes the redistribution of the temperature and concentration fields of the gas medium components and, accordingly, the heat and mass exchange conditions, which can lead to a deterioration of the thermal performance of the individual elements of the boiler [5].

When the air is enriched to 30% O_2, the length of the visible part of the flame is reduced by 4–6%, and the region with the maximum temperature approaches the burner by 25%, while enriching up to 35%, the length reduction reaches 9–12% [5].

One of the negative consequences of increasing the temperature of the flare may be an increase in the content of nitrogen oxides in the off-gas. Therefore, the emission of nitrogen oxides was calculated for the case of combustion of a blend of blast furnace gas and natural gas in boilers of hydroelectric power stations of PJSC "Magnitogorsk Iron and Steel Works": 67–33%, 75–25%, 80–20%, respectively, with a different proportion of oxygen in the air supplied for combustion. The results of the calculations are shown in Fig. 2.

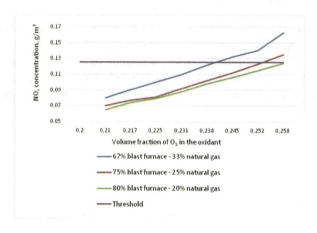

Fig. 2. The content of nitrogen oxides in combustion products of a mixture of blast furnace and natural gas at different degrees of air enrichment with oxygen.

Increase in the caloric content of the fuel mixture with an increase in the share of natural gas leads to a significant increase in nitrogen oxides in the combustion products and reduces the potentially permissible threshold of the oxygen blast enrichment threshold.

Another significant factor in the combustion process when oxygen is enriched with air is the decrease in the volume of combustion products, which affects the heat exchange in the tail surfaces of the boiler and the operation of the smoke exhaust.

Numerical calculations connected with the estimation of the potential for additional input of blast furnace gas into the fuel mixture in the case of enrichment of the blast with oxygen while maintaining the volume of the combustion products at the same level were performed for a boiler of the SBPS with a steam capacity of up to 125 t/h with P = 10 MPa and a superheated steam Tss = 5400 °C, which is able to work on blends of blast furnace, coke oven and natural gas (Fig. 3) [6].

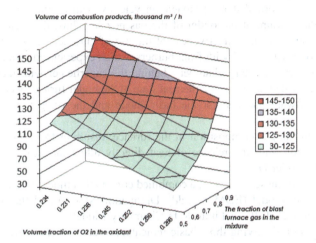

Fig. 3. Volume of combustion products depending on the fraction of blast furnace gas and O_2 content in air.

The analysis of Fig. 3, taking into account the data for calculating the concentrations of nitrogen oxides, it is possible to distinguish the composition range of mixtures that provide acceptable conditions for emissions, depending on the volume of combustion products:

- With a volume of combustion products up to 125000 m^3, the boiler operates with acceptable quality indicators with a ratio of natural gas and blast furnace gas of 1:2 respectively, and a steam load of up to 101 t/h
- With a volume of combustion products up to 100000 m^3, the boiler operates with acceptable quality indicators at a ratio of natural gas, coke oven gas and blast furnace gas of 1: 2:7 respectively, and a maximum capacity of 117 t/h
- To ensure the nominal steam capacity of the boiler at 125 t/h, the composition of the mixture of natural and blast furnace gases and the degree of oxygen enrichment of the air at which the nitrogen oxides in the combustion products do not exceed the permissible standards can be selected. Such conditions are satisfied by the fuel composition of 70% of blast furnace gas and 30% of natural gas, while enriching O_2 of combustion air up to 23,7% [7, 8]

When burning fuel, the main component of which is blast furnace gas, a non-luminous flame is produced. The addition of natural gas in an amount of 8–9% in heat creates the conditions for the formation of black carbon and increase the luminosity of the flame. However, the increase in the temperature of the flare in the range 1000–20000 °C causes a decrease in the degree of blackness. The use of oxygen, including when it is premixed with fuel, raises the temperature of the flare, which leads to a decrease in the degree of blackness of CO_2 and H_2O at a temperature of 15000 °C by 0,005–0,015. The change in the composition of gases increases blackness CO_2 and H_2O by approximately the same value [9]. With the use of enriched blast, the ratio of the air flow through the burners varies, which can lead to a reduction in the flame due to the growth of the normal flame propagation velocity with an increase in the average velocity of the mixture at the outlet of the gas nozzle. Thus, there can be no unambiguous assessment of the effectiveness of the use of enriched blast in the furnace.

For the boiler KGM-125 with front burners, which experienced problems of raising the temperature in the rotary chamber - behind the superheater with increasing steam load, numerical calculations of the thermal state of the boiler furnace were performed in the ANSIS software package for the basic mode and the oxygen-enriched air blast mode.

Changes in the gas composition in the furnace, resulting from the chemical reactions of oxidation with the release of heat directly affect the flow pattern and heat transfer in the boiler [10–12].

To account for these processes, a combined combustion model was used combining the vortex decay model (EDM or Eddy Dissipation Model) and the finite chemical reaction rate model (FRC or Finite Rate Chemistry Model).

In the process of solving three basic indicators of convergence level were tracked:

- Root-mean-square discrepancies of the computed variables
- Global imbalances of the solved equations
- Changes in the target parameters (temperatures, mass fractions of the components at several points in the calculated region, the velocity of gases averaged over the output section)

The quality control of the performed calculations was carried out by means of parallel control calculations of the radiation in the furnace [13, 14]. The radiant heat, perceived by the furnace, kJ/m^3 was estimated:

$$Q_\pi = \sigma_0 \cdot \frac{a_k \cdot H_l}{B} \cdot \left(T_f^4 - T_z^4\right), \qquad (1)$$

where σ_0 – the emissivity of an absolutely black body, $\sigma_0 = 5{,}67 \cdot 10^{-11}$, kW/($m^2 \cdot K^4$).

The reduced degree of blackness of the combustion chamber:

$$a_k = \frac{1}{\frac{1}{a_l} + \chi \cdot \left(\frac{1}{a_f} - 1\right)}, \qquad (2)$$

The effective degree of blackness of the torch was evaluated:

$$a_f = m \cdot a_{CB} + (1-m) \cdot a_g, \tag{3}$$

where a_{CB} – the degree of blackness of the glowing part of the torch; a_g – the degree of blackness of the gas (non-luminous) part of the torch; m – the coefficient of filling of the furnace with the luminous part of the torch.

Degree of blackness of the glowing part of the torch:

$$\alpha_{CB} = 1 - e^{-k_{CB}ps}, \tag{4}$$

The coefficient of attenuation of the rays by the luminous part of the flare is calculated from the temperature and composition of the gases at the exit from the furnace (in its determination, the radiation of water vapor and triatomic gases and suspended particles of soot are taken into account).

$$k_{CB} = k_g^0 \cdot r_p + k_c, \tag{5}$$

The coefficient of attenuation of rays by the gaseous medium was determined by the formula:

$$k = k_g^0 \cdot r_p = \left(\frac{7,8 + 16 \cdot r_{H_2O}}{\sqrt{10 \cdot p \cdot r_n \cdot s_T}} - 1\right) \cdot (1 - 0,37 \cdot 10^{-3} T_T'') \cdot r_p, \tag{6}$$

The coefficient of attenuation of rays by black particles was calculated by the formula:

$$k_c = \frac{1,2}{1+\alpha^2} \cdot \left(\frac{C^r}{H^r}\right)^{0,4} \cdot (1,6 \cdot 10^{-3} \cdot T_T'' - 0,5), \tag{7}$$

The effective thickness of the radiating layer of the furnace is determined by the formula:

$$s_T = 3,6 \cdot V_T / F_{CT}, \tag{8}$$

The degree of blackness of the non-luminous part of the flame is determined by the formula:

$$a_A = 1 - e^{-kps}, \tag{9}$$

Calculations showed that the supply of enriched blast (the degree of oxygen enrichment by 24%) allows to significantly change the geometry of the flame and improve the operation of the superheater by shifting the high-temperature zone of the torch to the upper part of the boiler furnace (Fig. 4).

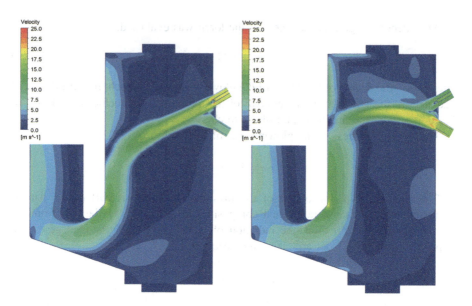

Fig. 4. Change in radiation intensity in the section passing through the burner block for the initial version of the boiler operation (on the left) and for the alternative with enrichment of 24% O_2 (on the right).

Calculations showed that, despite the fact that the local maximum temperature of the torch increased by 3,5%, the volume of gas with a temperature above 12000 °C decreased by 14%, which led to a change in the geometry of the torch and the zone of high temperatures from the rotary chamber moved to the combustion volume.

3 Conclusion

The use of oxygen-enriched air allows us to expand the boundaries of the use of secondary gases in power boilers and control the heat exchange processes in the furnace, however, this procedure must be used with some caution associated with the growth of local temperatures in the combustion chamber and the increase in nitrogen oxides in the off-gas.

References

1. Polyakov, V.V.: Resource saving in the iron and steel industry. Engineering (1999)
2. Mulder, M.: Basic principles of membrane technology. Center for Membrane Science and Technology. University of Twente, Enschede, Netherlands (1996)
3. Agapitov, E.B., Kartavtsev, S.V., Mikhaylovskiy, V.N., et al.: Technical and economic approaches to assessing the efficiency of using blast-furnace gas at a metallurgical plant. Ind. Power Eng. **3**, 15–22 (2016)

4. Agapitov, E.B., Mikhaylovskiy, V.N., Dautov, R.N., et al.: Improving the efficiency of TBS power plant of metallurgical enterprise in solving multi-objective problems. Electrotech. Syst. Complexes 3(32), 48–53 (2016). https://doi.org/10.18503/2311-8318-2016-3(32)-48-53
5. Agapitov, E.B., Mikhaylovskiy, V.N., Agapitov, A.E., et al.: Mathematical software of grid module-efficiency evaluation system for power stations at metallurgical enterprise. Electrotech. Syst. Complexes 4(29), 25–30 (2015)
6. Agapitov, E.B., Mikhaylovskiy, V.N., Nikolaev, A.A., et al.: The study of the influence of the volume use of the secondary energy resources for electricity generation at TBS power plant of metallurgical enterprise. In: Proceedings of the IEEE Russia Section Young Researchers in Electrical and Electronic Engineering Conference, pp. 1467–1470 (2017)
7. Revun, M.P., Granovskiy, V.I., Baybuz, A.N.: Intensification of the operation of heating furnaces. Technik, Kiev (1987)
8. Ryzhkov, A.F., Levin, E.I., Filippov, P.S., et al.: Increase in efficiency of use of blast-furnace gas at the metallurgical entities of Russia. Metallurgist 1, 26–34 (2016)
9. Reznikov, M.I., Lipov, U.M.: Boiler plants for power plants. Energoatomizdat (1987)
10. Munz, V.A.: Energy saving in energy and heat technologies. Ural State Technical University Publishing, Yekaterinburg (2006)
11. Sazanov, B.V., Sitas, V.I.: Industrial heat power installations and systems. Moscow Power Engineering Institute Publishing, Moscow (2014)
12. Makhmudov, S.A., Yakovlev, V.K., Nikiforov, G.V., et al.: The raising of efficiency and reliability of TEC – PVS boilers of metallurgical works when maximum using of blast-furnace and coke-oven gases. Ind. Power 5, 17–20 (2006)
13. Timoshpolskiy, V.I., Kabishov, S.M., Trusova, I.A.: Technique of assessment energy efficiency of enrichment air oxygen at combustion of gaseous fuel. Energ. Effi. 11, 32–34 (2013)
14. Shtym, K.A., Solovieva, T.A.: Conversion of the KVGM-100-150 boiler into cyclone-vortex gas combustion. Thermal Eng. 62(3), 202–207 (2015)

Renewable Sources of Energy for Efficient Development of Electricity Supplies for Agriculture in Chechen Republic

A. R. Elbazurov[✉] and G. R. Titova

Moscow Power Engineering Institute, 14 Krasnokazarmennaya Street,
Moscow 111250, Russian Federation
essame@yandex.ru

Abstract. Nowadays the industry and economy of the Chechen Republic are at the stage of recovery growth. Over the last years electric energy consumption has been growing in the region. But the energy system of the Chechen Republic lacks supply as there is no electrical generation except the Kokadoyskaya small water power plant on the Argun River. The lack of power in the energy system of the Chechen Republic is covered by the power flow from the neighboring energy systems which, in its turn, leads to the loss in capacity. After the events which took place in the Republic from 1994 to 2000 the energy system hasn't been restored yet. The factors mentioned above hold back the development of the region with the unique environmental conditions. Nowadays the utilization of renewable energy sources such as wind and solar energy must be the strategic goal for the economic development of the Chechen Republic, especially its agriculture.

Keywords: Renewable sources of energy · Agriculture · Power

1 Introduction

The foreseeable depletion of hydrocarbon fuel resources, the growing monopolization of this branch of economy and the rise in prices for power sources pose a threat to the energy security. The state must ensure economic growth irrespective of the availability of energy resources and the prices for them. Moreover, the environment must be protected. Environment protection and the agricultural growth require energy security which must be based on power efficiency as well as using new and renewable energy sources.

2 Analysis of the Power Sector Condition in the Chechen Republic

The branches of industry connected with fuel and energy sector have raw material orientation. The industrial and agricultural growth requires more electric power. The increasing output leads to the growth of thermal and electric power consumption, but electric energy consumption grows more rapidly that is shown in Fig. 1.

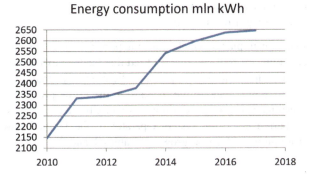

Fig. 1. Energy consumption dynamics in the enegy system of the Chechen Republic from 2010 to 2017.

The energy system of the Chechen Republic lacks supply as there is no electrical generation except Kokadoyskaya small water power plant on the Argun River which has the maximum rated power of 1,3 MW.

The energy demand of the Chechen Republic is covered by the power flow through the network of 110 kW from the substation in the Republic and through the network of 330 kW from the neighboring energy systems that in its turn leads to the loss in power [1]. The main providers of electricity are the energy systems of the North Ossetia and Dagestan. For example, at winter consumption peak on the 18th of December of 2016 at 6 p.m. the consumption of the Chechen energy system was 493 MW as shown in Fig. 2 and was covered as follows:

- 106 MW from Dagestan's energy system
- 327 MW from North Ossetian power grid

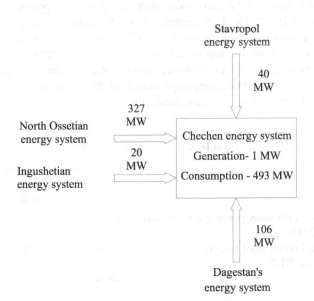

Fig. 2. Power flows in Chechen energy system in 18 of Dec. 2016.

- 20 MW from the energy system of Ingushetia
- 40 MW from Stavropol power grid

Most thermal power plants which transmit power to the Chechen Republic are natural gas fired, some are coal and fuel oil fired that leads to exhaustion of oil and gas which may be depleted within 70 years. Extraction and transportation of these resources rise in price that leads to the increase in the prime price of electricity.

«Realistic» estimate of power consumption and Chechen energy system capacity for a three-year period made by System Operator of the Unified Power System [2] shows that consumption will go on growing, Tables 1 and 2.

Table 1. Electricity consumption forecast for the period from 2018 to 2020.

	Period		
Year	2018	2019	2020
Electricity consumption, mln kWh	2714	2807	2838
Average annual growth,%	2,57	3,43	1,1

Table 2. Electricity consumption forecast for the period from 2018 to 2020.

		Period		
Year		2018	2019	2020
Summer peak	Power input, MW	360	363	365
Winter peak		530	534	537

The construction of a thermal power plant with generating power of 498,6 MW is underway in Grozny, Table 3. According to the electricity consumption forecast this thermal power plant must cover from 90 to 95% of the needed power.

Taking into account the significant rise of energy consumption in the Chechen Republic in the recent years and in future the current condition of the power facilities doesn't always meet the demanding requirements of the consumers, authorities and the System Operator to their quality and reliability.

Table 3. Electricity generation.

Power plant	Rated power, MW	Available power, MW	
		Winter peak	Summer peak
Kokadoyskaya small water power plant (put into operation in 2015)	1,3	1	1,3
Grozny thermal power plant (expected comissioning in 2018)	498,6	498,6	498,6
In total:	499,9	499,6	499,9

About a third of the substations 35, 110 kW were put into operation single-powered by a single-transformer scheme with no possibility of power reserve.

Most energy facilities are worn-out and outdated. The equipment of the substations for 110 and 35 kW is worn-out for 86%.

The condition of the equipment and grid doesn't meet reliability requirements.

All the factors mentioned above hold back the development of the region with the unique environmental conditions.

In the authors' opinion the development of the agriculture by means of renewable energy sources such as solar and wind energy must be the strategic goal of Chechen Economy.

3 Climatic and Energy Characteristics of the Sun in the Chechen Republic

The location of the Chechen Republic between 42 and 46° north latitude provides intensive inflow of solar radiation. Solar energy resources, expressed in terms of radiation balance, in plain and foothill areas consist 50 to 55 kcal/sm^2 per year. The duration of insolation is about 330 days per year. Days without sun are rare – 34 to 40 days in plain and foothill areas and 10–12 days in mountainous areas. The largest number of dull days is 61 in the plain areas. As a whole during a year the cloud cover reduces the inflow of direct radiation for 20 to 25% of the potential.

The total radiation is the sum of direct and scattered radiation on a horizontal surface. The total radiation in the Chechen Republic peaks from May to July. The total radiation intensity varies between 280 and 300 MJ/m^2 in the foothill areas. In the mountainous areas it ranges from 360 to 400 MJ/m^2 [3].

The NASA database [4] provides the average monthly size of the total solar radiation falling on a horizontal surface on the ground surface for each month for a period of 22 years from July 1993 to June 2015, Table 4. Each month the average value is the arithmetic mean of 3 h values for that month. The values are given for the village of Shelkovskaya in the Chechen Republic with an area of 167,21 km^2.

The advisability of combined power supplies can be shown using the example of the power supplies to a cattle-breeding complex for 800 head with the total peak load of 257,6 kWh and an annual power consumption of 875,84 MWh.

The average value of solar energy gross potential per year is calculated on the following formula:

$$E_g^y = \sum_{k=1}^{d} E_{gk}^y / d, \qquad (1)$$

d – number of estimated months, $\Im_g^y$ – inflow of solar radiation on a horizontal ground for k – a month.

The solar energy gross potential in the Chechen Republic is estimated at 3,67 kWh/m^2 per day:

$$E_g^y = (1,61 + 2,36 + 3,3 + 4,51 + 5,67 + 6,07 + 5,92 \\ + 5,05 + 3,94 + 2,69 + 1,67 + 1,3)/12 = 3,67 \text{ kW} \cdot \text{h/m}^2 \cdot \text{d}, \quad (2)$$

Then the value of an annual solar energy gross potential is:

$$E_g^y = 3,67 \cdot 365 = 1340 \text{ kW} \cdot \text{h}(\text{m}^2 \cdot \text{year}), \quad (3)$$

The solar energy gross potential in the village of Shelkovskaya taking into account its land area is:

$$E_g^y = E_g^y \cdot S \\ E_g^y = 1340 \cdot 167,21 \cdot 10^6 = 224,061 \cdot 10^9 \text{ kW} \cdot \text{h/year} \quad (4)$$

These calculations show that the village of Shelkovskaya has a significant solar energy potential which can be used as a source of electricity for a cattle-breeding complex power supplies.

4 Climatic and Energy Characteristics of the Wind in the Chechen Republic

According to the wind stream gross potential the Chechen Republic has a medium level of wind energy.

For efficient utilization of wind energy it's necessary to have full information about the wind as a natural phenomenon and a source of energy. The environmental characteristics needed to estimate wind energy gross potential are:

- Wind speed temporal variations
- An average wind speed V
- Wind speed recurrence $t(V)$, %
- Wind speed duration $P(V)$, %
- Correction factors which take into account changes in the wind on the territory due to heterogeneity of the underlying surface
- Average vertical wind speed profile
- Top wind speed F_{max}^e
- specific power Ns and specific energy Es of the wind, wind energy resources of the region [5].

According to the NASA database the multi-year average wind speed in the village of Shelkovskaya is 4,67 m/s at the altitude of 10 m and 5,91 at the altitude of 50 m, Table 5.

According to the multi-year average wind speed values at the altitude of the weathervane at the wind speed V > 5 m/s there are good conditions for wind energy utilization, V < 4 m/s utilization of wind energy isn't recommended, 4 < V < 5 m/s a business case for wind energy utilization is needed [6–8].

Along with the multi-year average wind speed the specific power of the wind flow is also an original characteristic of the wind intensity total level in the estimated region. It allows to indicate the prospects of wind resources utilization: with specific power less than 100 (Ns < 100 W/m^2) it's not recommended, more than 400 (Ns > 400 W/m^2) there are good conditions, 100 < Ns < 400 – a business case is needed [9].

Let's calculate the specific power and the specific energy of the wind flow at the altitude of 50 m to estimate the prospects of wind energy utilization.

Table 4 Long-term (22 years) monthly average values of insolation falling on a horizontal surface. (kWh/m^2 per day).

Month	I	II	III	IV	V	VI	VII	VIII	IX	X	XI	XII	Year
$Э_g$	1.61	2.36	3.30	4.51	5.67	6.07	5.92	5.05	3.94	2.69	1.67	1.30	3.67

Table 5. Long-term (10 years) monthly and multi-year average wind speed at the altitude of 10 and 50 m over a landscape like an airport.

Month	I	II	III	IV	V	VI	VII	VIII	IX	X	XI	XII	Year
$\overline{V}, \overline{V_0}$ (10 m)	5.11	5.14	4.79	4.59	4.43	4.12	4.26	4.41	4.51	4.61	4.93	5.17	4.67
$\overline{V}, \overline{V_0}$ (50 m)	6.47	6.50	6.06	5.81	5.61	5.22	5.39	5.58	5.72	5.84	6.24	6.55	5.91

In the calculation of the specific power by different means will take into account that the air density doesn't change and equates $\rho_0 = 1{,}226$ kg/m^3.

1. N_S is growing according to the multi-year wind speed value V_0:

$$N_s = \frac{1}{2}\rho V_0^3 = 126{,}54 \text{ W/m}^2, \tag{5}$$

2. Calculation of N_S according to a range of speed observations V_i (a range of 3-hour observation):

$$E_S = \sum_{i=1}^{8760} N_{Si} = \frac{1}{2}\rho \sum_{i=1}^{8760} V_i^3 \Rightarrow N_S = E_S/(8760/3) = \sum_{i=1}^{2920} N_{Si}/2920 =$$
$$= (\frac{1}{2}\rho \sum_{i=1}^{2920} V_i^3)/2920 = 126{,}92 \text{ W/m}^2, \tag{6}$$

3. Calculation of N_S according to multi-year recurrence $t(V)$:

$$E_s = \frac{1}{2}\rho T \cdot V_j^{gr3} t(\Delta V_j^{gr}), \quad N_S = \frac{1}{2}\rho V^3, \tag{7}$$

The results of the calculation are shown in Table 6.

Table 6. The results of the specific power and specific energy calculation.

V_j^{gr}	1	4,5	8,5	12,5	16,5	22
$t(\Delta V)_j^{gr}$	10	52	31	7	0	0
N_{sj}	0,61	55,86	376,46	1197,27	2753,67	6527,22
$Э_{sj}$	0,54	254,45	1022,31	734,16	0,00	0,00

Summing the specific energy values $Э_{sj}$ in the full range of the observed speeds we calculate the multi-year annual specific energy $Э_s$ and the specific power value N_s of the wind flow:

$$E_{Sj} = \sum E_{Sj}; \ N_{Sj} = \frac{E_S}{T} = 230 \text{ W/m}^2, \tag{8}$$

According to these calculations the power values range from 126 to 230 W/m². For the following calculations we will take the lowest value $N = 126,54$ W/m².

The power generated by a wind plant Fuhrlander FL100 with a rated power of 100 kW is calculated as follows:

$$N_{WPP}(v_{hb}) = N'_{g.p.}(v_{hb})\eta_{ww}\eta_r\eta_{gen} = 0,125\rho\pi D_{ww}^2 v_{hb}(t)^3 \eta_{ww}\eta_r\eta_{gen}, \tag{9}$$

where

$$N'_{g.p.}(v_{h6}(t)) = 0,5\rho F_{ww} v_{h6}(t)^3; \ F_{ww} = 0,25\pi D_{ww}^2,$$

Electricity generation by a wind plant in this region according to recurrence of the wind speed for period T is calculated as follows (kWh):

$$E_{WPP}(T) = \sum_{j=1}^{N^{gr}} \left[N_{WPP}(V^{-h6})t(\Delta V_j^{gr}) \right] T, \tag{10}$$

The results are shown in Table 7.

Table 7. Calculation of electricity generation by a wind plant Fuhrlander FL100.

Range of wind speeds, m/s	Average wind speed, m/s	Possibility of wind speed,%	Number of hours per year, h	Wind plant power, kW	Electricity generation, kWh
0–2	1	9,5	832,2	0,00	0
3–6	4,5	52,25	4577,1	7,94	36325
7–10	8,5	31	2715,6	53,49	145244
11–14	12,5	7	613,2	170,10	104306
15–18	16,5	0,25	21,9	391,23	8568
19–25	22	0	0	927,35	0
In total:		100	8760	622,75	294443

The gross potential of the wind flow in the village of Shelkovskaya:

$$\begin{aligned} E_g &= E_g \cdot \frac{S}{20} = N_g \cdot T \cdot \frac{S}{20} = 126,54 \cdot 8760 \cdot \frac{167,21 \cdot 10^3}{20} \\ &= 9267,534\,\text{MW} \cdot \text{h/km}^2 \cdot \text{year}, \end{aligned} \quad (11)$$

The annual electricity consumption of a cattle-breeding complex for 800 head:

$$W_y = P_{\max}T = 257,6 \cdot 3400 = 875,84\,\text{MW} \cdot \text{h}, \quad (12)$$

where $T = 3400$ h is the number of hours at the peak load [10].

The annual electricity consumption of the cattle-breeding complex is 875,84 MWh at the gross potential of the wind flow of 9267,53 MWh/km² per year.

These calculations show that the village of Shelkovskaya has a significant wind energy potential which can be used as a source of electricity for a cattle-breeding complex power supplies.

5 Conclusion

The analysis of the condition and development prospects of the power industry in the Chechen Republic shows the two best directions of its development: the wind and solar power industry.

On the basis of the data obtained we can state that renewable energy utilization allows to use local energy sources and guarantees energy security to the consumers. And the calculations on the electricity supplies to the cattle breeding complex show the possibility of providing electricity to a complex like this from renewable energy sources in the Chechen Republic.

References

1. Energy sector development Scheme and Agenda in the Chechen Republic for the period from 2010 to 2016 (2016). http://minpromchr.ru/investment/development-program. Accessed 17 Sept 2016
2. Russian Unified Energy System development scheme and agenda for the period from 2012 to 2018 (2016). http://so-ups.ru/fileadmin/files/laws/orders/sipr_ups/sipr_ups_12-18.pdf. Accessed 10 Oct 2016
3. Don's engineering Bulletin (2015). http://www.ivdon.ru/ru/magazine/archive/n1y2012/677. Accessed 17 Nov 2016
4. The NASA global meteorological database (2014). https://eosweb.larc.nasa.gov/cgi-bin/sse/grid.cgi?&email=na&step=1&p=&lat=43.509167&submit=&lon=46.327778. Accessed 3 Nov 2016
5. Derugina, G., Malinin, N., Pugachev, R.: The main characteristics of the wind. Wind resources and ways of their calculation. MPEI Publ, Moscow (2012)
6. Bezrukikh, P., Arbuzov, U., Borisov, G.: Sources and efficiency of renewable energy in Russia. Science, St. Petersburg (2002)

7. Fateev, E.: Methods of wind energy inventory elaboration. USSR Academy of Science, Moscow
8. RD. 52.04.275-89: Guidelines for work on evaluation of wind energy resources for proper wind plants location and design. USSR State's committee on meteorology, Moscow (1990)
9. Shtannikova, O.O.: Guidelines on Climatic Characterization of Wind Energy Resources. Gidrometizdat Publ., Leningrad (1989)

Renewable Energy Potential of Russian Federation

E. Solomin[✉], A. Ibragim, and P. Yunusov

South Ural State University, 76 Lenina Avenue,
Chelyabinsk 454080, Russian Federation
e.solomin@bk.ru

Abstract. The paper describes the potential of alternative and renewable energy in Russian Federation and opportunities of efficient usage of pollution-free environmentally friendly renewable energy sources as a one of the solutions for reduction of carbon dioxide emissions of which the main part will be produced by power plants running on the base of renewables – wind, solar, biomass, geothermal, small hydropower, and tide energy. Authors are working on the scientific, public and political issues in the promoting renewable energy sources on the Russian and International market as a prospective developing market. The main barriers and obstacles to renewable energy in Russia are psychological, economical, legislative, technical and organizational. However, all are not high enough and should be taken into account while developing the business model of renewable energy business. The structure of the thermal power generation in Russia on the base of pollution-free environmentally friendly renewable energy sources has been also considered, which showed that boilers and thermal power plant produce the main part of biomass.

Keywords: Renewable energy · Wind · Solar · Geothermal · Hydro · Biomass

1 Introduction

Global climate change had affected Russia among almost all other countries and moved the nations to save the planet for future generations [1–6]. In accordance with Kyoto Protocol, Russia should not exceed the greenhouse gas emissions of 1990 level (2372 Mt or 78% of total GHG emissions), for 2008–2012 period [6].

Predicted dynamics of CO2 emissions of Russian Power Industry in 1990–2030 shown in diagram of Fig. 1.

It shows that starting from 2025 the number of amount of emissions from fossil fuels combustion will overbalance the level of 1990 [7]. The main source of greenhouse gases emissions in Russia is energy segment, which represents more than 1/3 of total emissions in the atmosphere (36.5%) [8].

Since the main source of heating and electrical energy in the World today including Russia is the power plants running on fossil fuels, one of the solutions for reduction of CO2 emissions is the usage of pollution-free environmentally friendly renewable energy sources (RES). Today the share of RES in global energy segment is 3.3%, while in Russia even less than 1% [9–12].

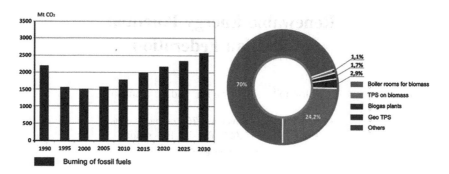

Fig. 1. Predicted dynamics of CO2 emissions of Russian Power Industry up to 2030 (top) and Structure of thermal power generation in Russia on the base of renewable sources of energy, % (lower).

Electric power generation in Russia on the base of RES including small Thermal Power Plant (TPP) in 2011 was 6320.1 million kW h [13]. Of that:

- TPP on biomass - 2995.0 million kW h (47.4%)
- Small TPP - 2846.0 million kW h (45%)
- Geothermal TPP (GTPP) - 474.9 million kW h (7.5%)
- Wind Power Plants (WPP) and others - 4.2 million kW h (0.1%)

The structure of the thermal power generation in Russia on the base of RES demonstrated in Fig. 1, which shows that boilers and TPP produce the main part of biomass, including burning of firewood in the countryside.

Why the development of renewable energy in Russia is still slow? The answer is because of the huge amounts of fossil fuels resources. Which is actually misleading. In accordance with various expert conclusions, the oil and gas will be over in about 100–200 years. The main barriers and obstacles to renewable energy in Russia are psychological, economical, legislative, technical and organizational as shown in of Fig. 2.

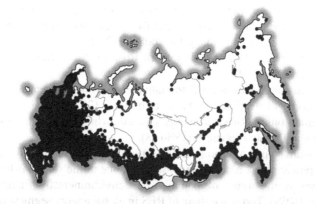

Fig. 2. Barriers and obstacles to RES development in Russia (top) and Centralized electric power supply (lower).

Some legal documents intended to promote renewables had started appearing from 2008 [14–17]. The most important of them is the Direction of Russian Federation Government #1p of April 8th, 2010, regulating the part of RES in electric energy production as follows: in 2010 the share of RES should be 1.5%; in 2015 – 2.5%; in 2020 – 4.5%.

Implementation of this Direction should start green energy development in Russia and by 2020, the contribution of RES should reach 25 GW, of which the main part will be produced by power plants running on biomass, small Hydropower Plants (HPP), wind, solar and tides [18].

The said development will focus primarily on the territories with no electric lines or decentralized power supply zones, which exceed 70% of the area representing over 10 million people as shown in Fig. 2.

Table 1, gives the economic potential of RES (as a part of the gross and technical potential) which is determined in the amount 320 million tons of conventional fuel. There is still no widespread development of renewable energy in Russia. However, there are many good examples of using this potential.

Table 1. Assessment of the potential of RES in Russia.

RES	Gross potential, million tons of conv., fuel/year	Technical potential, million tons of conv., fuel/year	Economical potential, million tons of conv., fuel/year
Wind energy	44326	2216	11
Small hydro energy	402	126	70
Sun energy	2 205400	9695	3
Bio energy	467	129	69
Geothermal energy	a	11869	114
Low-potential heat	563	194	53
Total	2 251158	24229	320

Here (a) is gross potential of hydrothermal energy is 29.2 trillion tons of conventional fuel.

2 Wind Energy

The installed capacity of wind power plants in Russia is planned in the amount of 4750 MW by 2020 as it is shown in Fig. 3.

Now wind power plants are operating mostly in Central and North-West Federal Districts of Russia as well as in Ural and Volga Federal Districts with less concentration [21]. There are about 50 key players on Russian wind power market but only the half of them fabricate the turbines. Almost all of them have the self-developed design. Less than 1% of the wind turbines made under the transfer of foreign technologies.

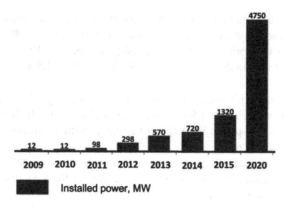

Fig. 3. The planned wind power capacity in Russia.

A family of small unique autonomous Vertical Axis Wind Turbines (VAWT) on 0.1–30 kW with high energy and operational parameters was designed and developed by South Ural State University in 2000–2017 [22]. Turbines are used in various combinations including hybrid applications as shown in Fig. 4.

3 Solar Energy

The most solated parts of Russia are the Northern Caucasus, Far East and South Siberia where the average annual influx of solar radiation is 4–5 kW h per square meter per day, which is comparable with South Germany and Northern Spain, leading countries of the implementation of photovoltaic systems (PVS).

Fig. 4. VAWT family and wind-solar hybrid plants.

The main manufacturers of solar panels in Russia are:

- NPP Kvant (Moscow)
- Solnechny veter, Ltd. (Krasnodar)
- NPC Krasnoe znamya (Ryazan)
- Mining and chemical combine RosAtom (Zheleznogorsk)
- Nitol Solar (Usolyc-Sibirskoe) involved in the production of polysilicon

The Scientific potential of solar energy is concentrated mainly in Joffe Institution, Rusnano, Moscow State University, VIESH, etc.

The market for solar energy today is represented rather poorly. Most of the solar panels arrive in Russia from other countries, primarily from China. The structure of solar panels imports in 2011 is shown in the diagram of Fig. 5 in accordance with the country of origin. Export from Russia represents the solar panels, assembled or not assembled on the base of monosilicic, and silicic cells. Germany and the Czech Republic are the main countries-importers of Russian solar products.

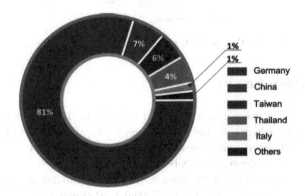

Fig. 5. Structure of solar panels imports in Russia in 2011, percent (Source: EPIA, AS MARKETING).

4 Bioenergy

It is the fast-growing segment of the renewable energy industry in Russia. The main areas of development are:

- Biofuel
- Energy mechanisms of nature
- Waste utilizing and ecology saving
- Biomass burning
- Bioenergy engineering

Among them, there are innovation projects of high investment priority:

- Agro-energy clusters
- Municipal bioheat supply
- Aviculture bioenergy supply

Biogas plants and biofuel production in Russia are shown on the map in Fig. 6.

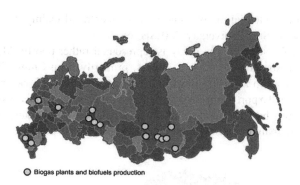

Fig. 6. Biogas plants and biofuel production in Russia.

5 Small Hydro Energy

The number of small rivers in Russia is over 2.5 million. The total drain is more than 1000 km^3 per year which give the annual energy potential of 500 billion kW h. The potential of small hydro energy resources in Federal Districts is shown in Fig. 7, represented mostly by the rivers of Siberia and the Far East.

INSET (Saint Petersburg) is the main manufacturer of hydro energy equipment for small HPP which is used at "Kiwi-Koivu" Mini-HPP (Karelia).

South Ural District is the location where the unique small hydropower plant was installed more than 100 years ago, providing the Porogi villagers with inexpensive energy. The author and designer of this project is well-known scientist and politician Boris Bakhmetev, who managed the development of HPP in Lappeenranta.

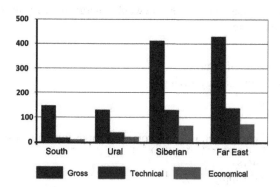

Fig. 7. The potential of small hydro energy resources of Russian Federal Districts (Billion kW h per year).

6 Geothermal Energy

Chukotka, Kuril Islands, Seaside region, Western Siberia, North Caucasus and the area near Lake Baikal, have significant geothermal resources. Heat pumps based on the low-potential heat could be used on two thirds of the territory of Russia. Geothermal energy potential in Russia 10–15 times exceeds the resources of fossil fuels and represents the heat in the range of 30…200 °C. There are more than 4,000 wells of 5 km depth had been drilled for the local usage of low potential heat.

Kamchatka and the Kuril Islands are the most "hot" points of Russia. Geothermal resources of Kamchatka exceed 5,000 MW.

There are three geothermal power plants installed in Kamchatka region: Pauzhetskaya, Verhne-Mutnovskaya and Mutnovskaya GTPP. The total capacity of these plants is about 80 MW. Pauzhetskaya GTPP was the first Russian geothermal power plant built in 1966 in the south of Kamchatka on Pauzhetka river valley on 5 MW power increasing up today to 11 MW.

7 Tides Energy

The tides energy is also included in the electric energy segment development program up to 2020. The plan is to build two tidal power plants (TiPP). Mezenskaya TiPP on 8 GW power in the Mezenskaya Bay of Arkhangelsk District, and Tugurskaya TiPP on 3640 MW power in the Tugursky Bay of Okhotskoye Sea in Khabarovsk District.

The project for the year 2012 is the construction of North TiPP in Murmansk District in Barentsevo Sea. Nominal calculated power is 12 MW; average power production is 19 million kW h per year.

8 Conclusion

Due to the efforts of scientists, ecologists, volunteers and progressively thinking people the situation in Russia is coming to the step-by-step increasing of the part of renewables in generating of power in the country. Some of the innovations of RF designers are promising in the development of the RES infrastructure [23, 24] especially involving the battery charging systems [25]. Authors are working on the scientific, public and political issues in the promoting RES on the Russian market.

References

1. Martinez-Cesena, E.A., Mutale, J.: Wind power projects planning considering real options for the wind resource assessment. IEEE Trans. Sustain. Energy 3(1), 158–166 (2012)
2. Popel, O.S., Frid, S.E., Kolomiets, Yu.G., et al.: Atlas of resources of solar energy on the Russian territory. OIVT RAN, Moscow (2010)
3. Solomin, E., Kirpichnikova, I., Martyanov, A.: Iterative approach in design and development of vertical axis wind turbines. Appl. Mech. Mater. **792**, 582–589 (2015)
4. Bezrukih, P.P., et al.: Resources and efficiency of the use of renewable energy sources in Russia. Science, St. Petersburg (2002)

5. Boute, A.: A promoting renewable energy through capacity markets: an analysis of the Russian support scheme. Energy Policy **46**, 68–77 (2012)
6. Boute, A.: A comparative analysis of the European and Russian support schemes for renewable energy: return on European experience for Russia. J. World Energy Law Bus. **4**, 157–180 (2011)
7. Boute, A.: The Modernization of the Russian electricity production sector: Regulatory risks and investment protection. Ph.D. thesis at the University of Groningen (2011)
8. Gore, O., Viljainen, S., Makkonen, M., et al.: Russian electricity market reform: deregulation on reregulation. Energy Policy **41**, 676–685 (2011)
9. Government of the Russian Federation. Decree 426 on the qualification of the renewable energy installation, 3 June 2008
10. Government of the Russian Federation. Resolution 1662-r on the Concept for long-term social and economic development to 2020, 17 November 2008
11. Government of the Russian Federation: Resolution 1-r on the main direction for the state policy for improvement of energy efficiency of the electricity sector on the base of renewable energy sources for the period up to 2020 (2009)
12. Government of the Russian Federation: Resolution 1715-r on Russia's energy strategy until 2030 (2009)
13. Government of the Russian Federation: Decree 238 on the determination of price parameters for the trade in capacity (2010)
14. Government of the Russian Federation: Decree 1172 approving the wholesale market rules (2010)
15. Government of the Russian Federation: Resolution 2446-r on the Federal energy efficiency program for the period to 2020 (2010)
16. President of the Russian Federation: Decree 861-rp on the climate doctrine of the Russian Federation (2009)
17. Kopylov, A.: New approach to the support of RES in Russia on the base of the payment of generating power. Energi Bill **1**, 35–42 (2011)
18. Keane, A.: Capacity value of wind power. IEEE PES. Trans. Power Syst. **26**, 564–572 (2011)
19. Solomin, E.V., Sirotkin, E.A., Martyanov, A.S.: Adaptive control over the permanent characteristics of a wind turbine. Procedia Eng. J. **129**, 640–646 (2015)
20. Martyanov, A.S., Solomin, E.V., Korobatov, D.V.: Development of control algorithms in Matlab/Simulink. Procedia Eng. J. **129**, 922–926 (2015). International conference on industrial engineering, 23 November 2015
21. Martyanov, A.S., Korobatov, D.V., Solomin, E.V.: Research of IGBT–transistor in pulse switch. In: 2nd International Conference on Industrial Engineering, Applications and manufacturing, Chelyabinsk, 19–20 May 2016 (2016). Procedia engineering Journal
22. Keller, A.V., Korobatov, D.V., Solomin, E.V.: Development of algorithms of rapid charging for batteries of hybrid and electric drives of city freight and passenger automobile transportation vehicles. In: Proceedings of the International Siberian Conference on Control and Communications (2015)
23. Louie, S., Miguel, A.: Lossless compression of wind plant data. IEEE Trans. Sustain. Energy **3**(3), 598–606 (2012)
24. Hatziargyriou, N., et al.: Energy management and control of island power systems with increased penetration from renewable energy sources. In: IEEE PES 2002 Winter power meeting, New York, January 2002 (2002)
25. Parsons, B., et al.: Grid impacts of wind power: a summary of recent studies in the United States. Wind Energy J. (2003)

Study of Spiral Air Accelerators for Wind Power Plants Using a Vertical Rotation Axis

A. A. Bubenchikov[✉], T. V. Bubenchikova, and E. Yu. Shepeleva

Omsk State Technical University, 11 Mira Avenue, Omsk 644050,
Russian Federation
antech-energo@mail.ru

Abstract. The article analyzes the use of the simplest designs of air flow accelerators. The best geometry of a multi-blade construction is determined for use as a flow accelerator for wind turbines with a vertical axis of rotation. For analysis, during operation, 12 models of concentrator plants were created: three designs with three, five, seven-blade and three with six-blade. The analysis of the use of the simplest structures representing the blades having the base of the Archimedes spiral and the analysis of the execution of the upper part of the structure to simplify installation and additional rigidity. The effect of the use of additional bridges in the channel for additional lifting and acceleration of the flow is studied, and boundary options for the geometry of the design of bridges that can be used if necessary to simplify the installation of the structure are selected. Optimum geometry with maximum efficiency for three five and seven-blade structures was chosen. To determine the dependence of acceleration on wind speed for optimal designs, experimental studies were carried out at various free-stream speeds. The best geometry was determined for its use as a flow accelerator for wind turbines with a vertical axis of rotation. The achieved increase in speed is 52%. Since the power depends on the wind speed in the cube, this design will increase the power generated by the wind power installation by 3 times.

Keywords: Wind power · Flow accelerator · Hub · Confuser

1 Introduction

In many parts of the world wind power engineering has already reached the level allowing it to become the main power source. The growth of wind power in developed countries, especially in Europe, for a long time was due to the problem of global climate change [1].

Wind power is the most attractive solution to global energy problems. It does not pollute the environment and does not depend on fuel. Moreover, wind resources are present in any part of the world and they are enough to meet the growing demand for electricity.

However to implement the electricity generation using wind power in regions with low wind flow velocities the application of classical wind turbines do not produce the expected results. In this connection the question arises about developing new structural designs of wind turbines deprived of these shortcomings.

According to the wind Atlas the average wind speed on the Central Russia territory does not exceed 3–5 m/s [2]. Most wind turbine available on the market are designed for the wind velocity of 10 m/s and at the flow rate of 3–5 m/s and don't produce more than 10–30% of their committed capacity.

For developing new sources of wind energy it is necessary to design wind turbines with a wind power hub able to work effectively in the regions with low wind flow velocities.

2 Development of Models

Various additional structures, enabling to increase speed, flow density or add rotation are developed to improve the efficiency of wind turbines [3, 4]. They are confuser-diffuser channels of the tower-like wind turbine with several wind turbines, stacked multi-blade designs, installations with different blade modifications and those with guide planes. Also they are hybrid plants: solar wind, hydro wind installations and horizontal-axis wind turbines with pneumatic way of transmitting air power [5–14]. Some installations allow increasing the flow velocity several time according to the authors.

Unfortunately, over 90% of these inventions have remained at the stage of patenting and haven't received its implementation. The probable causes may be the large size, and hence high cost of the proposed installations, vibration and noise unacceptable to people and animals, the non-conformance between the real design and claimed characteristics.

At present the niche of wind turbines with a vertical rotation axis [15–17] able to operate at low speeds of wind flow remains poorly developed, and new technical solutions that do not require developing complex devices, but at the same time allowing to improve significantly power parameters and characteristics are of great practical interest.

For the developed wind turbine with a flow accelerator the following conditions are made:

- Accelerating the air flow in 2–2,5 times more compared to 20–40% of counterparts
- Increased reliability
- Excluding dynamics items with backward motion in the wind orientation and more stability, if needed (in contrast to mast wind turbines)
- Environmental safety (reduced noise level, no adverse effect on birds), for this purpose it is proposed to place a wind wheel in a hub body
- Ability to work effectively in regions with low wind load (up to 3 m/s)
- Acceptable cost, size and appearance

It is very difficult to implement some conditions at the same time, because some vortex plants have a very complex form of confusing channels and the design in general, which not only increases the installation cost, but also complicates its operation.

An ideal hub is known to be a tapering tube, curved in a logarithmic helix with spiral guides inside [18, 19]. Designing, operating and repairing such an accelerator are very difficult, so during the operation it was proposed to undertake a study of some intermediate designs from the simple to the complex, indicating the effect of changing the geometry and introducing additional elements to installation angles. Designing, manufacturing and experimental investigations of a series of flow accelerator models were carried out.

For the set tasks theoretical and experimental methods were used.

Theoretical studies were conducted based on the methods of hydro - and aerodynamics using mathematical simulation by means of the ANSYS CFX software [20].

In experimental studies the optimal parameters of accelerator models with different configurations have been determined. The studies were conducted in a laboratory consisting of:

1. Testing laboratory premises 14 m in length for studies;
2. Air injection systems (1 kW vertical axis fans);
3. GS-0.5 low-speed generator for studies of wind turbines with an accelerator under load.

For measuring the flow velocity 4 wind instruments were used. They were installed at the accelerator inlet and outlet. Each measurement was carried out during 120 s in recording readings 2 times per second. A mean value was accepted as final.

The following parameters were taken as constant ones for the studied design:

- The base diameter is 1.4 m
- The outlet diameter in the top of the accelerator is 0.6 m
- The installation height is 1.5 m

In the experiment, the following parameters were changed:

- The radius of the blades curvature, affecting the blades length (the larger the radius, the smaller the length of the blade in the same dimensions)
- Blade number
- The air flow direction at the confuser inlet
- The swirl (linear) air rise tilt in the body of confusing channels by adding flow generating areas (bridges of various shapes)

In the first approximation, the possibility of using the logarithmic spiral as guides at the accelerator base was analyzed (Fig. 1).

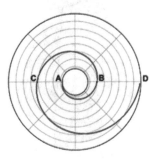

Fig. 1. An example of a logarithmic spiral at the accelerator base.

For the model base three versions with different length of the AB, AC and AD arc were offered. Three-, five- and seven-bladed designs were developed for each of them (Fig. 2).

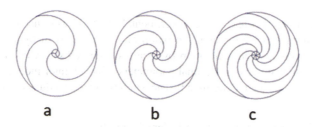

Fig. 2. Examples of the installation base (a) tree blades; (b) five blades; (c) seven blades.

As a result nine models, differing by the arc length and the number of blades, were received. An example of the geometry of the five-bladed design with AB and AC arc length is presented in Fig. 3.

Fig. 3. The inlet and outlet geometry of the five-bladed accelerator channel (a) the AB arc length design; (b) the AC arc length design.

Further all developed models were numbered for convenience according to Figs. 4, 5, 6, 7, 8, 9, 10, 11 and 12.

Fig. 4. Model No.1 – three-bladed with AB arc length.

Study of Spiral Air Accelerators for Wind Power Plants 481

Fig. 5. Model No.2 – three-bladed with AC arc length.

Fig. 6. Model No.3 – three-bladed with AD arc.

Fig. 7. Model No.4 – five-bladed with AB arc length.

Fig. 8. Model No.5 – five-bladed with AC arc length.

Fig. 9. Model No.6 – five-bladed with AD arc length.

Fig. 10. Model No.7 – seven-bladed with AB arc length.

Fig. 11. Model No.8 – five-bladed with AC arc length.

Fig. 12. Model No.9 – five-bladed with AD arc length.

It is notable that the principal dimensions for all models in all experiments remain the same:

- The height is 1.5 m
- The inscribed diameter of the base is 1.4 m
- The inscribed diameter of the top is 0.6 m

3 Result of Research

During the experiments it was found that under flow acceleration the ratio of the average velocity at the channel inlet and outlet decreases in reducing the input velocity, and in its increasing it becomes greater. Therefore, the analysis of the accelerator at low wind velocities approximate to our region average speed was of main interest. The velocity of 3 and 5 m/s was preset and design investigations were carried out in using the Ansys program. Examples of air distribution for a three-bladed model are presented in Fig. 13.

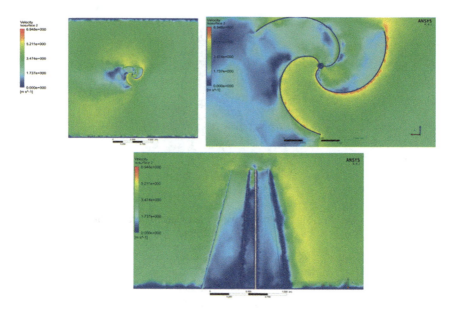

Fig. 13. The Ansys analytical simulation of the model No.1.

The measurement results at the velocity of 3 m/s when the flow is directed perpendicular to the inlet channel area (Fig. 14) are presented in Table 1.

Similar measurements were carried out at the flow velocity of 5 m/s (Table 2).

After the analytical simulation the models of the proposed designs were developed. Experimental studies under various wind velocities were conducted for them. The measurements were carried out for all channels to determine the operation in conditions close to real ones. In the first stage the velocity was measured on the installation approach and at the points where the accelerator channel inlets will be placed. After determining the average flow velocity from the source the main experiments were carried out.

Fig. 14. The flow direction in the Ansys model analysis.

Study of Spiral Air Accelerators for Wind Power Plants   485

Table 1. Measurement results of the three-bladed models 1–8 at the average ram air velocity of 3 m/s.

Model number	Model No. 1		Model No. 2		Model No. 4		Model No. 5		Model No. 7		Model No. 8	
	Top	Bottom	Top	Bottom	Top	Bottom	Top	Bottom	Top	Bottom	Top	Bottom
Average velocity, m/s	4.544	1.555	5.01	1.53	4.613	1.907	4.613	1.907	4.613	1.907	5.34	1.77
Velocity increase	2.92		3.29		3.26		3.08		2.42		3.01	

Table 2. Measurement results of the three-bladed models 1–8 at the average ram air velocity of 5 m/s.

Model number	Model No. 1		Model No. 2		Model No. 4		Model No. 5		Model No. 7		Model No. 8	
	Top	Bottom	Top	Bottom	Top	Bottom	Top	Bottom	Top	Bottom	Top	Bottom
Average velocity, m/s	7.52	2.58	7.57	2.87	8.34	2.36	8.58	2.69	7.616	3.17	9.06	2.64
Velocity increase	2.9		2.63		3.55		3.18		2.39		3.42	

It should be pointed out that in measuring a fan system unable to define correctly the laminar flow was used. But such conditions are enough to estimate the acceleration parameters of different installations, as the measurements were conducted simultaneously at the channel inlet and outlet by a group of the anemometers with the PC registration in real time. This allows assessing the situation close to real flow changes. All the measurements were taken in the same conditions with minimal experiment intervention.

Such measurements were carried out for five- and seven- bladed constructions.

The measurement results of the most acceleration received for these constructions are given in Tables 3 and 4.

Table 3. Measurement results of the five bladed models 4 and 5 at the average ram air velocity of 1,5 m/s.

Model number	5 blades (model No. 4)						5 blades (model No. 5)					
Rotation angle	$\alpha = 0$		$\alpha = 72$		$\alpha = 144$		$\alpha = 0$		$\alpha = 45$		$\alpha = 90$	
	Top	Bottom	Top	Bottom	Top	Bottom	Top	Bottom	Top	Bottom	Top	Bottom
Average velocity, m/s	1.312	0.574	0.89	0.368	0.289	0.0494	2.607	1.34	2.81	1	1.172	0.48
Velocity increase	2.286		2.418		5.85		1.946		2.81		2.442	

Table 4. Measurement results of the three-bladed models 1–8 at the average ram air velocity of 5 m/s.

Model number	7 blades (model No. 7)									7 blades (model No. 8)									3 blades (model No. 5)			
Rotation angle	$\alpha = 0$		$\alpha = 51{,}4$		$\alpha = 102{,}8$		$\alpha = 154{,}2$		$\alpha = 0$		$\alpha = 51{,}4$		$\alpha = 102{,}8$		$\alpha = 154{,}2$		$\alpha = 0$		$\alpha = 120$	
	Top	Bottom	Top	Bottom	Top	Bottom	Top	Bottom	Top	Bottom	Top	Bottom	Top	Bottom	Top	Bottom	Top	Bottom	Top	Bottom
Average velocity, m/s	1.017	0.394	1.075	0.678	0.364	0.253	0.446	0.104	1.306	0.524	1.573	0.689	0.388	0.155	0.193	0.069	1.37	0.575	0.279	0.269
Velocity increase	2.58		1.59		1.44		4.29		2.49		2.28		2.5		2.797		2.38		1.04	

After measuring the simplest designs, the models No. 3, 6, and 9 were dropped. They had a long redundant channel where a flow met the maximum deceleration. These models are superior to the others as to the cost variance.

Further two top versions were considered (Fig. 15).

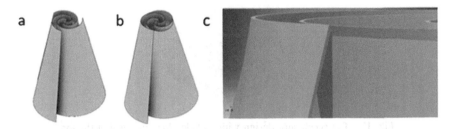

Fig. 15. Versions of an installation top (a) initial spiral design; (b) external view of the installation with a changed output design; (c) the contact of two blades in the changed design.

The second version (Fig. 15b, c) facilitates the option of the assembly in adding an upper block containing wind wheel blades (Figs. 16 and 17). Experimental studies have shown that the resulting acceleration obtained due to changes of the output geometry shape varies in minimum and any version can be selected without changing the beneficial properties of the design.

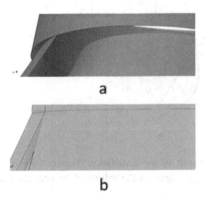

Fig. 16. Versions of the construction top in adding a blade block (a) contact of a blade block on the initial design; (b) close contact of a blade block on the changed design.

Fig. 17. The accelerator design with a blade block installed at the top.

Further the versions of adding bridges in contraction channels for supplementary and more gradual flow rise were considered. Three versions of bridges for each installation were simulated and investigated. They are shown in Fig. 18.

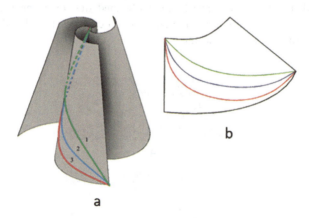

Fig. 18. Three bridge versions in the installation channel (a) the bridges position on a three-bladed installation; (b) bridge edges on the rolling of an acceleration blade.

Determining the best geometry of a bridge was done firstly in the ANSYS program complex, and then was checked and corrected using an experimental model. Change results in laboratory conditions are presented in Tables 5, 6 and 7.

In fact, the bridge version number was over 16. Versions No. 1 and No. 3 are presented as boundary ones of very flat and sharp bridge rise for analyzing their use with the possible design simplification. The results of the best geometry are included in Table 8, i.e. the model No. 4 bridge No. 2. The similar work was done with other models.

Table 5. Measurement results of three bridge versions of the five-bladed model No. 5 in laboratory conditions.

Model number	Model No. 5 bridge 1				Model No. 5 bridge 2				Model No. 5 bridge 3			
Rotation angle	$\alpha = 0$		$\alpha = 72$		$\alpha = 0$		$\alpha = 72$		$\alpha = 0$		$\alpha = 72$	
	Top	Bottom	Top	Bottom	Top	Bottom	Top	Bottom	Top	Bottom	Top	Bottom
Average velocity	1.196	0.596	1.021	0.471	1.166	0.461	1.166	0.461	1.2635	0.644	1.2635	0.644
Velocity increase	2.007		2.168		2.529		2.065		1.962		1.25	

Table 6. Measurement results of three bridge versions of the five-bladed model No. 4 in laboratory conditions.

Model number	Model No. 4 bridge 1				Model No. 4 bridge 2				Model No. 4 bridge 3			
Rotation angle	$\alpha = 0$		$\alpha = 72$		$\alpha = 0$		$\alpha = 72$		$\alpha = 0$		$\alpha = 72$	
	Top	Bottom	Top	Bottom	Top	Bottom	Top	Bottom	Top	Bottom	Top	Bottom
Average velocity	1.276	0.753	1.006	0.427	1.481	0.557	1.012	0.495	1.312	0.889	0.993	0.642
Velocity increase	1.693		2.351		2.65		2.263		1.475		1.545	

Table 7. Measurement results of three bridge versions of the best seven- and three-bladed installations in laboratory conditions.

Model number	Model No. 7 (seven-bladed)						Model No. 1 (three-bladed)			
Rotation angle	$\alpha = 0$		$\alpha = 120$		$\alpha = 102,8$		$\alpha = 0$		$\alpha = 120$	
	Top	Bottom	Top	Bottom	Top	Bottom	Top	Bottom	Top	Bottom
Average velocity, m/s	1.517	0.668	1.606	0.515	0.339	0.132	1.3867	0.531	0.325	0.123
Velocity increase	2.27		3.12		2.56		2.61		2.64	

Table 8. Results of measuring flow acceleration at a different orientation angle of the inlet area to the ram air.

Model number	Model No. 4 bridge 2 (five-bladed)							
Rotation angle	$\alpha = 0$		$\alpha = 45$		$\alpha = 90$		$\alpha = 75$	
	Top	Bottom	Top	Bottom	Top	Bottom	Top	Bottom
Average velocity	1.2195	0.459	1.314	0.3678	0.294	0.231	1.012	0.495
Velocity increase	2.66		3.58		1.27		2.263	

After finding the optimal geometry the analysis of the flow angle attack was conducted to determine the best acceleration. The experiment was carried out by turning the construction around the axis at a pitch of 15° are shown in Table 8. The best angle is 45° and the worst one is 90°.

At the next stage the acceleration coefficient analysis at different flow velocities for the optimal bridge designs of No. 4 and No. 5 models.

The air blast system was presented by a fan of 1 kW. The flow velocity is known to decrease in increasing distance from the source. A series of experiments was conducted to determine the dependence of the air flow velocity on the distance from the inlet area of a confused channel. Later experiments at a different flow velocity were carried out to confirm and correct initial results.

After analyzing all designs, optimal ones were chosen and acceleration measurements relative to the ram air were performed. As a result, it was confirmed on all installations and was in the range of 40–52%, which ultimately allows increasing power generated by wind turbines twofold.

4 Conclusion

The research results are the following:

1. Twelve models of hub power plants are developed, three of them having three, five and seven blades and three are six-bladed one;
2. The analysis of the use of the simplest installations which are blades having the Archimedean spiral at their base;
3. The analysis of the installation top was conducted for facilitating the assembly and extra rigidity;
4. The analysis of using additional bridges in a channel for additional rise and flow acceleration;
5. Boundary geometry versions of bridges which can be used in simplifying an installation assembly are analyzed and chosen;
6. Optimum geometry with maximum efficiency for the three five and seven-bladed installations is chosen;
7. For determining the dependence of acceleration on wind speed for optimal designs the experimental studies with different ram flow velocity are conducted;

8. Five-bladed turn design with the bridge No. 2 (model No. 4) is the best geometry to be used as a flow accelerator for wind turbines with a vertical rotation axis. The achieved speed increase is 52%. Since the power depends on wind speed cubed, such a design will increase the power produced by the wind turbine threefold.

References

1. Lisitsyn, A.N., Zadorozhnaya, N.Z.: Concerning the prospect of wind power in the modern world. In: Collection of Articles of V International Scientific-Practical Conference, pp. 36–42 (2017)
2. Klochko, A.A., Romanovskaya, M.A.: The national Atlas of Russia, vol. 2. Nature and Environment. Gosgistcentr, Moscow (2004)
3. Sokolovsky, Yu.B., Sokolovsky, A.Yu.: Improving the efficiency of wind power plants. Energy saving. Power engineering. Energy Audit 9(127), 28–37 (2014)
4. Biramov, F.D., Galimov, N.S., Ivanov, V.A.: Ways to improve the efficiency of the wind turbine rotor with a rotation vertical axis in Megapolis. Sci. Tech. Bull. Volga Region 2, 99–102 (2014)
5. Zanegin, L., Petro, Yu., Suchankova, E.: Wind power plant. RU Patent 2387871, 27 April 2010 (2010)
6. Tebuev, V.V.: Power generation with the placement of wind generator in vertical exhaust air duct in designing residential buildings. RU patent 2369772, 10 October 2009 (2009)
7. McDavid, Jr., William, K.: Fluid-powered energy conversion device. RU Patent 2694010, 08 July 2019 (2019)
8. Golovenkin, E.N., Zadorin, I.N., Khalimanovich, V.I., et al.: Carousel wind turbine. RU Patent 2432493, 16 February 2018 (2018)
9. Sigaev, V.P., Mihov, A.P.: Wind power. RU Patent 2422673, 27 June 2011 (2011)
10. Smolich, R., Buryak, A., Buryak, D.: Method of converting wind energy and device for its implementation. RU Patent 2403439, 10 November 2010 (2010)
11. Yarygin, L.A., Ermakov, I.G.: Wind turbine. RU Patent 2399789, 20 September 2010 (2010)
12. Sigaev, V.P., Mihov, A.P.: Wind power plant. RU Patent 2390654, 27 May 2010 (2010)
13. Kolomatsky, S.I., Kolomatskiy, D., Kolomak, E.S.: The vortex wind turbine. RU Patent 2386853, 20 April 2010 (2010)
14. Strebkov, D.S., Tanygin, V.V.: Wind turbine rotor. RU Patent 2383775, 10 March 2010 (2010)
15. Galas, M.I., Dimkovets, Yu.P., Kev, N.: Concerning the feasibility of vertical-axis electrical installations of a megawatt class. Energ. Constr. 3, 33–37 (1991)
16. Turian, K.J.: Power of wind turbines with a vertical axis of rotation. Aerosp. Eng. 8, 105–121 (1988)
17. Lee, Y.T., Lim, H.C.: Power performance improvement of 500 W vertical axis wind turbine with salient design parameters. Int. J. Mech. Aerosp. Ind. Mechatronic Manuf. Eng. 10(1), 84–88 (2016)
18. Bubenchikov, A.A., Gorlinsky, N.A., Tcherbinov, V.V., et al.: Flow hubs for wind turbines. Young Sci. 28(2), 10–14 (2016)
19. Taho-Gody, A.Z.: Air flow hubs for a wind power station, operated by a tracking system with an optimal angle control loop. Fundamental 2-14, 3056–3058 (2015)
20. ANSYS CFX. Website of the product company. http://www.ansys.com/products/fluids/ansys-cfx. Accessed 25 Sept 2017

Generating Gas from Wood Waste as Alternative to Natural Gas in Package Boilers

E. M. Kashin[1(✉)], R. R. Safin[2], and V. N. Didenko[1]

[1] Izhevsk State Technical University, 7 Studencheskaya Street,
Izhevsk 426069, Russian Federation
kawuh@yandex.ru
[2] Kazan National Research Technological University, 68 Karla Marksa,
Kazan 420015, Russian Federation

Abstract. The article is devoted to the perspective way of utilization of wood waste by gasification. The design of a gas generator with a rotating active zone of vertical type is offered. It allows increasing efficiency of the process of gasification and expanding a scope of gas generators as the devices for utilization of solid waste. A mathematical model of a rotor gas generator is developed allowing one to define constructive characteristics of the device and the indicators of the gasification process, depending on initial characteristics of fuel, regime parameters, physical and chemical parties of the process. The necessary quantity of gas generators with the cylindrical mine in the transfer of boilers from natural gas to waste wood is defined on the basis of the received results. According to the results one gas generator of a rotor type of reciprocal circuit with an extreme diameter of mine of 0.7 m and 2.47 high is capable to ensure a complete functioning of a copper of KV-GM-4.65–150. At the same time it is rather compact, that is caused by an internal arrangement of layers - active zones by the principle "one in another" and has a wide range of the given power varied depending on the speed of rotation of a rotor that gives the chance to use any grades of fuels including low-grade.

Keywords: Gas generator · Solid fuel · Waste · Generating gas

1 Introduction

Currently, in connection with rise in prices for gas and electricity, there is a renewal of interest in waste management issues, logging and timber processing, agricultural residues, used tires, household waste, oil sludge and processing of local low-calorie fuels: coal, peat to produce heat and electrical energy.

One of the ways of increase of efficiency of use of fossil solid fuels is the creation and implementation of clean, multi-purpose technologies in which organic and mineral mass is a valuable natural raw material for the production of electricity, heat and other types of products. One of these technologies is gasification of solid fuels, the purpose of which consists in transformation of fuel into gas generator by partial oxidation by the gasification agent (air, oxygen, steam or a mixture of these substances) and in the use of gas as an energy source or as a chemical feedstock [1–16].

Apparatus for producing generating gas - gas generators - are the devices, intended for gasification of raw materials of organic origin by means of reactions of interaction of carbon of raw materials with oxidizers, which final result is receiving of combustible gases SO, SN4, N2.

Currently there is an increased interest in gas generation in the world, as one of the most relevant methods for the use of renewable energy resources. The authors [1] believe that one of the most promising ways to produce green fuels is combination of synthesis biomass gasification and synthesis Fischer-Tropsch (FT), in which the biomass is gasified and is used to obtain hydrocarbons after cleaning.

The main characteristics of supercritical steam gasification of biomass are offered in the work of authors [2]. They point out that concentration of the diluted initial raw materials and high temperatures of reaction lead to highly effective gasification of carbon. However, influence of speed of heating on supercritical gasification of water isn't completely investigated.

According to Minou and Fang [3], the cellulose is first dissolved in water in the process of gasification in supercritical steam, and then is subjected to gasification and simultaneously produces a resinous material. Authors found out that resinous forms of material would never be gasified. Therefore, suppression of production of resinous material is a key to improving the efficiency of gasification.

Seo and others [4] determined the effects of gasification temperature (750–900 °C), the ratio of steam/fuel (0,5–0,8) and the ratio of biomass (0–1,0) to the composition of the gas in the double circulating gasifier with a fluidized layer. They found out that the gasification temperature played an important role in that process (higher temperature leads to a large amount of gas and less content of pitch).

Gasifiers with a fluidized layer are the later developments in which excellent characteristics of mixing and high speeds of reaction are used [5]. They work at the temperature up to 1000 °C and therefore are used only for gasification of biomass as a rule [6].

Aznar and others [7] conducted experimental work on joint gasification of biomass and coal to study the effect of equivalence ratio and mixing ratio on gas composition, pitch content and amount of carbon. They found out that the layer temperature increase caused a decrease in calorific value. The authors [8] carried out experimental study of gasification of biomass flow and coal to determine the effect of the ratio of fuel/air (2,5–7,5), temperature of gasification (750–1150 °C) and the ratio of biomass/coal to the composition and performance of the gas generator. They concluded that increasing the ratio of biomass/coal the relationship H_2/CO, CO/CO_2 and H_2/CO_2 increased, while the ratio of CH_4/H_2 followed the opposite trend.

Despite the numerous works in the field of gasification [9–16], the use of generating gas as an alternative energy source has not received wide distribution due to many reasons, one of which is a low power of gas-generating unit, as well as the lack of regulatory documentation for gasification. Having studied all the shortcomings of the existing generators [17–20], the authors propose a promising mathematical model of the gas generator flow solid fuel with a rotating active area.

2 Methods and Materials

The authors propose a design of the gas generator with a rotating active area of vertical type for the process of gasification of solid fuel, which is shown in Fig. 1. The principle of operation of the gas generator lies in the fact that the action of centrifugal forces, arising upon rotation of the inner chamber with a perforated side wall, allows changing the direction of movement of raw materials from top to bottom, due to the action of force of gravity of the raw material, on the radial direction, oriented from the center to the periphery. The zones of drying, pyrolysis, oxidation and reduction of solid fuel are in the form of cylindrical surfaces and are arranged one inside the other.

Under the influence of centrifugal forces, arising at rotation of the rotor (the gasification chamber with a perforated side wall), the direction of movement of the gasified fuel and the gasifying agent inside the rotor is changed and becomes directed from the center to the periphery. Thus, the active zones of gasification (drying, pyrolysis, oxidation and reduction of solid fuel) are sequentially arranged from the center to the periphery and have the shape of axially symmetric layers. The appearance of the gas generator is shown in Fig. 2.

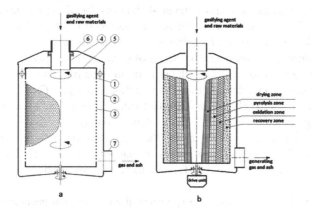

Fig. 1. Scheme of the gas generator of solid fuel with a rotating active area: a - design of gas generator: 1 - internal rotating chamber; 2 - outer stationary chamber; 3 - perforated side wall of the internal chamber; 4 - pipe for feeding of raw materials and gasifying agent 5, 6 - covers of the inner and outer bodies; 7 - pipe for collecting gas and ash; b - location of the active zone in the chamber of gas generator.

At the established process of gasification the following arrangement of active zones is formed: zone of warming up and drying, pyrolysis zone, burning zone and restoration zone. Such arrangement of active zones is explained by features of distribution of the gas agent in the gas generator of rotor type. As the gas agent moves only in the central part of a rotor 2 through a branch pipe 4, the conditions for completeness of implementation of reactions of burning exist in that part of a layer which is as close as possible to the center of a rotor (Fig. 1a). In the process of removal from the center

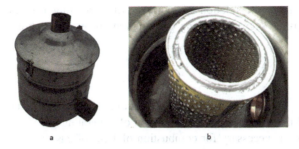

Fig. 2. The gas generator of solid fuel with a rotating active area: a - appearance; b - inner rotating chamber with a perforated side wall.

of a rotor to its periphery access of gas agent to raw materials decreases and its amount is minimum in close the perforated lattice. Exactly there restoration reactions can take place.

The gas-generating gas which is formed in a restoration zone under the influence of centrifugal forces is taken out through the perforated lattice of a rotor 2 in free space between a rotor and the internal surface of the chamber of a gas generator. Generating gas is forced out from free space by newly-formed gas in a branch pipe of collecting gas for the subsequent cleaning and cooling. Ash is carried away from the gas generator chamber with a stream of products of gasification.

The intensity of the gasification process depending on the type of raw material is adjusted by changing the value of the speed of rotation of the rotor.

3 Mathematical Model

The flow of any gaseous fuel, required to provide the specified capacity of the boiler, is determined from the energy balance:

$$Gg = \frac{Q_B \cdot 3600}{Q_L \cdot \eta_{EC}}, \qquad (1)$$

where: η_{EC} – efficiency coefficient of the boiler unit; Q_l – the lower calorific value of gaseous fuel.

$$Q_L = 127,5\ CO + 108,1\ H_2 + 358,8\ CH_4, \qquad (2)$$

The average elemental composition of wood raw material [9, 10]:

$W^w = 17\%$; $A^d = 0,7\%$; $C^c = 50,0\%$; $H^c = 6,0\%$; $O^c = 43,0\ \%$; $N^c = 1,0\ \%$,

The composition of generating gas from wood raw materials for the reciprocal circuit of gasification includes: CO_2 - 10; O_2 - 0,2; N_2 - 50,7; CO - 19,5; H_2 - 14,5; CH_4 - 3,0, and the general content of combustible substances is 37%.

At $\eta_{EC} = const$, the relation of expenses of generating gas (further «gg») and natural gas (further «ng»), providing the required power of any boiler, is:

$$\frac{G_{gg}}{G_{ng}} = \frac{Q_{l.ng}}{Q_{l.gg}}, \qquad (3)$$

where $G_{ng}, Q_{l.ng}$ – expense and the lowest heat of combustion of natural gas; $G_{gg}, Q_{l.gg}$ – expense and the lowest heat of combustion of generating gas.

Amount of air, necessary for combustion of 1 m³ of gas, is:

$$L_0 = [0,5(CO + H_2) + 2CH_4 - O_2]/21, \qquad (4)$$

Gas generator efficiency:

$$\eta_g = V_g \, Q_L / (Q_L^C)_T, \qquad (5)$$

where: $(Q_L^C)_T$ – the lowest calorific value of the gasified raw material, kJ/kg.

The yield of dry gas from 1 kg of gasified wood raw material was determined by the formula:

$$V_g = \frac{22,4(C^w - C_d)}{12(CO + CO_2 + CH_4)} = \frac{1,867(C^w - C_d)}{CO + CO_2 + CH_4} \, m^3/kg, \qquad (6)$$

$$C^w = C^c(100 - A^d)(100 - W^w)/100, \qquad (7)$$

where: C^w, C^c and C^d – the content of carbon in working, combustible and dry masses, %; C_d – the losses of carbon in the focal remains and in the form of dust (2%); W^w – the moisture content in the working mass, %; A^d – the ash content in dry weight, %.

The results of the calculation of the average values of the energy characteristics of a wood gas generator and gas generators according to the formulas (2–7): $Q_{l.gg}$ – 5130 kJ/m³; G_{gg}/G_{ng} – 6,811; L_0 – 1,088 m³/m³; V_g – 2,26 m³/kg; η_g – 58,0%.

It is accepted in calculations: for natural gas $Q_{l.ng} = 34,94 \, Mj/m^3$, for wood raw materials – $(Q_L^C)_T \approx 20,0 \, Mj/kg$.

4 Results

For the purpose of preservation of power of a copper upon transition from natural gas on generating one, the consumption of generating gas of the reciprocal circuit of gasification has to exceed a consumption of natural gas from c $Q_{l.ng} = 34,94 \, MW/m^3$ by 6,811 times.

The amount of air necessary for full burning of 1 m³ of generating gas is 1,088 m³ that is nearly 9 times less than for burning of 1 m³ of natural gas [9, 10]. Therefore, at conversion of a copper to generating gas with the expense, increased in comparison with natural gas by 5,8–6,9 times, the volume of products of combustion of generating gas is less, than of natural gas. Thus, for adaptation of the existing coppers to low-calorie generating gas, only modernization (or replacement) of the burner devices and fuel-supply lines without change of furnace and convective parts of a copper will be required [11].

The authors have received a formula for determination of quantity of the gas generators with the cylindrical mine, providing the required boiler power:

$$n = Q_B \cdot \frac{1}{\eta_g} \cdot \frac{1}{\eta_{EC}} \cdot \frac{1}{q_{TT}} \cdot \frac{4}{\pi \cdot D_M^2}, \qquad (8)$$

$$q_{TT} = \frac{B_T^I (Q_L^C)_T}{4\pi \cdot D_M^2/4}, \frac{K_{kJ}}{m^2 \cdot hour}, \qquad (9)$$

where: q_{TT} – «the thermal tension of cross section of the mine», the value which is integrally characterizing the mass speed of gasification and the valid form of a surface of an active zone in a gas generator with set value of raw materials and the scheme of gasification; B_T^I – raw materials consumption in a gas generator, kg/h; D_M – diameter of the mine of a gas generator, m.

Generally, for gas generators of any design [12]:

$$n = S_{ZG}/F_M, \qquad (10)$$

where:

$$S_{ZG} = Q_B \cdot \frac{1}{\eta_g} \cdot \frac{1}{\eta_{EC}} \cdot \frac{1}{q_{TT}}, m^2 \qquad (11)$$

where F_M – the area of a furnace grate lattice of the mine (for the cylindrical mine $F_M = \pi \cdot D_M^2/4$); S_{ZG} – is the area demanded for ensuring work of a copper of the set power, the area of a projection of the valid surface of an active zone of a gas generator to the plane of a furnace grate lattice.

The schedules, constructed on the dependences (9, 10, 11) for determination of quantity of gas generators with the cylindrical mine, are submitted in Fig. 3.

On the basis of works [13, 14] for gas generators with the reciprocal circuit of gasification of wood it is accepted:

$$q_{TT} = 12,72 \cdot 10^6, kJ/(m^2 \cdot hour),$$

At these values q_{TT} and η_g, the ratio for calculation of quantity of the gas generators, providing the required copper power, is revealed by formula (8):

$$n_{RC} = 487,8 \cdot 10^{-6} \cdot \frac{Q_B}{\eta_{EC}} \cdot \frac{1}{\pi \cdot D_{MN}^2/4}, \qquad (12)$$

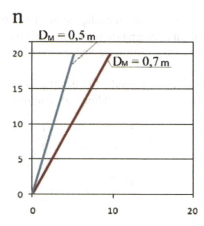

Fig. 3. The schedules for determination of quantity of gas generators with the cylindrical mine and reciprocal circuit of gasification.

At these values q_{TT} and η_g, the ratio for calculation of quantity of the gas generators, providing the required copper power, is revealed by formula (8):

$$n_{RC} = 487,8 \cdot 10^{-6} \cdot \frac{Q_B}{\eta_{EC}} \cdot \frac{1}{\pi \cdot D_{MN}^2/4}, \qquad (12)$$

The diameter of the mine of gas generators with the reciprocal circuit of gasification is limited $D_M = 0,7$ m [14].

The maximum power of a copper, provided with one gas generator, is equal:

$$Q_B = \eta_g \cdot \eta_{EC} \cdot q_{TT} \cdot \pi \cdot D_M^2/4, \qquad (13)$$

Therefore, one gas generator of the reciprocal circuit with an extreme diameter of the mine 0,7 m can ensure functioning of a copper with the maximum power $Q_B = 725,21$ kW.

The results of calculations of the required quantity of gas generators with the turned scheme of gasification for boilers of the KV-GM series are presented in the Table 1. In particular it is revealed, that for package boilers with a power up to 8 MW, the quantity of gas generators with the cylindrical mine and the reciprocal circuit of gasification of wood doesn't exceed $n = 3$, that is quite acceptable for practice.

As follows from (8), quantity of the gas generators, ensuring functioning of a copper with the required power, is in inverse proportion to the area of an active zone of gasification. Authors [15–18] have offered the new type of gas generator: the gas generator of rotor type, allowing increasing significantly the area of an active zone of gasification owing to creation of an active zone of gasification close to cylindrical, capable to increase on height up to gas generator rotor height.

Table 1. Calculation of quantity of gas generators with the cylindrical mine and the reciprocal circuit of gasification for conversion of boilers of the OGB (oil-gas boiler) series to wood generating gas.

Characteristics of boiler and gas generator	Boiler brand							
	KVE-0, 7–115Gn	KV-1, 6GM	KV-GM-2,9–150C	KV-GM-4,65–150	KV-GM-7,56–150	KV-GM-11,63–150	KV-GM-23,26–150	KV-GM-35–150
Q_B, MW	0.7	1.6	2.9	4.65	7.65	11.63	23.26	35
η_{EC}	0.86	0.91	0.92	0.92	0.92	0.92	0.92	0.91
η_{gRC}	0.58	0.58	0.58	0.58	0.58	0.58	0.58	0.58
q_{TT}, MW/m²	3.533	3.533	3.533	3.533	3.533	3.533	3.533	3.533
$4S/\pi$, m²	0.51	1.09	1.96	3.14	5.17	7.86	15.72	23.91
$D_{M\ RC}$, m	0.70	0.70	0.70	0.70	0.70	0.70	0.70	0.70
n_{RC}	1.03	2.23	4.00	6.41	10.55	16.04	32.07	48.79

For determination of quantity of gas generators of rotor type and for comparison it with quantity of classical gas generators with the reciprocal circuit the following formulas are used [19, 20]:

$$n_r = 1076,4 \cdot 10^{-6} \frac{Q_B}{\eta_{EC} F_{ZG}} = 1076,4 \cdot 10^{-6} \frac{Q_B}{\eta_{EC} \pi D_{SG} H}, \quad (14)$$

$$\frac{n_{RC}}{n_r} = \frac{1,81271 \cdot H}{D_M}, \quad (15)$$

where F_{ZG}, H – area and height of an active zone of gasification (here $F_{ZG} = F_M$); D_{SG} – diameter of a cylindrical surface of a zone of gasification equivalent to D_M.

Thus, the number of gas generators of rotor type in $(1,81271H/D_M)$ times are less than quantity of gas generators with the reciprocal circuit of gasification. The advantage of a gas generator of rotor type increases with increase in height. So at a rotor with $H = 2,47\,м$ and $D_{RG} = 0,7\,м$, one gas generator of rotor type is capable to ensure completely functioning of a copper KV-GM - 4, 65–150.

5 Conclusion

As a result of the conducted pilot researches it is established that one gas generator of rotor type of the reciprocal circuit with an extreme diameter of mine of 0,7 m and 2,47 high is capable to ensure completely functioning of a copper of KV-GM-4, 65–150. It is rather compact that is caused by an internal arrangement of layers - active zones by the principle "one in another", has the wide range of the given power which varies depending on the speed of rotation of a rotor that gives the chance to use any grades of fuels including low-grade.

References

1. Tijmensen, M.J.A., Faaij, A.P.C., Hamelinck, C.N., et al.: Exploration of the possibilities for production of Fischer Tropsch liquids and power via biomass gasification. Biomass Bioenergy **23**, 129–152 (2002)
2. Xu, X., Matsumura, Y., Stenberg, J., et al.: Ind. Eng. Chem. Res. **35**, 2522–2530 (1996)
3. Minowa, T., Fang, Z., Chem, J.: Eng. Jpn. **31**, 488–491 (1998)
4. Seo, M.W., Goo, J.H., Kim, S.D.: Gasification characteristics of coal/biomass blend in a dual circulating fluidized bed reactor. Energy Fuels **24**, 3108–3118 (2010)
5. Warnecke, R.: Gasification of biomass: comparison of fixed bed and fluidized bed gasifier. Biomass Bioenergy **18**, 489–497 (2000)
6. Boerrigter, H., Drift, A., Hazewinkel, J.H.O., et al.: Biosyngas: multifunctional intermediary for the production of renewable power, gaseous energy carriers, transportation fuels, and chemicals from biomass. Energy research centre of the Netherlands, Netherlands (2004)
7. Aznar, M.P., Caballero, M.A., Sancho, J.A.: Plastic waste elimination by co-gasification with coal and biomass in fluidized bed with air in pilot plant. Fuel Process. Technol. **87**, 409–420 (2006)
8. Hernandez, J.J., Aranda-Almansa, G., Serrano, C.: Co-gasification of biomass wastes and coal-coke blends in an entrained flow gasifier: an experimental study. Energy Fuels **24**, 2479–2488 (2010)
9. Dhole, V.R., Linnhoff, B.: Total site targets for fuel, co-generation and emissions. Comput. Chem. Eng. **17**, 101–109 (1993)
10. Gandiglio, M., Lanzini, A., Leone, P., et al.: Thermoeconomic analysis of large solid oxide fuel cell plants: atmospheric vs. pressurized performance. Energy **55**, 142–155 (2013)
11. Spiegel, R.J., Preston, J.L.: Test results for fuel cell operation on anaerobic digester gas. J. Power Sources **86**, 283–287 (2000)
12. Abashar, M.E.E.: Coupling of steam and dry reforming of methane in catalytic fluidized bed membrane reactors. Int. J. Hydrog. Energy **29**, 799–808 (2003)
13. Sandelli, G.C., Trocciola, J.C., Spiegel, R.J.: Landfill gas pre-treatment for fuel cell applications. J. Power Sources **49**, 143–149 (1994)
14. Lehman, A.K., Russel, A.E., Hoogers, G.: Renewable fuel cell power from biogas. Renew. Energy World **4**(76) (2001)
15. Yan, R., Liang, D.T., Tsen, L.: Kinetics and mechanism of H2S Adsorption by alkaline activated carbon. Environ. Sci. Technol. **36**, 4460–4466 (2002)
16. Lin, Y.C., Lee, W.J., Hou, H.C.: PAH emissions and energy efficiency of palm-biodiesel blends fueled on diesel generator. Atmos. Environ. **40**, 3930–3940 (2006)
17. Tsai, J.H., Chen, S.J., Huang, K.L., et al.: PM, carbon, and PAH emissions from a diesel generator fuelled with soy-biodiesel blends. J. Hazard. Mater. **179**, 237–243 (2010)
18. Durbin, T.D., Collins, J.R., Norbeck, J.M.: Effects of biodiesel, biodiesel blends, and a synthetic diesel on emissions from light heavy-duty diesel vehicles. Environ. Sci. Technol. **34**, 349–355 (2000)
19. Kim, S.B., Kim, K.J., Cho, M.H., et al.: Micro- and nanoscle energetic materials as effective heat energy sources for enhanced gas generators. ACS Appl. Mater. Interfaces **8**, 9405–9412 (2016). https://doi.org/10.1021/acsami.6b00070
20. Hasue, K., Yoshitake, K.: The mixture of the phase stabilized ammonium nitrate containing potassium nitrate and 1HT as the new gas generate composition. Sci. Technol. Energy Mater. **74**(3–4), 66–72 (2013)

Investigating the Effectiveness of Solar Tracking for PV Facility in Chelyabinsk

A. A. Smirnov[✉], A. G. Vozmilov, and O. O. Sultonov

South Ural State University, 76 Lenina avenue,
Chelyabinsk 454080, Russian Federation
tolix2007@gmail.com

Abstract. For the last decade energy production of PV facilities has greatly increased. Due to the problem of the low actual efficiency of PV panels, many researchers conduct experiments with the sun trackers at different locations to obtain the maximum and actual efficiency gain. The purpose of our study is to investigate the effectiveness of one-axis sun tracker for a PV panel located in Chelyabinsk. The four experimental sets are compared to obtain the efficiency gain during the period from May to September of 2016. The experimental sets comprise identical PV panels (Topray solar) and differ with the mount angle of inclination (90°, 0°, 55°, and 55° with the one-axis sun tracker). An average monthly and total solar irradiation is calculated for the collected data. The main result of the study shows that the usage of the one-axis solar tracking system can increase the efficiency of the PV panel up to 33% during a season compared to the fixed at the optimal angle PV panel. Further research is surely perspective in Chelyabinsk, in particularly, for investigating the maximum efficiency gain for a dual-axis sun tracker and designing a sun tracker with the low energy consumption.

Keywords: Photovoltaic · PV panel · Solar tracking system · One-axis sun tracker · BIPV · Chelyabinsk · Effectiveness gain

1 Introduction

The usage of solar energy persistently grows in different spheres of human activity. It is caused mostly by the developed countries policy to reduce their dependence on fossil fuels and to lower greenhouse gas emissions [1]. Photovoltaic (PV) plants are actual and perspective in production of electricity [2]. Therefore, total installed capacity of the world PV plants dramatically increased during last decade, from 2.6 GW to 177 GW [3].

However, PV panels have a major problem of low efficiency. The maximum efficiency of the most productive PV panels is about 20%, more commonly it is around 14–18% or lower [4]. As a rule, the actual efficiency of the PV panels statically mounted at the optimum angle of inclination is noticeably decreased at the morning and evening. Using of the sun tracker, or solar tracking system, is the most effective way to gain the actual efficiency of the PV panels [5]. PV panel equipped with the dual-axis sun tracking system is considered to be up to 40% more efficient compared to the fixed at the optimal angle PV panel [6].

Though, the efficiency gain for the PV panel with sun tracker strongly depends on the latitude of the location, its climate and the season [7]. Thus, many of researchers conduct the experiments for their geographic location to obtain the possible maximum value of the efficiency gain. Particularly, Serhan et al. [8] conducted the experiment using two-axis sun tracker in Lybia (33° N) and obtained the efficiency gain between 20% and 28%, unfortunately, researchers did not mention the season and the period. Another experiment was made by Carvalho et al. [9] during only 3-day periods in November and December in Brazil, Vicosa (21° S). They stated more than 52% of the efficiency gain for the PV panel with a sun tracker. Yao et al. [10] conducted the test with the one-axis sun tracker from 17 of July to 12 of September in Harbin, China (45° N) and obtained almost 32% of the efficiency increase. Recent research of Hoffman et al. [11] conducted in Santa Cruz do Sul, Brazil (29,7° S) from June to November with the two-axis sun tracker has showed 17%–31% of the efficiency gain. Despite of plenty research in this area, no one to the best of our knowledge has conducted this experiment in Chelyabinsk (55° N).

The aim of this study is to investigate the effectiveness of PV facility with one-axis sun tracking (from East to West) in Chelyabinsk. In order to achieve the goal, the four experimental sets were compared for five months, from May to September of 2016.

As a result, the usage of sun tracker for increasing of the PV panel efficiency in Chelyabinsk was approved. This experiment was mentioned in the earlier research [12], but it was not described and discussed properly. Thus, the other goal of this paper is to completely describe the conducted experiment.

The following section describes the methodology and used materials for the experiment. This is followed in third section by the obtained results. In the final section a discussion of the results is given and conclusions are made.

2 Methodology

2.1 Photovoltaic Panels

The four identical PV panels were chosen for conducting the experiment. The units TPS-936M by Topray solar were selected for its low cost. The output power and operating voltage of the panels are 28 W and 17.5 V, respectively. Due to these units are similar to the ones that used by Hoffman et al. [11], it allowed us to use the suggested by the authors measuring system.

2.2 Simple One-Axis Sun Tracker

The simplest one-axis sun tracker was assembled to investigate the effectiveness of the sun tracking in Chelyabinsk. It consisted of an AC motor YE2-(H71-355) (Fig. 1), connected to the electrical grid, and a simple gearbox for reducing the rotation speed. The AC motor of Qizhimotors brand was chosen because of low-rate RPM (660) and affordable price [13].

Investigating the Effectiveness of Solar Tracking 503

Fig. 1. AC motor used for the simple one-axis sun tracker.

The rotation speed of the tracker should be 15° per hour, or 0.25 RPM (equals to the rotation speed of the Earth around a polar axis), but a gearbox with the required gear ratio was not found. Thus, the plain worm gearbox was made by ourselves.

2.3 Location and Various Orientation of PV Panels

The PV panels were located at an open place of a garden plot near Chelyabinsk (55° 14′ N, 61° 22′ E). The surfaces of the panels were not shaded during a day, i.e. shadows of building construction or plants did not interfere the results.

The reference panels were oriented directly to the south and set up at the different fixed angles of inclination to the surface of the earth:

- 90° (Fig. 3a), corresponding to the common case of wall-mounted panel at BIPV facility
- 0° (Fig. 3b), corresponding to the case of measuring Global Horizontal Irradiation (GHI) [14]

Fig. 2. The reference fixed PV panel was set up at the optimal angle of inclination (55°).

- 55° (Fig. 3c), corresponding to the common case of a stand-alone PV facility. This angle equals to the latitude of the location (Chelyabinsk), i.e. it is close to the optimal angle [15]. Figure 2 represents the fixed PV panel mounted at the optimal angle

The fourth panel with the simple one-axis sun tracker were set up at the angle of 55° and started to rotate around the polar axis synchronously with the sun moving (Fig. 3d).

2.4 Measuring System

Solar irradiation was measured for the installed PV panels during 5 months, from May to September of 2016. The measuring method suggested by Hoffman et al. [11] was adopted in the current study. According to the method, each panel were connected to A/D inputs of the PIC18F452 microcontroller via bias resistors.

Collected data were saved to a MMC card, connected to the microcontroller via SPI interface, for further analysis. This method of data collection were suggested by Nadjah et al. [16].

2.5 Data Analysis

The collected data were analyzed using MS EXCEL. Total solar radiation for each of the PV panel was calculated. As a result, the percent of the effectiveness of using the one-axis solar tracker was calculated in comparison with the fixed panels.

3 Results

There was a significant efficiency gain for the PV panel with the one-axis solar tracker compared to the fixed panel at the same optimal angle of inclination.

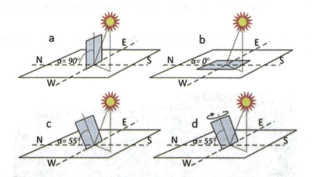

Fig. 3. Various orientation of the PV panels of the four different experimental sets (a) vertical position ($\alpha = 90°$); (b) horizontal position ($\alpha = 0°$); (c) optimum angle of inclination ($\alpha = 55°$); (d) optimum angle of inclination ($\alpha = 55°$) and polar rotation.

It is interesting to note that May, June and July of 2016 were the months with the maximum solar irradiance in Chelyabinsk during the measuring period. September of 2016 was the least sunny month in Chelyabinsk (Fig. 4).

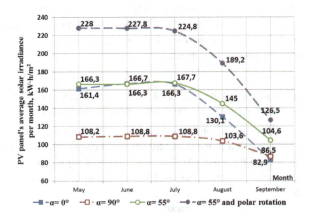

Fig. 4. Monthly solar irradiation of the fixed PV panels at the different angle of inclination and the PV panel at the optimal angle (55°) with the one-axis sun tracker rotating around polar axis (filled purple circles). Measuring values were gathered between May and September of 2016 near Chelyabinsk.

Total solar irradiation had maximum value for the PV panel with the one-axis sun tracker and minimum value for the fixed vertically PV panel (Fig. 5).

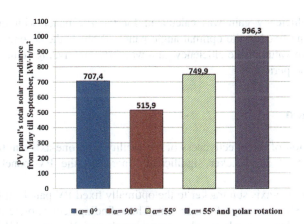

Fig. 5. Total solar irradiation of the fixed PV panels at the different angle of inclination and the PV panel at the optimal angle (55°) with the one-axis sun tracker rotating around polar axis (purple cross-hatched bar). Measuring values were gathered between May and September of 2016 near Chelyabinsk.

Efficiency gain was calculated for the fixed horizontally PV panel (0°), for the fixed at the optimal angle PV panel (55°) and for the PV panel with the one-axis sun tracker compared to the fixed vertically PV panel considered as the control (Table 1, the first group of columns). There was a significant increase of the efficiency (maximum was 111%, average was 93%) for the panel with the sun tracker compared to the control. Negative value for the fixed horizontally PV panel in September could be caused by decreasing of the efficiency due to the dirty surface of the panel (horizontal surface tended to be dirty, especially in autumn).

Table 1. Comparison of the monthly and total solar radiation gain for the differently oriented fixed PV panels and the PV panel with the simple one-axis sun tracker (source: authors).

Angle of inclination/month	First group: efficiency gain compared to the vertically fixed PV panel (90°)			Second group: efficiency gain compared to the fixed PV panel at the optimal angle (55°)
	0°	55°	55° and 1-axis sun tracker	55° and 1-axis sun tracker
May	49%	54%	111%	37%
June	53%	53%	109%	37%
July	53%	54%	107%	34%
August	26%	40%	83%	31%
September	−4%	21%	46%	21%
Average	37%	45%	93%	33%

Measuring values were gathered between May and September of 2016 near Chelyabinsk

Besides, efficiency gain was calculated for the PV panel with the sun tracker compared to the fixed at the optimal angle panel (Table 1, the second group). In that case, maximum value of the efficiency gain was 37% and average efficiency gain was 33% during the period.

4 Discussion

The investigation of the effectiveness of PV facility with one-axis sun tracking shows that the usage of sun tracker significantly increases the PV panel efficiency in Chelyabinsk.

Adding the one-axis sun tracker to the optimally fixed PV panel could increase the efficiency to 33% in average for a season. These findings corroborate the experiment results of the other researchers. The closest result had Yao et al. [10] (32% of the efficiency increase). It seemed to be connected with the closest location (45° N for Harbin, China and 55° N for Chelyabinsk, Russia) and the similar methodology (they also used one-axis sun tracker).

Interestingly that the maximum average efficiency gain was 93% compared to the vertically fixed PV panel. Excluding unexpected results of September, when the lowest efficiency had the fixed horizontally PV panel. As we said in the Results section, it is probably due to dirtiness of the horizontal surface of the panel in the rainy season. Nevertheless, we suggest that the vertically fixed PV panels have the lowest efficiency, and such installations used in BIPV plants possibly have the room for improvements. Besides, we could assume that the using of the sun tracker in Chelyabinsk is the most effective from May to July. Most likely it is due to summer solstice that almost bisects this period.

We confirm the limitations of our experiment due to short period of time (not whole year), and due to using the one type of the PV panels (polycrystalline silicon). However, we consider that the obtained results are significant and could be used for the further research. Further research is needed to investigate the maximum efficiency of the dual-axis sun tracker, and it should be also studied possible designs of a sun tracker with minimal energy consumption to maximize actual PV panel efficiency. Further research could help in solving the main problem of low efficiency of PV panels.

Acknowledgment. The work was supported by Act 211 Government of the Russian Federation, contract № 02.A03.21.0011.

References

1. Vozmilov, A.G., Malugin, S.A., Malugina, A.A.: Increasing efficency of sun energy use at autonomous systems of energy supply of agriculture. Bull. CSAA **69**, 10–13 (2014)
2. Uskov, A.E., Girkin, A.S., Daurov, A.V.: Solar power: condition and prospects. Sci. J. KubSAU **98**, 364–380 (2014)
3. Vozmilov, A.G., Malugin, S.A., Malugina, A.A., et al.: Analytical review of PV plants' types and constructions. Technics - from theory to practice, pp. 33–41 (2015)
4. Vissarionov, V.I., Deryugina, G.V., Kuznetsova, V.A.: Solar Energy: A Teaching Aid for Universities. Publishing House MPEI, Moscow (2008)
5. Prinsloo, G.J., Dobson, R.T.: Solar Tracking. SolarBooks, Stellenbosch (2015)
6. Kitaeva, M.V., Yurchenko, A.V., Skorohodov, A.V.: Solar tracker. Sib. J. Sci. **3**(4), 61–66 (2012). https://cyberleninka.ru/article/n/sistemy-slezheniya-za-solntsem/viewer
7. Yurchenko, A.V., Kozlov, A.V.: The long-term prediction of silicon solar batteries functioning for any geographical conditions. In: Proceedings of XXII European PV Solar Energy Conference and Exhibition, pp. 3019–3022 (2007)
8. Serhan, M., El-chaar, L.: Two axes sun tracking system: comparison with a fixed system. In: International Conference on Renewable Energies and Power Quality, vol. 10 (2010)
9. Carvalho, D.R., Lacerda Filho, A.F., Resende, R.C., et al.: An economical, two axes solar tracking system for implementation in Brazil. Appl. Eng. Agric. **29**, 123–128 (2013)
10. Yao, Y., Hu, Y., Gao, S., et al.: A multipurpose dual-axis solar tracker with two tracking strategies. Renew. Energy **72**, 88–98 (2014)
11. Hoffmann, F.M., Molz, R.F., Kothe, J.F., et al.: Monthly profile analysis based on a two-axis solar tracker proposal for photovoltaic panels. Renew. Energy **115**, 750–759 (2018)
12. Smirnov, A.A., Malugin, S.A., Bakanov, A.V.: Designing integrated PV facility with dual-axis solar tracking system mounted on the south building face. In: 3rd International Conference on Industrial Engineering, Applications and Manufacturing (2017)

13. Alibaba (2018). https://www.alibaba.com/product-detail/Iron-Cast-220v-high-torque-low_60686625467.html. Accessed 25 Sept 2018
14. Solargis (2018). https://solargis.com/products/maps-and-gis-data/free/download/world. Accessed 18 Oct 2018
15. Breyer, Ch., Schmid, J.: Global distribution of optimal tilt angles for fixed tilted PV systems. In: 25th European Photovoltaic Solar Energy Conference and Exhibition, pp. 4715–4721 (2010)
16. Nadjah, A., Kadri, B., Sellam, M.: New design of dual axis sun tracker with DSPIC microcontroller. In: 16th International Power Electronics and Motion Control Conference and Exposition, vol. 6980644, pp. 1030–1034 (2014)

Optimization of Diesel-Driven Generators in Continuous Power Systems of Essential Consumers

V. V. Karagodin, K. A. Polyansky[✉], and D. V. Ribakov

Military Space Academy, 13 Zhdanovskaya Street, Saint Petersburg 197082, Russian Federation

Abstract. Today's trends in the development of engineering and technical systems of various facilities stipulate for higher requirements to the reliability of essential consumers' power supply and the electrical power quality in all operation modes. At the moment, to execute these requirements, continuous power systems including, as a rule, static uninterrupted power supply sources and diesel-driven generators are used. The latter are necessary to ensure the required durability of the system autonomous operation. Resolving the problem of diesel-driven generators optimization in the continuous power systems of essential consumers providing for high reliability indicators is impossible without taking into account the peculiarities of the combined operation of DDGs and static UPSes, both in transient and steady operation modes. This problem can be stated as a discrete programming problem the resolution of which will help to avoid unreasonable economic costs as well as possible technical failures occurring when the parameters of the considered system components do not match. To resolve it, the authors developed a mathematical model where the capital investment minimum or the minimum of the reduced annual costs taking into account operating costs are chosen as optimality criteria. The obtained research results are implemented by means of the software and brought up to the engineering level.

Keywords: Optimization · Electromagnetic compatibility · Diesel-driven generator · Uninterrupted power supply source · Continuous power system · Essential consumer

1 Introduction

At the moment the development of engineering and technical systems at various industrial and special facilities along with the mainstreaming of continuous engineering processes, microcontroller and industrial controller application to control engineering equipment, application of PC-based and microprocessor-based automated control systems stipulates for higher requirements set forth by essential consumers (consumers involved in engineering processes) to the electrical power quality, power supply reliability and uninterrupted operation. At the same time, the unresolved problem of a set reliability level provision pertaining in the electrical power engineering in the Russian Federation [1] (external power supply systems cannot guarantee the required power supply reliability) results in a significant increase of the importance of the stand-alone (backup) power supply, uninterrupted and continuous power supply systems for the provision of a required power supply reliability level and the quality of such power supply. Thus, it is necessary to develop new approaches towards such system building with the use of modern equipment of enhanced operation reliability as well as providing for environmental and engineering safety [2].

An integral part of modern autonomous power complexes are continuous power systems (CPS). A continuous power system is a power plant intended for the stand-alone power supply of essential consumers in the cases when the unified power quality index of the main sources power deviates beyond accepted values. A subsystem of a modern CPS is an uninterrupted power system (UPS) capable of providing power supply to essential consumers in case of power failures (electric power break, voltage falls, etc.) for a period determined by the storage power amount in the accumulator [3].

The key structural components of a modern UPS determining its performance indicators are static uninterrupted power supply sources (UPSes) of double conversion.

The problem of UPS building on the basis of static double conversion UPSes can be resolved as an optimization problem [4]. On the basis of its mathematical model [5] which allows taking into account the most important UPS performance indicators, including operating costs, the authors developed a method of UPS structural and parametric optimization [6].

Non-standardized cutouts in terms of the duration and recurrence in the external power supply mains induce the necessity to integrate stand-alone power plants in the CPSes to provide or the necessary stand-alone operation (longer than the period determined by the UPS accumulator batteries). Combined with UPSes, the stand-alone power supply sources make a CPS. Diesel-driven generators (DDGs) are most often used as stand-alone power supply sources for CPSes.

It is explained by their easy servicing, large operational experience, completed units and high automation degree, quick start and other advantages of their use [7]. Besides, the application of stand-alone sources in conjunction with a UPS allows one to reduce the time of the network support with the help of accumulator batteries (AB) by several minutes, in other words, to reduce their capacity and, consequently, reduce the UPS cost.

At the DDG operation significant problems occur with the double conversion UPS [8]; such problems necessitate the resolution of electromagnetic compatibility issues in a CPS. However, the method developed by the authors [6] did not consider such issues.

2 Peculiarities of DDG and UPS Joint Operation

The considered peculiarities of DDG an UPS joint operation in the CPS are predetermined by the nature of energy exchange processes taking place in the systems containing power electronics devices which cause significant disturbances in the harmonic form of currents and voltages due to the operation discrete nature. Higher current harmonics induced by the UPS being the component with a non-linear input characteristic can pose serious problems for CPS electromagnetic compatibility.

The issues of interaction of the DDG and UPS functioning in the CPS shall be considered both for the transient (for example, during the UPS transition to the power supply from the DDG) and steady operation modes.

The transient mode, in particular, the load connection mode, negatively influences the DDG operation [9]. Such influence can lead to the generator voltage reduction. The dependences of the voltage change at the generator output on the simultaneously connected load are provided by the manufacturers for the specific DDG models.

The CPS including a UPS and DDG parameters of which, nominal rating powers, in the first place, were chosen without the account of considered peculiarities, the synchronous generator voltage at the moment of load connection can reach a minimum acceptable voltage value (equal to $U_{ac}^{min} = 0,85 U_{nom}$ in the UPS produced by most manufacturers) at which the UPSm starts taking power supply from accumulator batteries. This process is illustrated in Fig. 1. After that, the generator will pass into an idle mode and restore nominal voltage in some time. Due to this the UPSm will again get the supply from the generator instead of ABs. The load connection will cause a recurrent generator voltage drop - the cycle will recur.

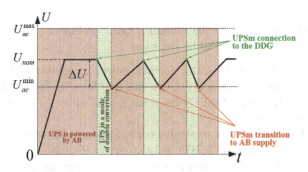

Fig. 1. Diagram of the generator voltage changing at the UPS connection to the DDG in the CPS with unmatching component parameters.

In the CPS steady operation mode, as it has been aforementioned, another problem of the electromagnetic compatibility of its main functional components occurs: it is characterized by the presence of high-frequency harmonics of the UPS input current that consume non-sinusoidal current which leads to a significant increase in the level of conductive electromagnetic disturbances.

Non-sinusoidal currents participate in the creation of a machine magnetic field and, consequently, influence the value and the form of output voltage. The output voltage of the generator working with a rectifier load becomes non-sinusoidal.

Such voltage can cause the failures or incorrect operation of electric equipment connected to the synchronous generator (SG) output and precondition false response of the SG emergency protection, an incorrect operation of the generator voltage automatic regulator, etc.

For a synchronous unit operating with a powerful UPS the condition on the phase current sinusoidal nature assumed at it designing becomes inapplicable.

The currents distorted by a UPS rectifier block (Fig. 2) [10] are closed on the generator windings and cause the following negative phenomena: additional power loss in the generator windings and cores, reduction of the load power, potentially dangerous winding and core overheating. All these can lead to the DDG emergency stops.

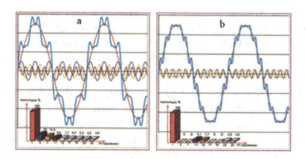

Fig. 2. Form of input currents characteristic for thyristor-controlled rectifiers (a) 6–pulse; (b) 12–pulse.

It is possible to eliminate the aforementioned problems occurring at the joint operation of the UPSm and DDG by means of:

- Significant increase of the DDG nominal power in relation to the load power. According to some estimations [11], at 6–pulse UPS rectifier circuit the DDG power multiplicity in relation to the CPS can reach 3, at 12–pulse UPS rectifier circuit this value can reach 1.9
- Using the CPS with a more expensive configuration under which one understands the type (thyristor type or IGBT) and the circuit (6–pulse or 12–pulse) of UPS rectifiers included in the system as well as the presence or absence of additional devices such as higher harmonic filters and the standardized load connection devices (SLCD)

Thyristor-based rectifiers applied in static UPSes generate higher harmonics the order of which is determined by the expression [12]:

$$n = k\rho \pm 1,$$

where k - counting numerals, $k = 1, 2, 3...$; ρ - number of rectifier impulses.

That is why a 6–pulse rectifier ($\rho = 6$) is the source of harmonics with the numbers: 5, 7, 11, 13, 17, 19, 23, etc./while a 12–pulse rectifier ($\rho = 12$) is characterized by the harmonics with the following numbers: 11, 13, 23, 25 etc.

As it is seen from Fig. 2, the input current form at the UPS with a 12 – pulse rectifier resembles a sinusoidal wave, harmonic distortion coefficient has a smaller value and thus, the negative influence on the DDG will be smaller as well.

Modern UPSes can be equipped with SLCDs with a smooth or step-wise input. This device allows for avoiding inadmissible generator voltage reduction in the transient mode by means of a gradual load switching from the accumulator batteries to the

DDG and thus, there will be no need to significantly increase the DDG power comparing to the UPSm input power.

As it has been mentioned above, the resolution of the problems occurring at the joint operation of the DDG and UPS in the CPS is possible by means of the increase of the DDG nominal rating power and/or using the UPS with a more expensive configuration. The technical and economic analysis of these solutions proved that the UPS with a more sophisticated configuration results in the rise of economic costs for the UPSm but allows for DDG cost reduction. Another solution for the problem of matching the UPSm and DDG parameters (by means of the nominal rating power increasing) will result in the rise of the DDG costs but, at the same time, one can use a less expensive UPSm configuration,

Therefore, it is clear that resolution of the problem of DDG and UPSm parameters optimization for essential consumers to provide high reliability indicators is impossible without the account of the joint operation of the DDG and UPSm both in transient and steady operation modes. Moreover, taking into consideration that the UPSm can have various configurations and depending on such configuration the DDG operation is influenced in different ways, and the degree of such influence also depends on the DDG nominal rating power, this problem can be viewed as an optimization problem.

3 Optimization Problem Mathematical Statement

When the optimization problem is mathematically stated, one shall determine the variables the fixed values of which define the specific problem solution; the target function representing the optimality criterion mathematical notation and a set of limiting conditions.

Let us introduce the following boolean variables as the variables to be determined within the optimization problem solving:

$$w_{ij} = \begin{cases} 1, & \text{if the DDG with } i \text{ nominal rating power supplies} \\ & \text{the power to the UPSm with the } j \text{ configuration;} \\ 0, & \text{otherwise.} \end{cases} \quad (1)$$

The target function represents economic (capital or reduced annual) costs for the CPS building:

$$F_{CPS} = \sum_i \sum_j w_{ij}(F_{DDG_i} + F_{UPS_j}), \quad (2)$$

where F_{DDG_i} – economic costs for the DDG of the i nominal rating power; F_{UPS_j} – economic costs for the UPSm with the j configuration.

If the capital cost minimum (without the costs or further operation) is taken as an optimality criterion, the target function takes the following form:

$$K_{CPS} = \sum_i \sum_j w_{ij}(c_{DDG_i} + K_{UPS_j}), \quad (3)$$

where c_{DDG_i} – cost of the DDG with the i nominal rating power, rub.; K_{UPS_j} – capital costs for the CPS of the j configuration, rub.

Taking into account that the DDG serves as a standby power supply, the cost of power losses in the DDG can be neglected and the expression (2) will take the following form in case the minimum of reduced costs is taken as an optimality criteria:

$$F_{CPS} = Z_{CPS} = \sum_i \sum_j w_{ij}(p_\Sigma c_{DDG_i} + Z_{UPS_j}), \qquad (4)$$

In the expression (4) the authors assume: p_Σ – aggregate estimated coefficient of deductions depending on the total of capital investments into the power supply system taking into account operation costs (for the DDG up to 500 rpm $p_\Sigma = 0,212$, higher than 500 rpm. $p_\Sigma = 0,252$ [13, p. 21]); Z_{UPS_j} – reduced annual costs for the CPS with the j configuration, rub./year.

A set of admissible alternatives is formed by the limitations representing the following mathematical and technical taken into consideration during the optimization problem solving:

1. The variables can take only one value:

$$w_{ij} = \{0, 1\}, \qquad (5)$$

2. The condition providing for the voltage deviation on the generator terminals at load connection within admissible limits:

$$\sum_i w_{ij}(P_{DDG_i} - k_{r_j}^{tm} P_{UPS_j}^{in}) \geq 0, \qquad (6)$$

where P_{DDG_i} – nominal rating power of the DDG, i configuration, kW; $k_{r_j}^{tm}$ – redundancy rate of the DDG power over the UPSm power with the j configuration, p.u.; $P_{UPS_j}^{in}$ – input consumed power of the UPSm with the u configuration, kW.

The redundancy rate of the DDG power over the CPS power has various values depending on the type of the drive diesel engine and the SLCD presence.

At the application of the UPS without software input devices, the redundancy rate can be calculated as follows [14]:

$$k_r^{tm} = \frac{100 k_s}{q_{ad}},$$

where k_s – safety factor (for the diesel engines with gas turbine charging is assumed as being equal to 1.3–1.4, without such gas turbine charging 1.1–1.2), p.u.; q_{ad} – admissible instantaneous load rise in the percentage points of the DDG nominal rating power (taken according to the technical documents for a specific DDSG or in compliance with [15]).

For the UPS with the SLCD k_r^{tm} is determined in accordance with the data provided by the manufacturer or is taken as being equal to 1.2–1.3 for the DDG with diesel engines equipped with a gas turbine charging and equal to 1.05–1.1 without such charging.

The UPSm input power can be significantly higher than the load power (depending on the UPS load) and depends on the power consumption for the accumulator batteries charge and UPS internal losses.

For the case when the vital load is powered by z twin UPSes the input power can be obtained with the help of the formula:

$$P_{UPS}^{in} = \sum_z P_z^{in} = \sum_z \frac{100 P_{c_z}}{\eta_z(k_{us_z})} + \sum_z P_{ch_z},$$

where P_{UPS}^{in} – input power consumed by the z UPSes, kW; P_{c_z} – calculated power of the load connected to the z UPS, kW; η_z – efficiency factor of the z UPS,%; k_{us_z} – usage coefficient of the z UPS,%; P_{ch_z} – power consumed for the charge of accumulator batteries, kW.

3. DDG minimum load condition in a steady mode:

$$\sum_i w_{ij}(k_{min}^{sm} P_{DDG_i} - P_{UPS_j}^{in}) \leq 0, \qquad (7)$$

Here k_{min}^{sm} – DDG minimum load coefficient in the steady mode, p.u.

4. The unique CPS building variant:

$$\sum_i \sum_j w_{ij} = 1, \qquad (8)$$

5. Condition providing the UPSm and DDG electromagnetic compatibility:

$$\sum_i w_{ij}(P_{DDG_i} - k_{r_j}^{sm} P_{UPS_j}^{in}) \geq 0, \qquad (9)$$

The redundancy rate $k_{r_j}^{sm}$ in the expression (9) can be calculated as the ratio of losses in a synchronous generator at the non-linear load supply to the power losses at the DDG operation under a linear load [11]:

$$k_r^{sm} = \frac{\Delta P}{\Delta P_1} = 1 + \frac{\sum_{n>1} \Delta P_n}{\Delta P_1} = 1 + \sum_{n>1} \Delta P_n^* = 1 + \sum_{n>1} \left(K_{I(n)}^2 + K_{U(n)}^2 (2n^{1,3} + 1) \right),$$

where: ΔP_n – losses in the synchronous generator at the n harmonic wave, W; ΔP_n^* – relative value of losses at the n harmonic wave, p.u.; $K_{I(n)}$ – coefficient of the current n harmonic component, p.u.; $K_{U(n)}$ – coefficient of the voltage n harmonic component, p.u.

The calculation of voltage higher harmonics' levels is conducted on the basis of the equivalent circuit applicable for each harmonic component separately (Fig. 3). Non-linear loads (UPS) on the circuit are represented in the form of the current sources [16].

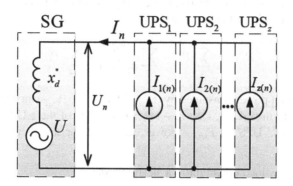

Fig. 3. CPS equivalence circuit for the calculation of higher harmonics.

At the neglect of the overlap angle of the UPS rectifier:

$$I_n = \sum_z I_{z(n)},$$

where: I_n – current equivalent value of the n harmonic component; $I_{z(n)}$ – current of the n harmonic component, z UPS.

In this case the coefficient of the voltage n harmonic component $K_{U(n)}$ can be calculated according to the following formula:

$$K_{U(n)} = \frac{U_n}{U_1} = \frac{nx_d'' I_n}{x_g I_1} = nx_d''^* K_{I(n)},$$

where x_d'' – subtransient inductive generator resistance along the longitudinal axis, Ω; x_g – generator resistance, Ω; $x_d''^*$ – relative subtransient inductive resistance of the generator along the long axis for the n harmonic component, p.u.; $K_{I\,(n)}$ – coefficient of the n harmonic component of the stator current.

In its final form the expression (9) will be the following

$$\sum_i w_{ij} \left(P_{DDG_i} - P_{UPS_j}^{in} - P_{UPS_j}^{in} \sum_{n>1} K_{I(n)_j}^2 \left(1 + x_{d_i}''^{*2}(2n^{3,3} + n^2)\right) \right) \geq 0. \quad (10)$$

Therefore, the authors obtained a mathematical model for the optimization problem which includes: variables (1), target function (2) which, depending on the choice of the economic indicator, can be expressed (3) or (4), and limitations: (5)–(8), (10).

4 Problem Resolving

The obtained mathematical model of the considered optimization problem belongs to the class of linear boolean mathematical models. The boolean linear optimization problems represent a specific case of discrete optimization problems as to resolve them, one shall apply special methods among which the methods of the type "branches and bounds" and dynamic programming became the most wide-spread [5]. To find out the solution of a considered optimization problem, the authors used a combinatorial technique which is a modified method "branches and bounds".

The solution for the optimization problem for the DDG parameters in the CPS of essential consumers with the help of the presented mathematical mode and the combinatorial method to resolve the linear programming problems with boolean variables is implemented in a software developed in the MatLab mathematical package intended for the determination of the DDG nominal rating power and the UPS configuration in the UPSm of essential consumers [17].

The validity of the developed mathematical approach is confirmed by the solution of a large number of practical problems for various input data with the subsequent check of the obtained solution validity.

Acknowledgements. DDG parameters' optimization in the power supply systems of essential consumers is impossible without taking into account the peculiarities of their joint operation with static UPSes. The developed mathematical model application to solve this problem will help to avoid unreasonable economic costs as well as possible technical failures occurring when the parameters of the developed CPSes do not match.

References

1. Voropai, N.I., Kovalyov, G.F., Kucherov, Yu.N.: Reliability provision concept in electrical power engineering. Energy, Moscow (2013)
2. Karagodin, V.V., Polyanskiy, K.A., Zverev, A.V.: Approaches to power supply provision to essential consumers of special facility with set reliability level. Theoretical and applied problems of development of internal and stand-alone power supply systems for special facilities, pp. 113–119 (2015)
3. Gerasimov, A.N., Orlov, A.V., Petrushin, V.F.: Uninterrupted power systems: study guide. Ministry of defense of the Russian Federation, St. Petersburg (1997)
4. Karagodnin, V.V., YeM, R., Polyanskiy, K.A., et al.: Analysis of power supply provision to essential consumers with set level of reliability ensuring. Bull. SPBO AIN **11**, 162–171 (2015)
5. Karagodin, V.V., Polyanskiy, K.A., Rybakov, D.V.: Selecting optimum option to build uninterrupted power supply system for essential consumers of special facilities with set reliability level. Ind. Eng., 295–299 (2016)

6. Karagodin, V.V., Polyanskiy, K.A., Groin, V.A.: Structural and parametric optimization of uninterrupted power system for essential consumers. News High. Educ. Inst. Instrum. Eng. **60**(1), 14–24 (2017)
7. Nye, P.: Special complex power supply. Military Space Academy, St. Petersburg (2011)
8. Karagodin, V.V., Polyanskiy, K.A., Rybakov, D.V.: Account of joint operation of diesel-driven generators and static uninterrupted power systems at building of continuous power systems. Relevant issues of stand-alone power supply systems for the facilities of the Ministry of Defense of the Russian Federation: book of the reports of the round table conducted within the scientific and business program of the International Scientific and technical forum ARMY, pp. 226–231 (2017)
9. Garganyeev, A.G.: Emergency Power Supply for AC Essential Consumers. TPU Publishing House, Tomsk (2010)
10. Kuzmina, O.A.: On the joint operation of DDG and UPS. Netw. Bus. **2**(3), 18–21 (2002)
11. Oblakyevich, S.V.: To Calculation of DES power jointly operated with UPSU. Ind. Electr. Power Ind. Electr. Eng. **2**, 35–40 (2005)
12. Kusko, A., Thompson, M.: Power supply networks. Power quality in electrical systems. Dodeka-XXI, Moscow (2011)
13. Revyakov, B.A., Karagodin, V.V., Nye, P.: Designing of power supply systems of special facilities. Part 2. Power Grids Over 1,000 V. Military Space Academy, St. Petersburg (2012)
14. Goldinyer, A.Ya.: Static and diesel redundant electrical power units. Galeya, St. Petersburg (2006)
15. GOST R 55231: Automatic rotational frequency control systems of marine, locomotive and industrial reciprocating internal combustion engines. General specifications. Standartinform, Moscow (2012)
16. Zhezhelenko, IV: Electromagnetic compatibility of consumers. Machine construction, Moscow (2012)
17. Polyanskiy, K.A.: Software for determination of capacity of DGS for CPS. RU certificate of registration of the software application 2016663288 (2016)

Study of Operating Characteristics of Pellets Made from Torrefied Wood Raw Material

R. R. Khasanshin[✉], R. R. Safin, and A. R. Shaikhutdinova

Kazan National Research Technological University, 68 Karla Marksa Street, Kazan 420015, Russian Federation
rus12881@mail.ru

Abstract. The tray unit for heat treatment of reduced wood, which provides uniform torrefaction of raw materials with different fractional composition without a prior fractionation, is developed. Fuel pellets derived from thermally heated reduced wood using the unit are studied. The research showed the effectiveness of thermal processing of wood raw material prior to production of solid fuel to reduce requirements for storage conditions and energy efficiency. It is noted that the equilibrium moisture content of torrefied pellets is more than twice as low as of standard pellets. The rise in combustibility of wood pellets with increasing temperature of torrefaction is observed which can be explained by the increase in a released heat of combustion and reduction of the initial humidity, which in total allows torreficating and maintaining self-combustion longer. It is established that the increase in productivity and economic efficiency of the torrefaction process of wood raw material can be achieved by completion of processing at the time of significant reduction in a stream of gaseous decomposition products without a substantial loss of energy quality of fuel.

Keywords: Wood · Torrefication · Biofuel · Fuel pellets

1 Introduction

Heat-treated reduced wood has recently been widely used when producing fuel pellets with reference to which the heat treatment process of raw wood is called "torrefaction" (from the English word "torrefy" - frying), and the finished product is termed "biochar" [1]. The process was first developed in France in 1930s for preliminary preparation of wood fuel [2, 3]. It is established that the thermal destruction of the organic component of biomass and a decrease in the equilibrium moisture content of the finished product (which takes place during torrefaction) allow obtaining a solid hydrophobic product with high calorific characteristics: depending on the conditions of the process high combustion value can reach 22 MJ/kg, while of dry wood raw material doesn't exceed 18 MJ/kg [4].

The process of heat treatment of wood, carried out to change the physical properties and chemical composition of wood raw materials, occurs at temperatures of 180–300 °C in an oxygen-free environment. The lack of oxygen in the reactor is necessary to prevent self-ignition of the wood during processing. The various inert media can be used for this, such as flue gases, nitrogen, and water vapor in superheated or saturated states. The process can be carried out in vacuum apparatuses [4] with the supply of heat by the contact method.

Active stage of development of bioenergy in recent decades has led to the establishment of such methods of preparation of solid fuel as the granulation and briquetting, which allowed not only increasing energy density, but also greatly simplifying the regulation of combustion process of solid biofuels [5–9]. However, the granular fuel is not devoid of such shortcomings characterizing the biomass, as the hygroscopicity, swelling and smouldering. That is why the researches came to torrefaction again as not only one of the most effective methods of increasing energy intensity of wood fuel pellets, but as to the method of preparation of solid biofuels, which can reduce transportation and storage requirements [3–7]. The greatest volume of research in this area belongs to the scientists from the USA, Canada and Western Europe [8–14]. Russian scientists are not less interested today in the research [15, 16].

Operating parameters of the process of torrefaction of reduced wood, which affect the degree of heat treatment, and, as a result, the properties of the final product are the temperature and duration of thermal effects [13–18]. However, if we are talking about a relatively low degree of heat treatment, it can be achieved by using different combinations of temperature and duration of exposure: the modes with low processing temperature and relatively long duration of the process can afford to obtain the same degree of processing that the modes with high temperature and relatively short duration of exposure [12–14].

Because of this, a number of researchers use the amount of loss of mass or density of the feedstock when processing as the main characteristics of the degree of torrefaction of biomass [17–21]. In particular, in work [4] heat treatment is recommended to proceed up to weight loss 30% to increase heat of combustion of the processed raw materials by not less than 20% and double reduction of the equilibrium moisture content.

In the production process of the solid fuel, the operation of the torrefaction can be used when preparing raw wood for pelletizing, processing after the pelletizing or briquetting as well [1–4]. The second option received a larger industrial use as a preliminary granulation or briquetting allows sending bulk material, excluding fractional variance, to the reactor for thermal processing and, consequently, obtaining equal degree of torrefaction for the entire batch of product. At the same time, the preliminary heat treatment of raw materials before granulation allows obtaining pellets with the maximum energy intensity, which in certain conditions allows improving significantly the competitiveness of products.

In this connection, the researches focused on receiving pellets from thermally treated raw materials are still topical.

2 Methods and Materials

The authors developed a tray unit (Fig. 1) for heat treatment of reduced wood, which provides uniform torrefaction of raw materials with different fractional composition without preliminary fractionation. A peculiarity of this unit is the existence of

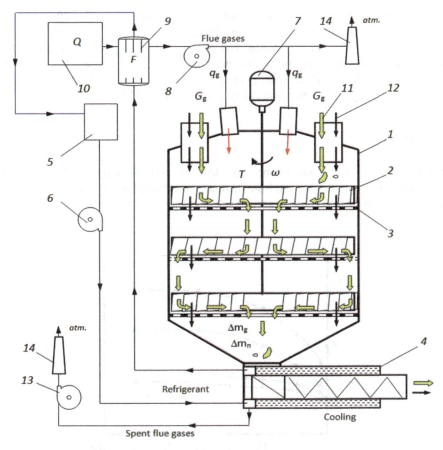

Fig. 1. Tray type reactor of clean burn with different time of processing raw materials: 1 – reactor; 2 – impeller with spades; 3 – tray with perforation; 4 – desiccator with cooling facing element; 5 – heat consumption device; 6, 8, 13 – fan; 7 – engine; 9 – heat exchanger; 10 – gasifier; 11 – large feedstock; 12 – fine feedstock; 14 – device for mechanical cleaning of gases.

perforations in the trays, the dimensions of which are gradually changed according to the height of the apparatus: from large on the top trays to smaller ones in the central and lower parts of the column. This scheme allows smaller fractions of raw materials, not lingering on the upper trays, spilling out to the underlying trays through large perforations of top trays; this ensures uniformity of processing the whole batch of raw materials [22–24].

For testing fuel pellets, derived from thermally treated reduced wood in the proposed unit, the pellets were produced, which were subsequently analyzed for main operating characteristics: hygroscopicity and swelling, affecting storage conditions, and fire parameters (calorific value, and smouldering time). To study the process of swelling and water absorption, a desiccator with mesh inserts was used. The obtained samples were kept in saturated steam on desiccation inserts for 35 days.

The study of the combustibility of fuel granules created according to the proposed scheme was carried out on an installation designed to determine refractory and combustible solid non-metallic substances and materials. The purpose of the study is to determine the combustibility of the test material in the created temperature conditions and study the behavior of the material in these conditions. As a result, the mass loss of the sample during combustion and decay was determined [25, 26].

3 Results

Figure 2 presents the graphs of the dependence of changes in relative humidity and volume of fuel granules obtained from thermally modified biomass and untreated wood raw materials over the period of their treatment in a humid environment.

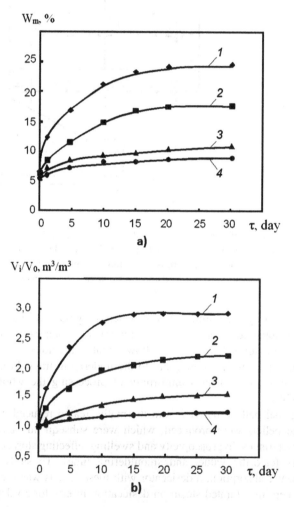

Fig. 2. Change in humidity (a) and volume (b) of torrefied solid fuel (1) from untreated raw materials; (2) T = 473 K; (3) T = 523 K; (4) T = 573 K.

As follows from the graphs (Fig. 2), a longer heat treatment provided a slightly greater effect of increasing water-repellent properties. It is found that the larger the diameter of the pellets, the less its swelling and water absorption. This is due to the lower penetration of water into the inner layers of the material.

The curves of equilibrium moisture content of torrefied pellets, held under normal conditions, which are presented in Fig. 3, characterize a significant reduction of this parameter with increasing temperature of heat treatment of raw materials, which ultimately affects the calorific value of the fuel, changing its moisture during storage and transportation.

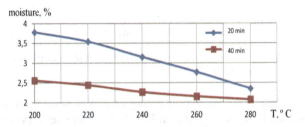

Fig. 3. The equilibrium moisture content of the fuel when held indoor depending on the temperature of torrefaction.

Regarding the calorific value of torreficate, it was found that with increasing temperature of treatment there is a significant increase in the heat of combustion of wood pellets (Fig. 4). Meanwhile, a double increase in the duration of treatment does not cause a significant increase in calorific value of the fuel, especially at relatively low temperatures of treatment (up to 240 °C). Thus, if it is necessary to increase productivity and economic efficiency of the torrefaction process of wood raw material, this can be done in the rough by finishing the processing at the time of significant reduction in flow of gaseous products of decomposition (in our case 20 min) without significant loss of energy quality of fuel.

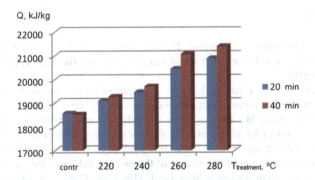

Fig. 4. High heat value of pellets depending on the temperature and time of torrefaction.

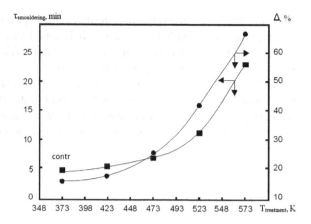

Fig. 5. The dependence of the time smouldering and loss of mass of pellets on the temperature of torrefaction of wood raw material.

The obtained results on the combustibility of pellets made from thermally treated wood are presented in form of graphs in Fig. 5. The increase in the combustibility of wood pellets with increasing temperature of torrefaction is explained by the increase of the released heat of combustion and reduction of initial humidity, which together allows maintaining self-combustion longer.

4 Conclusion

The research demonstrates the effectiveness of thermal processing of wood raw material prior to production of solid fuel for the purpose of reducing requirements for storage conditions and energy efficiency. It is established that increasing of productivity and economic efficiency of the torrefaction process of wood raw material can be done by finishing the processing at the time of significant reduction in flow of gaseous products of decomposition without significant loss of energy quality of fuel.

References

1. Barcik, S., Ugryumov, S., Razumov, E., et al.: Studies of component interconnection in a plywood structure with internal layers of veneer chips. Acta facultatis xylologiae zvolen **61**, 121–129 (2019)
2. Arias, B., Pevida, C., Fermoso, J., et al.: Influence of torrefaction on the grindability and reactivity of woody biomass. Fuel Process. Technol. **89**, 169–175 (2008)
3. Koleda, P., Korcok, M., Barcik, S., et al.: Effect of temperature of heat treatment on energetic intensity of flat milling of picea abies. Manag. Syst. Prod. Eng. **26**, 151–156 (2018)
4. Khasanhin, R.R., Safin, R.R., Kainov, P.A.: Effect of process pressure on the yield of products and the duration of the process of thermochemical processing of wood waste. Int. Multidiscip. Sci. GeoConference Surv. Geol. Min. Ecol. Manag. **19**, 145–151 (2019)
5. Timerbaev, N.F., Almohammed, O.A., Ali, A.K.: Longitudinal fin effect on effectiveness of double pipe heat exchanger. Lecture Notes in Mechanical Engineering, vol. 10, pp. 605–614 (2019)

6. Hughes, M., Constant, B.: The water vapour sorption properties of thermally modified and densified wood. J. Mater. Sci. **47**, 3191–3197 (2012)
7. Militz, H., Lande, S.: Challenges in wood modification technology on the way to practical applications. Wood Mater. Sci. Eng. **4**, 23–29 (2009)
8. Koleda, P., Barcik, S., Nascak, L., et al.: Cutting power during lengthwise milling of thermally modified oak wood. J. Wood Research **64**, 537–548 (2019)
9. Zabelkin, S., Bikbulatova, G., Grachev, A., et al.: Modification of bitumen binder by the liquid products of wood fast pyrolysis. Road materials and pavement design **20**, 1182–1200 (2019)
10. Tumuluru, J.S., Sokhansanj, S.: Review on biomass torrefaction process and product properties and design of moving bed torrefaction system model development. Annual International Meeting sponsored by ASABE, pp. 758–771 (2011)
11. Hosseinaei, O., Wang, S., Taylor, A.M.: Effect of hemicellulose extraction on water absorption and mold susceptibility of wood-plastic composites. Int. Biodeter biodegr **71**, 29–35 (2012)
12. Mukhametzyanov, S.R., Safin, R.R., Kainov, P.A.: Alternative energy in the processes of drying of thermolabile materials. Int. Multi-Conf. Ind. Eng. Mod. Technol. **4**, 860–865 (2019)
13. Chubov, A.B., Matjushenkova, E.I., Tsarev, G.I.: Rationale modes of oil heat treatment of plywood. J. News of the St Petersburg forestry Academy **194**, 129–137 (2011)
14. Jukka, A.: The activities of finnish thermowood association to commercialize thermowood. In: The Third European Conference on Wood Modification, pp. 3–9 (2007)
15. Khasanshin, R.R., Safin, R.R., Semushina, E.Y.: Technology of creating decorative panels made of thermomodified veneer. IOP Conf. Ser.: Mater. Sci. Eng. **463**, 699–706 (2018)
16. Vanco, M., Mazan, A., Barcik, S., et al.: Influence of thermal modified pinewood on the quality of machining at plain milling. J. MM Sci. 1171–1177 (2016)
17. Mohebby, B., Ilbeighi, F., Kazemi-Najafi, S.: Influence of hydrothermal modification of fibers on some physical and mechanical properties of medium density fiberboard. J. Holz Roh Werkst **66**, 213–218 (2008)
18. Hosseinaei, O., Wang, S., Rials, T.G., et al.: Effect of hemicellulose extraction on physical and mechanical properties and mold susceptibility of flakeboard. J. Forest Prod. **61**, 31–37 (2011)
19. Prins, M.J., Ptasinski, K.J., Janssen, F.J.: Torrefaction of wood. Part 2. Analysis of products. J. Anal. Appl. Pyrolysis **77**, 35–40 (2006)
20. Ayrilmis, N., Jarusombuti, S., Fueangvivat, V.: Effect of thermal-treatment of wood fidres on properties of flat-pressed wood plastic composites. J. Polym degrade Stabil **96**, 818–822 (2011)
21. Zhou, X., Segovia, C., Abdullah, U.H., et al.: A novel fider-veneer-laminated composite based on tannin resin. J. Adhes. **93**, 1–7 (2015)
22. Thek, G., Obernberger, I.: Wood pellet production costs under Austrian and in comparison to Swedish framework conditions. Biomass Bioenergy **27**, 671–693 (2004)
23. Gandiglio, M., Lanzini, A., Leone, P., et al.: Thermoeconomic analysis of large solid oxide fuel cell plants: atmospheric vs. pressurized performance. Energy **55**, 142–155 (2013)
24. Alakangas, E., Paju, P.: Wood pellets in Finland technology, economy, and market 64 OPET Report 5. VTT Processes (2002)
25. Wiinikka, H., Gebart, R.: Critical parameters for particle emissions in small scale fixed bed combustion of wood pellets. Energy Fuels **18**, 897–907 (2004)
26. Melin, S.: Wood pellets-material safety data sheet. Delta research, Canada (2006)

Development and Research of Automated Compact Cogeneration Plant

A. M. Makarov[✉], P. V. Dikarev, and V. V. Lazarev

Volgograd State Technical University, 28 Lenina Avenue,
Volgograd 400005, Russian Federation
amm34@mail.ru

Abstract. An automatic control system for an automated compact cogeneration plant has been developed, in which the secondary energy generated, which produced when electricity is obtained, is used to generate thermal energy. Generated electricity is redistributed using an adaptive control system to additional heating of water if it is not necessary for external consumers to use it. A functional automation circuit, an electrical circuit diagram and a block diagram of a water heating algorithm have been developed. A mathematical model for calculating the efficiency of the installation has been developed. For experimental research of the developed device, a prototype was designed and manufactured, consisting of a gasoline-driven electric generator, heat exchanger, heating element, pumps, and also a control system based on a microcontroller and sensors. A program has been developed for receiving and processing data from installation sensors. A preliminary assessment of the performance of the developed installation has been carried out.

Keywords: Cogeneration · Mobile plant · Water heating · Thermal energy · Adjustment · Hot water supply · Electric power supply

1 Introduction

Cogeneration is a process of joint production of electric and thermal energy, which allows increasing the energy efficiency of the installation. The primary drive is a mechanical source of generated electricity, and thermal energy is generated through the use of heat loss (the use of heat from the coolant, lubricating oil, compressed air-gas mixture and carbon monoxide exhaust) of the primary drive motor [1].

The main privilege of cogeneration over other types of electricity production using renewable energy sources, such as the sun or wind, is the ability to supply electricity in a known amount [2]. Thus, cogeneration can be attributed to regulated sources of energy.

In recent years, cogeneration plants called mini-CHP plants have been used more and more widely in Russia. Mini-CHPs are used in oil and gas refineries and chemical plants, in the glass, paper, machine-building, textile and food industries, in large administrative and shopping centers and hospitals. They are successfully operated to cover long or peak loads in stand-alone or emergency operation [3]. Nowadays, hot water systems are being developed in which cogeneration plants are used, including steam heating plants [4]. But all the above cogeneration plants are stationary. The majority of these devices cannot be used non-stationarity. It is possible to use the principle of cogeneration to receive heat power in non-stationary conditions. Cogeneration plants allow using the total thermal energy of an internal combustion engine (ICE) of a car or power plant as a source of additional energy, while it is possible to obtain electricity [4]. World manufacturers of power plants produce cogeneration units with electric power from 20 to 2000 kW, equipped with engines of MAN, Perkins, MTU, Doosan manufacturers with low levels of environmental pollution. These installations, as autonomous energy sources, are widely used in the housing and communal and agricultural sectors of Russia, cottage villages, sanatorium and medical complexes, etc.

Now the heat-availability factor of fuel combustion in petrol cogeneration plants on the basis of internal combustion engines is about 87% and only 13% of heat are lost in production [6]. In comparison with separate generation of the same amount of electric and thermal energy, savings of fuel in cogeneration installations reach 40% [7].

Using the heat of the coolant, exhaust and lubricating oil of an internal combustion engine with a capacity of about 500 kW for heating, it is possible to provide an area of 4.3 thousand square meters with thermal energy, and the installation will maintain a normal temperature in premises [8].

2 Mathematical Model

To evaluate the efficiency of using cogeneration plants on the ICE basis, a mathematical model thereof is developed.

The ECE (energy conversion efficiency) of the cogeneration plant consists of the ECE for electric energy and the ECE for thermal energy:

$$\eta = \eta_{el} + \eta_{th}, \qquad (1)$$

The ECE for electric energy is determined by the formula:

$$q_g = f(n_g, P_g); \eta_{el} = \frac{P_{el}}{P_f}, \qquad (2)$$

where P_{el} is the electric power supplied to the consumer for his needs; P_f is the power given by fuel, [W].

The ECE for thermal energy is determined by the formula:

$$\eta_{th} = \frac{P_{th}}{P_f}, \qquad (3)$$

where P_{th} is the thermal power received by the water, [W].

The thermal power received by the water includes heating by the exhaust gases, cooling fluid, oil in the lubrication system, and the TEH, and is calculated by the formula:

$$P_{th} = \frac{Q_{th}}{t_{heat}}, \qquad (4)$$

where Q_{th} is the thermal energy received by the water, [J]; t_{heat} is water heating time, [s].

The thermal energy received by the water [9] is calculated by the formula:

$$Q_{th} = cm_w(T_a - T_b), \qquad (5)$$

where c = 4187 [J/kgK] is specific water heat capacity; m_w is the volume of water pumped through the heat exchanger and the tubular electric heater in the course of operation, [kg]; T_a and T_b is the water temperature before and after heating, respectively, [K].

We determine the volume of water which is pumped in the course of operation by the formula:

$$m_w = M_w t_{heat}, \qquad (6)$$

where M_w is the mass consumption of water, [kg/s].

The relation of the mass consumption of water with the volumetric one:

$$M_w = \rho_w q_w, \qquad (7)$$

where q_w is the volumetric water consumption, [m^3/s]; ρ_w = 1000 [kg/m^3] is the water density.

Substituting expressions (5), (6) and (7) into formula (4), we shall obtain:

$$P_{th} = \frac{c\rho_w q_w t_{heat}(T_a - T_b)}{t_{heat}} = c\rho_w q_w (T_a - T_b).$$

Thus, the thermal power is calculated by the formula:

$$P_{th} = c\rho_w(T_a - T_b) \cdot q_w. \qquad (8)$$

The power given by the fuel is calculated by the formula:

$$P_f = \frac{Q_f}{t_{heat}}, \qquad (9)$$

where Q_f is the thermal energy given by the fuel.

The thermal energy given by the fuel can be calculated by the formula:

$$Q_f = \alpha m_g, \qquad (10)$$

where $\alpha = 4.4 \cdot 10^7$ [J/kg] is specific combustion heat of gasoline AI-92; m_g is the volume of gasoline consumed in the course of heating [kg].

The volume of gasoline, consumed in the course of heating is calculated similar to the volume of water by the formula:

$$m_g = t_{heat} M_g = \rho_g q_g t_{heat}, \tag{11}$$

where M_g is the mass consumption of gasoline, [kg/s]; $\rho_g = 725...780$ [kg/m^3] is the density of gasoline AI-92 at $t = 15$ °C; q_g the volumetric consumption of gasoline, [m^3/s].

Substituting expressions 10 and 11 in formula 9, we shall obtain:

$$P_f = \frac{\alpha \rho_g q_g t_{heat}}{t_{heat}} = \alpha \rho_g q_g.$$

Thus, we have received formula 12 to calculate the power of fuel:

$$P_f = \alpha \rho_g q_g. \tag{12}$$

Substituting expressions (8) and (12) in formulas (2) and (3), we shall receive the expressions to determine the ECE for the electric and thermal power:

$$\eta_{el} = \frac{1}{\alpha \rho_g} \cdot \frac{P_{el}}{q_g}, \tag{13}$$

$$\eta_{th} = \frac{c \rho_w}{\alpha \rho_g} (T_a - T_b) \cdot \frac{q_w}{q_g}. \tag{14}$$

Let us make a replacement for convenience:

$$k_1 = \frac{1}{\alpha \rho_g} = \frac{1}{4,4 \cdot 10^7 \cdot (725 \div 780)} = (2.914...3.135) \cdot 10^{-11} \text{m}^3/\text{J};$$

$$k_2 = c \rho_w = 4187 \cdot 1000 = 4.187 \cdot 10^6 \text{J}/(\text{m}^3 \cdot \text{K}).$$

Then dependences (13) and (14) will be recorded as:

$$\eta_{el} = k_1 \frac{P_{el}}{q_g}, \tag{15}$$

$$\eta_{th} = k_1 k_2 (T_a - T_b) \cdot \frac{q_w}{q_g}. \tag{16}$$

When determining the general ECE of the plant, one should consider that the gasoline consumption depends on the load power of the electric generator and the frequency of its rotation. That is:

$$q_g = f(n_g, P_g),$$

$$\eta = k_1 \frac{P_{el}}{q_g} + k_1 k_2 (T_a - T_b) \cdot \frac{q_w}{q_g}.$$

Then the general ECE of the plant is:

$$\eta = \frac{k_1}{f(n_g, P_g)} (P_{el} + k_2(T_a - T_b)q_w). \tag{17}$$

Let us carry out a check of the formula by the dimensions of the input values:

$$\frac{\frac{m^3}{J}}{\frac{m^3}{s}} \left(\frac{J}{s} + \frac{J}{m^3 \cdot K} K \cdot \frac{m^3}{s} \right) = \frac{m^3}{J} \cdot \frac{s}{m^3} \cdot \frac{J}{s} = 1$$

The ECE is a dimensionless value, therefore, the dependence obtained is correct.
The analysis of the dependence received allows to draw the following conclusions:
The required and initial water temperatures are basic data for the calculations, that is (in the flowing mode of heating) $T_a - T_b = const$.

Coefficients k_1 and k_2 depend on the physical parameters which characterize the gasoline and water. Let us accept them as constant $k_1 = k_2 = const$.

The electric power to be supplied to the consumer depends on the load created by the consumer:

$$P_{el} = U_c I_c. \tag{18}$$

The variables in the formula are the rotating speed of the generator n_g, load of the generator P_g and water consumption through the heat exchangers q_w. Where the ECE is directly dependent on q_w.

As a result, we obtain that for receiving the final formula to calculate the ECE of the plant it is necessary to explore the dependence $q_g = f(n_g, P_g)$.

3 Materials and Methods of Study

To evaluate the adequacy of the mathematical model obtained, as well as the practical use of the cogeneration principle for non-stationary obtaining and application of heat and electric power, the automated compact cogeneration plant [10] has been developed. It includes a converter of direct cycle – engine 1 with generator of electricity 2 on a single shaft, heat exchanger-utilizer of exhaust gases heat from engine 3 which is linked to engine 1 by highway of exhaust gases 4 (Fig. 1) [1].

System of outer heat transfer 5 to users of heat is performed, e.g., like a radiator or capacity with water linked to heat exchanger-utilizer 3 by supply line 6 to additional line 7 and valve 8 for immediately hot water deliver to the user. Also this system includes supply line of cool water 9 with circulatory pump 10, which is connected with heat exchanger-cooler 11 and the cooling shirt of engine 12 by highway of coolant 13 with another circulatory pump [11].

The heat exchanger-utilizer 3 is linked to the pump 17 by cold water supply line, equipped with a control valve 16. The pump can be installed in a pond or reservoir of cold water 18. Inside the heat exchanger-utilizer 3 is equipped with an electric heating element 19 and temperature sensor 20, which are connected to the microprocessor unit 21.

An electric generator 2, circulatory pump 10, another circulatory pump 14, control valve 16, pump 17 and socket 22 also are connected to the microprocessor unit 21, which included an information input device 23 and an information output device 24.

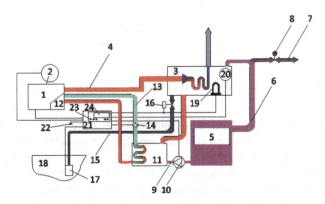

Fig. 1. Diagram of automated compact cogeneration plant.

The efficiency of the installation can be increased (higher than that of a gas and electric portable heater) by using the generated electricity for additional heating of water, provided that external consumers do not need electricity at the moment. More efficient utilization of thermal and electric energy of fuel combustion allows achieving better results of the functionality system in comparison with analogues [12].

The cogeneration plant is presented by a petrol power plant which consists of an internal combustion engine with an electric generator on a shaft. There are two sockets of 220 V and 12 V for a user.

The exhaust gases, which are emitted during the operation of the power plant, as well as the cooling fluid, which washes the ICE, are used to obtain thermal energy (water heating) [13]. The heat exchange of the heat sources with the water takes place respectively in heat exchangers 1 and 2. Also, the surplus of the electric energy produced is brought to the tubular electric heater thermal electric heater.

The microcontroller is responsible for collecting data from the process of operation of the cogeneration unit and controlling this process. Based on the gathered information, we build an automation scheme for the automated compact cogeneration plant. Functional diagram of the plant is shown in Fig. 2.

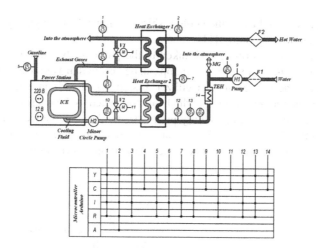

Fig. 2. Function automation chart of automated compact cogeneration plant.

Lock valves K1 and K2 serve to shutdown the corresponding heating circuit. Coarse filter F1 is intended to prevent large mechanical particles from getting in; the latter can get into the system, for example, at the water intake from the tank. Filter F2 is intended to purify the water from possible pollution in the system, for example, from scum. Manual gate MG is intended to dump air out of the system at the beginning of operation of the cogeneration plant. The temperature control is carried out by thermocouples 0 ... 400 °C Sensor-M6Screw and Dallas DS18B20 temperature sensors.

Water temperature regulation at the outlet is carried out by regulating its consumption in the system by means of pump 1 where a PWM (pulse-width modulation) signal comes from the microcontroller when heating in the flowing mode. In case of heating water in the recirculation mode, regulation can be carried out by means of turning on/off the heating circuits by closing/opening lock valves K1 and K2, as well as by regulating the flow passing through the tubular electric heater. When heating water up to the required temperature, lock valves K1 and K2 open, thereby, the heat sources (the exhaust gases and cooling fluid) pass by the heat exchangers. Also, in this case, the tubular electric heater is disconnected.

Control and recording of the following parameters is carried out in this diagram:

- Exhaust gas temperature at the inlet of the heat exchanger 1 and at its outlet
- Exhaust gas temperature at the inlet of the heat exchanger 2 and at its outlet
- Temperature of the cold water
- Temperature of heated water after each heating level: after the heater, heat exchanger-utilizer 2 and heat exchanger-utilizer 1
- Water flow in the system using the flow sensor Sabocn SBS-HZ 21WA
- Petrol flow using the Micro Low Sensor OF-201

To research the dependence of the efficiency of water heating on various factors, an experimental setup was built on the basis of the Huter HT1000L electric generator (Fig. 2).

In addition to the ability to produce thermal energy for heating water, the prototype can generate electrical energy up to 1000 W, which can be used by users, or for additional heating to grow the efficiency of the system as a whole [14].

The heat exchanger serves to transfer heat from the heat carrier of the primary contour to the heat carrier of the secondary contour through the plates, which avoids the mixing of heat carrier flows with each other [15].

4 Results and Discussion Thereof

To check and assess the efficiency of developed device (Fig. 3) and adequacy of the mathematical model obtained, a range of experiments have been conducted. The average fuel (gasoline) consumption was about 0.4 l/h in the process.

Fig. 3. Picture of the experimental plant: 1 – electric generator; 2 – heat exchanger-utilizer; 3 – control unit; 4 – capacity with water; 5 – tubular electric heater; 6 – circulatory pump; 7 – water flow sensor; 8 – temperature sensor.

Based on the results of the experiments, the dependences of the temperature of the water at the outlet of the heat exchanger (water for the users) on time were found (Fig. 4).

The first experiment (the blue diagram), water heating was carried out only by the exhaust gases, the external load of the generator was 600 W, the volume of the water pumped was 10.5 l. The average water consumption was 7.59 l/min.

The second experiment (the red diagram), water heating was carried out by the exhaust gases and the tubular electric heater (capacity 224 W) simultaneously, the volume of the water pumped was 10.5 l. The average water consumption was 7.59 l/min.

The third experiment (the green diagram), heating was carried out by the exhaust gases and the tubular electric heater (capacity 1000 W) simultaneously, the volume of the water pumped was 10.5 l. The average water consumption was 7.59 l/min.

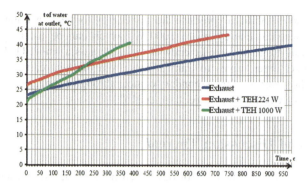

Fig. 4. Diagram of dependences of the temperature of the water at the outlet of the heat exchanger on time were found.

The results of the experiment have demonstrated that when the water is heated from 20 to 40 degrees centigrade, the heating time was:

- When heated only by the exhaust gases (the blue diagram) – 16 min
- When heated with the tubular electric heater on 224 W (the red diagram) – 10 min
- When heated with the tubular electric heater on 1000 W (the green diagram) – 6 min

Thus, when an additional source of heat, such as TEH 1000 W, was introduced, heating time decreased by 2.67 times. As the reduction in heating time indirectly influences the reduction in the volume of gasoline consumed by the plant, then, proceeding from formula (17), the ECE of the plant has increased. Therefore, the hypothesis of an increase in the ECE of a cogeneration plant is confirmed by the experimental data.

5 Conclusion

To more efficiently obtain thermal and electric energy under non-stationary provisions, automated compact cogeneration plant has been developed, which allows increasing the efficiency of obtaining thermal energy, since for this purpose, in addition to the energy of the exhaust gases and coolant, excess generated electric energy is used. At the same time, the installation is controlled from the microprocessor control unit, this allows setting various operating modes of the plant depending on the type and amount of energy consumed and effectively control them by controlling the installation in automatic mode. All this increases the efficiency and efficiency of the installation as a whole.

Also, the plant can be manufactured sufficiently compact and equipped with an additional line for continuous hot and cold-water supply to the consumer. It will increase the universality of the plant and allow using it in the form of an additional source of electric and thermal energy (for example, when energy supply is interrupted or used non-stationarity).

A mathematical model of operation of a cogeneration plant has been developed. For experimental research of the developed device, a prototype has been designed and manufactured. The results of mathematical modeling agree with the results of the experiment, which may indicate the effectiveness of the use of automated compact cogeneration plants.

References

1. Dikarev, P.V.: Development of technical solution for automated compact cogeneration plant using search engineering In: Proceedings of International Ural Conference on Green Energy South Ural State University, 4–6 October 2018, pp. 253–258 (2018)
2. Bystritsky, G.F., Borodich, E.A.: Autonomous and cogeneration installations of power supply (reference materials). NTF Energo-progress, Moscow (2014)
3. Dikarev, P.V.: Calculation of the efficiency of a cogeneration electromechanical system with an adaptive control system. Energy Resour. Saving Ind. Transp. 2, 31–36 (2019)
4. Tikhomirov, K.V., Sergeenko, E.S.: Heat Engineering, Heat Supply and Ventilation: Textbook for High Schools. Stroiizdat, Moscow (1991)
5. Dikarev, P.V., Makarov, A.M., Volkov, I.V.: Development and study of a compact mobile cogeneration plant. In: Instrumentation and the Automated Electric Drive in the Fuel and Energy Complex, Housing and Utilities Infrastructure: a Digest of the III Volga Region Scientific and Practical Conference, vol. 2, pp. 103–105 (2017)
6. Barkov, V.M.: Cogeneration technologies: opportunities and prospects. ESCO 7 (2004)
7. Sivukhin, D.V.: Thermodynamics and Molecular Physics. Fizmatlit, Moscow (2005)
8. Esterkin, R.I.: Boiler Installations. Energoatomizdat, Leningrad (1989)
9. Kondratyev, D.A.: Cogeneration. SPB Grafike publishing house (2012)
10. Makarov, A.M., Volkov, I.V., Lazarev, V.V., et al.: A mobile cogeneration plant. RU Patent 174173, 5 Oct 2017
11. Lazarev, V.V., Dikarev, P.V.: Compact cogeneration field installation. Review-competition of scientific, design and technological works of students of Volgograd State Technical University, pp. 32–33 (2016)
12. Dikarev, P.V., Makarov, A.M., Volkov, I.V. et al.: A compact independent cogeneration plant with a microprocessor-based control system. VolgGTU News. Progressive Technologies in Mechanical Engineering Series, vol. 8(187), pp. 72–74 (2016)
13. Saydanov, V.O., Agafonov, A.N., Antipov, M.A., et al.: Power plant. RU Patent 2280777, 27 June 2006
14. Dikarev, P.V., Makarov, A.M., Volkov, I.V., et al.: Automatic compact cogeneration device. In: Instrumentation and the Automated Electric Drive in the Fuel and Energy Complex, Housing and Utilities Infrastructure: a Digest of the II Volga Region Scientific and Practical Conference, vol. 1, pp. 140–142 (2016)
15. Kasatkin, A.G.: The Main Processes and Apparatus of Chemical Technology: A Textbook for Universities. Alliance, Moscow (2004)

CAD in Electrical Engineering: New Approaches to an Outdoor Switchyard Design

E. A. Panova[✉], A. V. Varganova, and N. T. Patshin

Magnitogorsk State Technical University, 38 Lenina Avenue,
Magnitogorsk 455000, Russian Federation
ea.panova@magtu.ru

Abstract. The outdoor switchyard design is one of the most time-consuming stage of electric substation engineering. It requires the designer to know a large number of design standards, safety requirements for the switchgear and standard design decisions. To date, there is no CAD system that allows you to automate the design process of plans for outdoor switchgear substations taking into account their economic indicators and offer the designer not only a technically feasible solution, but also carry out its economic assessment on the basis of which designer is able to choose the optimal layout option. The aim of this study is to elaborate the algorithmic basis for electric power substation CAD software. The algorithm of economically reasonable scheme and open switchyard arrangement design had been developed. The algorithm makes it possible to adopt a scheme option, choose electrical equipment and compose switchyard single-line diagram and layout. An example of engineering and economical comparison results form had been given. This indicates that the elaborated CAD system could be used in the work of architect engineer.

Keywords: CAD · Outdoor switchyard · Single-line diagram · Graph editor · Electrical swithgear · Technical-economic calculation · Short-circuit current calculations

1 Introduction

Since 1980s CAD systems are widely implemented in the construction of technical objects of diverse complexity. Initially CAD was graphical software enabling a drawing making process. The next stage of its elaboration was production of a full pack of project documents. Today industry also demands CAD to make computations and decision optimization.

CAD is widely exploited in architecture and construction, design of machine detail and other engineering systems. In the field of electric and power engineering the software for electrical machine CAD [1, 2] are developed. Thus [3] and [4] describe an application of genetic algorithm modification to a construction optimization which provides required performance characteristics and minimizes costs of item production and operation. This algorithm is also effective for radioelectronics design [5]. The task of electric machine optimal design is also solved in [6–8].

The independent line of CAD development is a software for 3D modeling of a designed item, which is a separate element [9] or even a whole substation [10]. In [10] 3D modeling is not the main function of the software. It only complements the system and helps designing engineer to judge the solutions. The main tasks of this software are outdoor switchgear composition and cross section views of its units, and also bus system mechanical analysis.

Also CAD software is applied to virtual simulation of complex technical objects such as thermoelectric power station [11] which enables not only construction project drawings and specifications preparation, but also simulation of a whole object life cycle up to its shut down.

A range of papers is dedicated to CAD systems for some local problems of an electric power system design process such as wires choice and verification [12], building wiring design [13], electric lighting design [14], earthing system and lightning protection computation [15–17], corona protection system design [18].

As soon as distributed generation design and operation is an urgent problem CAD systems are also exploited to solve it. For instance [19] describes a CAD system for an effective operation of photovoltaic system in the grid, the authors of [20] offer a CAD of complex solar cell metallization patterns, in [21] CAD of a compact doubly-fed induction generator for small wind power application is considered.

CAD is also widely integrated in electric power supply schemes designing process [22], but this software is directed towards project documentation forming according to unified system of engineering drawings. Thus it sets the designer free from routine tasks, but it doesn't simplify decision making process. Similar approach is implemented in secondary control wiring CAD [23, 24].

More advanced are CAD systems which combine construction project drawings and specifications preparation and computation functions. E.g. [25] describes the software for electric system mode parameters computation, its graphical representation and the choice of conductors and transformers. Also there are CAD systems able to solve circuit equation [26].

The reported study was funded by RFBR according to the research project № 18-37-00115.

Up-to-date CAD systems should not only simplify projectors work in drawing and carry out some computations but also it should optimize project decision making process. It means that out of a range of variants it should pick up economically reasonable one beyond all permissible. Thus [27] gives an application of coordinate descent method and its probabilistic version, gradient-based algorithm and two-stage probabilistic algorithm to project decision optimization.

The authors of [28] had made an approach towards electrical energy system CAD similarly to electronic circuits, but there is a doubt that such a method is applicable to complex multiloop systems.

Apparently, the trend of CAD system developments is the integration of graphical software and computational programs and sub-scripts. In project making process the most time-consuming stage is the project of an outdoor switchgear. Currently there is no CAD software for outdoor switchgear projects which allows to prepare drawings and compose project documentation according to standards together with switchgear layout optimization in terms of technical boundary conditions and construction and operation expenses minimization.

2 CAD Algorithm

The principals of man-machine interaction are used to design the CAD of electrical switchgear. CAD includes the main workflow phase of architect engineer of electrical department: regulations and specifications work management and statement of work, grounds for a variant decision, and generation of planning documentation.

Developed engineering soft makes it possible to automates the process of engineering design of switchyard single-line diagram and layout, adoption and control the parameters of equipment, to calculate technical-economic indexes, using the criteria of discount costs minimum, taking into account maintenance cost, depreciation expenses, supply interruption costs.

CAD algorithm is performed on Fig. 1.

Bases of design are contained (block 1, Fig. 1):

- Hand input: rated high-voltage side U_{nomHVS} (kV) and low-voltage side U_{nomLVS} (kV) of electrical swithgear, number of feeders (N), maximum of power load S_{max} (kVA), the value of symmetrical ampere I_{n0} (kA), electrical substation type
- Data basis (DB) which makes it possible to correct the such information as: technical-economic indexes of equipment, equipment cost (C) and reliability targets (w - failure rate, (1/year), recovery time of distribution substation, (h)), conventional representation of elements on single-line diagram (in keeping with GOST) and mechanical drawings on layouts and sections (equipment certificate), space limitations of unifying electrical swithgear, Fig. 2, Table 1 [29].

Using U_{nomHVS}, U_{nomLVS}, мвн, $U_{номНН}$ and S_{max} the number (M) and capacity of transformers are selected (block 2, Fig. 1) and controlled at operating emergency and normal conditions..

CAD is selected the variable schemes of electrical swithgear taking into account terms of reference, normative and technical documentation, the number and capacity of transformers (block 3, Fig. 1). Technical-economic calculation of variable variants of scheme are also generated by the CAD (block 3, Fig. 1), using the methods allows at [30].

Scheme projecting (block 5, Fig. 1) and displaying (block 6, Fig. 1) are realized using specific graph editor KOMPAS.

The technical-and-economic indexes of schemes variants are displayed (block 7, Fig. 1) at *MS Excel*, to ratchet that result up of technical-economic calculation. The main criteria of choosing the scheme is the minimum of discounted costs. As an example (Table 2) of technical-economic calculation of outdoor switchyard of 220 kV with number of feeders - 2, type of electrical substation - tail station.

CAD in Electrical Engineering 539

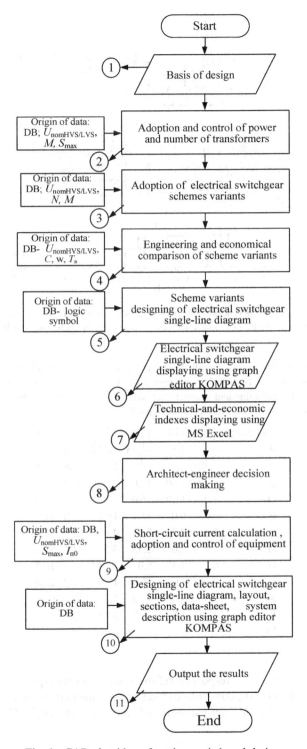

Fig. 1. CAD algorithm of outdoor switchyard design.

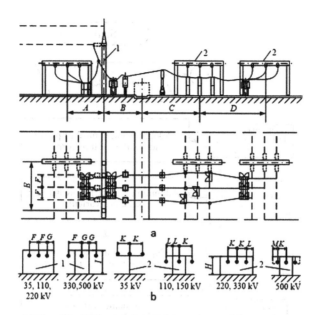

Fig. 2. Unify model lie of outdoor switchyard (a) section and layout (as an example outdoor switchyard 220 kV); (b) line (1) and selector (2) gantry towers schemes 35–500 kV.

Table 1. Elements size range of unified outdoor switchgear 35–500 kV, m (1) transfers bars doesn't used at schemes of distribution substation of 35 kV; (2) it is necessary taking into account that size between line and gantry towers equal B + C = 7,5 m at distribution substation of 35 kV; (3) gantry towers of voltage 35 kV has T- form.

Size range	35	110	150	220	330	500
A	—(1)	7,5	11,75	11,75	18,0	29,0
B	—(2)	9,0	9,0	11,8	19,6	26,8
C	—(3)	13,0	15,0	19,45	20,4	29
D	7,0	11,0	16,0	17,5	31,5	45,0
E	6,0	9,0	11,1	15,4	22,0	31,0
F	1,6	2,5	3,0	4,0	8,0	11,0
G	1,4	2,0	2,55	3,7	4,0	5,5
H	6,1	7,85	8,0	11,35	12,8	16,5
I	7,85	11,35	13,0	17,0	20,0	26,0
K	2,0	3,0	4,25	4,0	4,5	6,0
L	—(3)	1,5	2,13	3,25	3,5	–
M	–	–	–	–	–	5

The "Scheme name" is responded off-the-shelf solutions report Company's Code 56947007-29.240.30.010-2008 "Electric schematic diagram of substation electrical swithgear of 35–750 kV. Off-the-shelf solution" и Company's Code 56947007-

29.240.30.047-2010 "Recommendation of using of electric schematic diagram of substation electrical swithgear of 35−750 kV".

Table 2 is given three variants: 1 - scheme "220-3H" (block of power line - transformer - circuit-breaker HVS (n. 3.3.2 [29])); 2 scheme "220-5AH" ("bridge circuit" (n. 3.3.4 [29])); 3 - scheme "220-4H" (block of power line - transformer - circuit-breaker HVS with jumper (n. 3.3.3 [29])).

Table 2. Engineering and economical comparison results.

Variant	1	2	3
Scheme name	220-3H	220-5AH	220-4H
Number of transformers, pcs	2	2	2
Transformer power, kVA	40	40	40
Size of a land plot, m^2	5400	6800	6200
Discounted costs	41850	43000	42500
Equipment capital contributions	28000	34000	29000
Land values	12 000	14 000	13 500
Maintenance cost	3360	4080	3480
Depreciation expenses	2520	3060	2610
Supply interruption costs	32610	31780	32930
Number of switches, pcs	2	3	2
Failure rate, 1/year	0,04	0,03	0,035
Recovery time of distribution substation, h	34	26	33

The number and power of transformers are chosen by the mean of terms of reference on the first stage of projecting (block 2, Fig. 1). The "Size of a land plot" is counted using the footprint of outdoor switchyard taking into account space for indoor switchgear. The "Discounted costs", the "Equipment capital contributions", the "Land values", the "Maintenance cost", the "Depreciation expenses" are counted using the methods of [30]. The number of switches are counted using the single-line diagram of schemes. The "Supply interruption costs", the "Failure rate", the "Recovery time of distribution substation" are computed with the methods of [31, 32].

To high up the motivation of architect-engineer using the result of CAD it is possible to make: off-the-shelf solution, modified solution or own decision, taking into account specific features of terms of reference. It is important that engineer could use own background to meet the nontypical customer requirements (block 8, Fig. 1).

For taking variant short-circuit current calculations are computed using the value of symmetrical ampere (block 9, Fig. 1). Adoption and control of equipment are designed: switching devices, voltage measuring/measuring current transformers, excess-voltage suppressor and conductors.

Using the results of automated projecting, the CAD is designed single-line diagram, layout, chosen sections, date-sheet and generate the description with the content of short-circuit current and technical-economic calculations of finalize of electrical switchgear (block 10−11, Fig. 1) using graph editor KOMPAS.

3 Conclusion

Existing works and electrical engineering CAD systems don't have a complex approach to substation design. This paper gives CAD system which makes it possible to design substation outdoor switchyard taking into account economical and technical indexes, reliability factor, object specific features, etc. More over CAD application will enhance architect engineer performance, reduce routine operations and minimizes possible mistakes. The reported study was funded by RFBR according to the research project № 18-37-00115.

References

1. Rivera, C.A., Poza, J., Ugalde, G.: A knowledge based system architecture to manage and automate the electrical machine design process. In: IEEE International Workshop of Electronics, Control, Measurement, Signals and Their Application to Mechatronics, Donostia-San Sebastian, pp. 1–6 (2017). https://doi.org/10.1109/ecmsm.2017.7945875
2. Aravind, C.V., Grace, I., Rozita, T., et al.: Universal computer aided design for electrical machines. In: IEEE 8th International Colloquium on Signal Processing and its Applications. Melaka, pp. 99–104 (2012). https://doi.org/10.1109/cspa.2012.6194699
3. Tikhonov, A.I., Zaitsev, A.S., Stulov, A.V., et al.: Elaboration of power transformers CAD optimization. Bull. IGEU **6**, 87–91 (2017)
4. Ojaghi, M., Daliri, S.: Analytic model for performance study and computer-aided design of single-phase shaded-pole induction motors. IEEE Trans. Energy Convers. **32**(2), 649–657 (2017). https://doi.org/10.1109/TEC.2016.2645641
5. Pavlushin, V.A., Mikheev, F.A., Markov, M.V.: Genetic algorithm in frequency characteristics optimization. In: Proceedings of Saint Petersburg Electrotechnical University, vol. 11, pp. 33–38 (2006)
6. Ghorbanian, V., Salimi, A., Lowther, D.A.: A computer-aided design process for optimizing the size of inverter-fed permanent magnet motors. IEEE Trans. Ind. Electron. **65**(2), 1819–1827 (2018). https://doi.org/10.1109/TIE.2017.2733460
7. Zhengming, Z., Ahmed, E.A.: Advanced computer-aided design and analysis for synchronous reluctance-permanent magnet machines. Tsinghua Sci. Technol. **3**(3), 1143–1148 (1998)
8. Halvaii, A.E., Ehsani, M.: Computer aided design tool for electric, hybrid electric and plug-in hybrid electric vehicles. In: IEEE Vehicle Power and Propulsion Conference, pp. 1–6 (2011). https://doi.org/10.1109/vppc.2011.6043005
9. Agapov, A.A., Chernykh, T.E.: Application of CAD to electric drive wires modelling. Applied problems of electromechanics, power engineering, electronics. Engineer ideas of XXI century, pp. 15–18 (2016)
10. Orelyana, I.A., Vorobyev, S.A.: Model studio CS. Outdoor switchgear. CAD and graphics (2008). http://sapr.ru/article/18709. Accessed 23 Sept 2017
11. Osika, L.K., Zhuravlev, V.S.: Virtual modeling as a single tool for managing the life cycle of thermal power plants. Power Technol. Eng. **4**(981), 2–10 (2013)
12. Eliseev, D.S.: CAD algorithms for wires and cables choice. Volgograd State Agrarian University, p. 184 (2012)

13. Pionkevich, V.A.: CIM-TEAM E3 Series – CAD for complex electric projects. In: Promotion of Production and Energy Utilization Potential in Siberia, Conference Proceedings, pp. 458–463 (2010)
14. Kovalev, A.A., Golovin, A.A.: Application for CAD lighting calculations. Electron. Sci. Pract. J. Mod. Sci. Res. Innov. **6**–1(38), 37 (2014)
15. Shishigin, D.S.: AutoCAD software for electrical earthing system and lightning protection computation. In: XII Meeting on Management Issues, pp. 9374–9380 (2014)
16. Kokorus, M., Mujezinovic, A.: Computer aided design of the substation grounding system. In: ICHVE International Conference on High Voltage Engineering and Application, pp. 1–4 (2014). https://doi.org/10.1109/ichve.2014.7035394
17. Duta, M.I., Diga, S.M.M., Ruşinaru, D.G., et al.: Computer aided design of the earthing installations for the substations. In: 3rd International Symposium on Electrical and Electronics Engineering, pp. 34–38 (2010). https://doi.org/10.1109/iseee.2010.5628483
18. Staubach, C., Kempen, S., Pohlmann, F.: Computer aided design of an end corona protection system for accelerated voltage endurance testing at increased line frequency. In: Electrical Insulation Conference, pp. 170–174 (2011). https://doi.org/10.1109/eic.2011.5996140
19. Bagre, A., Ikni, D., Dakyo, B., et al.: Computer aided design and complex power control effectiveness of large-scale photovoltaic system integrated into a grid. In: Pointe-Aux-Piments, pp. 1–5 (2013). https://doi.org/10.1109/afrcon.2013.6757624
20. Wong, J.: Griddler: intelligent computer aided design of complex solar cell metallization patterns. In: IEEE 39th Photovoltaic Specialists Conference, pp. 0933–0938 (2013). https://doi.org/10.1109/pvsc.2013.6744296
21. Vakil, G.I., Rajagopal, K.R.: Computer aided design of a compact doubly-fed induction generator for small wind power application. In: Joint International Conference on Power Electronics, Drives and Energy Systems and Power India, pp. 1–4 (2010). https://doi.org/10.1109/pedes.2010.5712574
22. Akhtulov, A.L., Akhtulova, L.N., Leonov, E.N.: Statement of problem of synthesis of industrial electric supply basic schemes by means of modern CAD. Bull. Kalashnikov ISTU **1**, 110–113 (2011)
23. Trofimov, A.V.: Computer-aided designing of secondary circuits of electrical power stations and substations. Power Technol. Eng. **10**, 46–49 (2009)
24. Tselishchev, E.S., Kotlova, A.V., Kudryashov, I.S., et al.: Improvement of the effectiveness of CAD application in auxiliary mechanisms control schemes. Ind. Autom. **9**, 10–15 (2015)
25. Fursanov, M.I., Dul, I.I.: Basic principles of industrial electric power network computer aided design and engineering. In: Energetika Proceedings of CIS Higher Education Institutions and Power Engineering Associations, vol. 5, pp. 18–24 (2012)
26. Ngoya, E., Rousset, J., Obregon, J.J.: Newton-Raphson iteration speed-up algorithm for the solution of nonlinear circuit equations in general-purpose CAD programs. IEEE Trans. Comput. Aided Des. Integr. Circ. Syst. **16**(6), 638–644 (1997). https://doi.org/10.1109/43.640621
27. Lvovich, Ya.L., Chernysheva, G.D., Kashirina, I.L.: Project decisions optimization on the basis of equivalent transformations of the task of minimal cover. In: Science and Education: Scientific Edition of Bauman MSTU, vol. 1, p. 4 (2006)
28. Kim, Y., Shin, D., Petricca, M., et al.: Computer-aided design of electrical energy systems. In: IEEE/ACM International Conference on Computer-Aided Design, pp. 194–201 (2013). https://doi.org/10.1109/iccad.2013.6691118
29. Igumenshev, V.A., Olenikov, V.K., Malafeev, A.V., et al.: Step down substation projecting of industrial plant. Magnitogorsk (2014)
30. Karapetian, I.G., Faibisovich, D.L.: Reference-book on design of electric power systems. ENAS, Moscow (2012)

31. Shemetov, A.N.: ERCOT electric reliability. NSTU, Magnitogorsk (2007)
32. Panova, E.A., Irikhov, A.S., Dubina, I.A., et al.: Calculation of economic components of target function of the algorithm for determining the optimal option of scheme of substations distribution device with the high voltage of 35 kV and above. Electrotech. Syst. Complexes **1**(42), 4–11 (2019). https://doi.org/10.18503/2311-8318-2019-1(42)-4-11

Increasing Efficiency of Water and Energy Supply Systems in Southern Region Environment of Russian Federation

D. V. Kasharin[✉]

Platov South-Russian State Polytechnic University, 132 Prosveshcheniya Street, Novocherkassk 346428, Russian Federation
dendvkl@mail.ru

Abstract. The article deals with the issues of increasing the energy efficiency of water and energy supply technologies in the southern region environment of the Russian Federation through the recuperation of energy in reverse osmosis units. An analysis was made of the use of water resources in the southern regions. Qualitative indicators of surface and groundwaters of the Rostov, Volgograd, Astrakhan regions, Krasnodar Territory and the republics of Kalmykia and Adygea are determined, based on which water treatment technologies are proposed for their use for the needs of decentralized drinking water supply in small settlements. A comparative analysis of technologies for water treatment of surface and sea water is carried out using ultrafiltration methods, nanofiltration, depending on the possibility of recovering the energy of the concentrate stream. A justification is given for the planned laboratory studies to determine the efficiency of using membrane element-based units on decentralized water and energy supply to small settlements.

Keywords: Southern region · Drinking water · Water resources · Water supply systems · Ultrafiltration · Nanofiltration · Energy recovery · Laboratory installation · Small settlements

1 Analysis of the Municipal – Domestic Water Supply of the Southern Region of the Russian Federation

There is an insufficient quantitative and qualitative water supply to municipal-domestic facilities within most of the Southern region territory of the Russian Federation, which is due to the following factors: a decrease in the quantitative and qualitative indicators of water supply sources; the level of productivity and quality of water treatment in water treatment facilities; unsatisfactory sanitary-technical condition of drinking water transportation systems and irrational use of drinking water. Insufficient drinking water being replenished by transported water is indicative, especially in small settlements. The distribution of transported water to subjects of the Russian Federation is shown in Fig. 1 [1–8].

One of the main factors hampering the creation of autonomous water supply systems for small settlements is the significant energy consumption and low energy efficiency of existing water supply technological schemes associated with the most promising membrane technologies at the moment.

Surface and groundwater sources are used to supply water in the Southern region. Let's consider their characteristic features.

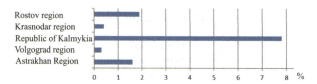

Fig. 1. Percentage of transported water in the Southern region from total water supply.

1.1 Surface Water Sources

Although large rivers are present in Russia - the Volga, the Don and the Kuban, it is difficult to supply the water for drinking purposes to water treatment facilities of small settlements (Fig. 2) due to the high level of water contamination, insufficiently treatment of sewage waters in Category I reservoirs, eutrophication and algal blooming of water bodies.

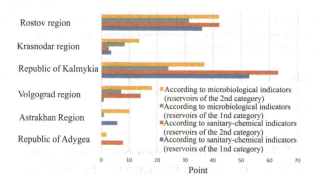

Fig. 2. Data on sanitary-chemical and microbiological indicators for the Southern region in a point system.

Considering the development of recreational use of the coasts of the Azov and Black Seas, sea water for water supply is potentially perspective, provided there are no surface and underground sources of water supply, indicators for the Black and Azov Seas are given in Table 1 [9–20].

Table 1. Average characteristics of the chemical composition of water from the Black and Azov seas.

№	Average characteristics of sea water		
	Concentrations	Black sea	Azov sea
1	Mineralization, g/l	18,60	11,8
2	Water Hardness, mg · equiv/l	68,00	42,9
3	HCO_3^-, g/l	0,17	0,16
4	Na^+, g/l	6,20	3,49
5	K^+, g/l	0,19	0,13
6	Ca^{2+}, g/l	0,20	0,17
7	Mg^{2+}, g/l	0,70	0,42
8	Cl^-, g/l	10,00	6,53
9	SO_4^{2-}, g/l	1,44	0,92

1.2 Underground Sources for Water Supply

On average, only 10.9% of underground waters of the Southern region are being used, so increase of its use can cover the need for domestic and drinking water supply to small settlements. Operational reserves of groundwater are shown in Fig. 3.

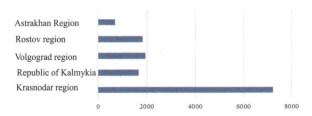

Fig. 3. Operational reserves of groundwater in the southern regions of the Russian Federation, thousand m^3.

Currently, groundwater supply in populated areas is used to a greater extent than water from surface water bodies in the Krasnodar Krai, the Volgograd Region and the Republic of Adygea [1, 2].

The total mineralization of groundwater varies from 487.5 to 380038 mg/l, which requires special methods of water treatment, including the use of nanofiltration methods. In spite of its lower value as compared to reverse osmosis its energy consumption is higher.

2 Analysis of the Energy Efficiency of Water Treatment Methods for the Southern Region Environment of the Russian Federation

2.1 Justification of Methods and Technological Water Treatment Plans

According to analysis of the existing water treatment methods, reagent methods are the most common at present with their lowest specific energy consumption per m^3 of purified water at 0.05–0.07 (kWh/m^3), however, considering better efficiency due to energy recovery and automation, membrane technologies based on reverse osmosis have great prospects for small settlements. It should be noted that for drinking water

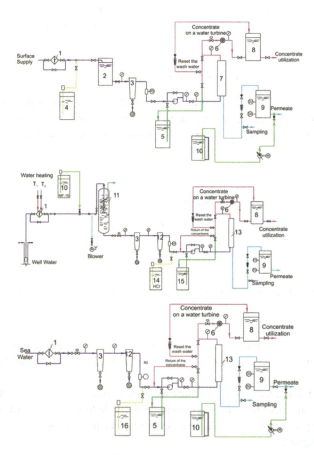

Fig. 4. The Technological schemes of treatment plants for the purification of: a – surface waters based on ultrafiltration; b – underground waters of nanofiltration and reverse osmosis; c – sea water: 1- sump; 2 – contact capacity; 3 – mechanical filter; 4 – capacity of coagulant; 5 – capacity of regeneration solution; 6 – water turbine; 7 – ultrafiltration membrane; 8 – tank concentrate; 9 – storage tank; 10 – tank with sodium hypochlorite; 11 – air stripping tower; 12 – pre-sediment filter; 13 – membrane module; 14 – capacity of hydrochloric acid; 15 – capacity of regeneration solution; 16 – capacity of antiscalants.

supply it is promising to use nanofiltration technologies with a lower working pressure (3.5–16 bar) compared to reverse osmosis and a selectivity that allows to obtain water with a salt content corresponding to human biological needs [4, 5].

The following figure shows a combined plan of water clarification methods (ultrafiltration in combination with a reagent method) for surface water bodies, desalination of underground (nanofiltration and reverse osmosis) and seawater (reverse osmosis), including the recovery of energy from the concentrate, using the Pelton hydraulic turbine (Fig. 4).

2.2 Comparative Analysis of Water Treatment Methods with Metering

Using the NanoTechPRO program in accordance with the average indicators of surface, groundwater and sea water, a comparative analysis of the technological plan for water treatment was carried out (Fig. 4), taking into account the possibility of energy recovery, shown in Fig. 5.

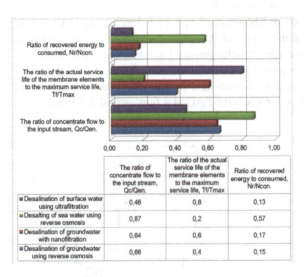

Fig. 5. Comparative analysis of water treatment methods taking into account the possibility of energy recovery in the NanoTechPRO program.

As seen from Fig. 5 the best indicators for the desalination of groundwater compared to reverse osmosis belong to nanofiltration units that provide a longer service life of membrane elements (with a ratio of estimated service life extending to a maximum of 5 years for nanofiltration 0.6 and for reverse osmosis 0.4) and energy efficiency ratio of recovered energy to consumed energy 0.17 to 0.15 respectively) (Fig. 4).

To determine the ranges of operation and ensure more efficient energy consumption of the technological plans for water and energy supply, we envisage to conduct laboratory studies at a stand that models the initial water in accordance with the qualitative indicators of underground and sea waters of the Southern region and the operation of a reverse osmosis unit with energy recovery.

3 Laboratory Stand

3.1 Schematic Diagram of the Laboratory Stand

To carry out further research, a schematic diagram of the laboratory stand investigating membrane technologies for the purification of underground and sea water was developed for a wide temperature range from 5 to 30 °C, which will allow us to specify the application ranges and optimal parameters of nanofiltration units for water supply systems in the southern region of the Russian Federation (Figs. 6 and 7).

3.2 Methodology

1. Tap water at a flow rate of 20 l/min preheated to 30 °C in flow-through water heater EVPN-7,5 (position 1) is supplied to tank DK-500 with a paddle stirrer (position 2) to prepare the model water. Chemical reagents necessary for the preparation of a solution with a specific concentration of salts, are added to the tank to a volume of 500 L and mixed with a stirrer for 10–15 min.

2. Then, a drum laboratory pump with a flow rate of 27 l/min is supplied from tank DK-500 (position 2) to tank DK-500 (position 10) where the main hydro-chemical indicators of the model water prepared are measured (pH, TDS, t).

3. Then the model water solution from tank DK-500 (position 10) is fed to the reverse osmosis unit, if necessary, pre-cooled to the set temperature in flow-through water cooling unit ZXA 2019 PAC at a flow rate of 8.3 l/min.

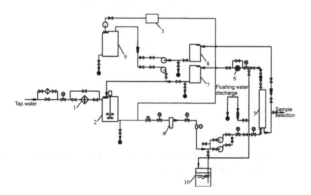

Fig. 6. Schematic diagram of the laboratory stand for the study of membrane treatment technologies: 1 – water heating unit; 2 – tank with a stirrer; 3 – cooler; 4 – mechanical filter; 5 – membrane module; 6 – turbine; 7 – tank with concentrate; 8 – tank with permeate; 9 – intermediate tank; regeneration tank.

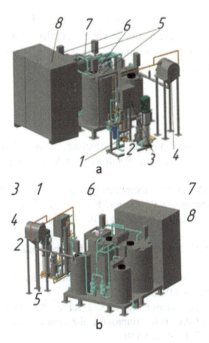

Fig. 7. Laboratory stand for the study of membrane treatment technologies: 1 – the membrane element; 2 – pumps; 3 – shield of control; 4 – water turbine; 5 – dosing tank; 6 – chiller; 7 – the heater; 8 – the regulator.

4. After preliminary mechanical filtration on a cartridge filter at a flow rate of 8.3 l/min, a high-pressure pump supplies the prepared model solution to the membrane module (position 6) with a roll filter element (for seawater, for groundwater, for surface water) where the flow is separated into a concentrate and a permeate.

5. Purified water (permeate) accumulates in tank DK-200 (position 9). The concentrate at high pressure, is directed to bucket turbine PLT-1 (position 7) to recover the energy necessary to compensate the power of membrane separation consumed by the unit. After that, the concentrate is gravity fed to tank DK-200 (position 8).

6. To reduce the water and chemical reagents consumption required to model the initial solution, the drum laboratory pumps mix and supply water with the experimentally determined proportions from the concentrate tanks (position 8) and permeate (position 9) to tank DK-500 (position 10).

4 Conclusion

It is recommended to use units with nanomembrane filtration elements and the possibility of energy recovery using hydro turbine plants as fully automated compact units. In the future, laboratory studies are planned for a temperature range of 5 to 30 °C to develop recommendations and a rationale for the use of nanofiltration units in the Southern Region.

This article was executed as per assignment №13.1236.2017/4.6 on the topic: "Development of energy-efficient and environmentally safe systems of decentralized water and power supply for recreational facilities in the Southern Region environment of the Russian Federation".

References

1. Kasharin, D.V.: Protective engineering structures made of composite materials in the construction of water construction. South-Russian State Technical University, Novocherkassk (NPI) (2012)
2. Giroud, J.P.: Geotextiles and geomembranes. Geotext. Geomembr. **1**, 5–40 (1984)
3. Kasharin, D.V., Kasharina, T.P., Godin, P.A., et al.: Use of pipelines fabricated from composite materials for mobile diversion hydroelectric power plants. Power Technol. Eng. **6**(48), 448–452 (2015)
4. Kasharin, D.V.: Integral mobile diversion conduit and method of construction. RU Patent 22607650 (2016)
5. Leaf, J.A.V.: Woven fabric tensile mechanics. In: International Conference on Mechanics of Flexible Fibre Assemblies (1980)
6. Mefferl, B.: Mechanische eigenschaften PVC-gesehichteter polyestergewebe. Aachen (1978)
7. Pervov, A.G., Yurchevsky, E.B.: Improving the structures of membrane devices. Water Supply Sanitary Eng. **7**, 62–68 (2009)
8. Kasharin, D.V., Thai, T.T.K.: System of mobile dams and a method of its erection. RU Patent 2539143, 10 January 2015
9. Kasharin, D.V.: Mobile hydropower structure of composite materials. Reclam. Water Manag. **4**, 45–46 (2007)
10. Kasharin, D.V., Godin, M.A.: Scope and justification of the parameters of the mobile micro hydro sleeve-type for small streams. Bull. Volgograd State Univ. Architect. Civil Eng. Ser. Constr. Architect. **20**, 142–148 (2010)
11. Kasharin, D.V., Kasharina, T.P., Godin, M.A.: Mobile derivational micro-HPP for reserve water supply and standby power service of recreation facilities and harbour installations of Russky Island. In: Nase more, vol. 62, no. 4, pp. 272–277 (2015)
12. Kasharin, D.V.: Calculation analysis of parameters of mobile water retaining structures made of composite materials. Bull. Volgograd State Univ. Archit. Civ. Eng. Ser. Constr. Archit. **30**(49), 128–138 (2013)
13. Kasharin, D.V.: Numerical simulation of the structural elements of a mobile microhydroelectric power plant of derivative type 423 of the series. Advances in Intelligent Systems and Computing, pp. 51–61 (2016)
14. Kasharin, D.V.: Optimization of selection and calculation of structures made of composite materials for engineering of protection of water during construction. RU certificate of state registration of computer program 201011300 (2010)
15. Kasharin, D.V.: Reliability assessment of lightweight hydrotechnical structures formed from composite materials. Power Technol. Eng. **4**(43), 230–236 (2009)
16. Kasharin, D.V.: Optimizing the parameters of water-retaining protective structures under small stream conditions. In: 2nd International Conference on Industrial Engineering, pp. 1804–1810 (2016)
17. Mikheeva, I.V.: Water bodies and the state of water supply in the territories of the Southern Federal district. In: Technologies of Water Treatment, TECHNOVOD, pp. 144–155 (2017)
18. Quality of surface waters of the Russian Federation. Rosgidromet

19. Kasharin, D.V.: Using biopositive constructions made of composite materials in water and electricity supply for recreational facilities. In: International Conference on Industrial Engineering, Applications and Manufacturing (2017)
20. Pervov, A.G.: Modern high-efficiency technologies for the purification of drinking and industrial water using membranes: reverse osmosis, nanofiltration, ultrafiltration. ASV, Moscow (2009)

Electron-Ion Technology as Protection of Solar Modules from Contamination

I. M. Kirpichnikova[✉] and V. V. Shestakova

South Ural State University, 76 Lenina Avenue, Chelyabinsk
454080, Russian Federation
ionkim@mail.ru

Abstract. In this paper, the problem of dependence of a state of solar modules' surface on the external environmental conditions: dust, precipitation, sand, pollen of plants, various technical suspensions in megacities is investigated. It is concluded that such an impact significantly reduces the efficiency of solar installations and leads to a decrease in electricity generation and the using of famous technologies for cleaning solar panels is associated with the attracting human resources and leads to increasing costs. An overview of possible methods and devices for solving this problem is given. The use of processes of electron-ion technology for the protection of solar modules from pollution is proposed, the theoretical justification of this method and the prospects for its use are presented. The main forces acting on a dust particle in an electric field during its deposition are considered. The cleaning technology developed does not require large investments, maintenance and does not pollute the environment with harmful chemicals.

Keywords: Solar energy · Contamination of the solar module · Purification system · Ionization · Electron-ion technology

1 Introduction

One of the main strategic directions of development of any state is the increase of energy efficiency and energy saving. The transition to an innovative development path is impossible without increasing the share of renewable energy resources in the future.

The government of our country at the legislative level is supporting the introduction of renewable energy sources (RES) [1–3]. The world development of the last two decades has been accompanied by an increase in the negative impact on the environment, a violation of the biosphere balance, and therefore it becomes necessary to write down the ever-increasing needs of the world community for the natural opportunities of the planet, observing harmonious relations in the system "resources - society – people" [4, 5].

In the conditions of limited and non-renewability of traditional energy sources (gas, coal, wood), humanity entered the era of using alternative energy sources and, in particular, solar energy.

For the conversion of solar radiation into other types of energy, various technologies and devices are used, the continuous improvement of which is a process of continuous increase in the efficiency of their operation [6–8].

2 Effect of Atmospheric Precipitation on the Efficiency Working of Solar Modules

The efficiency of working the solar panels and collectors depends largely on the location of their installation. In most cases, they are installed at an angle to the horizon, corresponding to the geographical latitude of the terrain.

To be able to use solar radiation, it is necessary to determine in what quantity it is delivered to a location over a certain period. This is called the «radiation balance", which is regulated by astronomical relations (daily and seasonal cycles) on the one hand and atmospheric conditions (turbidity and cloudiness), on the other [9–12].

Of course, not everything depends only on the choice of successful installation site in accordance with the radiation balance. The solar units themselves must be ready to absorb solar energy, especially their glass working panels. It is important to consider and take into account a number of issues related to electricity generation, for example, blackout, pollution and snow cover of solar photovoltaic installations [13].

When the surface of the glasses, the constituent elements of the solar modules, becomes dirty, their transmission coefficient decreases, and accordingly the amount of radiation entering the plates decreases. This leads to decreasing the efficiency of the installation. Because of pollution, the transmission coefficient of the glass decreases by 2–3% [14].

Solar modules (panels) and the entire solar power plant works best in a clear, clean, cloudless day. But the presence of one of these factors does not mean that the work of solar modules will be most effective all the time. Due to weather conditions such as snowfall, rain, fog, sandstorms, the operation of solar installations can be significantly hampered [15]. In such conditions, solar panels produce electrical energy from scattered solar radiation, since the arrival of any direct sunlight on the glass of the solar module is difficult. In the United States, Germany, Japan, Britain, Russia, Ukraine and other countries where there is always snow, the effect of the presence of precipitation on solar modules becomes a tangible problem, especially in solar power plants. Solar modules can work quite successfully, for example, under light snow. But when there is heavy snow and there is accumulation of precipitation on the surface of the solar module (when the solar modules are blocked), this can reduce the generation of electricity or even completely stop the electric generation. The more snow that accumulates on the surface of the solar module, the more decrease of electric generation of the solar power plant. But we also need to understand that snow will not remain on the roof and surface of solar modules for a long time even in the heaviest snowfall. Since solar panels are usually installed at an angle, this factor can help the discharge the snow [16]. For this reason, industrial solar power plants and installed solar modules can be tilted at a steeper angle to the ground to help the snow avalanche reliably slide off its surface.

Rain leaves sediment and divorce on the surface of the solar module (pollution caused by industrial impurities in droplets in industrial cities), which reduces the effectiveness of interaction between solar radiation and photocells. The same effect is caused remained on the surface of modules dust, soot, pollen, sand in the sandy areas.

3 Ways and Methods of Cleaning Solar Modules from Pollution

Pollution by precipitation, dust, soot and pollen significantly reduces the performance of photovoltaic batteries and solar power systems - energy losses can reach 20% [17]. The decline in efficiency is only a consequence. It is necessary to start with a reason to eliminate losses, which is why various large enterprises develop different methods and technological complexes for cleaning glass from solar installations.

The most common method of cleaning the surface of aggregates is manual. Human resource must be involved for this. Manual cleaning of working surface is not temporary, it required constantly. The employee must manually clean the surface of the glass from external contamination or snow cover, dust, sand with using the detergents and special accessories. This method consumes much energy, because often it is necessary to clean more than one panel, but the whole complex of batteries, which also makes this method economically inefficient and quite long in time. To improve the efficiency of cleaning, certain purification systems and technologies are required, which will support the generation of electricity at the required level, but will not significantly increase the cost of the module.

Large industrial companies are engaged in decision of this task, creating the whole purification complexes.

For example, Kärcher [18] offers for this purpose its own solution - an accessory system for high-pressure devices iSolar, consisting of telescopic rods and attached devices with rotating brushes. Various rods and brush attachments make it possible to economically clean up solar modules on area of up to 1.500 m^2.

The brushes of the devices are driven into the rotation by a jet of water, formed by a high-pressure apparatus, and is used simultaneously for washing off the mud (Fig. 1).

The cleaning effect is provided by the mechanical action of brushes, whose nylon bristles of which do not leave scratches on the surface of the modules (Fig. 2). Operation in low pressure mode eliminates the risk of damage the modules. To ensure a long service life, the rotating brushes are fixed to the sturdy ball bearings.

If hard water is used, limescale may remain on the surface of the cleaned modules, which significantly reduces their energy output. A reliable protection against stains lime is guaranteed by the Kärcher mobile water softeners and cleaner of solar battery RM 99.

The ion-exchange resin used in water softeners absorbs dissolved lime and reduces the hardness of water to a normal level. And if water alone is not enough for effective cleaning, RM 99 is used, removing even heavy grease and simultaneously binding lime, which eliminates limescale formation [18].

Fig. 1. iSolar system.

Fig. 2. The iSolar system brush.

The disadvantage of this device is that water with the product can cause environment pollution, because it is a chemical compound that will flow down the surface of solar module and soak into the ground.

Therefore, now it is still necessary to search perfect device for cleaning the surface of solar modules.

4 Use of Electron-Ion Technology to Protect the Solar Module from Contamination

There is an opinion in medicine that it is easier to prevent the disease than to treat it later. Something similar can be said about the processes of cleaning solar modules. We suggest not to clean already dusty glass surfaces of modules, but to prevent their contamination. This can be done by using methods of electron-ion technology (EIT). The essence of method is that dust particles, having their natural bipolar electric charge or obtained artificially in a corona discharge field, are deposited under the action of electric field forces on the electrodes of the opposite sign. The work of electrical and electrostatic filters for air purification is based on this principle.

To properly design a device for the deposition of dust particles, it is necessary to know the characteristics of the dust in the terrain (dimensions, dielectric permittivity, roughness, dispersion, density, wettability, adhesive properties, etc.) and the origin of the dust.

The sources of the natural electric charge of aerosol particles are:

- Radiation of radioactive substances contained in the earth's crust and air
- Cosmic rays
- Balloelectric effect (crushing and splitting of water)
- Electric discharges in the atmosphere
- Triboelectric effect (mutual friction of particles of dust, grains of sand, snow, etc.).

The dust particles suspended in the air, as a rule, have either a positive or negative electrical charge, and can be neutral. The ratio of the number of neutral and positively or negatively charged particles in the same volume of air is not constant. Signs of charge and their magnitude are determined by the properties of the initial material and their time being in the air (Fig. 3).

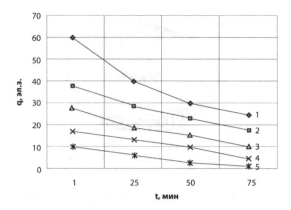

Fig. 3. Dependence of the average natural charge of particles on the time of being in air for different sizes [19]: 1 – r = 5, 6... 11, 2 μm; 2 – r = 2, 8... 5, 8 μm; 3 – r = 1, 4... 2, 8 μm; 4 – r = 0, 7...1, 4 μm; 5 – r = 0, 45... 0, 7 μm.

As we can be seen from the figure, the dependence of the charge on time has a transitional character and is mainly determined by the particle size and ionic air composition. The steady-state distribution of charges on particles of given sizes in calm air comes in after 10, 15, 20, 35 and 65 min, respectively. Preservation of charge is observed for several hours, and in most particles.

This circumstance allows us to conclude that we should not be neglected the natural charge of particles and while developing a device that captures dust particles it is necessary to keep in mind the magnitude and polarity of their natural charge.

Air purification in such a device occurs as follows: dust particles which are in the air stream, approaching to the glass panel of the solar module fall into the electric field of the interelectrode gap and under the influence of this field they settle on one or another precipitating plate in accordance with the polarity and magnitude of the natural charge. Since there is no zone for artificial charging of particles in the device under development, it was customary to call it the electrostatic device (ESD).

To ensure the effectiveness of the ESD work, while maintaining its economy, it is necessary either to increase the particle deposition rate, or to reduce the distance between them. One of the factors affecting the rate of particle deposition is the electric field strength E, which, as known, determined by the ratio:

$$E = \frac{U}{d}, \qquad (1)$$

where U – the voltage supplied to the electrodes of filter, B; d – the interelectrode distance, m.

A voltage of the order of 9–12 kV is used for the operation of the ESD. From a high voltage source, it is supplied to the precipitating plates of the device, creating there a negative potential. The second group of precipitating plates has a positive potential.

5 Analysis of Forces Acting on the Particle in the ESD

Precipitation of dust particles on ESD plates occurs due to several forces. If the solar module is horizontal or inclined at a small angle, the dust particles will settle on the glass surface due to gravity forces F_T. This force is determined by the multiplication of the mass of the particle m by the acceleration of gravity g. The time for which the particle of aerosol flies a distance equal to the width or length of the module at air flow velocities from 0.1 to 1.0 m/s does not exceed 0–0.5 s. During this time, a particle up to 1 μm under the action of gravity falls by several micrometers, however, during the day nevertheless reaches the surface of the glass and the precipitation electrodes. Considering that not only suspended particle of aerosol with the size of micrometer size are in the air, but particles with much more size, this force should not be neglected, since heavier and larger particles, having their own specific charge, will be directed to the electrode of the opposite sign.

Another force acting on the particle when deposited on the precipitation electrodes is the Coulomb force Fk. Its value is determined by the magnitude of the electric field strength E and the natural charge of a particles q_{ecm}:

$$F_K = q_{ecm} \cdot E, \qquad (2)$$

where E – the electric field strength in the filter, V/m; q_{ecm} – the natural charge of a particle, Cl.

Obviously, the larger charge, the greater the impact of this force. The value of the natural charge of particles from data of different sources [20, 21] varies from 1 to 10% of the charge, which it could obtain by artificial charging. However, even such a small electrical charge can affect the value of the Coulomb force and it must be considered in the calculations.

The natural charge of dust particles is bipolar, that is, it can have both a positive and a negative value. The particle, falling into the interelectrode gap (between positively and negatively charged plates), behaves accordance with the laws of physics: a positively charged particle repelled by a positively charged plate, settles on a negative plate. Similarly occurs with negatively charged particles.

Particles moving in the air stream can have a wide variety of shapes. In the general case, they can be spherical and nonspherical. The magnitude of the force acting on them depends on this. If the particle is spherical, that is, it has the shape of a ball, then, in addition to the Coulomb force Fk, another panderomotive force acts on it, determined by the non-uniformity of the distribution of the electric field strength [20].

$$F_E = 2\pi\varepsilon_0 a^3 \frac{\varepsilon - 1}{\varepsilon + 2} grad E^2, \qquad (3)$$

where a – the size of particle, m; ε_0 – the electric constant, F/m; ε – the permittivity of the particle; $gradE$ – the gradient of the electric field strength, V/m².

The component $2\pi\varepsilon_0 a^3 \frac{\varepsilon-1}{\varepsilon+2}$ is a polarization charge that appears on the particle when it is in an electric field. Consequently, a dielectric particle of radius a is considered as a dipole.

However, often the particles have a shape different from spherical. In this case, such a particle, for example, of an ellipsoidal shape, falling into an electrostatic field, also polarizes. But, unlike a spherical particle, it is oriented along the force lines of the field along its long axis. In this case, the charge polarization does not occur uniformly around the circle, as in the sphere, but concentrated along the ends of the ellipsoid (Fig. 4).

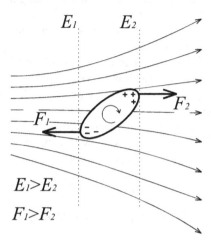

Fig. 4. Polarization of a particle in an electrostatic field.

Around these charges appears a field that distorts the ground field of the inter-electrode gap. The main field, striving to regain its original position, acts on the particle, pushing it to the side with more tension. This force with which the field acts to the particle is the ponderomotive force. Therefore, the more particles of non-spherical form fall into the zone of action of the ground field, greater its non-uniformity and, correspondingly, and greater the force F_E, acting on these particles.

A significant influence on the motion of particles is exerted by the resistance force of the environ F_C. It determines the steady-state velocity of the particles. In the case of sufficiently slow motion of small size particles, i.e. for small values of the Reynolds number (Re < 5), the resistance force of the environ to the motion of particles is expressed by the Stokes formula [20]:

$$F_C = -6\pi \cdot \mu \cdot a \cdot v_c, \qquad (4)$$

where μ – the coefficient of dynamic viscosity of the air, N s/м^2; a – the particle size, m; v_c – the drift velocity of a particle moving under the action of electric field forces, m/s.

Thus, the analysis of the forces acting on the particle in the electrostatic trap for dust showed that when calculating and designing the ESD, it is necessary to take into account the processes occurring in the electrostatic field of the device with particles of dust during their trapping and deposition.

6 Conclusion

The efficiency of solar modules depends to a large extent on the environmental conditions and contamination to which the module's working surface is exposed. The using of famous technologies for cleaning solar panels is associated with the attracting human resources and leads to increasing costs. The cleaning technology developed by us not only allows to effectively protect the panel from dust, but also does not require large investments, maintenance and does not pollute the environment with harmful chemicals.

The proposed electron-ion method has no analogues in the market of purification systems, which means that consumers and manufacturers of solar power plants can be interested in it.

Acknowledgements. The work was supported by Act 211 Government of the Russian Federation, contract № 02.A03.21.0011.

References

1. Federal Law of the Russian Federation of March 12 35-FZ. On the electric power industry (2007)
2. Rustamov, N.A.: Standardization for the development of energy on renewable sources. Stand. Qual. **6**(936), 38–40 (2015)
3. Rustamov, N.A.: Issues of technical regulation of the development of renewable energy in Russia. Energetik **2**, 48–49 (2014)
4. Golitzin, M.V., Golitzin, A.M., Pronina, N.V.: Alternative energy sources. Science, Moscow (2004)
5. Anikeev, V.V.: Ecological and economic aspects of the implementation of the provisions of the Kyoto Protocol in Russia. Bull. Cent. Environ. Policy Russia Towards Sustain. Dev. Russia **32**, 31–32 (2005)
6. Bushuev, V.V.: Energy of Russia. Potential and implementation strategy 1. Energy, Moscow (2012)
7. Petlevanaya, E.V., Levshov, A.V.: Information control systems and computer monitoring. In: Proceedings of the II International Scientific and Technical Conference of Students, Graduate Students and Young Scientists, vol. 2, pp. 132–136 (2011)
8. Shafranic, U.K.: Fuel and energy and the economy of russia: yesterday, today, tomorrow. Energy, Moscow (2011)
9. Butuzov, V.A.: Solar Heat Supply in Russia, Design, Construction Operation. Lambert Academic Publishing, Saarbrücken (2012)
10. Shestakova, V.V.: Trends in the development of solar photovoltaic installations. In: 65th International Youth Scientific and Technical Conference Youth. Science. Innovations, Vladivostok (2017)
11. Bakirov, A.A., Karkach, D.V.: Numerical model of dynamics flux of solar radiation and its comparison with experiment. In: Abstract of Third International Forum Renewable Energy: Towards Raising Energy and Economic Efficiency, pp. 28–34 (2015)
12. Elistratov, V.V., Kobysheva, N.B., Sidorenko, G.I.: Climatic factors of renewable energy sources. Science, St. Petersburg (2010)

13. BAPV and BIPV Installation Trends in Sweden. In: Conference Proceedings, Chalners University of Technology, Gotherburg, 16–19 September 2014 (2014)
14. Strebkov, D.S., Polyakov, V.I., Panchenko, V.A.: High-voltage silicon modules for the conversion of concentrated solar radiation. Period. Sci. Tech. J. Minor Energy **1–2**, 116–122 (2014)
15. Popel, O.S., Frid, S.E., Kiseleva, S.V., et al.: Climatic data for renewable energy in Russia. Tutorial. MPTU Publications, Moscow (2010)
16. State standard of Russian Federation 51595: Nontraditional power engineering. Solar power engineering. Solar collectors. General specifications (2000)
17. Shestakova, V.V., Kirpichnikova, I.M.: The study of unforeseen consequences of the production of solar photovoltaic panels. In: Electrotechnical Complexes and Systems: Materials of the International Scientific-Practical Conference, vol. 2, p. 221 (2017)
18. iSolar Karcher: System solving for cleaning solar batteries. https://www.kaercher.com/by/professional/apparaty-vysokogo-davlenija/isolar-ehffektivnaja-ochistka-solnechnykh-batarei-i-povyshenie-ikh-ehnergootdachi.html. Accessed 25 Oct 2017
19. Kirpichnikova, I.M.: Energy-saving air-cleaning systems in the s. premises with increased requirements for air purity, p. 38 (2001)
20. Vereschagin, I.P., Levitov, V.I., Mirzabekyan, G.Z., et al.: The fundamentals of the electrogas dynamic of disperse systems. Energy, Moscow (1974)
21. Grin, H., Lein, V.: Aerosols - dusts, fogs and smokes. Himia, p. 428 (1972)

Plating Technology for Improvement of Moveable Contact Joint Performance in Power-Distribution Equipment

V. Goman[✉] and S. Fedoreev

Ural Federal University, 19 Mira Street,
Yekaterinburg 620002, Russian Federation
vvg_electro@hotmail.com

Abstract. The article proposes a method for better characteristics of moveable contact joint performance. In particular, the improvement is related to reliability, operating economy, and durability. The method supposes application of protective metal coatings made of special materials on the work surfaces of the contacts. The method suggests applying protective coatings on contact surfaces without using complex equipment and requires temperatures not higher than 80–90 °C. The method is based on the contact melting process, which important property is the interaction of solid and liquid metals or an alloy below the autonomous melting point of the solid metal. The article describes how to choose the alloys and how the technology of their application was developed. As a result, two surface-active indium alloys were selected. The article shows that the coated contact joints demonstrate stable and almost constant contact electrical resistance, while the resistance of the uncoated contact joints significantly increases in the two years of test. The technology is applicable to real operation of electrical equipment.

Keywords: Coating · Electrical contact · Contact joint · Contact electrical resistance · Power transmission and distribution · Energy efficiency

1 Introduction

Power supply systems extensively use moveable contact joints to ensure controllable electrical circuit closing/opening and continuous (sliding) contact. This type of connections is used in various electric devices: relays, contactors, circuit breakers, knife switches, electric motors, current collectors etc.

Such features of moveable contact joint operation as friction between the members, sparkling, and electric arcs result in mechanical wear and electrical erosion. Moreover, moveable contact joints share the common problems of electrical contacts [1–11]:

- The contact surfaces get welded together with strong current impulses
- Oxidation and corrosion of the contact surfaces as a result of the impact of such factors as sharp fluctuations in temperature, high humidity, aggressive substances, present in the ambient air

Exposure to the above-mentioned factors leads to an increase in contact electrical resistance and results in an increase in energy losses [1–11].

Emergency power failures cause production equipment outage, danger to the people in the in workshops and public facilities, and power deficiency. Owners would need extra time and funds to rectify such power failures. Failures resulting from degradation of electrical circuit contacts often end in fire in the electrical equipment and buildings.

The application of metal coatings [12–18] solves the above problems, reduces losses and improves the contact joints reliability. Unlike the conductive lubricants [19, 20] not fitting to moveable contact joints, the metal coatings are resistant to ambient factors and provide long-term oxidation protection for the contact surfaces.

There are several conventional methods for metal coating application: electroplating, electro sparking, gas-dynamic etc.

Electroplating is an energy-intensive process; it requires special equipment and causes environmental risks [18]. Contact surface metallization through gas-thermal spray (in particular, plasma spray) needs complex special equipment as well [21]. The above methods are useful only for of a stationary plant producing electrical equipment.

When a thin layer of heated solder is manually applied on the contact surfaces (e.g., with a soldering iron), it requires much time and temperatures of 300–400 °C, even though the process is functional in the environment of electrical equipment operation and maintenance. Therewith, the worker's skills and sense of responsibility influence the coating quality considerably.

Unlike other known solutions, the technology, developed by the authors, suggests applying protective coatings on contact surfaces without using complex equipment and requires temperatures not higher than 80–90 °C. This is achieved by using low-melting alloys. In addition, the coating provides stable characteristics of the contact joints in the long term [22–24].

The product is used in contact joints without requirements for electric arc effect limitation. For the high-load contacts producing an electric arc when open or closed, special measures for contact surface protection are required (it may be magnetic blowout, build-up melting of high-melting metals etc.).

The research has been intended to:

- Select the optimal compounds of the low-temperature alloys
- Develop and optimize the technology for production of the protective coating on the contact surfaces
- Test the contact joints with the new type of protective coating both in lab environment and in actual operational conditions

2 Theory

During the process of the coating production, the contact surface hosts the well-known contact melting process, its important property being the interaction of solid and liquid metals or an alloy below the autonomous melting point of the solid metal [25].

At the first stage of the process, the surface of the solid metal is wetted with the liquid metal. During the second stage the localized melting of the solid metal surface occurs. At the third stage diffusion takes place, i.e. the solid body atoms from the melted volume move to the remaining volume of the liquid phase [22, 24]. This process may also be characterized as partial dilution of the solid metal in the liquid one, occurring in the localized surface layer. At the fourth stage hardening of the molten mass takes place. At the fifth stage a layer of metal coating forms on the solid metal surface. This layer differs from the solid metal in its chemical and physical properties. Meanwhile, its properties differ from the solder metal or alloy applied on its surface [23].

Depending on the properties of the metals involved in the contact melting process, the interaction results in either solid solutions or chemical compounds (intermetallic metals). These processes are usually examined by means of experiment [23].

3 Materials and Methods

The alloys for protective coating on contact joint surfaces were selected on the basis of the following criteria:

- Low melting point, below 100 °C
- Mutual diffusibility with the wide-spread contact member materials
- Good wettability of the contact member surfaces
- Corrosion resistance
- Lack of toxic, expensive or noble metals
- Lack of negative effect on the contact member material
- Low resistance of the alloy and its oxide films
- Plasticity of the produced coating allowing to increase the effective contact area and preventing electrical erosion

For the alloys making protective coatings on the working surface of moveable contact joints, there is an extra requirement: the alloy must contain a component preventing mechanical wear of the coating (decreasing friction ratio).

The required antifriction properties can be secured with indium demonstrating such properties as [26]:

- High adhesion to many metals
- High antifriction and lubricating properties (e.g., the dry friction ratio of indium per steel is 0.05–0.07)
- Resistance to alkali, sea water, and corrosion
- Being one of the low-melting alloys (indium easily dissolves the metals standing near in the periodic system: gallium, thallium, stannum, plumbum, bismuth, cadmium, and mercury)

To test the hypothesis, 12 alloys with melting point at 44 to 96 °C were examined after which 2 surface-active indium alloys with the necessary properties were selected.

We examined comparative sample applications of the alloys on the surfaces of different materials used for production of contact members (copper and copper-bearing alloys, aluminum, and steel). When applying, we also measured the sample wettability and evaluated the labor intensity of the application process. Then we put the ready samples through mechanical bend and compression tests, climate effects, high-temperature heating (up to 500 °C), and corrosion resistance tests. To find the contact electrical resistance, we used the tested samples to construct fixed dismountable contact joints. Nominal current heating and cyclic heating tests were also performed. The content and structure of the coatings were examined with an electronic microscope. The choice of alloys, the technology, and the alloy production technology should be considered in an individual publication.

The target technology criteria proceeded from the following characteristics of the metal coating production process:

- Low consumption of alloy per unit area during application
- Low energy consumption of application process (important for field work without stationary power supply)
- Low time input resulting in highly productive work
- Understandability of the technology for network facilities personnel; consequently, fewer requirements for personnel qualification and shorter training periods
- Lack of negative influence on the contact member material
- Production of a uniform metal coating on the contact member surface resistant to external factors

The proposed technology includes the following key elements:

- Contact surface degreasing
- For heavily soiled surfaces, cleaning of the aluminum contact members with alkaline flux materials and cleaning of the copper contact members with acid flux materials
- Dressing of the contact surface with a rotary wire brush to fresh metal
- Heating of the contact member up to a substrate point (Ts) higher than the alloy melting point ($Tamp$) $Ts>Tamp$; therewith, $Tamp<<Tsmp$, where $Tsmp$ – substrate melting point). Use of a heat gun (hot air gun) or a gas burner (for field conditions) to heat the contact members
- Applying of the alloy by touching the contact member surface with an ingot of the solid alloy or by applying the preheated liquid alloy with a doser. As a result, one or more pools of the melted alloy are formed on the contact surface
- Rubbing in the alloy to the contact member surface with a soft metal brush (Fig. 1)

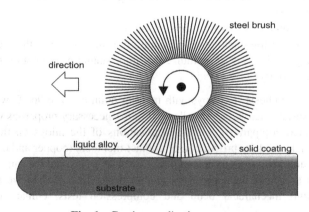

Fig. 1. Coating application process.

The advantage of this technology for coating aluminum and copper contact members is that when the alloy is rubbed in with a brush, the oxidation film on the member surface is destructed under the liquid alloy layer; therefore, no oxidation film is formed on the member surface. Mechanical impact on the contact solid-liquid melting area stimulates the diffusion processes.

Eventually, the solid-coating area (Fig. 1) is covered with a coating consisting of the solid solution of the metals being components of the low-melting alloy on the surface and of the contact member metal. The physical and chemical properties of this layer ensure necessary contact joint performance still being different from those of the applied alloy and the contact member material.

The technology and the applicable alloys can be used for production of protective coatings on the surfaces of both fixed (dismountable) and moveable contact joints.

4 Laboratory Tests and Results

The test worked with the contact joints of aluminum bars 20 × 40 mm fastened with a steel M8 bolt with steel washers 24 mm in diameter. The contact member width was 5 mm. The cable lugs were fastened to the aluminum buses with M8 bolts and d24 washers. Photos of the contact joints are shown in Fig. 2.

The working contact surfaces were covered with the protective coating according to the proposed technology. The surfaces of the other joints were dressed. Resistances were measured at ambient temperature with a micro ohmmeter.

According to Fig. 3, the coated samples independently demonstrate stable and almost constant contact electrical resistance, while the resistance of the uncoated samples increases (approximately twofold in the two years of test). As soon as in six months, the contact resistance of the uncoated aluminum contact joints increases considerably. The initial contact electrical resistance is approximately 1.3 times less for the coated contact joints.

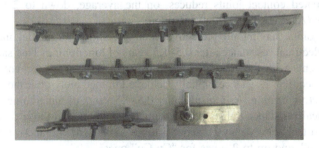

Fig. 2. Samples of fixed contact joints.

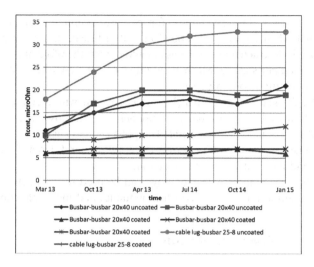

Fig. 3. Test results for laboratory test objects.

5 Conclusion

The research has provided the following main characteristics and advantages of the technology:

- The protective metal coatings are applicable to the contact surfaces of the materials used for contact electrical joints (copper, aluminum, steel and corresponding alloys)
- Production of the protective coatings poses no danger to health; the relevant low-melting alloys contain no toxic or precious metals
- Application of the coating does not require temperatures over 100 °C; no complex equipment is necessary
- As compared to the uncoated but dressed contact joints, the initial contact resistance of the coated contact joints reduces, on the average, down to 2 times (Al-Al), 1.5 times (Al-Cu), and 1.3 times (Cu-Cu)
- Contact joints, having no protective coatings, show substantial increase in the contact electrical resistance after 0.5–1.5 years of functioning. That said, the coated contact joints with the coating, applied using the developed technology, keep the level of electrical resistance, initially achieved immediately after the application of the protective coating. As a result, after 0.5–1.5 years of functioning the difference in electrical resistance values between uncoated joints and those coated using the developed technology may be about 5 times for "Al-Al" contacts, up to 2.5 times for "Al-Cu", and up to 2 times for "Cu-Cu" ones

To develop this technology further, it is reasonable to:

- Improve the application process efficiency (needed by electrical equipment manufacturers)
- Improve technology performance (lower prime cost, better usability)

References

1. Braunovic, M., Myshkin, N.K., Konchits, V.V.: Electrical Contacts: Fundamentals, Applications and Technology. CRC Press, London (2006)
2. Braunovic, M.: Reliability of power connections. J. Zhejiang Univ. **8**(3), 343–356 (2007)
3. Slade, P.G.: Electrical Contacts: Principles and Applications. CRC Press, London (2013)
4. Preston, P.F.: An industrial atmosphere corrosion test for electrical contacts and connections. Trans. Instit. Metal Finish. **50**, 125–130 (1972)
5. Braunovic, M.: Aluminum connections: legacies of the past. In: Proceedings of 40th International Conference on Electrical Contacts, pp. 1–31 (1994)
6. Timsit, R.: Some fundamental properties of aluminum-aluminum electrical contacts. IEEE Trans. Compon. Hybrids Manufact. Technol. **3**(1), 71–79 (1980)
7. Johnson, J.L., Moberly, L.E.: Separable electric power contacts involving aluminum busbars. Electric Contacts **21**, 53 (1975)
8. Timsit, R.S.: Electrical contact resistance: properties of stationary interfaces. IEEE Trans. Compon. Packag. Technol. **22**(1), 85–98 (1999)
9. Naybour, R., Farrell, T.: Connectors for aluminum cables: a study of the degradation mechanisms and design criteria for reliable connectors. IEEE Trans. Parts Hybrids Packag. **9**(1), 30–36 (1973)
10. Antler, M.: Survey of contact fretting in electrical connections. Electr. Contacts **3** (1984)
11. Aronstein, J.: An updated view of the aluminum contact interface. In: Proceedings of the 50th IEEE Holm Conference on Electrical Contacts and the 22nd International Conference on Electrical Contacts, pp. 98–103 (2004)
12. Krumbein, S.: Contact properties of tin plates. Electric Contacts **3** (1974)
13. Shao, C.B., Zhang, J.G.: Electric contact behavior of Cu-Sn intermetallic compound formed in tin platings. In: Proceedings of the 44th Holm Conference on Electrical Contacts, pp. 26–33 (1998)
14. Misra, P., Nagaraju, J.: Electrical contact resistance in thin gold-plated contacts: effect of gold plating thickness. IEEE Trans. Compon. Packag. Technol. **33**(4), 830–835 (2010)
15. Antler, M.: Gold plated contacts: effects of thermal aging on contact resistance. In: Proceedings of 43rd IEEE Holm Conference on Electrical Contacts, pp. 121–131 (1997)
16. Braunovic, M.: Evaluation of different platings for aluminum-to-copper connections. IEEE Trans. CHMT **15**, 205 (1992)
17. Abbott, W.H.: The role of electroplates in contact reliability. In: Proceedings of 48th IEEE Holm Conference on Electrical Contacts, pp. 163–164 (2002)
18. Shlesinger, M., Paunovic, M.: Modern Electroplating. Wiley, New Jersey (2010)
19. Chudnovsky, B.H.: Lubrication of electrical contacts. In: Proceedings of the Fifty-First IEEE Holm Conference on Electrical Contacts, pp. 107–114 (2005)
20. Timsit, R.S.: Effect of surface reactivity of lubricants on the properties of aluminum electrical contacts. In: Proceedings of the Forty-third IEEE Holm Conference on Electrical Contacts, pp. 57–66 (1997)
21. Fauchais, P.L., Heberlein, J.V.R., Boulos, M.: Thermal Spray Fundamentals: From Powder to Part. Springer, London (2014)
22. Goman, V.V., Fedoreev, S.A.: Plating technology for contact joint performance improvement in electrical equipment. Mater. Sci. Forum **870**, 271–275 (2016). https://doi.org/10.4028/www.scientific.net/MSF.870.271
23. Goman, V.V., Fedoreev, S.A.: Experimental study of contact joint characteristics in electrical equipment. Mater. Sci. Forum **870**, 276–281 (2016). https://doi.org/10.4028/www.scientific.net/MSF.870.276

24. Perelshtein, G.N.: High-reliability high-efficiency demountable electrical contact joints. Ind. Energetics J. **5**, 30–33 (2010)
25. Lashko, N.F., Lashko, S.V.: Contact metallurgical processes during soldering and brazing. Metallurgy, Moscow (1987)
26. Yacenco, S.P.: Indium. Properties and application. Science, Moscow (1987)

Refinement of Transmission Line Equivalent Circuit Parameters at Power System State Estimation by PMU Data

I. Kolosok and E. Korkina[✉]

Melentiev Energy Systems Institute, 130 Lermontova Street,
Irkutsk 664033, Russian Federation
korkina@isem.irk.ru

Abstract. The significance of the electric power system state estimation is getting increasingly relevant for control of the power industry facilities. The quality of state estimation results depends on the quality of input data, i.e. SCADA and PMU measurements, a properly constructed equivalent circuit based on the signals about position of switching equipment, and accuracy of equivalent circuit parameters (resistance, reactance, conductance, and susceptance of transmission lines). The evolution of WAMS alters the understanding of characteristics of the measurements involved in the state estimation procedure and the properties of the procedure itself. Therefore, special attention is paid to the improvement in quality and increase in accuracy in all stages of the measurement cycle, the requirements of failure-free synchronization of phasor measurements, and development of algorithms for SPM validation. The paper is concerned with the refinement of the equivalent circuit parameters, by using PMU measurements, and the influence of such refinement on the power system state estimation results. The validity of PMU measurements is preliminarily checked by the method of test equations and wavelet analysis. Results of calculations for two 750 kV transmission lines based on 10-min archives of PMU measurements prove the suggested approaches are practical and efficient.

Keywords: Synchrophasor measurements · State estimation · Equivalent circuit parameters

1 Introduction

State estimation methods are used for development of electric power system analytic model. The state estimation procedure results in the calculation of steady state (current state) of power system on the basis of current configuration of network equivalent circuit, its parameters and measurements of operating conditions.

WAMS technology development within UPS of Russia involves mainstreaming of computer-aided phasor measurement system using Phasor Measurement Units (PMU) [1]. To ensure that the information received from the PMU is accurate the computer-aided systems control state settings overrun (the same way SCADA telemetry data was checked before) [1], and signals are verified [2], which is very important in regards to state estimation. Computer-aided phasor measurement systems analyze data from the PMU with the help of visualization apps, which contribute to state bumping detection. However, during the adjustment of phasor measurement system state estimation procedure becomes even more important. Its mathematical tools based on pretest statistical and logic analysis of measurement reliability [3] helps to both detect measurement errors and specify equivalent circuit parameters, which are normally set according to equipment nominal data.

A method to identify equivalent circuit parameters at EPS SE was first proposed by [4]. However, as it was demonstrated by [5], at that time errors in SCADA telemetry data used for state estimation significantly exceeded error limits of specified equivalent circuit parameters, so refinement of these parameters had no considerable effect on SE results. As a consequence, suggested methods were not implemented into state estimation systems.

The situation changed when high-precision PMU measurements were introduced into state estimation process. In the context of WAMS, the problem of accurate specification of equivalent circuit parameters for steady-state calculation was emphasized, which provoked rather a lot of associated research.

One of the early researches of equivalent circuit parameters identification based on PMU measurements (active and reactive series impedance and shunt conductance R, X, G, B) was done by [6] using two major formulas to calculate R, X, G, B for a two-port terminal network. The paper highlights the following: due to PMU errors being significantly less substantial than errors of a metering circuit (instrument transformers and secondary circuits), it is impossible to avoid errors when defining equivalent circuit parameters.

When calculating equivalent circuit parameters in two states and at four ranges of errors on the basis of module and angle of voltage and current instrument transformers in [6] it was concluded that this method could be used with practical accuracy only to identify reactive components of equivalent circuit parameters (X and B). Later in [7], equivalent circuit parameters were identified based on PMU measurements in two transmission lines connected in series within a 10-min interval. First, to define the errors of test instrumentation (metering circuit and PMU) the authors used nonlinear optimization approach, at which the imposed limitations correspond to the instrument transformers accuracy of 0.5 class. Then the equivalent circuit parameters were identified considering these errors.

In [8] it is noted that the authors of [7] take the state to be balanced for all phases, though the case is not the same in real-life conditions. In [8] the identification of equivalent circuit parameters is considered separately for active (R and G) and reactive (X and B) elements. A method for simultaneous identification of equivalent circuit parameters and error negating coefficients for instrument transformers is suggested.

Study of how systematic errors in PMU measurements influence the identification of actual values of equivalent circuit parameters was made by [9]. It is demonstrated that R estimations are more sensitive to systematic errors in voltage module calculations and X estimations are more sensitive to phase angle deformations.

A method of equivalent circuit parameters identification based on phasor measurements is implemented by [10]. The author sets up a matrix for derived measurements of power and voltage based on components of equivalent circuit parameters and uses several cutoffs to specify values of defined R, X, G, B. For simplicity, when analyzing a certain transmission line where equivalent circuit parameters are identified, difference of voltage angle values is used instead of voltage angle value as such. Reference data of equivalent circuit parameters obtained at 20 °C is used as initial approximation.

The paper reviews an algorithm used to identify equivalent circuit parameters of the transmission line using PMU measurements based on state estimation. The algorithm is based on the approach suggested by [4], when the identified equivalent circuit parameters are included in the state vector at EPS SE. To check if equivalent circuit parameters need to be refined test equations [3] are applied to detect measurement errors. An experience of state estimation based on actual PMU measurements [11] has shown that verification of these measurements does not guarantee successful state estimation procedure. If PMU measurements are verified, but state estimation provides significantly different estimates of measurements, then is it conditioned by the deviation of equivalent circuit parameters from nominal values? A 10-min archive of data from four PMU installed on 750 kV transmission line ends was used for calculations.

2 State Estimation of Electric Power System

2.1 State Estimation Procedure

State estimation procedure [12] involves computation of current state of EPS based on the SCADA-data or PMU-data for actual analytical model generated according to the signals about the state of switching equipment. For state estimation of n-node analytical model a state vector x = (δ, U) with length of (2n–1) is introduced. It includes U modules and angles δ of voltage, but excludes the angle of reference bus. State vector uniquely determines both defined y and undefined $z = f(y)$ state parameters. Thus stated, the problem of state estimation is reduced to minimization of criterion:

$$J(x) = (\bar{y} - y(x))^T R_y^{-1} (\bar{y} - y(x)) \tag{1}$$

where R – covariance matrix of measurement errors with measurement variances $r_{ii} = \sigma_i^2$ on its diagonal.

Due to y(x) nonlinear relation, the problem is solved iteratively with re-evaluation of each i iteration:

$$\Delta x_i = (H_i^T R^{-1} H_i)^{-1} H_i^T R^{-1} (\bar{y} - y(x_i)) \tag{2}$$

where $H_i = \frac{\partial y}{\partial x_i}$ – Jacobian matrix; $\bar{y}$ – vector of measured parameters; x – state vector.

2.2 Test Equations Method

When solving a problem of state estimation it is important to know that the initial data received from PMU is accurate. In this study, a priori bad data detection (BDD) of measurements is accomplished using a method of test equations (TE) [3] developed in

Melentiev Energy Systems Institute of Siberian Branch of the Russian Academy of Sciences. *Test equations* are the steady-state equations, which only include measured state variables

$$w_k(y) = 0 \qquad (3)$$

or unmeasured variables calculated through measured $w_k(y, f(y)) = 0$.

A priori BDD using TE method is based on plugging in raw measurements into test equations and comparing the derived discrepancies (due to measurement errors) of test equations w_k with statistical thresholds d_k, defined by measurement accuracy (variances) included in TEs. To validate measurement accuracy the condition is verified

$$|w_k| < d_k \qquad (4)$$

If the condition (4) is met, then accurate and quality measurements are provided for state estimation procedure. Experience in telemetry and study of PMU measurements [13] suggest that prior to verification of equivalent circuit parameters these measurements should be validated.

3 Identification of Equivalent Circuit Parameters at Electric Power System State Estimation

The approach to the identification of equivalent circuit parameters at EPS SE proposed by [4] requires the identified parameters to be included into general set of state vector components. In this case, state estimation is reduced to minimization of objective function on T time intervals:

$$\phi = \sum_{t=1}^{T} [(\bar{y}(t) - y(x, D)]^T R_{y(t)}^{-1} [(\bar{y}(t) - y(x, D)] \qquad (5)$$

where D – identified parameters. As demonstrated in [14], to obtain estimations of state vector components and equivalent circuit parameters, criterion (5) is minimized according to x and D, leading to solution of two systems of linear equations:

$$\frac{\partial \phi}{\partial x(t)} = \left(\frac{\partial y(x, D)}{\partial x}\right)^T R_y^{-1} [(\bar{y}(t) - y(x, D)]$$

$$\frac{\partial \phi}{\partial D} = \sum_{t=1}^{T} \left(\frac{\partial y(x, D)}{\partial D}\right)^T R_y^{-1} [(\bar{y}(t) - y(x, D)]$$

To monitor change in circuit parameters, correction module D may be occasionally briefly activated during state estimation. Due to slow change in equivalent circuit parameters the operation of this module is not required at the rate of the estimation process.

To define actual values of R, X, G, B using methods of state estimation we introduce them into the state vector. State vector already includes modules and angles of node voltage, so introduction of R, X, G, B to the vector may create more unknowns, which would exceed existing measurements. In this case, to find a solution we need to increase the redundancy of given data. To increase data redundancy we simultaneously use several consecutive cutoffs: since the amortization of equivalent circuit parameters does not happen immediately, the refined parameter values will remain within new limits for a certain period. On this basis, to identify actual values of equivalent circuit parameters multiple cutoffs could be used. This method was applied to identify actual transformation ratio of instrument transformers [15].

Equivalent circuit parameters are refined using linear SE method and a matrix of derived current measurements for all equivalent circuit parameters (R, X, G, B):

$$\hat{Z}(t) = \left(\frac{\partial I}{\partial z}\right)^{-1} I(t) \qquad (6)$$

where Z – identified parameters.

Actual values of R, X, G, B were calculated using formula (6) for the intervals of 3000 measurements (1–3000, 3001–6000, 6001–9000 and so on up to 30000) and adaptable calculation method (recurrence formula from [3, 14]):

$$\hat{Z}(t) = (1-\alpha)\hat{Z}(t-1) + \alpha\hat{Z}(t) \qquad (7)$$

where $\alpha = 1/t$ on the first 100 cutoffs ($t = 1,2...100$), then $\alpha = 0.01$.

If the equivalent circuit parameters substantially change within reviewed time segment when compared with nominal values, the calculated values of R, X, G, B are added to analytical model parameters.

4 Case Study

To test the method we used 30000 cutoffs of 10-min data archive from two PMU (#3, #4) installed on 750 kV transmission line ends (Fig. 1), including the information about modules, phase angles of voltage and currents and total power. This sampling was sufficient to complete a statistical analysis of observational data and draw conclusions about the quality of measurements.

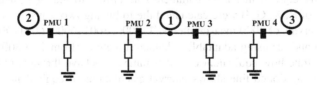

Fig. 1. 3-node circuit connecting of two 750 kV overhead lines [11].

The suggested method consists in the following:
1. Analysis of PMU measurement accuracy using test equations method.

To analyze PMU measurements accuracy of active and reactive power flows in test circuit the TE method is used. Balance equations of active and reactive power in line were used as test equations for branches 1–3:

$$w_P = P_{1-3} + P_{3-1} - \Delta P_{1-3} \tag{8}$$

$$w_Q = Q_{1-3} + Q_{1-3} - \Delta Q_{1-3} + Q_{gen} \tag{9}$$

In (8), (9) ΔP_{1-3}, ΔQ_{1-3} and Q_{gen} – are active and reactive power losses and line charging capacity of branch 1–3 respectively. They are calculated using power flow measurements, nodal voltage and preset values of equivalent circuit parameters.

While TE discrepancy is an algebraic sum of normally distributed errors of measurements included in the equation, it also has normal distribution with zero expectation and variance defined by variances of measurements included in it [3].

Discrepancies of test Eqs. (8) and (9) were calculated using nominal values of equivalent circuit parameters for branches 1–3. For the first cutoff we received the following values of TE discrepancies, which exceeded respective threshold values $d_P = 10\,\text{MW}$, $d_Q = 15,5\,\text{MVAr}$:

$$|w_P = 13.55\,\text{MW}| > d_P$$

$$|w_Q = -78.4\,\text{MVAr}| > d_Q$$

Then we calculated and analyzed TE discrepancies for remaining cutoffs. As a result it was found that the expectation of discrepancies of both test equations on all cutoffs were not equal zero. It could be either caused by systematic errors in PMU measurements or by incorrectly set equivalent circuit parameters used to calculate losses in (8), (9)

To check for systematic errors in PMU measurements methods of dynamic state estimation [16] were used along with wavelet analysis of stochastic processes [17] for measurements of currents, voltage and active and reactive powers calculated thereon. Both methods have shown that these measurements are accurate throughout the sampling: that is there are no systematic errors. That is why we need to specify equivalent circuit parameters for the transmission line in this 3-nodal network.

2. Calculation of equivalent circuit parameters using PMU data.

To identify equivalent circuit parameters according to the suggested method the actual values of R, X, G, B were calculated the following two ways: (1) using formula (6) for the intervals of 3000 measurements (1–3000, 3001–6000, 6001–9000 and so on up to 30000) and (2) using adaptable calculation method (from 1 to 30000). Figure 2 (a–d) demonstrate how equivalent circuit parameters behave throughout the whole of 10-min sampling: dotted line shows interval calculations of equivalent circuit parameters, solid line shows calculations based on formula (7). Figure 2(a–d) diagrams support the use of recurrence formula, which reflects the adaptive process.

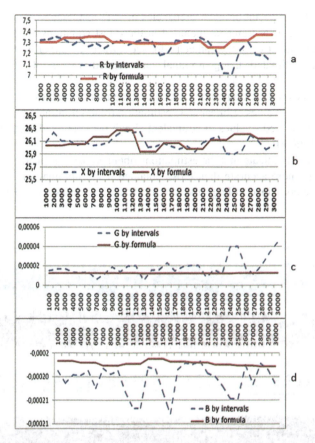

Fig. 2. (a) Active resistance value; (b) Reactive resistance value; (c) Corona conductance; (d) Capacity susceptance.

3. State estimation based on refined equivalent circuit parameters.

As a result, instead of nominal values of equivalent circuit parameters R = 2.66 Ω, X = 35.3 Ω G = 0 μS, B = 245 μS, new values R = 7.3 Ω, X = 26.06 Ω, G = 12 μS, B = 205 μS were obtained. Linear state estimation was made using these values (before the corrector of equivalent circuit parameters was activated). Once the equivalent circuit parameters were identified, the discrepancies of test equations were estimated based on the calculated values of equivalent circuit parameters. For the first cutoff we received the following values of test equation discrepancies:

$$|w_P = -0.6\,\text{MW}| < d_p$$

$$|w_Q = -1.46\,\text{MVAr}| < d_Q$$

It is evident that values of TE discrepancies of active and reactive power flows significantly decreased. Then we calculated TE discrepancies for remaining cutoffs to discover that the expectation of discrepancies of both test equations is close to zero. It indicates that the measurements included in these test equations do not contain gross errors and equivalent circuit parameters are set correctly. Further, state of the 3-nodal network was estimated using nominal and refined values of equivalent circuit parameters.

Figure 3 shows deviations of estimated voltage module values on the model nodes from the measured values at linear state estimation (blue - nominal values of equivalent circuit parameters, red - values of equivalent circuit parameters identified at an interval (cutoffs 3001–6000).

Figure 3(a, b) indicate that the estimations obtained using refined equivalent circuit parameters are more accurate than those obtained using nominal values of equivalent circuit parameters, since voltage measurements for this transmission line contain no errors.

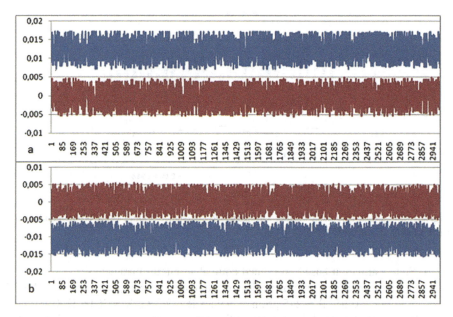

Fig. 3. Difference between measurements of voltage module and estimated voltage for (a) node 1; (b) node 3: blue - difference between measurements and estimations of U by nominal equivalent circuit parameters; red - difference between measurements and estimations of U by identified equivalent circuit parameters.

5 Conclusion

This paper analyzed the issues that occur when estimating the current state of EPS based on the PMU data. State estimation of transmission lines equipped with PMU show a mismatching between nominal values of equivalent circuit parameters and values calculated based on actual PMU measurements. To find a solution, methods of statistical data processing were suggested: a priori bad data detection using test equations method; wavelet analysis of measurements flow received from PMU; identification procedure of equivalent circuit parameters during linear state estimation process based on accurate PMU data; estimation of current state parameters based on obtained values of equivalent circuit parameters and PMU measurements after their substantiation. Results of calculations for two 750 kV transmission lines based on 10-min archives of data from four PMU prove the suggested approaches are practical and efficient and contribute to improved accuracy of estimations of current states of electric power systems using state estimation methods.

Acknowledgements. The study was performed within the Scientific Project III.17.4.2 of the Fundamental Research Program of SB RAS, reg. № AAAA-A17-117030310438-1.

References

1. Zhukov, A.V., Dubinin, D.M., Utkin, D.N., et al.: Development and implementation of automatic system for information acquisition from recorders of transient state monitor system in UPS of Russia. Releishchik **3**, 18–23 (2013)
2. Nebera, A.A., Shubin, N.G., Kazakov, P.N., et al.: The implementation in JSC "SO FGC" of software for transition processes monitoring. In: Proceedings of International Conference Actual Trends in Development of Power System Relay Protection and Automation, vol. 5, pp. 1–9 (2015)
3. Gamm, A.Z., Kolosok, I.N.: Bad data detection in electric power system telemetry. Science, Novosibirsk (2000)
4. Gamm, A.Z.: Methods of estimation and identification in electric power systems. Power Engineering Institute of Siberian Branch of the Science Academy of the USSR, Irkutsk (1974)
5. Idelchik, V.I., Novikov, A.S., Palamarchuk, S.I.: Equivalent circuit parameters errors in power system steady-state calculations. Statistical processing of immediate information in electric power systems, pp. 145–152 (1979)
6. Berdin, A.S., Kovalenko, P.Y., Plesnyaev, E.A.: Influence of errors in PMU measurements in defining the parameters of a power transmission line equivalent circuit. In: STC of Unified Power System Proceedings, vol. 1, no. 66, pp. 29–38 (2012)
7. Berdin, A.S., Kovalenko, P.Yu.: Defining the parameters of a power transmission line equivalent circuit on the basis of phasor measurements. In: STC of Unified Power System Proceedings, vol. 2, no. 71, pp. 29–34 (2014)
8. Rybasova, O.S., Kononov, Yu.G., Kostyukova, S.S.: Defining the parameters of 750 kV overhead line equivalent circuit on the basis of phasor measurements. In: Proceedings of int. sc. and pr. conf. Younger Generation Perspective on Power Generating Industry, pp. 201–206 (2015)
9. Kakovskiy, S.K., Nebera, A.A., Rabinovich, M.A., et al.: Estimation of power transmission line parameters on the basis of electric system model. Electric Power Stations **2**, 42–53 (2016)
10. Nikolaev, R.N.: Verification of power supply equipment parameters with transient state monitor system. In: International Science and Technology Conference Actual Trends in Development of Power System Relay Protection and Automation, p. 1 (2013)

11. Kolosok, I.N., Korkina, E.S., Buchinsky, E.A.: WAMS Data processing to address the issue of automatic control of power system state. In: Proceedings of International Conference Actual Trends in Development of Power System Relay Protection and Automation, vol. 5, pp. 2–6 (2015)
12. Gamm, A.Z.: Static methods to estimate electric power system state. Science, Moscow (1976)
13. Kolosok, I.N., Korkina, E.S., Buchinsky, E.A.: State estimation of digital substation based on phasor measurements. In: Proceedings of International Conference Actual Trends in Development of Power System Relay Protection and Automation, vol. 2, pp. 2–11 (2013)
14. Gamm, A.Z., Gerasimov, L.N., Golub, I.I., et al.: State estimation in electric power industry. Science, Moscow (1983)
15. Kolosok, I.N., Korkina, E.S., Buchinsky, E.A.: Instrument transformer error evaluation based on processing algorithms for synchrophasor measurements. Releishchik **3**, 24–29 (2013)
16. Glazunova, A.M., Kolosok, I.N., Semshchikov, E.S.: Bad data detection by the methods of dynamic state estimation for on-line control of electric power system operation. Electricity **2**, 18–27 (2017)
17. Kolosok, I.N., Gurina, L.A.: Credibility improvement of phasor measurements data flow. In: Proceedings of International Science and Technology Conference Relay Protection and Automation of Power Systems, pp. 915–921 (2017)